# Solid Edge 2020 Black Book

By
**Gaurav Verma**
**Matt Weber**
**(CADCAMCAE Works)**

Edited by
**Kristen**

ISBN # 978-1-77459-005-8

## NOTICE TO THE READER

## DEDICATION

To teachers, who make it possible to disseminate knowledge
to enlighten the young and curious minds
of our future generations

To students, who are the future of the world

## THANKS

To my friends and colleagues

To my family for their love and support

# Training and Consultant Services

At CADCAMCAE WORKS, we provide effective and affordable one to one online training on various software packages in Computer Aided Design(CAD), Computer Aided Manufacturing(CAM), Computer Aided Engineering (CAE), Computer programming languages(C/C++, Java, .NET, Android, Javascript, HTML and so on). The training is delivered through remote access to your system and voice chat via Internet at any time, any place, and at any pace to individuals, groups, students of colleges/universities, and CAD/CAM/CAE training centers. The main features of this program are:

## Training as per your need

Highly experienced Engineers and Technician conduct the classes on the software applications used in the industries. The methodology adopted to teach the software is totally practical based, so that the learner can adapt to the design and development industries in almost no time. The efforts are to make the training process cost effective and time saving while you have the comfort of your time and place, thereby relieving you from the hassles of traveling to training centers or rearranging your time table.

## Software Packages on which we provide
## basic and advanced training are:

CAD/CAM/CAE: CATIA, Creo Parametric, Creo Direct, SolidWorks, Autodesk Inventor, Solid Edge, UG NX, AutoCAD, AutoCAD LT, EdgeCAM, MasterCAM, SolidCAM, DelCAM, BOBCAM, UG NX Manufacturing, UG Mold Wizard, UG Progressive Die, UG Die Design, SolidWorks Mold, Creo Manufacturing, Creo Expert Machinist, NX Nastran, Hypermesh, SolidWorks Simulation, Autodesk Simulation Mechanical, Creo Simulate, Gambit, ANSYS and many others.

Computer Programming Languages: C++, VB.NET, HTML, Android, Javascript and so on.

Game Designing: Unity.

Civil Engineering: AutoCAD MEP, Revit Structure, Revit Architecture, AutoCAD Map 3D and so on.

We also provide consultant services for Design and development on the above mentioned software packages

For more information you can mail us at:
cadcamcaeworks@gmail.com

# Table of Contents

Training and Consultant Services                                                iv
Preface                                                                         xv
About Authors                                                                   xvii

## Chapter 1 : Starting with Solid Edge

**Introduction**                                                                **1-2**
**Starting Solid Edge**                                                         **1-2**
**Starting a new Part document**                                                **1-3**
**User Interface Overview**                                                     **1-5**
Customizing Ribbon                                                              1-6
Recording Solid Edge Screen                                                     1-9
**Application Menu Tools**                                                      **1-10**
Learn Options in Application Menu                                               1-10
New Options in Application Menu                                                 1-11
Opening a File                                                                  1-13
Saving File                                                                     1-13
Save As Tools                                                                   1-14
Saving Model in Image Format                                                    1-15
Exporting Solid Edge Model                                                      1-16
Saving Model for Tablet Devices                                                 1-18
Print File                                                                      1-19
Preparing File for 3D Printing                                                  1-21
Share                                                                           1-24
Settings                                                                        1-25
Tools                                                                           1-42
Info                                                                            1-45
Property Manager                                                                1-48
Exit Solid Edge                                                                 1-51
**Mouse Button Functions**                                                      **1-51**
**Workflow in Solid Edge**                                                      **1-51**
**Self-Assessment**                                                             **1-53**

## Chapter 2 : Sketching

**Basics for Sketching**                                                        **2-2**
**Planes**                                                                      **2-2**
Relation between sketch, plane and 3D model                                     2-3
**Starting Sketch**                                                             **2-3**
Sketching Environment and Tools                                                 2-3
**View toolbar**                                                                **2-5**
**Sketch Creation tools**                                                       **2-6**
Line Tools                                                                      2-7
Rectangle Tools                                                                 2-9
Circle Tools                                                                    2-12
Ellipse Tools                                                                   2-14
Tangent Arc Tools                                                               2-15
Curve Tool                                                                      2-17

Fillet Tools                                                    2-18
Split drop-down                                                2-21
Trim Tool                                                      2-22
Trim Corner Tool                                              2-22
Offset Tools drop-down                                        2-23
Move Tools                                                    2-25
Mirror drop-down                                              2-28
Construction Tool                                            2-31
Create As Construction Tool                                  2-32
**Relate Tools**                                             **2-33**
Connects                                                      2-33
Horizontal / Vertical                                        2-34
Tangent                                                      2-34
Parallel                                                      2-35
Equal                                                        2-35
Symmetric                                                    2-36
Concentric                                                  2-36
Perpendicular                                                2-37
Collinear                                                    2-37
Lock                                                        2-38
Rigid Set                                                    2-38
Symmetry Axis                                                2-38
Maintain Relationships                                      2-38
Relationships Handle                                        2-38
Relationship Assistant                                      2-39
**Dimensions**                                              **2-40**
Smart Dimension                                              2-40
Creating Distance Between Dimension                          2-44
Creating Angle Between Dimension                            2-44
Creating Coordinate Dimension                              2-45
Angular Coordinate Dimension                                2-46
Symmetric Diameter                                          2-47
Creating and Using Dimension Axis                          2-47
Activating Auto Dimensioning                                2-48
**Ordered Tools**                                          **2-48**
Starting 2D Sketch in Ordered Environment                  2-49
Draw Group                                                  2-49
IntelliSketch Tools                                        2-53
Intellisketch Options                                      2-57
Relationship Colors                                        2-58
Alignment Indicator                                        2-58
Styles                                                      2-58
**Practical - 1**                                          **2-64**
**Practical - 2**                                          **2-67**
**Practice 1**                                              **2-71**
**Practice - 2**                                            **2-71**

## Chapter 3 : 3D Sketch and Solid Modeling

**Introduction**                                            **3-2**
**3D Sketch**                                              **3-2**

**Extrude**                                                                  **3-3**
Project To Sketch                                                        3-10
Copying or Moving Sketch Elements to New Sketch            3-11
**Cut Feature**                                                              **3-12**
**Creating Planes and Coordinate System**                    **3-15**
Creating Coincident Plane                                            3-15
Creating Parallel Plane                                                3-15
Creating Plane at Angle                                                3-16
Creating Perpendicular Plane                                        3-17
Creating Coordinate System (Ordered Environment)        3-18
**Revolve**                                                                  **3-19**
**Revolve Cut**                                                            **3-21**
**Hole Tools**                                                              **3-21**
**Hole**                                                                        **3-22**
Creating Simple Hole                                                  3-23
Creating Threaded Hole                                                3-24
Creating Counterbore Holes                                          3-25
Thread                                                                         3-25
**Slot**                                                                        **3-27**
Recognizing Holes                                                       3-28
**Round**                                                                     **3-29**
Creating Constant Radius Round                                    3-29
Creating Variable Radius Round                                    3-31
Creating Blend Round                                                  3-32
Creating Surface Blend                                                3-33
**Chamfer**                                                                  **3-33**
**Draft**                                                                      **3-34**
**Thin wall drop-down**                                              **3-36**
Thin Wall (Shell)                                                        3-36
**Thin Region**                                                            **3-38**
**Rib**                                                                          **3-39**
**Web Network**                                                          **3-41**
**Lip**                                                                          **3-43**
**Vent**                                                                        **3-43**
**Mounting boss**                                                        **3-46**

## Chapter 4 : Advanced Solid Modeling

**Introduction**                                                            **4-2**
**Add drop-down**                                                        **4-2**
**Creating Sweep feature**                                            **4-2**
Sweep with Single Path and Single Cross-section            4-2
Multi Section Sweep                                                     4-4
Twist Sweep                                                                4-6
**Solid sweep**                                                            **4-7**
**Loft**                                                                        **4-8**
**Helix**                                                                      **4-9**
**Projecting Sketch Curves on Surface/Face**                 **4-10**
**Normal**                                                                    **4-11**
**Thicken**                                                                  **4-12**

**Creating Fillet Weld**                                      **4-13**
**Creating Groove Weld**                                      **4-14**
**Creating Stitch Weld**                                      **4-15**
**Creating Label Weld**                                       **4-17**
**Cut Features**                                              **4-17**
Sweep Cut                                                      4-17
Solid Sweep Cut                                               4-18
Loft Cutout                                                   4-19
Helix Cut                                                     4-20
Normal cut                                                    4-20
**Add body**                                                  **4-21**
Enclosure                                                     4-22
Union                                                         4-23
Subtract                                                      4-23
Intersect                                                     4-25
Split                                                         4-25
Scale Body                                                    4-26
**Pattern Features**                                          **4-27**
**Creating Pattern**                                          **4-27**
Rectangular Pattern                                           4-27
Circular Pattern                                              4-28
Along Curve Pattern                                           4-29
Pattern by Table                                              4-30
Duplicate Pattern                                             4-32
**Mirror**                                                    **4-33**
Mirror Copy Feature                                           4-33
Mirror Copy Part                                              4-34
**Moving faces**                                              **4-34**
**Rotating faces**                                            **4-35**
**Offset faces**                                              **4-36**
**Delete**                                                    **4-37**
Deleting Faces                                                4-37
Deleting Regions                                              4-38
Deleting Holes                                                4-38
Deleting Rounds                                               4-39
Re-sizing Holes                                               4-39
Resize Round                                                  4-40
Creating Sketches for Hook                                    4-42
Creating Sweep Feature                                        4-44
Applying Round at End                                         4-45

## Chapter 5 : Selection Tools, 3D Printing, and Views

**Introduction**                                             **5-2**
**Select drop-down**                                         **5-2**
**Selection Filter**                                         **5-4**
**Overlapping Selection Box**                                **5-4**
**3D Printing**                                              **5-4**
**Part Preparation for 3D Printing**                         **5-5**
3D Printing Processes                                        5-5

Part Preparation for 3D Printing                                        5-10
**3D Print Part Preparation in Solid Edge**                            **5-11**
Converting Cosmic Threads to Physical                                  5-11
Deleting Internal Voids                                                5-12
Reorienting Model for 3D Printing                                      5-12
Wall Thickness Validation                                              5-13
Checking Overhangs                                                     5-14
Exporting Model to STL File                                            5-14
**View Tab**                                                          **5-15**
**Showing Panes**                                                     **5-15**
**Construction Display Options**                                      **5-15**
**Setting planes**                                                    **5-16**
**Views**                                                             **5-17**
Front View                                                             5-17
Back View                                                              5-17
Left View                                                              5-17
Right View                                                             5-18
Top View                                                               5-18
Bottom View                                                            5-18
Dimetric View                                                          5-18
ISO View                                                               5-18
Trimetric View                                                         5-18
Saving Current View                                                    5-18
View Manager                                                           5-18
**Orient**                                                            **5-19**
Look at Face                                                            5-19
Previous View                                                           5-19
Spin About                                                             5-19
Wire Frame                                                             5-19
Visible and Hidden Edges                                               5-20
Visible Edges                                                          5-20
Shaded                                                                 5-20
Shaded with Visible Edges                                              5-21
**Floor reflection**                                                  **5-21**
**Floor shadow**                                                      **5-22**
**High-quality rendering**                                            **5-22**
**Edge Color**                                                        **5-22**
**Sharpen**                                                           **5-22**
**Perspective View**                                                  **5-22**
**Color manager**                                                     **5-22**
**View overrides**                                                    **5-23**
Rendering Tab                                                          5-24
Lights Tab                                                             5-24
Background Tab                                                         5-24
Reflection Box Tab                                                     5-25
**Applying Appearances to Part**                                      **5-25**

## Chapter 6 : Surfacing

**Surfacing**                                                          **6-2**

**Key point curve**                                                                 **6-2**
**Curve by table**                                                                  **6-3**
**Intersection**                                                                    **6-4**
**Project**                                                                         **6-5**
**Helical Curve**                                                                   **6-5**
**Cross Curve**                                                                     **6-8**
**Contour**                                                                         **6-9**
**Isocline**                                                                        **6-10**
**Derived**                                                                         **6-11**
**Split**                                                                           **6-12**
**Intersection point**                                                              **6-13**
**Creating Surfaces**                                                               **6-13**
Creating Surface using BlueSurf tool                                                6-13
Creating Surface using Bounded tool                                                 6-16
Creating Surface using Redefine tool                                                6-18
Creating Surface using Swept tool                                                   6-19
Creating Surface using Extruded tool                                                6-20
Revolved                                                                            6-21
Offset                                                                              6-22
Copy                                                                                6-23
Ruled                                                                               6-23
Blank Surface                                                                       6-25
**Modifying Surfaces**                                                              **6-26**
Intersect                                                                           6-26
Replace Face                                                                        6-29
Trim                                                                                6-29
Extend                                                                              6-30
Split                                                                               6-31
Stitched                                                                            6-31
Parting Split                                                                       6-33
BlueDot                                                                             6-35
**Surface Visualization Settings**                                                  **6-36**
**Displaying Section Curvature**                                                    **6-37**

## Chapter 7 : Product Manufacturing Information

**Introduction**                                                                    **7-2**
**Locking Dimension Plane**                                                          **7-2**
**Setting Dimension Axis**                                                           **7-2**
**Creating Dimension**                                                              **7-3**
Nominal Dimension Type                                                              7-7
Unit Tolerance Dimension Type                                                       7-7
Alpha Tolerance Dimension Type                                                      7-7
Class Dimension Type                                                                7-7
Limit Dimension Type                                                                7-8
Basic Dimension Type                                                                7-8
Reference Dimension Type                                                            7-8
Feature Callout Dimension Type                                                      7-9
Blank Dimension Type                                                                7-9
**Dimension Styles**                                                                **7-10**

**Model Size and Pixel Size PMI**                                           **7-10**
**Creating Callout Annotation**                                             **7-10**
General Tab                                                                  7-10
Text and Leader Tab                                                          7-13
Smart Depth Tab                                                             7-14
Feature Callout Tab                                                         7-14
Border Tab                                                                  7-14
**Creating Balloon Annotation**                                            **7-16**
**Creating Surface Texture Symbol**                                        **7-17**
**Creating Weld Symbol**                                                   **7-18**
**Creating Edge Condition**                                                **7-19**
**Creating Datum Frame and Feature Control Frame Annotation**              **7-20**
Creating Datum Frame                                                       7-22
Creating Feature Control Frame                                             7-23
**Creating Datum Target**                                                  **7-24**
**Creating Leader**                                                        **7-25**
**Creating Section by Plane**                                              **7-25**
**Creating User Defined Section**                                          **7-26**
**Creating Model Views**                                                   **7-28**

## Chapter 8 : Assembly-I

**Assembly**                                                               **8-2**
**Inserting Ground component**                                             **8-2**
**Inserting components in assembly**                                       **8-3**
**Assembly Constraints**                                                   **8-5**
FlashFit                                                                    8-5
Mate                                                                        8-5
Planar Align                                                                8-6
Axial Align                                                                 8-6
Insert                                                                      8-7
Connect                                                                     8-7
Angle                                                                       8-8
Tangent                                                                     8-9
Path                                                                        8-9
Cam                                                                         8-10
Parallel                                                                    8-10
Gear                                                                        8-11
Match Coordinate Systems                                                    8-11
Centre-Plane                                                                8-13
Rigid Set                                                                   8-13
Assembly Relationships Manager                                             8-13
Assembly Relationships Assistant                                          8-14
**Capture Fit**                                                            **8-16**
**Move on select**                                                         **8-17**
**Drag component**                                                         **8-17**
**Move component**                                                         **8-18**
Linear Move                                                                 8-19
Rotational Move                                                             8-19
**Replacing Part**                                                         **8-20**

Replace Part                                                              8-20
Replace Part With Standard Part                                           8-21
Replace Part with New Part                                               8-21
Replace Part with Copy                                                    8-23
**Transferring Part to Another Subassembly**                             **8-23**
**Dispersing Sub-assembly**                                              **8-23**
**Rotational motors**                                                    **8-23**
**Linear motor**                                                         **8-24**
**Simulate motor**                                                       **8-25**
**Variable table motor**                                                 **8-26**
**Pattern**                                                              **8-26**
**Pattern along curve**                                                  **8-27**
**Mirror Components**                                                    **8-28**
**Duplicate Pattern**                                                    **8-29**
**Inspection Tools**                                                     **8-29**
Smart Measure                                                             8-29
Measure                                                                   8-30
Measure Distance                                                         8-31
Measure Minimum Distance                                                 8-31
Measure Normal Distance                                                  8-32
Measure Angle                                                            8-32
Inquire Element                                                          8-33
**Properties**                                                           **8-33**
**Properties manager**                                                   **8-34**
**Check interference**                                                   **8-35**
**Assembly statistics**                                                  **8-36**
**Geometry Inspector**                                                   **8-36**
Inserting Ground Component                                               8-38
Inserting and Constraining Component                                     8-39

## Chapter 9 : Assembly Design - II

**Tools Tab**                                                            **9-2**
Peer Variable                                                            9-2
**Component Tracker**                                                    **9-4**
**Update Tools**                                                         **9-5**
**Limited Update and Limited Save**                                      **9-5**
**Reports**                                                              **9-5**
**ERRORS**                                                               **9-6**
**Inter-Part Manager**                                                   **9-7**
**Component Structure Editor**                                           **9-7**
**Publishing Virtual Components**                                        **9-8**
**Publishing Terrain Models**                                           **9-9**
**Update Structure**                                                     **9-9**
**Engineering Reference**                                                **9-9**
Shaft Designer                                                           9-9
Cam Designer                                                            9-14
Spur Gear Designer                                                      9-16
Bevel Gear Designer                                                     9-21
**Explode Render Animate Environment**                                   **9-23**

**ANIMATION EDITOR**     **9-23**

Animation drop-down     9-24

Creating New Animation     9-24

Saving Animation     9-24

Deleting Animation     9-24

Animation Properties     9-24

Save as Movie     9-24

Creating Camera Path and Enabling Camera Movement     9-25

**Show Camera Path**     **9-26**

**Working with Animation**     **9-27**

**Creating KeyShot Animation**     **9-27**

**Keyshot Render**     **9-27**

**Auto explode**     **9-27**

**Explode**     **9-29**

**Reposition**     **9-29**

**Removing Component from Exploded State**     **9-30**

**Collapsing Exploded Component**     **9-30**

**Unexplode**     **9-31**

**Binding Sub-Assembly**     **9-31**

**Unbinding Sub-Assembly**     **9-31**

**Flow lines**     **9-31**

Drop     9-32

Draw     9-32

Modify     9-33

**Configurations**     **9-33**

Saving and Retrieving Display Configurations     9-33

Configuration Manager     9-34

Configuration Options     9-35

Copying Components to Display Configuration     9-36

Taking Snapshot and Restoring     9-36

Unloading Hidden Parts     9-36

**Modes**     **9-36**

## Chapter 10 : Assembly - III

**Introduction to XpresRoute**     **10-2**

**Creating Path for Route**     **10-2**

**Creating Line Segment of Route Path**     **10-3**

**Creating Arc Segment**     **10-4**

**Moving Line Segment**     **10-4**

**Splitting Path Segment**     **10-5**

**Creating Curve Segment**     **10-5**

**Creating Keypoint Curve Segment**     **10-6**

**Creating Route using Paths**     **10-6**

**Creating Tube Along Path**     **10-7**

**Creating Piping Route**     **10-8**

**Introduction to Electrical Routing**     **10-10**

**Creating Paths**     **10-10**

**Creating Wires**     **10-10**

**Creating Cable**     **10-11**

**Introduction to Frame Design**     **10-12**

Creating Frame Members                                          10-12
Publishing Model to HTML Web page                              10-14

## Chapter 11 : Drawing

**Introduction**                                               11-2
**Starting Drawing Environment**                               11-2
Starting a New Drawing                                         11-2
Starting a Drawing using Model or Assembly                     11-2
**Placing Views with View Wizard**                             11-3
**Updating Views**                                             11-8
**Creating Principal Views**                                   11-8
**Creating Auxiliary View**                                    11-8
**Creating Detail View**                                       11-9
**Creating Broken View**                                       11-9
**Creating Section View Cutting Plane**                        11-10
**Creating Section using Cutting Plane Line**                  11-11
**Creating Broken-Out Drawing View**                           11-11
**Creating Parts List**                                        11-12
**Creating Hole Table**                                        11-12
**Creating Bend Table**                                        11-13
**Dimensioning Tools**                                         11-14
Creating Symmetric Diameter Dimension                          11-14
Applying Chamfer Dimension                                     11-15
Retrieving Dimension                                           11-15
Line Up Text                                                   11-15
Copying Attributes                                             11-16
Removing Dimensions from Alignment Set                         11-16
Stacking Dimensions                                            11-16

## Chapter 12 : Sheetmetal Design

**Introduction**                                               12-2
**Starting Sheetmetal Part Document**                          12-2
**Creating Tab Feature**                                       12-2
**Creating Flange Feature**                                    12-3
**Creating Contour Flange**                                    12-5
**Creating Lofted Flange**                                     12-7
**Creating Hem Feature**                                       12-8
**Creating Dimple Feature**                                    12-8
**Creating Louver Feature**                                    12-9
**Creating Drawn Cutout**                                      12-11
**Creating Bead Feature**                                      12-12
**Creating Gusset Feature**                                    12-13
**Creating Cross Brake**                                       12-14
**Creating Etch Feature**                                      12-14
**Creating Emboss Feature**                                    12-15
**Closing 2-Bend Corner**                                      12-16
**Closing 3-Bend Corners**                                     12-17
**Creating Ripped Corners**                                    12-17
**Creating Bend**                                              12-18

**Unbending Sheet Metal**                                              **12-19**
**Rebending**                                                          **12-20**
**Creating Jog Feature**                                               **12-20**
**Applying Bend Bulge Relief**                                         **12-21**
**Applying Break Corner**                                              **12-22**
**Sheet Metal Modification Tools**                                     **12-22**
Modifying Bend Angle                                                   12-23
Modifying Bend Radius                                                  12-23

## Chapter 13 : Simulation Study

**Introduction to Simulation**                                         **13-2**
**Types of Analyses performed in Solid Edge Simulation**              **13-2**
Linear Static Analysis                                                 13-2
Normal Modes Analysis                                                  13-3
Linear Buckling Analysis                                               13-3
**Starting a Study**                                                   **13-4**
**Switching between Studies**                                          **13-7**
**Selecting Material of Model**                                        **13-7**
**Selecting Geometry for Analysis**                                    **13-7**
**Uniting Bodies for Simulation**                                      **13-8**
**Recovering Bodies**                                                  **13-8**
**Applying Structural Loads**                                          **13-8**
Applying Force Load                                                    13-8
Applying Pressure Load                                                 13-9
Applying Torque Load                                                   13-10
Applying Displacement                                                  13-10
Applying Bearing Load                                                  13-11
Applying Body Temperature                                             13-11
Applying Centrifugal Load                                             13-11
Applying Gravity Load                                                  13-12
**Applying Constraints**                                               **13-12**
Applying Fixed Constraint                                             13-13
Applying Pinned Constraint                                            13-13
Applying No Rotation Constraint                                       13-13
Applying Slide Along Surface Constraint                               13-13
Applying Cylindrical Constraint                                       13-14
Applying User-Defined Constraint                                      13-14
**Meshing**                                                            **13-15**
Creating Mesh                                                          13-15
Setting Mesh Size for Edges                                           13-16
Specifying Surface Mesh Sizes                                         13-17
Specifying Body Mesh Size                                             13-18
**Solving Analysis**                                                   **13-18**
**Simulation Results**                                                 **13-19**
Probing Results                                                        13-19
Display Options                                                        13-20
Equilibrium Check                                                      13-21
Contour Style                                                          13-21

Deformation                                                      13-21
Animating Results                                                13-21
**Creating Report**                                              **13-22**
**Data Selection for Results**                                   **13-23**
Optimization Study                                               13-24
**Performing Generative Study**                                  **13-27**

**Index**                                                        **I-1**

# Preface

Solid Edge 2020 is a parametric (history based) and synchronous technology solid modeling tool that not only unites the three-dimensional (3D) parametric features with two-dimensional (2D) tools, but also addresses every design-through-manufacturing process. The Mechanical Designers functionality of the software makes it faster to design mechanical components for manufacturing. The software is capable of performing analysis with an ease. Due to a long journey of development in software, this solid modeling tool is remarkably user-friendly and it allows you to be productive from day one.

The **Solid Edge 2020 Black Book** is the first edition of our series on Solid Edge. This book is written to help beginners in creating some of the most complex solid models. The book follows a step by step methodology. In this book, we have tried to give real-world examples with real challenges in designing. We have tried to reduce the gap between university use of Solid Edge and industrial use of Solid Edge. The book covers almost all the information required by a learner to master the Solid Edge. The book starts with sketching and ends at advanced topics like Sheetmetal, Rendering, and Simulation Studies. Some of the salient features of this book are :

## In-Depth explanation of concepts

Every new topic of this book starts with the explanation of the basic concepts. In this way, the user becomes capable of relating the things with real world.

## Topics Covered

Every chapter starts with a list of topics being covered in that chapter. In this way, the user can easily find the topics of his/her interest easily.

## Instruction through illustration

The instructions to perform any action are provided by maximum number of illustrations so that the user can perform the actions discussed in the book easily and effectively. There are about 1350 illustrations that make the learning process effective.

### Tutorial point of view

At the end of concept's explanation, the tutorial make the understanding of users firm and long lasting. Almost each chapter of the book has tutorials that are real world projects. Moreover most of the tools in this book are discussed in the form of tutorials.

### Project

Free projects and exercises are provided to students for practicing.

### For Faculty

`If you are a faculty member, then you can ask for video tutorials on any of the topic, exercise, tutorial, or concept.`

## Formatting Conventions Used in the Text

All the key terms like name of button, tool, drop-down etc. are kept bold.

### Free Resources

Link to the resources used in this book are provided to the users via email. To get the resources, mail us at *cadcamcaeworks@gmail.com* with your contact information. With your contact record with us, you will be provided latest updates and informations regarding various technologies. The format to write us mail for resources is as follows:

Subject of E-mail as *Application for resources of .......................... book*.
Also, given your information like
*Name:*
*Course pursuing/Profession:*
*Contact Address:*
*E-mail ID:*

Note: We respect your privacy and value it. If you do not want to give your personal informations then you can ask for resources without giving your information.

## About Authors

The author of this book, Matt Weber, has written more than 15 books on CAD/CAM/CAE available in market. He has coauthored SolidWorks Simulation, SolidWorks Electrical, SolidWorks Flow Simulation, and SolidWorks CAM Black Books. The author has hands on experience on almost all the CAD/CAM/CAE packages. If you have any query/doubt in any CAD/CAM/CAE package, then you can contact the author by writing at cadcamcaeworks@gmail.com

The author of this book, Gaurav Verma, has written and assisted in more than 16 titles in CAD/CAM/CAE which are already available in market. He has authored Autodesk Fusion 360 Black Book, AutoCAD Electrical Black Book, Autodesk Revit Black Books, and so on. He has provided consultant services to many industries in US, Greece, Canada, and UK. He has assisted in preparing many Government aided skill development programs. He has been speaker for Autodesk University, Russia 2014. He has assisted in preparing AutoCAD Electrical course for Autodesk Design Academy. He has worked on Sheetmetal, Forging, Machining, and Casting designs in Design and Development departments of various manufacturing firms.

## For Any query or suggestion

If you have any query or suggestion, please let us know by mailing us on *cadcamcaeworks@gmail.com*. Your valuable constructive suggestions will be incorporated in our books and your name will be addressed in special thanks area of our books on your confirmation.

Page left blank intentionally

# Chapter 1

# Starting with Solid Edge

## Topics Covered

The major topics covered in this chapter are:

- *Starting Solid Edge 2020*
- *Starting a New Document*
- *User Interface Overview*
- *Setting User Interface*
- *Application Menu Tools*
- *3D Builder*
- *Settings*
- *Mouse Button Functions and Workflow*
- *Assessment*

# INTRODUCTION

Solid Edge is a 3D CAD software which works in both Parametric (History based) and Synchronous (direct) modeling style. You can create part, assembly, and 2D views using the software. You can integrate Team Viewer and Sharepoint with Solid Edge for collaborating with others. The software also provides support for FEA and CAM.

# STARTING SOLID EDGE

- To start **Solid Edge** in **Window 10** from **Start** menu, click on the **Start** button in the Taskbar at the bottom left corner. The **Start** menu will be displayed.
- Click on the **Solid Edge** folder. In this folder, select the **Solid Edge** icon; refer to Figure-1.

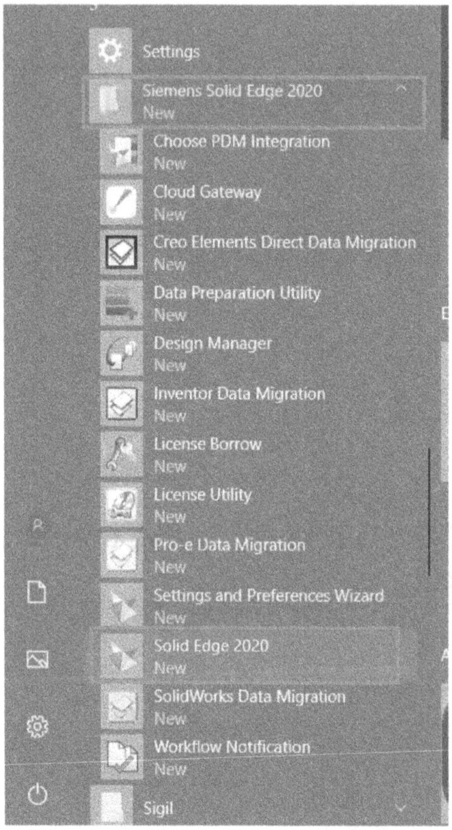

*Figure-1. Start menu*

- Before finishing the installation of this software, if you have selected the check box to create a desktop icon of this software then an icon of Solid Edge will be displayed on desktop.
- If you have not selected the check box to create the desktop icon but want to create the icon on desktop later, then drag and drop the **Solid Edge** icon from **Start** menu on the desktop.

After you have performed the above steps, the **Solid Edge** application window will be displayed with learn page; refer to Figure-2. Select desired link button from this page to access beginner's resources.

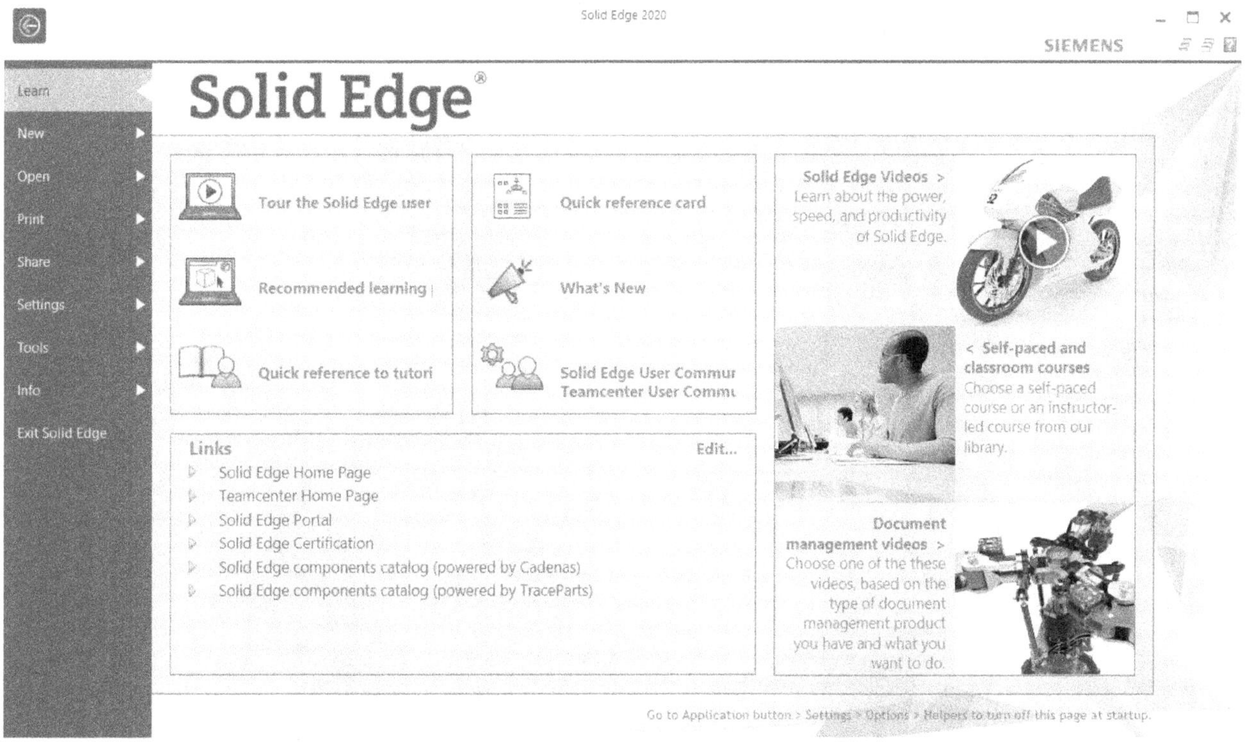

*Figure-2. Solid Edge application first view*

## STARTING A NEW PART DOCUMENT

**Application Menu: New cascading menu->New**

You can create a new document in **Solid Edge** by two ways: by using **New** menu or by **CTRL+N** shortcut key:

1. Click on **New** option from **Application** menu. The options to create new documents will be displayed in **New** cascading menu; refer to Figure-3. Click on the **New** button to create a new document. The **New** dialog box will be displayed; refer to Figure-4.

*Figure-3. Options to create new documents*

2. Press **Ctrl + N** from the keyboard. The **New** dialog box will be displayed; refer to Figure-4.

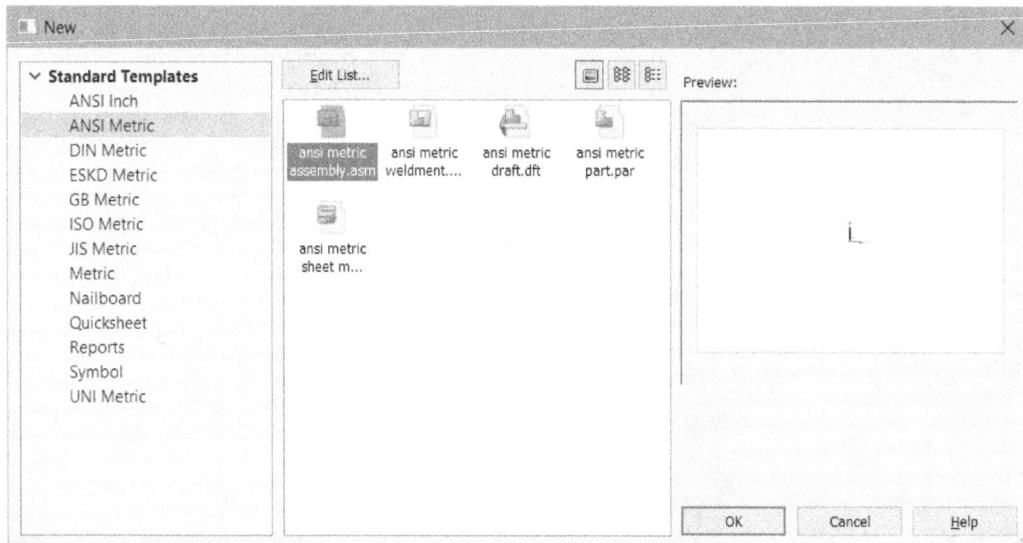

*Figure-4. New dialog box*

The template whose name ends with `assembly.asm` is used for starting new assembly files.

The template whose name ends with `weldment.asm` is used for starting new assembly files with weldment options.

The template whose name ends with `draft.dft` is for starting new drawing (drafting) file.

The template whose name ends with `part.par` is for starting new part files. 3D models are created using this template.

The templates whose name ends with `sheet metal.psm` are for starting new sheet metal design files.

- Select the desired template from the left area of the dialog box. ANSI Inch template uses inch as unit for length. ANSI Metric uses mm as unit for length. In the same way, other templates represent their standard unit system.
- On selecting the template from left area of the dialog box, the templates to create different type of files are displayed on the right in the dialog box. Double-click on the `ansi metric part.par` template file to start a part file with ANSI Metric standard. The interface will be displayed as shown in Figure-5.

*Figure-5. Solid Edge Interface for Part Modeling*

## USER INTERFACE OVERVIEW

Various common elements of user interface of Solid Edge are shown in Figure-6. These elements are discussed next.

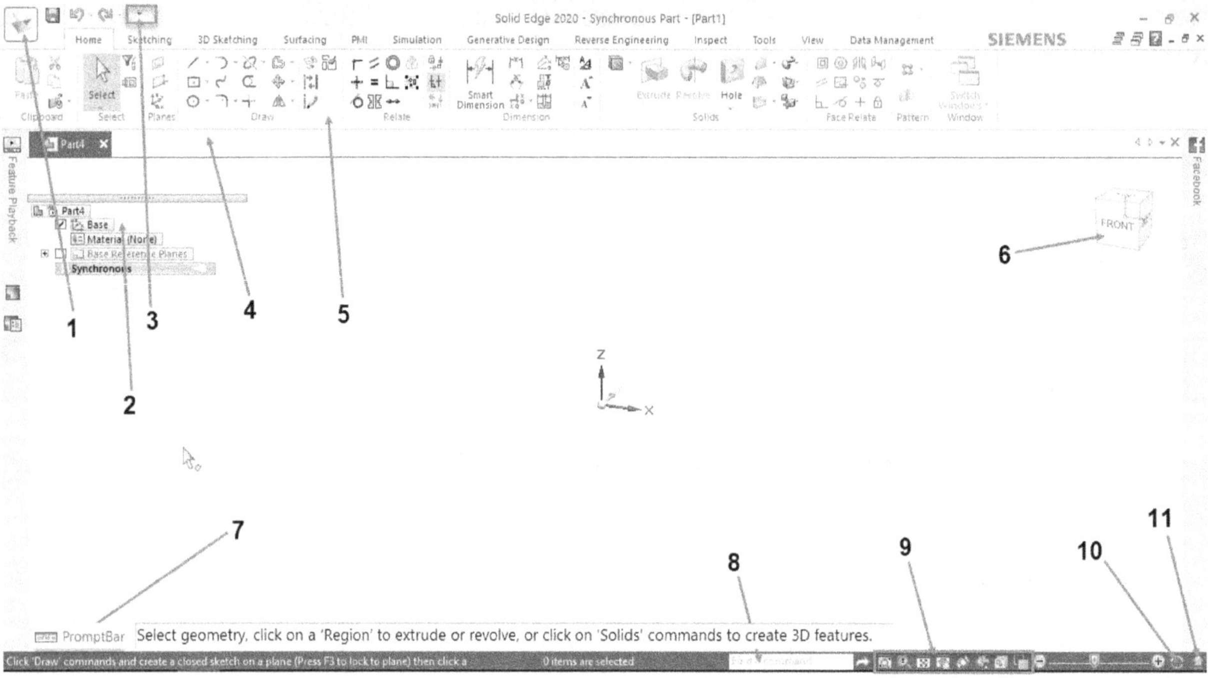

*Figure-6. User Interface Overview*

1. **Application button** : Displays the **Application** menu which provides access to all document level functions, templates, and standards.
2. **Path Finder** : Provides a quick way to identify and select model element.
3. **Quick Access toolbar** : The options in Quick Access toolbar are used to access common tools like Save, Undo, Redo etc. Use the down arrow in **Quick Access toolbar** to check additional options.
4. **Tabbed Document** : It shows all the document in the form of tiles which you have opened.
5. **Ribbon** : It is the group of tools from where you can use tools to create different features. It is classified into **tabs** and **groups**. It can be customized according to user. The procedure to add commands in **Ribbon** is discussed next.

## Customizing Ribbon

• Click on the **Arrow** button from **Quick Access Toolbar**, the drop-down will be displayed; refer to Figure-7.

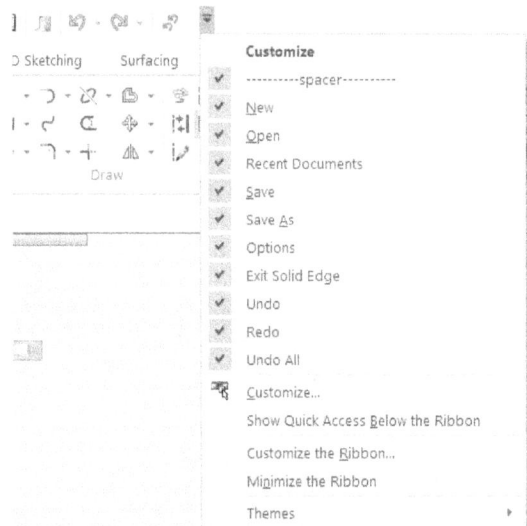

*Figure-7. Quick Access Toolbar drop-down*

- Select **Customize** option from the drop-down. The **Customize** dialog box will be displayed; refer to Figure-8.

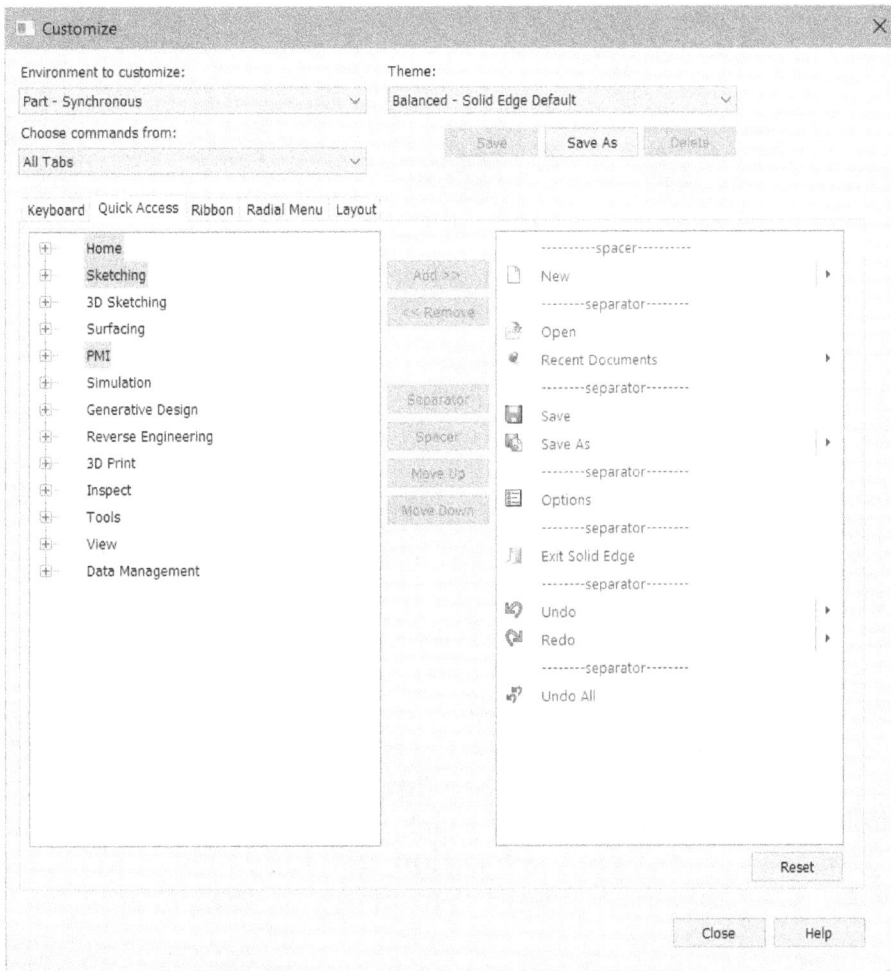

*Figure-8. Customize dialog box*

- Now, select **Ribbon** tab from **Customize** dialog box. After that select tools from left list box and location in right list box where you want to add the tool, and then click on the **Add>>** button. The added tools will be shown in the right list box; refer to Figure-9.

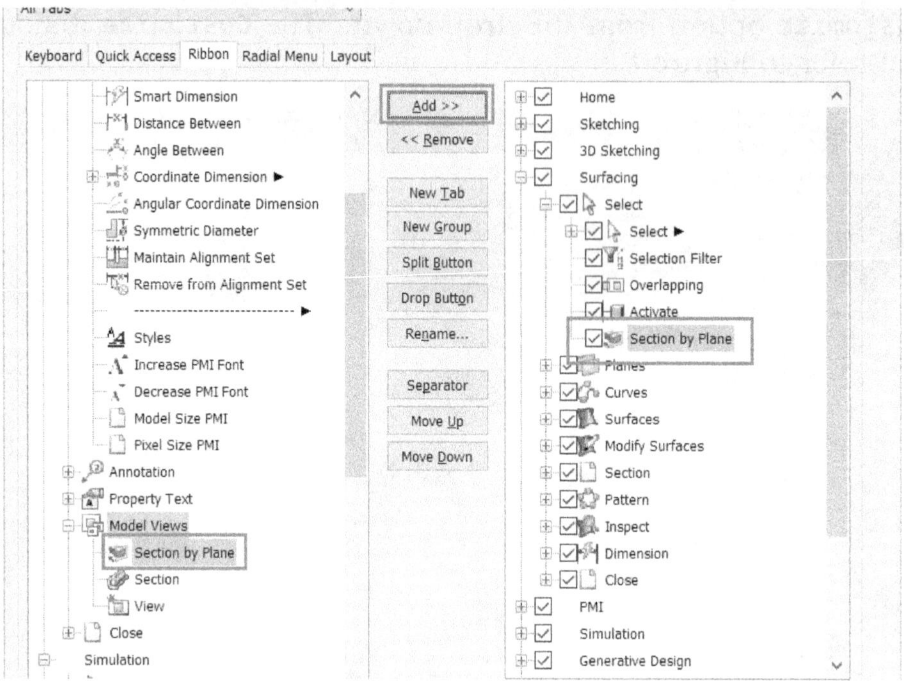

*Figure-9. Tool added in Ribbon by customizing*

- After setting desired parameters, click on the **Close** button.

6. **Quick View Cube**: It is used to change the orientation of object by rotating it or selecting the face. You can set parameters related to **Quick View Cube** by using **Quick View Cube Settings** tool in the **Orient** panel of **View** tab in the **Ribbon**. On clicking this tool, the **Quick View Cube Settings** dialog box will be displayed; refer to Figure-10. You can set the location, color, opacity, and other parameters of **Quick View Cube** in this dialog box. After specifying parameters, click on the **OK** button.

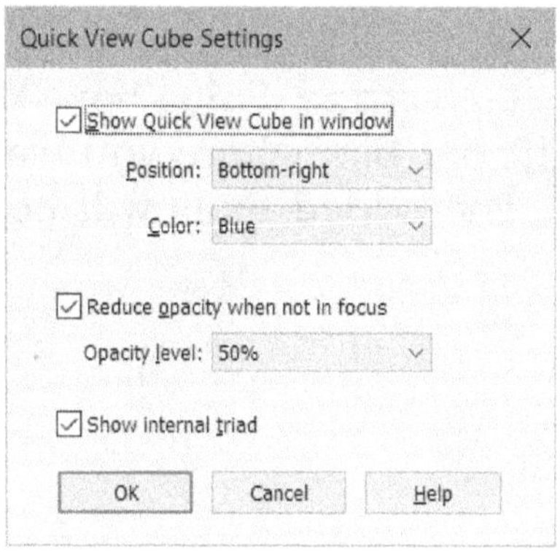

*Figure-10. Quick View Cube Settings dialog box*

7. **Prompt Bar**: It shows tips for the command you have selected.
8. **Command Finder**: It is used to find a command in the **User Interface**.
9. **View Tools**: These tools provide quick access to **Zoom**, **Fit**, **Pan**, **Rotate**, **View**, **Styles**, and **Saved View** options.
10. **Record**: User can record video while making any features in **Solid Edge 2020**. The steps are given next:

## Recording Solid Edge Screen

- Click on the red colored **Record** button at the bottom right corner of Application window to record video. The **Record Video** dialog box will be displayed; refer to Figure-11.

*Figure-11. Record Video dialog box*

- Click on **Record** button or press hot keys **Shift+F9** from keyboard to start recording the Solid Edge screen.
- After starting the recording, you can pause the recording anytime by pressing **Shift+F9** from keyboard.
- Click on **Stop** button or press hot keys **Shift+F10** to stop the video
- User can also record his/her voice. To do so, select the **Record audio** check box from the dialog box. You can select the device for recording audio from the drop-down below this check box.
- You can select the codec for video recording from the **Compression Codec** drop-down.
- After recording the video, click on **Save** button to save the video. The **Save Video** dialog box will be displayed; refer to Figure-12.

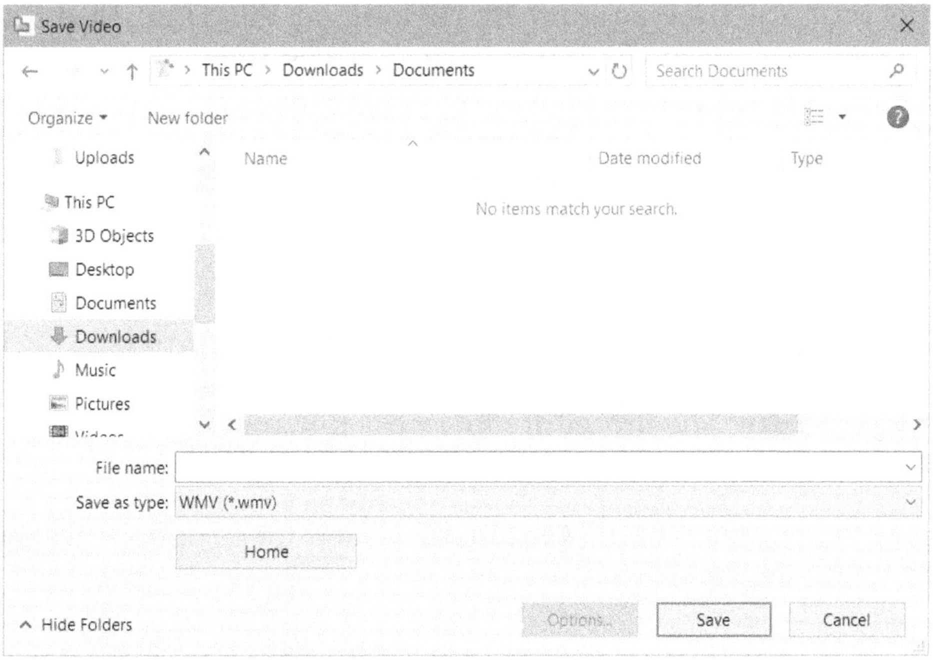

*Figure-12. Save Video dialog box*

- Specify desired name of file in the **File name** edit box and click on the **Save** button to save the recording.

## Uploading Recording on YouTube

- If you want to upload the recorded file on YouTube then click on the **Upload** button from the dialog box. You can also use the **Upload to YouTube** button from the bottom right corner of the **Application** window to upload video on YouTube. The **Upload to YouTube** dialog box will displayed; refer to Figure-13.

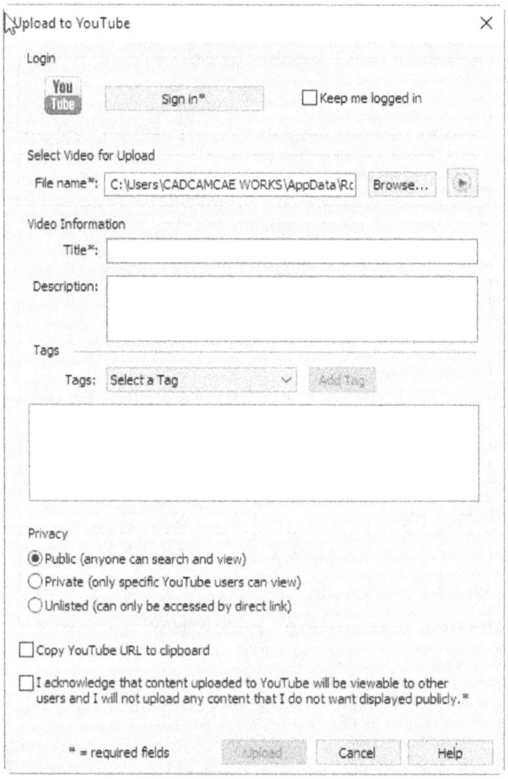

*Figure-13. Upload to YouTube dialog box*

- Login into your YouTube account by clicking on the **Sign In** button. Set the parameters as desired in the **Upload to YouTube** dialog box and click on the **Upload** button.

# APPLICATION MENU TOOLS

The tools in the **Application** menu are used to manage application and file related functions. These tools are discussed next.

## Learn Options in Application Menu

The options in **Learn** cascading menu are used to access various resources to learn about **Solid Edge**. The procedure to access different options in this menu are discussed next.

- Click on the **Learn** option in the **Application** menu. The options will be displayed as shown in Figure-14.

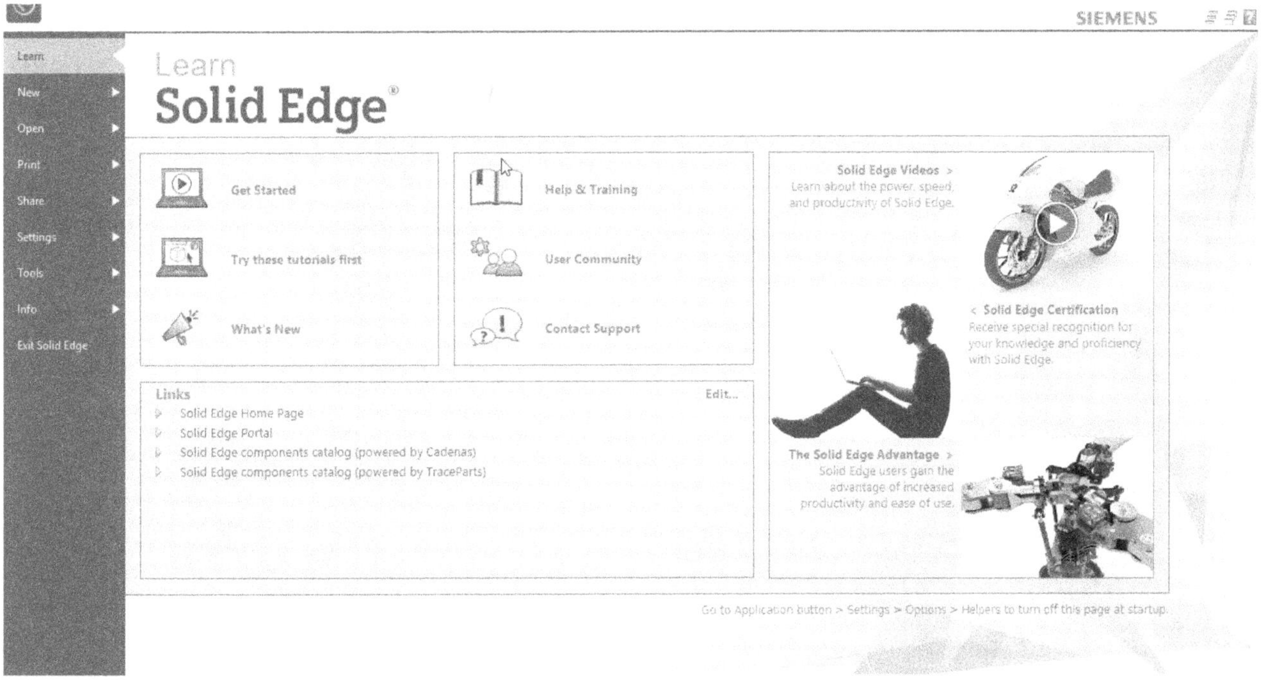

*Figure-14. Learn cascading menu*

- Click on the **Get Started** option in the Learn page to check beginners videos on **Solid Edge**.
- Click on the **Try these tutorials first** to know how to create basic objects.
- Click on the **What's New** to find the latest feature added to this version of software.
- Click on **Help & Training** to learn Solid Edge.
- Click on **User Community** to share ideas on Solid Edge.
- Click on **Contact Support** for help on Solid Edge software.

## New Options in Application Menu

The options in **New** cascading menu are used to create new documents. The procedure is discussed next.

- Click on **New** option in the **Application** menu. The options to start new documents will be displayed as shown in Figure-15.

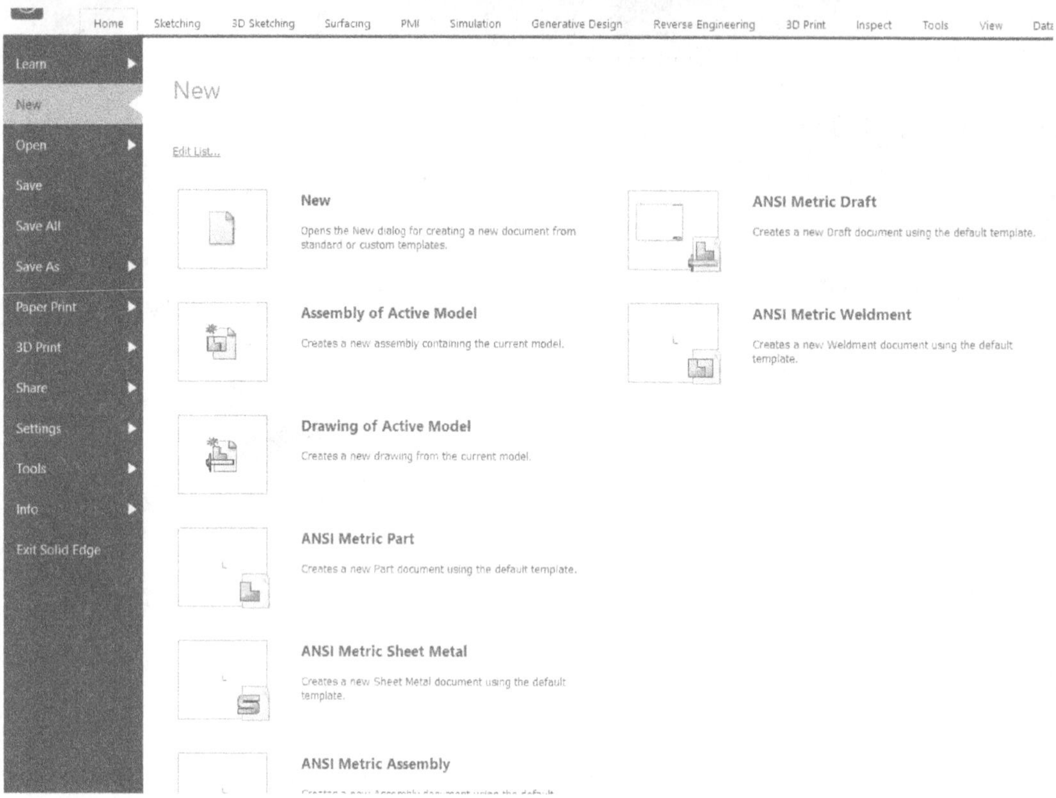

*Figure-15. New cascading menu*

- Click on **New** tool from **New** cascading menu. The **New** dialog box will be displayed with list of templates; refer to Figure-16.

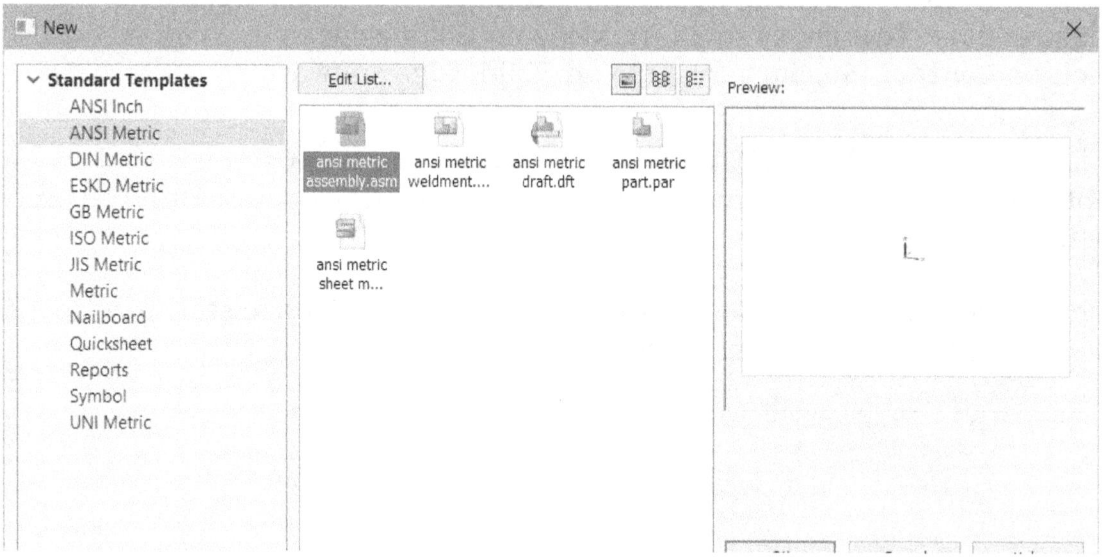

*Figure-16. New dialog box*

- Select `ansi metric part.par` template to create a new part document using default template.
- Select `ansi metric sheet metal.psm` template to create a new Sheet Metal document using the default template.
- Select `ansi metric assembly.asm` template to create a new assembly using the default template.
- Select `ansi metric draft.dft` template to create a new draft (Drawing) document using default template.
- Select `ansi metric weldment.asm` template to create a new welding design document using default template.

NOTE : You can also edit templates by using **Edit List** button from the **New** dialog box displayed on clicking **New** tool from **New** cascading menu of **Application** menu.

## Opening a File

The options in **Open** cascading menu are used to open files from local drive.

*   Click on **Browse** button 📂 from the **Open** cascading menu to browse files. The **Open File** dialog box will displayed; refer to Figure-13.

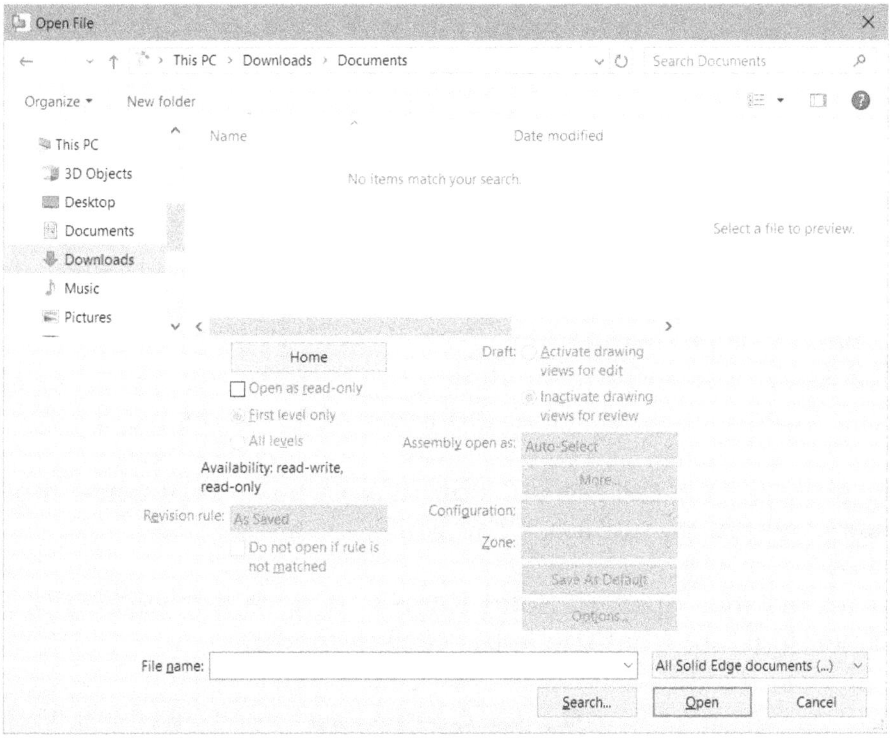

*Figure-17. Open file dialog box*

*   Select the format of file which you want to open in Solid Edge from the **File Type** drop-down.
*   Browse to the desired location and select the file you want to open in Solid Edge.
*   After selecting file, click on the **Open** button from the dialog box to open the file. The file will be displayed in Solid Edge.

## Saving File

The **Save** tool in **Application** menu is used to save the model file. If you are saving the file for first time then **Save As** dialog box will be displayed and once you have saved the file then no dialog box will be displayed later after selecting this tool. The procedure to use this tool is given next.

*   Click on the **Save** tool from the **Application** menu. The **Save As** dialog box will be displayed; refer to Figure-18.

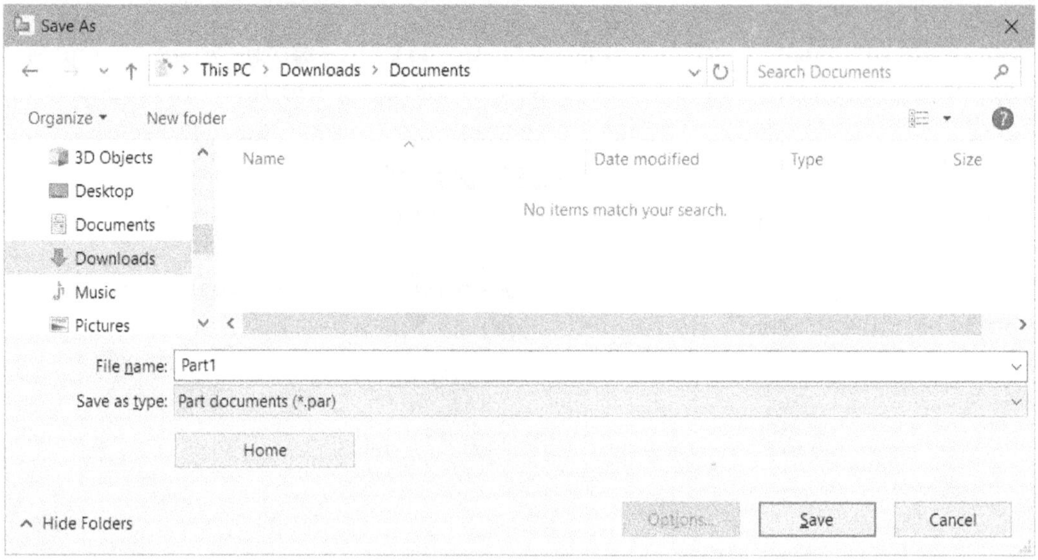

*Figure-18. Save As dialog box*

- Specify desired name of file in the **File name** edit box. If you want to save the file in format different from native format of Solid Edge then select the desired option from the **Save as type** drop-down.
- Click on the **Save** button from the dialog box to save the file. After saving the file once, modify the model as needed and press **CTRL+S** to save the file.

Note that you should save the file as soon as you have performed modifications in model to make sure you do not loose the work.

If you have multiple files open in Solid Edge and have not saved the files yet then click on the **Save All** tool from the **Application** menu.

## Save As Tools

The tools in the **Save As** cascading menu are used to save the model in different formats with different names; refer to Figure-19. These tools are discussed next.

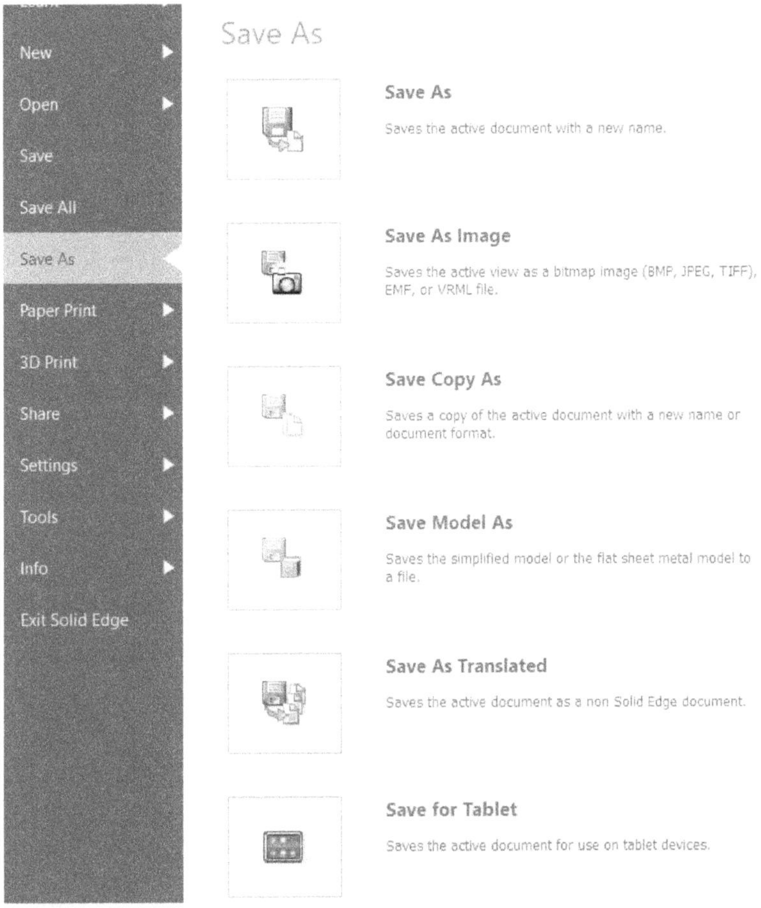

New
Open
Save
Save All
Save As
Paper Print
3D Print
Share
Settings
Tools
Info
Exit Solid Edge

Save As

**Save As**
Saves the active document with a new name.

**Save As Image**
Saves the active view as a bitmap image (BMP, JPEG, TIFF), EMF, or VRML file.

**Save Copy As**
Saves a copy of the active document with a new name or document format.

**Save Model As**
Saves the simplified model or the flat sheet metal model to a file.

**Save As Translated**
Saves the active document as a non Solid Edge document.

**Save for Tablet**
Saves the active document for use on tablet devices.

*Figure-19. Save As cascading menu*

The procedure to use **Save As** tool of **Save As** cascading menu is same as discussed for **Save** tool earlier.

## Saving Model in Image Format

The **Save As Image** tool is used to save a file in image format like JPEG, TIFF, and so on. The procedure to use this tool is given next.

- After creating model, set it in center of viewport using View tools. You can use zoom and pan tools of **View** toolbar to orient model.
- After orienting the model, click on the **Save As Image** tool from the **Save As** cascading menu of the **Application** menu. The **Save As Image/VRML** dialog box will be displayed; refer to Figure-20.
- Specify desired name of file in **File name** edit box and select the desired format from the **Save as type** drop-down.
- After setting desired parameters, click on the **Save** button. The file will be saved in location selected in the dialog box.

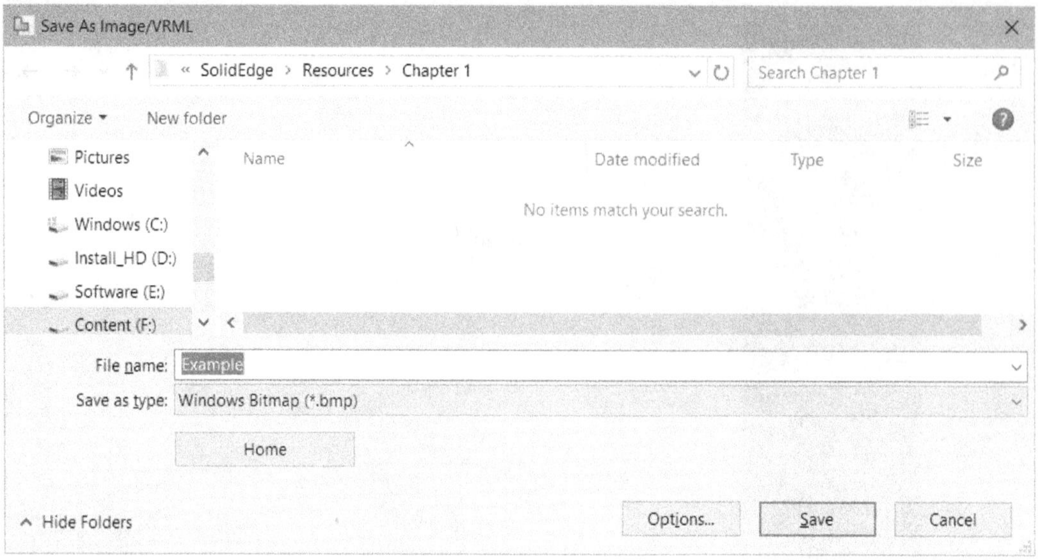

*Figure-20. Save As Image/VRML dialog box*

Similarly, the **Save Copy As** tool in the **Save As** cascading menu is used to store copy of the current active model by a new name in local drive and the **Save Model As** tool is used to save simplified model with no design history of current file.

## Exporting Solid Edge Model

The **Save As Translated** tool is used to export Solid Edge model in non-native CAD formats. The procedure to use this tool is given next.

* Click on the **Save As Translated** tool from the **Save As** cascading menu of the **Application** menu. The **Save As Translated** dialog box will be displayed; refer to Figure-21.

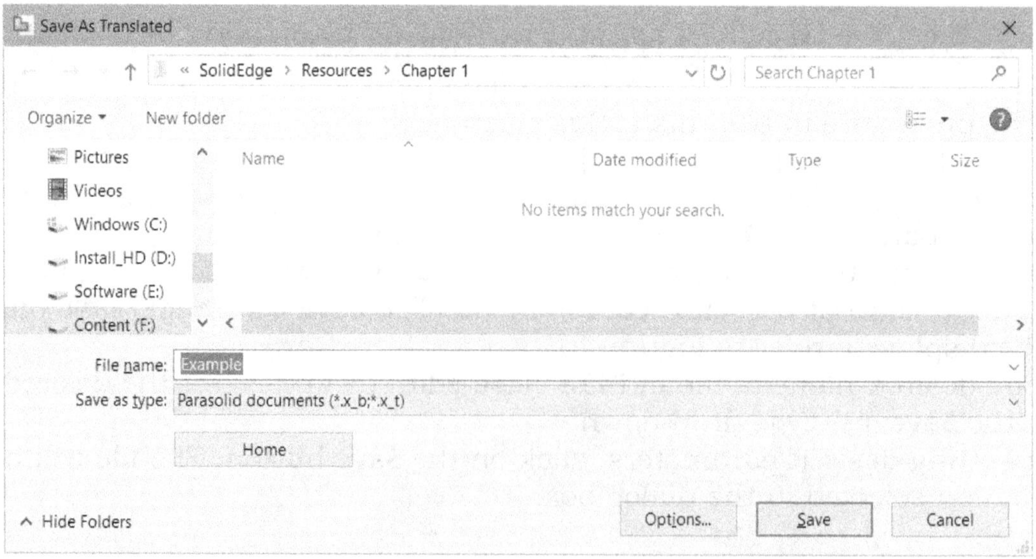

*Figure-21. Save As Translated dialog box*

* Specify the name and format of file in the dialog box.
* Click on the **Options** button to modify parameters related to translation in selected format. In our case, we have selected **IGES Document** option from the **Save as type** drop-down so the **IGES Export Wizard** dialog box will be displayed; refer to Figure-22. (IGES is one of the most common translation format for CAD models)

## IGES Export Options

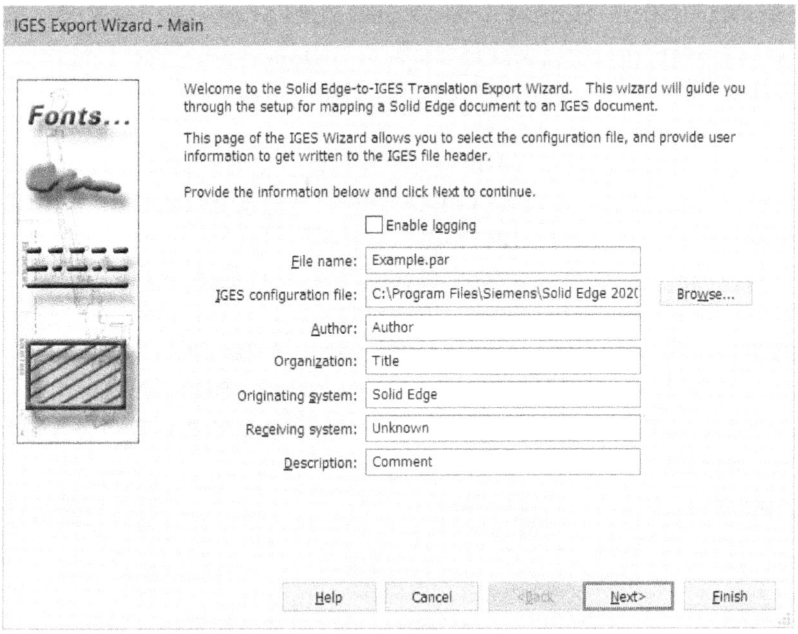

*Figure-22. IGES Export Wizard dialog box*

- Select the **Enable logging** check box if you want to generate a log file of translation.
- Click on the **Browse** button and select the desired configuration file for translation if you want to change the translation definition.
- Specify the desired user information in other edit boxes of the dialog box. Click on the **Next** button from the dialog box. The **Types** page of **IGES Export Wizard** will be displayed; refer to Figure-23.

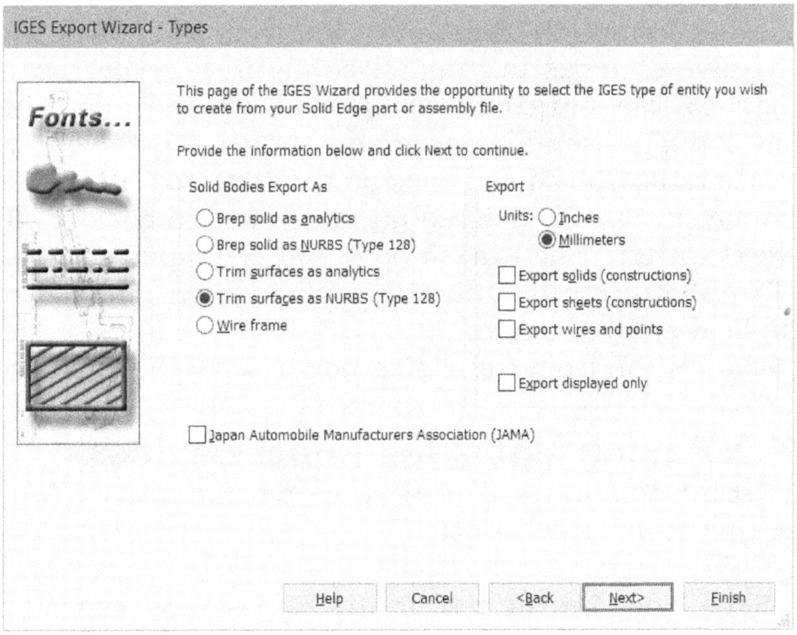

*Figure-23. Types page of IGES Export Wizard dialog box*

- Select the desired radio button from the **Solid Bodies Export As** area to define how solid bodies will be exported in iges format. Select the **Brep solid as analytics** radio button to use basic analytic surfaces (planar and cylindrical) wherever possible in exported file. Select the **Brep solid as NURBS (Type 128)** radio button to export the model as B-spline surface wherever possible. Select the **Trim surfaces as analytics** radio button to convert solid faces into analytic surfaces in exported file. Select the **Trim surfaces as NURBS (Type 128)** radio button to convert solid faces into B-spline surfaces in the exported file. Select the Wire frame radio button to export solid as skeleton of model comprised of lines, arcs, and spline curves.

- Select the desired radio button for unit from the **Export** area of the dialog box.
- Select the **Export solids (constructions)** check box to export constructions geometries in solid form to exported file. Similarly, you can use the **Export sheets (construction)** check box to export surfaces and **Export wires and points** check box to export curves used in construction.
- Select the **Japan Automobile Manufacturers Association (JAMA)** check box to export a trimmed version of IGES file as per Japanese manufacturer standards.
- Click on the **Next** button from the dialog box. The **Save** page of **IGES Export Wizard** dialog box will be displayed; refer to Figure-24.

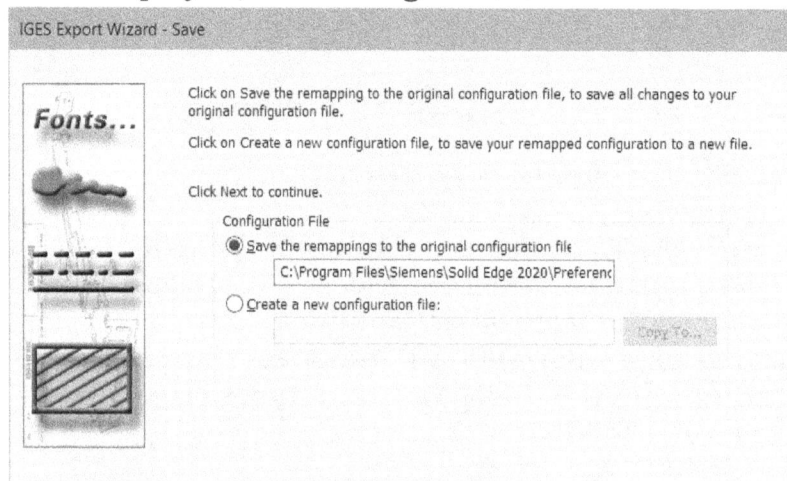

*Figure-24. Save page of IGES Export Wizard dialog box*

- If you want to save changes in original configuration file then select the **Save the remappings to the original configuration file** radio button. If you want to create a new configuration file based on changes then select the **Create a new configuration file** radio button and click on the **Copy To** button. The **Save As** dialog box will be displayed. Specify desired name of file and click on the **Save** button.
- Click on the **Next** button. The **Finish** page will be displayed.
- Click on the **Finish** button to apply the configuration. The **Save As Translated** dialog box will be displayed again.
- Click on the **Save** button from the dialog box to save the exported file.

## Saving Model for Tablet Devices

The **Save for Tablet** tool is used to save model for use in Tablet devices. The procedure to use this tool is given next.

- Click on the **Save for Tablet** tool from the **Save As** cascading menu of the **Application** menu. The **Save As** dialog box will be displayed as shown in Figure-25.

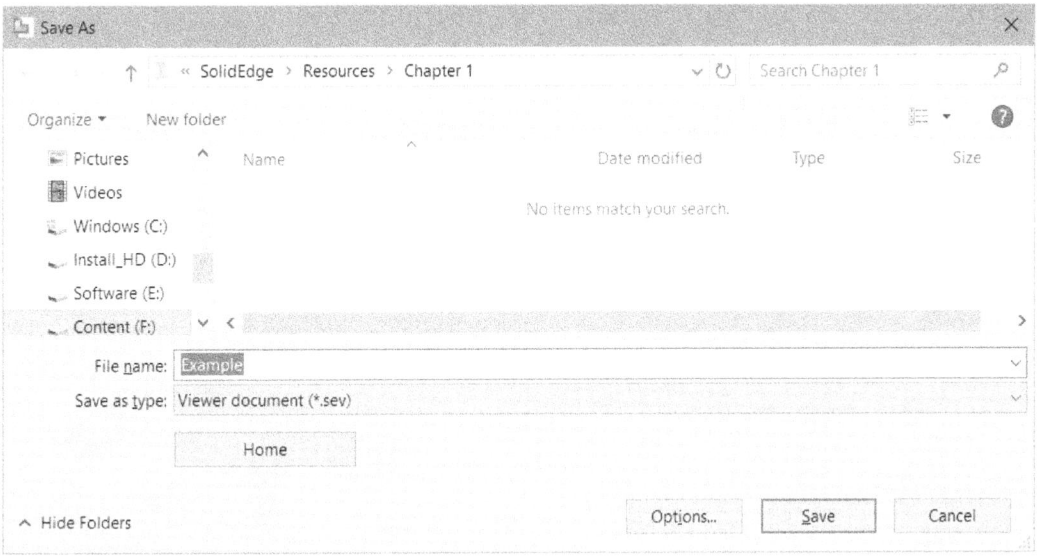

*Figure-25. Save As dialog box*

- Specify the desired name of file and click on the **Save** button. You can transfer this file to your tablet for viewing.

# Print File

The options in **Paper Print** cascading menu are used to print the model on paper; refer to Figure-26. The procedure to print model is given next

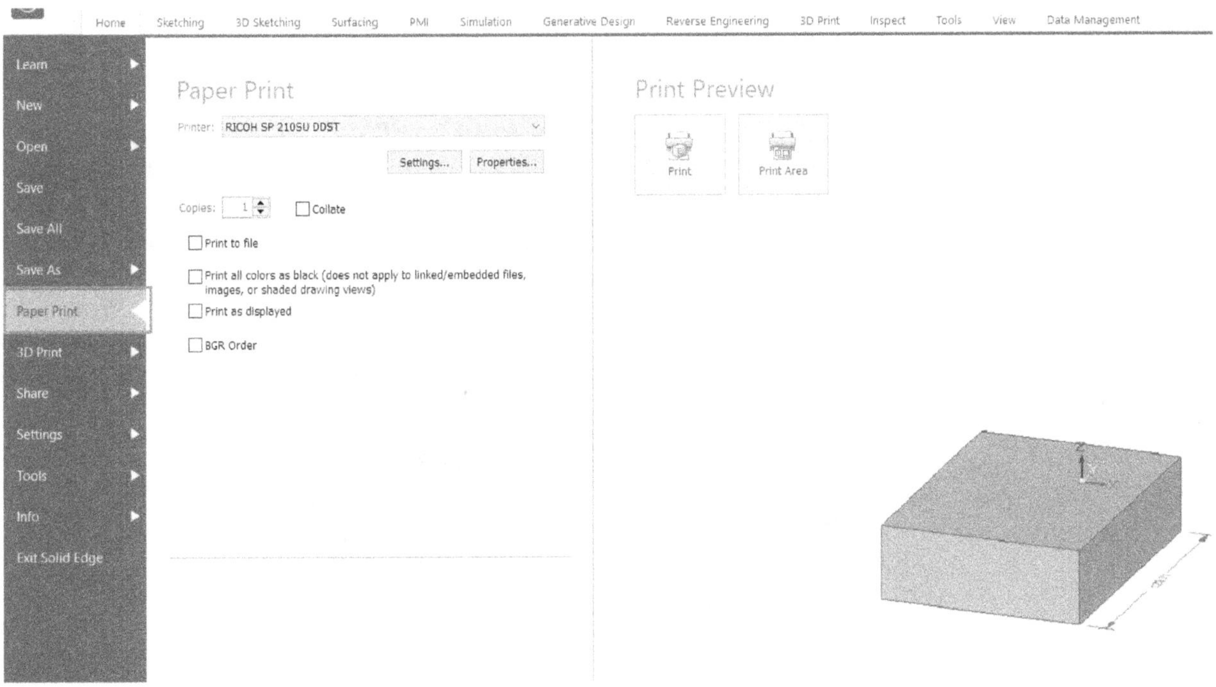

*Figure-26. Paper Print cascading menu*

- Select the desired printer to be used for printing from the **Printer** drop-down. If you cannot find your printer in the list then install the drivers of your printer.
- Using the **Settings** button, you can adjust the print range and scale of model for printing. Using the **Properties** button, you can define printer properties like orientation of paper, type of paper, print quality, and so on.
- Set the number of copies you want to print in the **Copies** spinner. Select the **Collate** check to print in sequence if there are more than one type of prints.

- Select the **Print to file** check box to print the model as a digital file. If you now click on the **Print** button from the cascading menu then the **Print to File** dialog box will be displayed; refer to Figure-27.

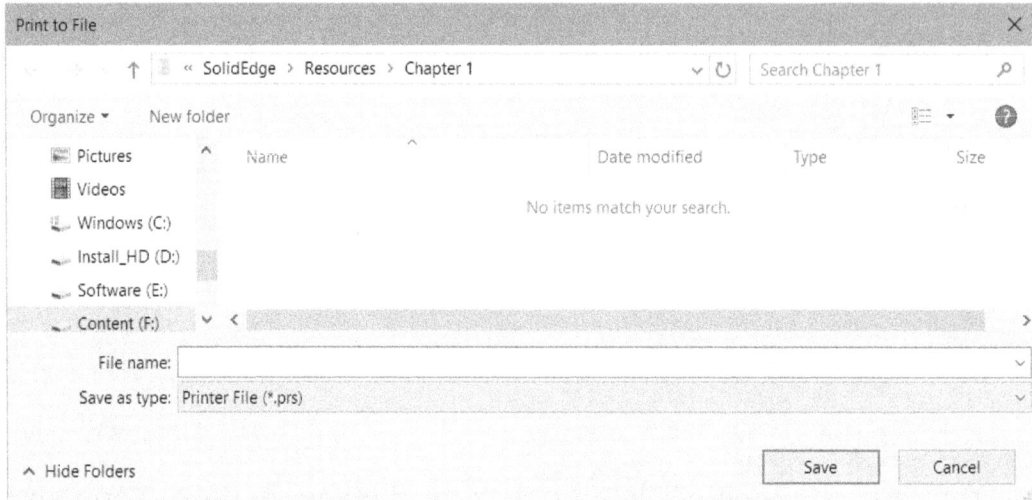

*Figure-27. Print to File dialog box*

- Select the **Print all colors as black** check box to print all the colored objects in Black and white format.
- Select the **Print as displayed** check box to print all the annotations and model as they are being displayed in the viewport.
- Select the **BGR Order** check box to print using Red, Blue, and Green color inks of printer.
- Click on the **Print** tool from the **Paper Print** cascading menu to print everything being displayed in the viewport. If you want to print specific portion of the viewport then click on the **Print Area** tool from the **Paper Print** cascading menu. You will be asked to define start point and end point of rectangle to define boundary of print area.
- Create the boundary enclosing the objects to be printed on paper; refer to Figure-28. The **Print Area** dialog box will be displayed; refer to Figure-29.

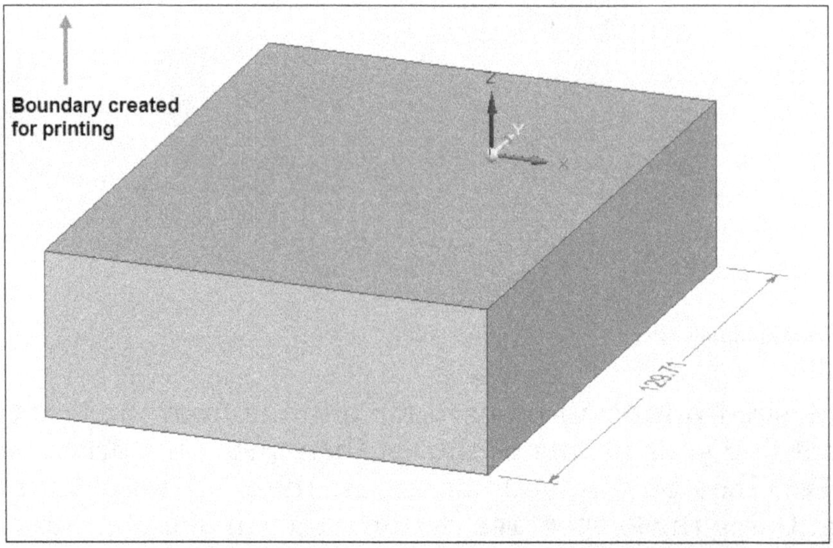

*Figure-28. Creating boundary for print*

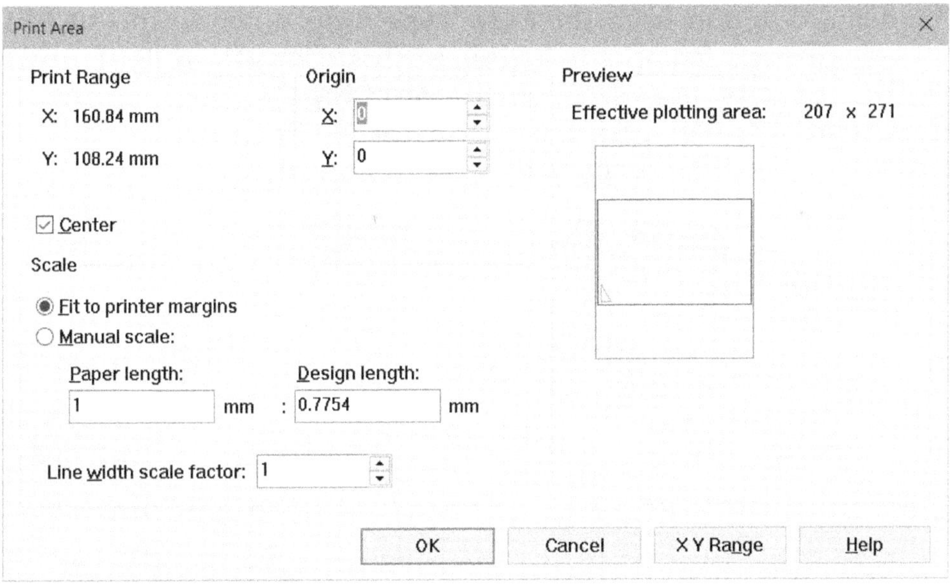

*Figure-29. Print Area dialog box*

- You can modify the orientation, location, and scale factor for printing in the dialog box. If you want to modify the print area then click on the **X Y Range** button
- After setting desired parameters, click on the **OK** button from the dialog box. The model will be printed in the file or paper based on your selections in the **Paper Print** cascading menu.

## Preparing File for 3D Printing

The options in the **3D Print** cascading menu are used to create file ready for 3D printing or post processing by 3D Printing software; refer to Figure-30. These options are discussed next.

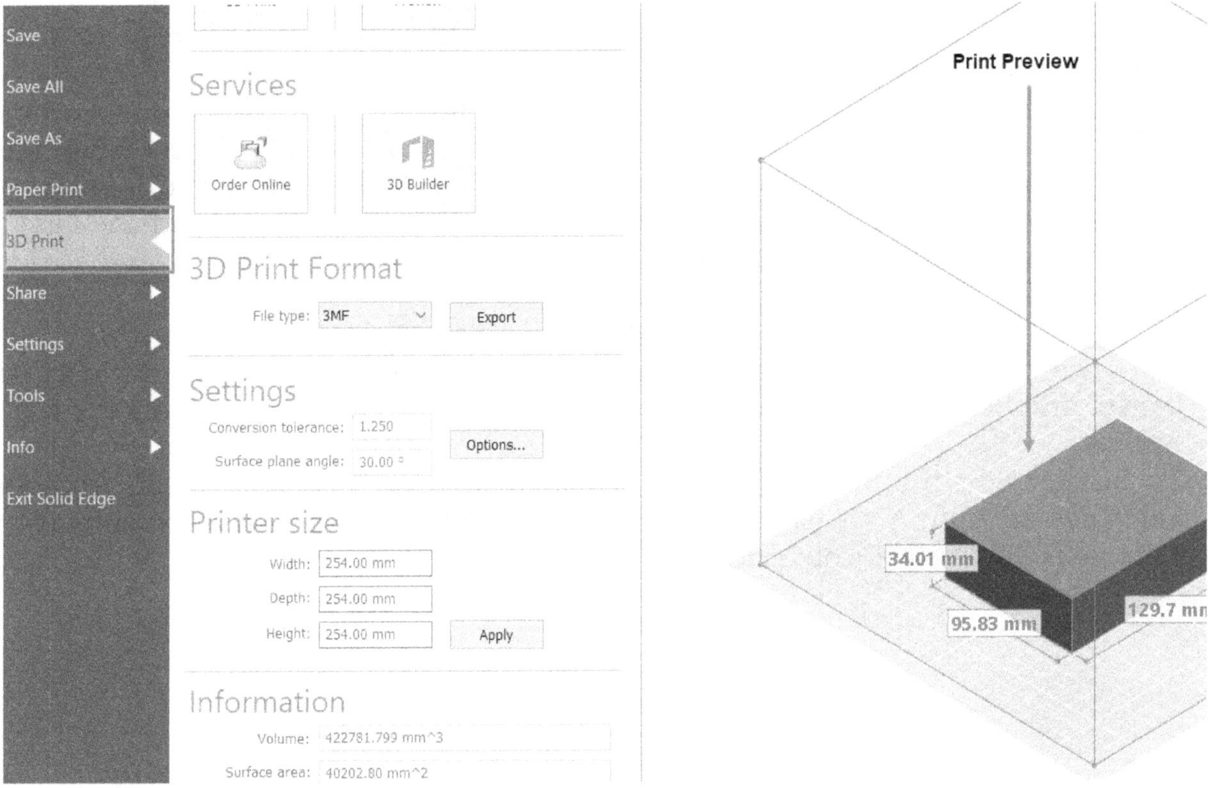

*Figure-30. 3D Print cascading menu*

- Select the desired option from the **File type** drop-down in the **3D Print Format** section. There are two formats available in this drop-down; 3MF and STL.
- Click on the **Options** button from the **Settings** section to define tolerance parameters related to conversion of model for 3D print ready model. The **Export Options** dialog box will be displayed; refer to Figure-31.

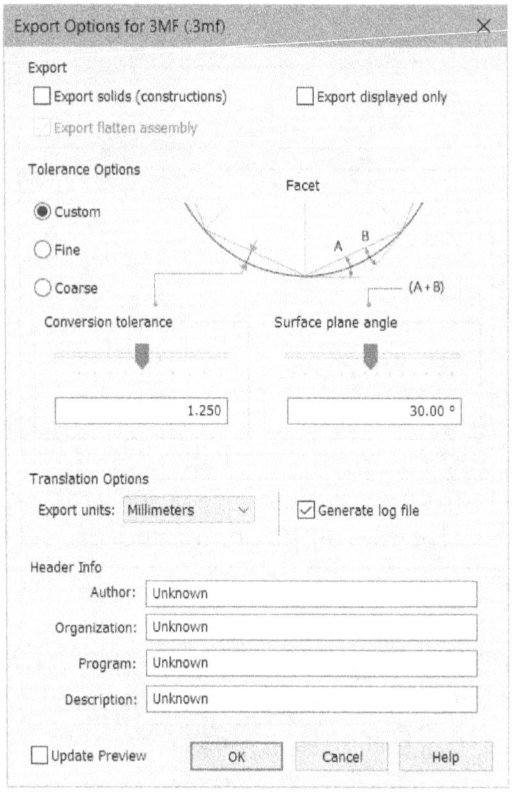

*Figure-31. Export Options dialog box*

- Select the check boxes from the **Export** area to define how elements of solid model will be exported in the 3D print ready model.
- Select desired radio button from the **Tolerance Options** area of the dialog box to define tolerance within which model will be converted to facets. Select **Fine** radio button to get better finish in converted model. Select the **Coarse** radio button to get faster conversion of model with lower finish. If you have selected the **Custom** radio button then you can manually define the conversion tolerance and surface plane angle.
- Set the other parameters as discussed earlier and click on the **OK** button.
- If you want to modify X, Y, and Z range of printer then specify the values in **Width**, **Depth**, and **Height** edit boxes of the **Printer Size** section and click on the **Apply** button.
- To further edit the 3D print model, click on the **3D Builder** button from the **Services** section. A message box will be displayed asking you to download the Microsoft 3D Builder app; refer to Figure-32.

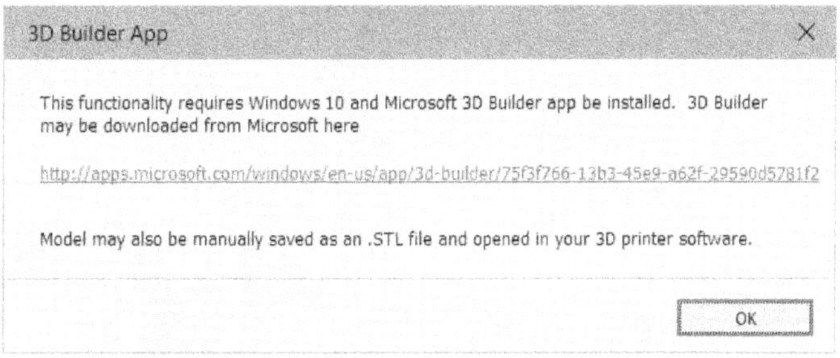

*Figure-32. Message box*

- Click on the link in the message box to download the app. A web page will be displayed in default web browser. Click on the **Get** button from the web page to download and then click on the **Install** button.
- Now, if you again click on the **3D Builder** button from the **3D Print** cascading menu. The 3D Builder application will be displayed; refer to Figure-33.

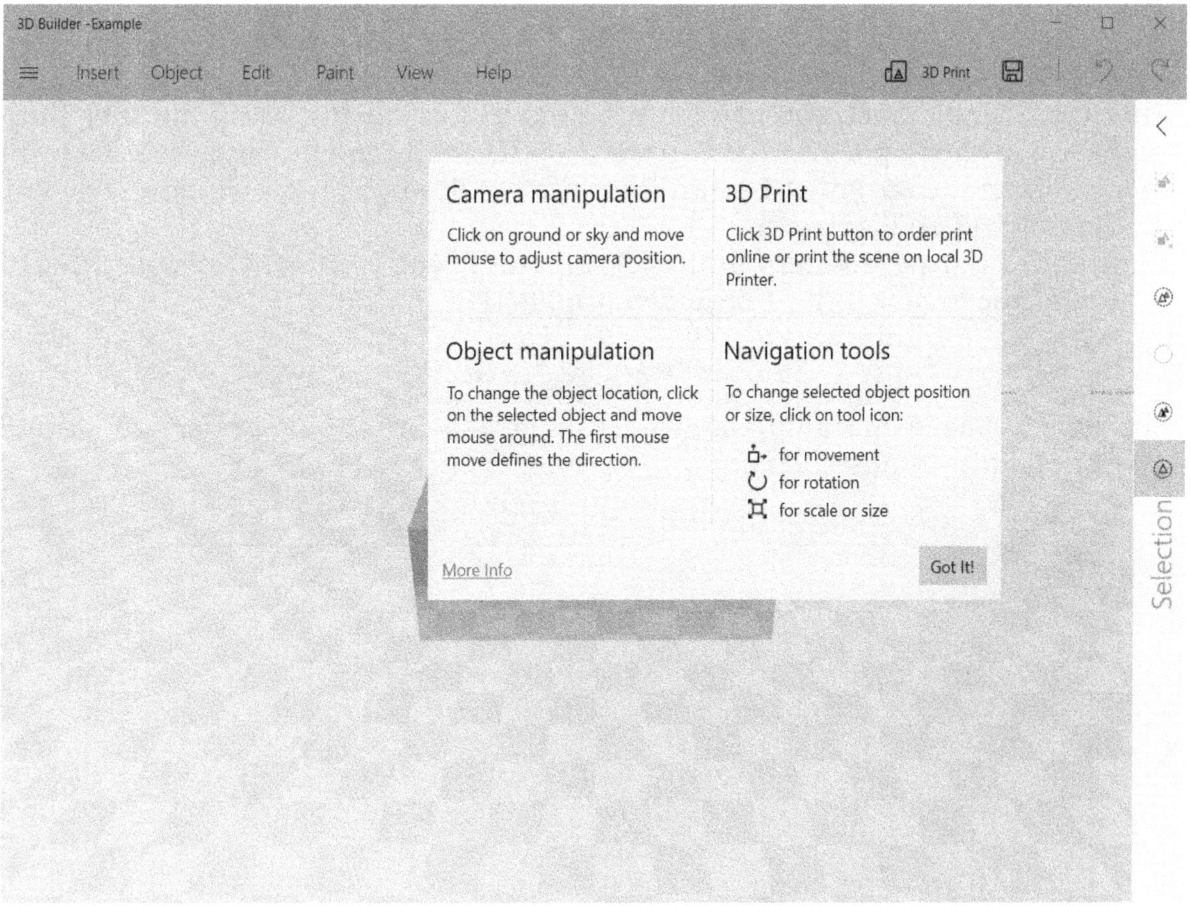

*Figure-33. 3D Builder application*

- Click on the **Got It!** button and select the model to be modified. The arrow handles to change location of model on printer bed will be displayed. Drag the arrow handles to change location; refer to Figure-34. Using the tools in the toolbar, you can rotate and scale the model.

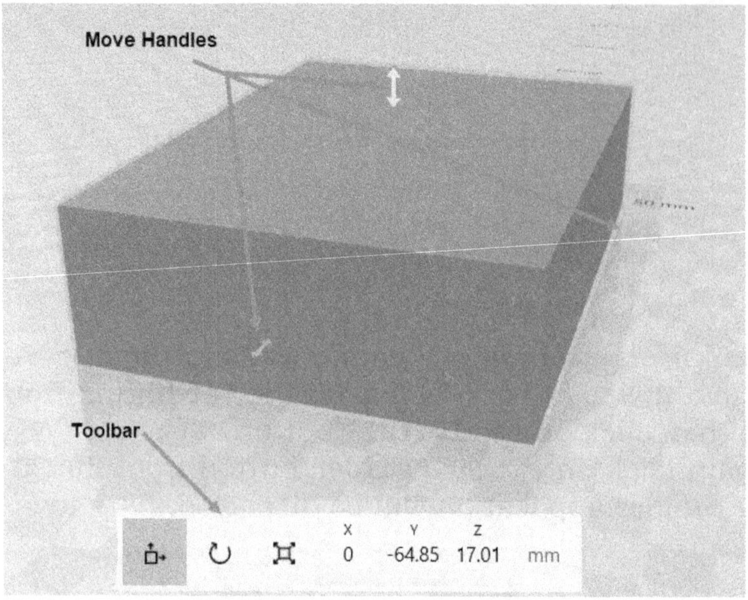

*Figure-34. Tools and arrow handles in 3D Builder*

- After modifying model, you can save the model for sharing using the **Save** button at the top in the application. If you want to 3D print model using your 3D printer then click on the **3D Print** button from the top in application window. The **Print 3D** window will be displayed.
- Set the desired parameters and click on the **Print** button. Click on the **Close** button at top-right corner to exit the application.

## Share

The options in **Share** cascading menu are used to share the file; refer to Figure-35. These options are discussed next.

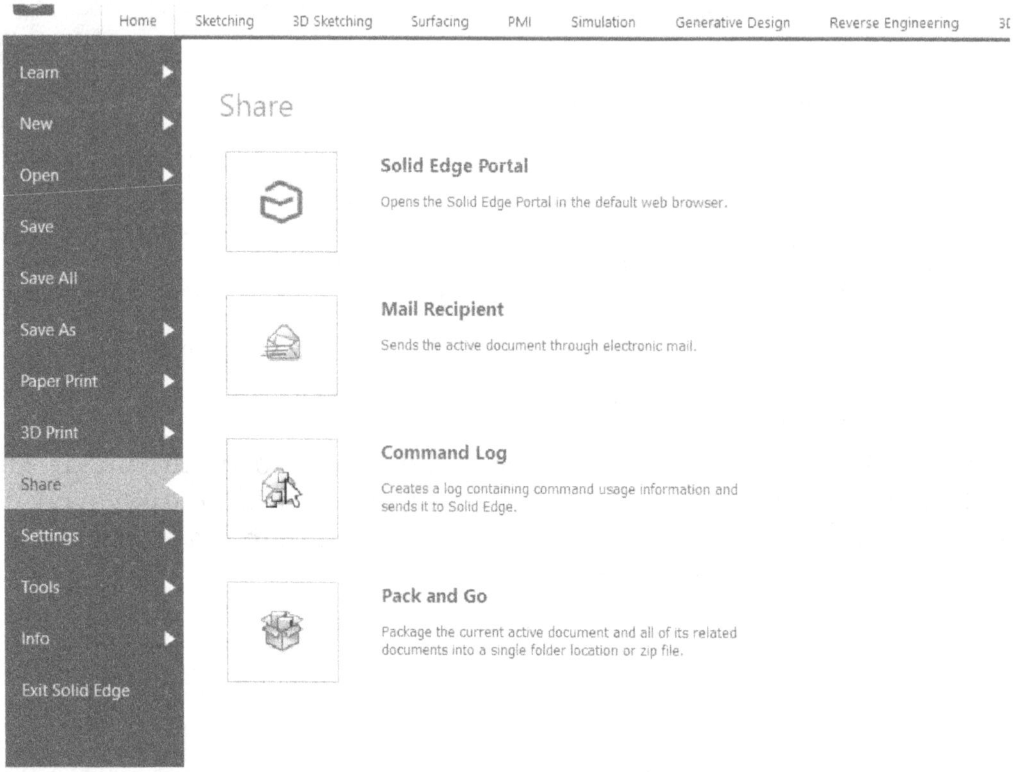

*Figure-35. Share cascading menu*

## Solid Edge Portal

The **Solid Edge Portal** tool is used to open the Solid Edge portal in the default browser where you can collaborate with friends and colleagues. You need to Sign In to your siemens account for collaboration by this method.

## Mail Recipient

The **Mail Recipient** tool is used to send the current document via E-mail.

## Command Log

The **Command Log** tool is used to creates a log containing command usage information and sends it to Solid Edge.

## Pack and Go

The **Pack and Go** tool is used to package the current active document and all of its related documents into a single folder location or zip file.

# Settings

The options in **Settings** cascading menu are used to customize Solid Edge as per your need.

* Click on **Settings** option from **Application** menu. The **Settings** cascading menu will be displayed; refer to Figure-36.

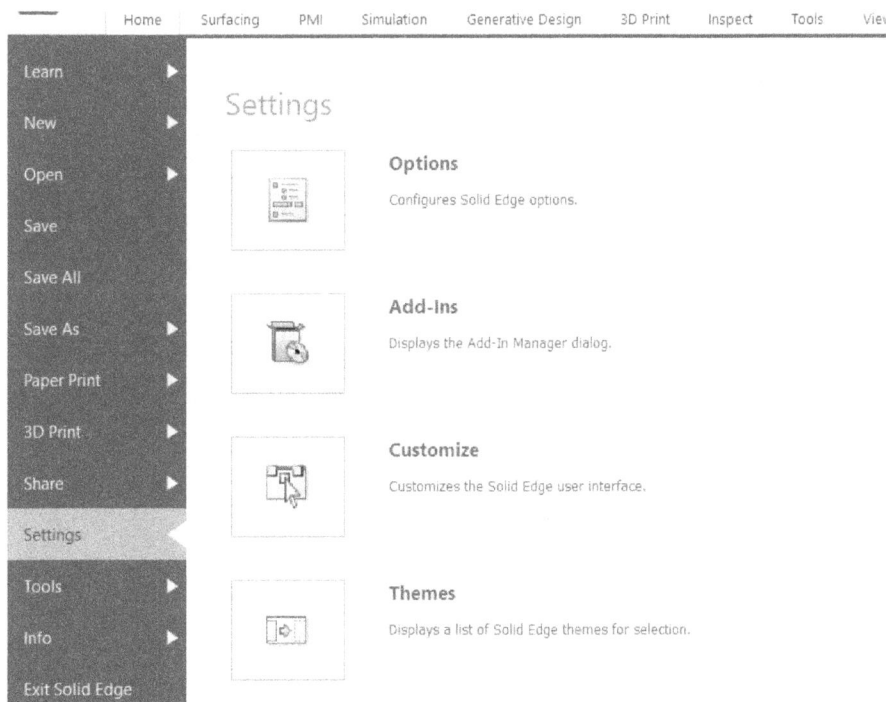

*Figure-36. Settings cascading menu*

# Option

The **Options** tool in **Settings** cascading menu is used to define various default parameters for Solid Edge.

* Click on the **Options** tool, the **Solid Edge Options** dialog box will be displayed; refer to Figure-37.

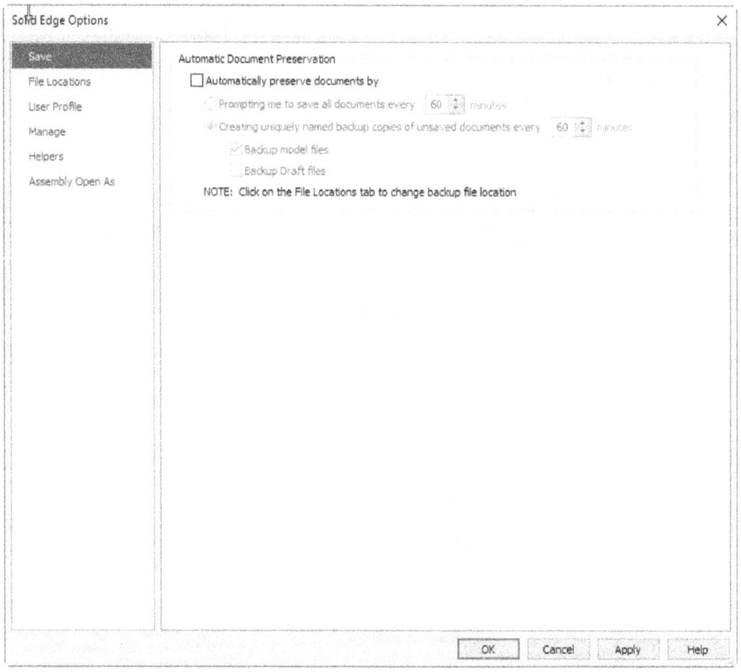

*Figure-37. Options dialog box when no document is open*

When a part model is already open in the software then **Solid Edge Options** dialog box will be displayed as shown in Figure-21. Some important options of Solid Edge are discuss next.

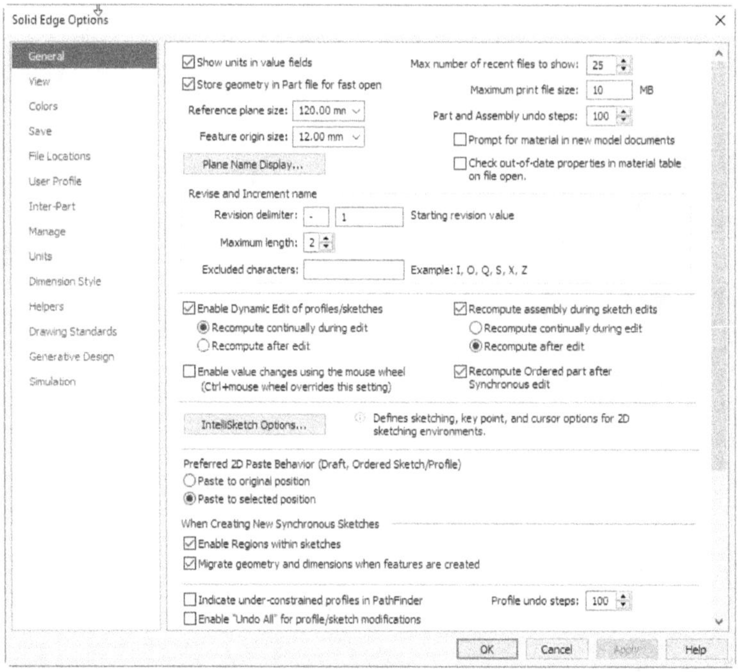

*Figure-38. Solid Edge Options dialog box when a document is open*

## General

The options in the **General** page of **Options** dialog box are used to set parameters related to general functions of software like number of file that can be shown in recent list, reference plane size, paste behavior, and so on. Some important options in this page are discussed next.

- Select the **Show units in value fields** check box to display the units when you are typing value of a parameter in dynamic edit box.
- You can specify the maximum number of documents displayed in the recent list by specify respective value in the **Max number of recent files to show** edit box.
- The value specified in the **Maximum print file size** edit box is used to define the file size of print ready file of model.
- You can specify the maximum number of undo steps that can be stored in volatile memory in **Part and Assembly undo steps** edit box. Note that a higher value specified here will cause more usage of RAM.
- In the **Reference plane size** and **Feature origin size** edit boxes, you can specify the size of reference plane and origin being displayed when starting a new model file.
- The **Plane Name Display** button is used to change font, color, text height of reference planes.
- Select the **Prompt for material in new model documents** check box to ask for selecting material when you are starting a new part model file.
- Select the **Check out-of-date properties in material table on file open** check box to check the model for an material data which has not been updated as per new material database.
- The options in the **Revise and Increment Name** area are used to define the sequence in which revision of model will be generated while collaborating with others on same file using the **Data Management** tools.
- Select the **Enable Dynamic Edit of profiles/sketches** check box to allow modification of sketches by simple mouse gestures like dragging an end point of a line to increase its length or location. On selecting this check box, two radio buttons below it will become active. Select the **Recompute continually during edit** radio button if you want to check the effect of dynamic operation in real-time. Select the **Recompute after edit** radio button if you want the effect of dynamic modification to be displayed after you have performed the operation. The later option is recommended if you do not have sufficient RAM and processing power.
- Select the **Recompute assembly during sketch edits** check box to automatically recompute the assembly when a sketch of part or assembly model is modified. The radio buttons available for this check box are same as discussed earlier.
- Select the **Enable value changes using the mouse wheel** check box to modify value in spinner edit box using the mouse wheel.
- Select the **Recompute Ordered part after Synchronous** edit check box to update the ordered part when it is modified by synchronous operation.
- Click on the **IntelliSketch Options** button from the dialog box to modify parameters related to automatic snapping and constraining of sketch. The **IntelliSketch** dialog box will be displayed; refer to Figure-39. By default, the **Relationships** tab is selected in the dialog box. Select check boxes for snap points to which relations will be applied automatically based on selection while creating sketch. You will learn more about this feature later in this book. Click on the **Auto-Dimension** tab from the dialog box and select the **Automatically create dimensions for new geometry** check box to automatically create dimensions while creating sketch; refer to Figure-40. After selecting the check box, select desired radio buttons to define when the dimensions will be created automatically. Click on the **Cursor** tab in the dialog box to define the size of cursor. After setting desired parameters, click on the **OK** button.

*Figure-39. IntelliSketch dialog box*

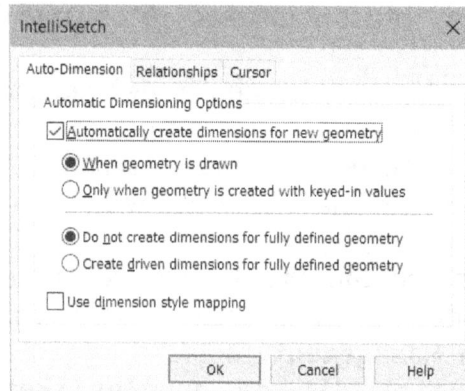

*Figure-40. Auto-Dimension tab*

- Select the desired radio button from the **Preferred 2D Paste Behavior** section to define how 2D objects will be pasted after copying in Draft and Ordered sketch environment.

- Select the **Enable Regions within sketches** check box from the **When Creating New Synchronous Sketches** area to enable selection and creation of regions in sketches.

- Select the **Migrate geometry and dimensions when features are created** check box to place the geometry and dimensions of features in respective categories after creating the feature.

- Select the **Indicate under-constrained profiles in PathFinder** check box to display whether profile is fully defined or under-defined.

- Select the **Enable "Undo All" for profile/sketch modifications** check box to display options for undoing all the modifications performed on sketch/profile.

- Set the other parameters as desired in the **General** page.

### View Options

Click on the **View** option from the left in the **Solid Edge Options** dialog box to modify view related parameters; refer to Figure-41.

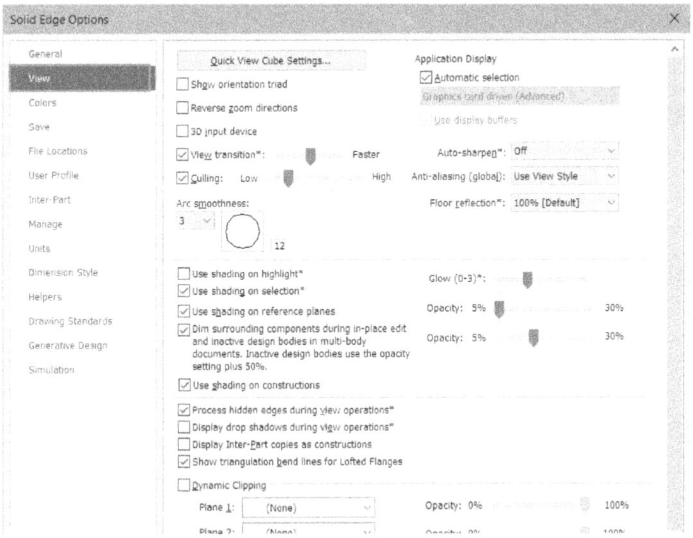

*Figure-41. View options*

- Click on the **Quick View Cube Settings** button to modify location, color, and transparency parameters of Quick View Cube. The **Quick View Cube Settings** dialog box will be displayed. Set the parameters as discussed earlier.
- Select the **Show orientation triad** check box to display triad in viewport while orienting the model.
- Select the **Reverse zoom directions** check box to reverse mouse gestures for zoom in/out.
- Select the **3D input device** check box to enable use of 3D input devices for modeling.
- Set desired value in the **View transition** slider to define how fast Solid Edge transitions from one orientation to another.
- Select the **Culling** check box and using the slider specify how aggressive the software will skip small features in the large assembly or complex model.
- Select desired options in the **Auto-sharpen**, **Anti-aliasing (global)**, and **Floor reflection** drop-downs to set quality of model in viewport.
- Set desired value in the **Arc smoothness** drop-down to define how fine the arcs will be generated in sketch and model.
- Select the **Use shading** check boxes to display shading on selection, highlight, reference planes, and so on.
- Similarly, you can set other view parameters.

## Color

In **Solid Edge**, you can use various color schemes for displaying the objects in application window. The process is discuss next.

- Click on the **Colors** option from left in the **Solid Edge Options** dialog box. The options in **Solid Edge Options** dialog box will be displayed as shown in Figure-42.
- Set the desired colors in various drop-downs to define colors for respective objects and click on the **Apply** button.

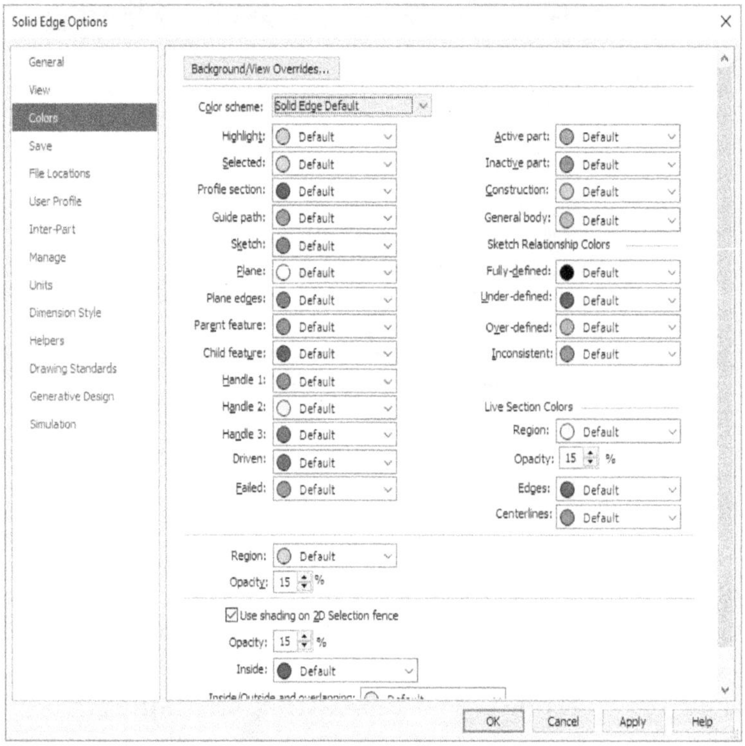

*Figure-42. Color options in Solid Edge Options dialog box*

## Save Options

The options in the **Save** page of **Solid Edge Options** dialog box are used to specify auto save duration and related parameters; refer to Figure-43.

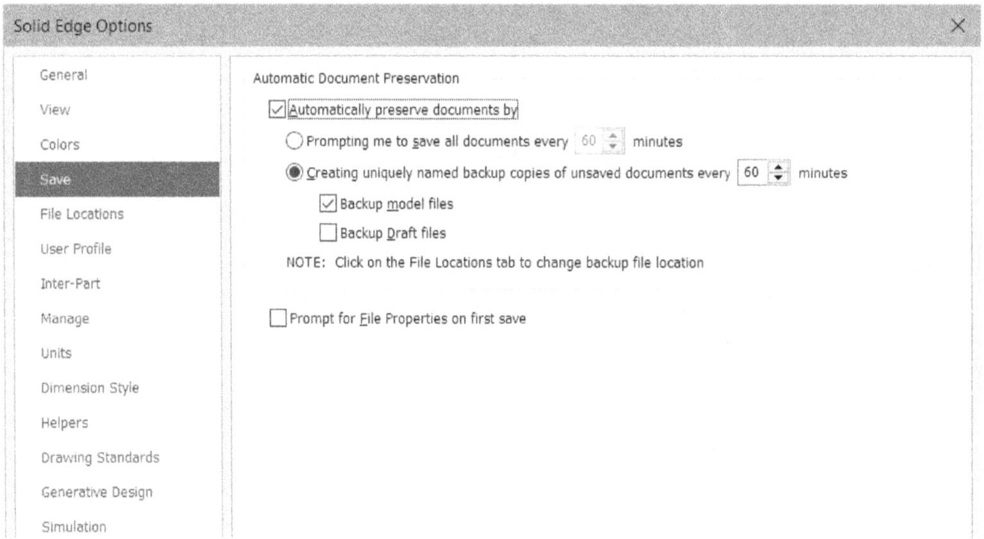

*Figure-43. Save options in Solid Edge Options dialog box*

- Select the **Automatically preserve documents by** check box to save or prompt you to save the active model. After selecting this check box, the related options will be displayed below it.
- Select the **Prompting me to save all documents every** radio button and specify the time in minutes in related edit box so that software will prompt you to save file after specified duration.
- Select the **Creating uniquely named backup copies of unsaved documents every** radio button to save backup copy of the document after specified duration. Select the **Backup model files** check box to backup model files. Select the **Backup Draft files** check box to create backup file of 2D drawings of the model.

- Select the **Prompt for File Properties on first save** check box if you want software to specify model properties also when you are saving the document for the first time.

### File Locations

The **File Locations** tab is used to modify the locations of Solid Edge system files. The process is discuss next.

- Click on the **File Locations** option from the left area in the dialog box. The options in the dialog box will be displayed; refer to Figure-44.

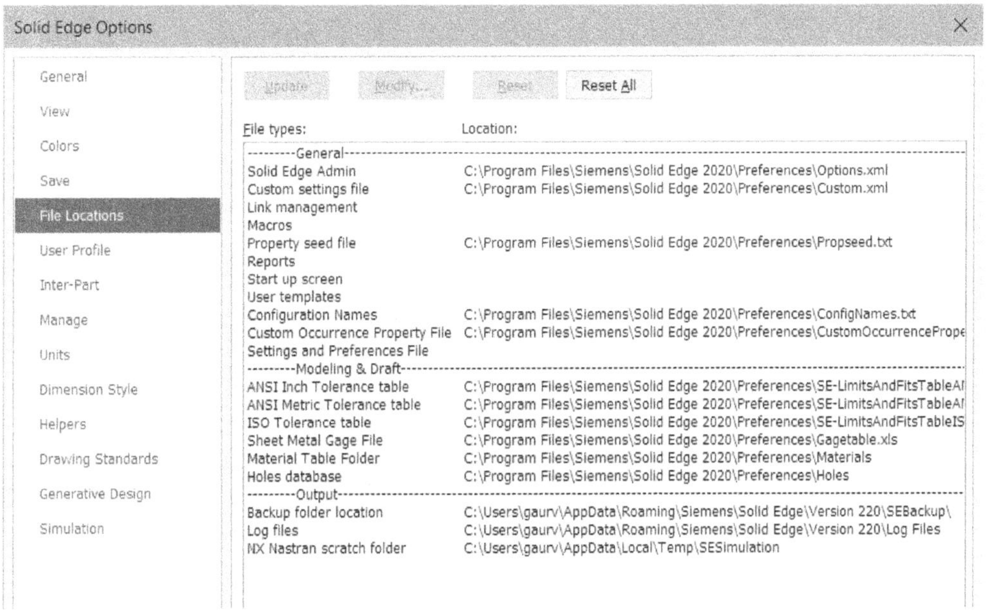

*Figure-44. File Locations options in Solid Edge Options dialog box*

- If you want to change any file then select it from the list and click on the **Modify** button. The **Select document** dialog box will be displayed; refer to Figure-45.

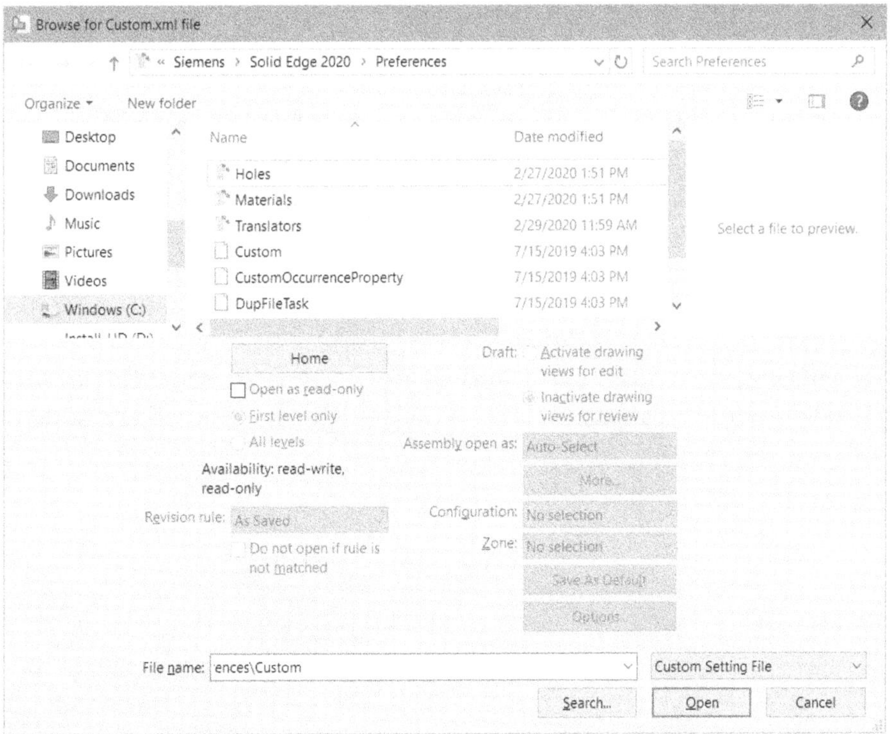

*Figure-45. Select Document dialog box*

*   Select the desired document and click on the **Open** button from the dialog box.

Similarly, you can select folders for various objects like macros, reports, startup screens and so on.

## User Profile Options

The options in the **User Profile** page of the dialog box are used to specify information about user of the software; refer to Figure-46.

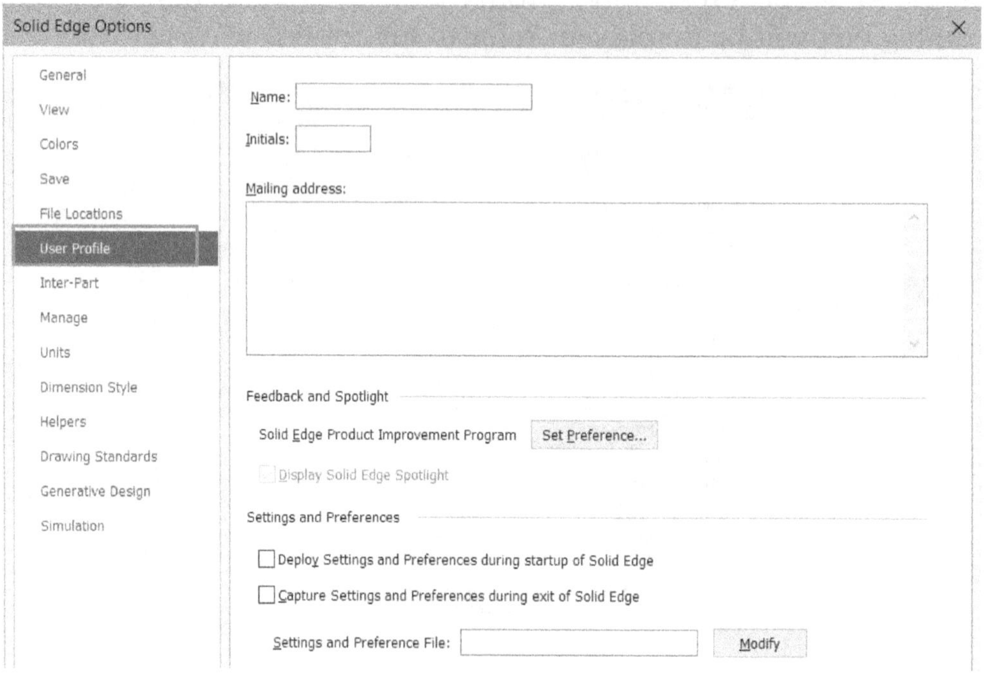

*Figure-46. User Profile page in Solid Edge Options dialog box*

*   Specify desired name and initials in the **Name** and **Initials** edit boxes.
*   In the **Mailing address** edit box, you can type your Email address.
*   Click on the **Set Preference** button to specify whether you want to participate in Solid Edge improvement or not. You can select **Yes** or **No** radio button from the dialog box displayed on clicking this button.
*   Select the **Deploy Settings and Preferences during startup of Solid Edge** check box to deploy file of settings and preferences at startup.
*   Select the **Capture Settings and Preferences during exit of Solid Edge** check box to deploy file of settings and preferences when you exit the software.
*   You can modify the preference file by clicking on the **Modify** button.

## Inter-Part Options

The options in **Inter-Part** page are used to activate and specify parameters related to inter-part linking; refer to Figure-47. You will learn about these options later in this book.

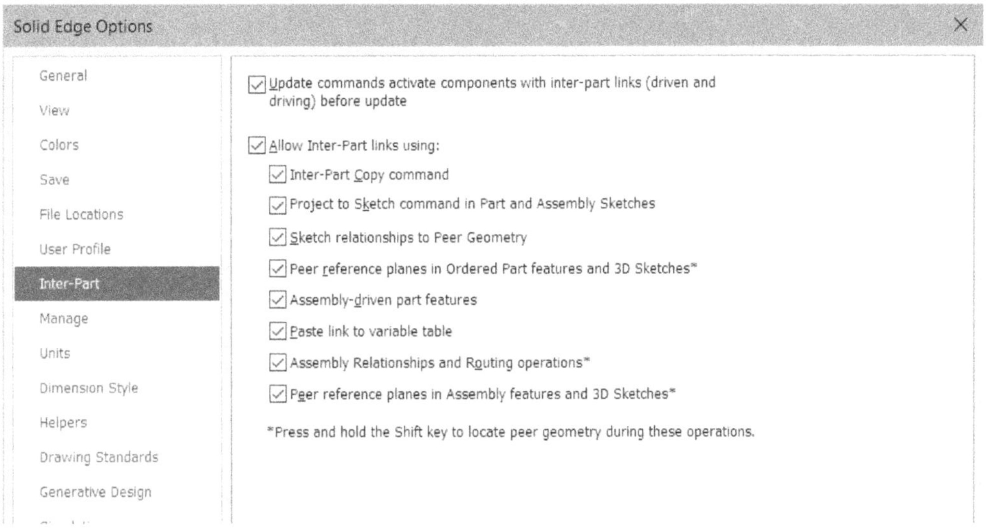

*Figure-47. Inter-Part page of Solid Edge Options dialog box*

## Units

The options in the **Unit** page are used to set the preference for units in Solid Edge. The process to set the units in Solid Edge is discuss next.

- Click on the **Units** option from the left area of the dialog box. The options to define units will be displayed in the dialog box; refer to Figure-48.
- Select the desired radio button from the **Units System** area to define unit system for creating model in Solid Edge.

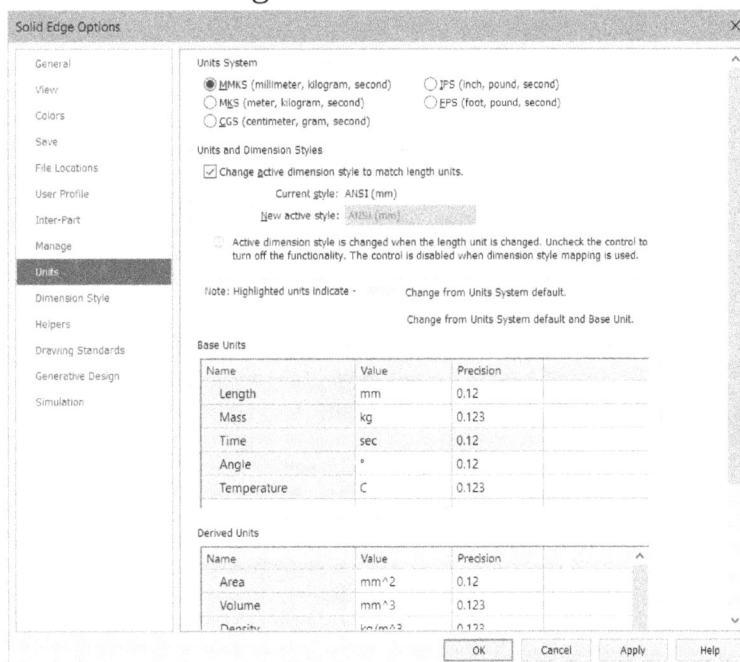

*Figure-48. Units option in Solid Edge Options dialog box*

- If you want to create custom unit system then set desired options in the **Base Units** and **Derived Units** tables.

## Dimension Style

The options in the **Dimension Style** page of **Solid Edge Options** dialog box are used to define dimensioning style for drafting; refer to Figure-49.

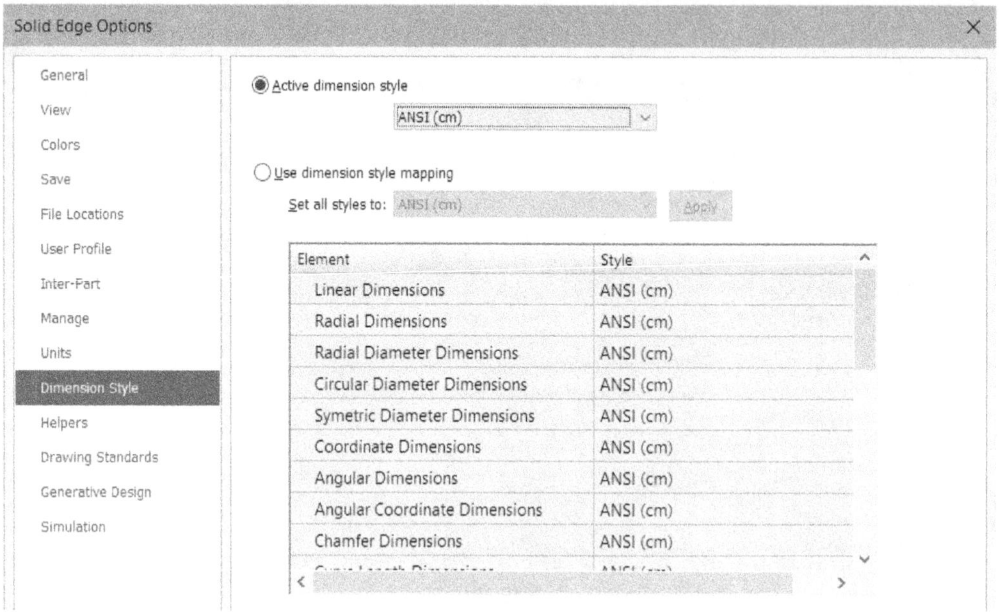

*Figure-49. Dimension Style page of Solid Edge Options dialog box*

- Select the **Active dimension style** radio button and select the desired option from the drop-down below it.

## Helpers Options

The options in the **Helpers** page are used to set parameters for help options while working on the software; refer to Figure-50. These options are discussed next.

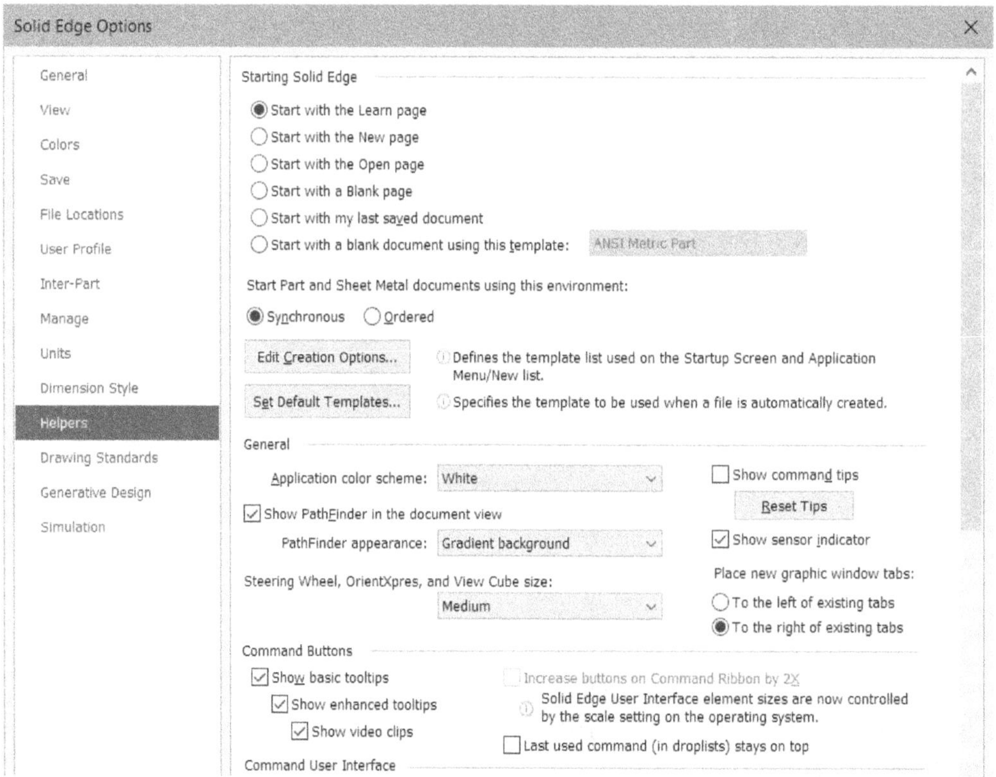

*Figure-50. Helpers page of Solid Edge Options dialog box*

- Select the desired radio button from the **Starting Solid Edge** area to define what type of starting page will be displayed in first interface of Solid Edge after starting.
- Select the desired radio button from the **Start Part and Sheet Metal documents using this environment** area to define whether you want to start Part and Sheet Metal model in **Ordered** or **Synchronous** environment.

- Click on the **Edit Creation Options** button from the dialog box to modify list of templates being displayed when you start new document. The **Template List Creation** dialog box will be displayed; refer to Figure-51. You can move the templates up/down or add new template in the list by using the options in this dialog box.

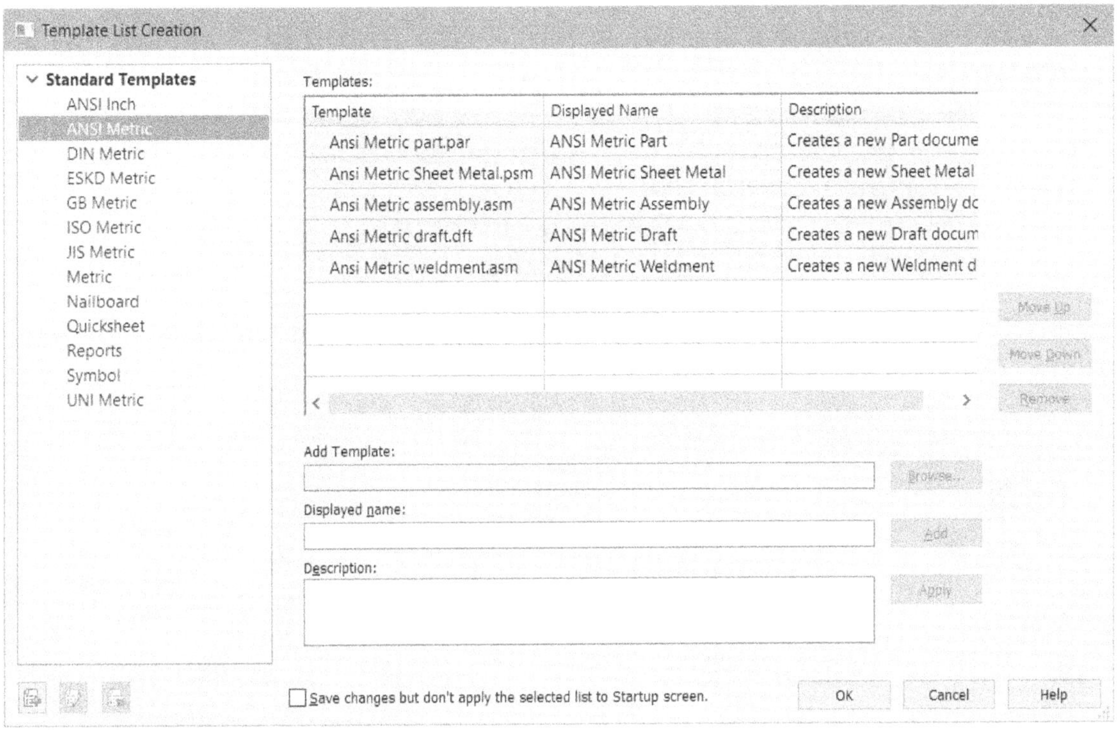

*Figure-51. Template List Creation dialog box*

- Click on the **Set Default Templates** button from the dialog box to define default templates for various documents. The **Default Templates** dialog box will be displayed; refer to Figure-52. Select the document type for which you want to change the template and click on the **Browse** button. The **Add Template** dialog box will be displayed where you can select a new template. After selecting template, click on the **OK** button from **Add Template** dialog box and **Default Templates** dialog box.

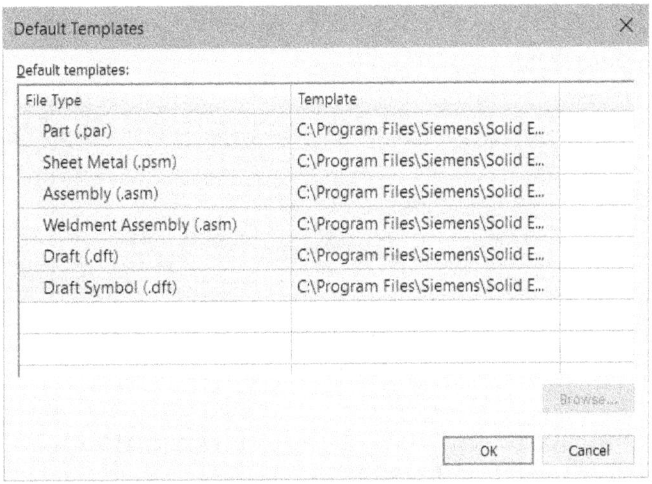

*Figure-52. Default Templates dialog box*

- Select desired option from the **Application color scheme** drop-down to modify color scheme for interface of application.

- Select the **Show PathFinder in the document view** check box to display PathFinder at the bottom in interface.
- Select the desired option from the **Steering Wheel, OrientXpress, and View Cube size** drop-down to define size of aforementioned objects.
- Select the **Show command tips** check box to display tips when you hover cursor on any command in interface.
- Select the **Show sensor indicator** check box to display sensor warnings in graphics area.
- Select the desired radio button from the **Place new graphics window tabs** section to define where next document tab will be added against existing document tab when a new document is started.
- Similarly, you can set other parameters in this page.

### Drawing Standards

The options in the **Drawing Standards** page are used to define standards for geometric design features; refer to Figure-53.

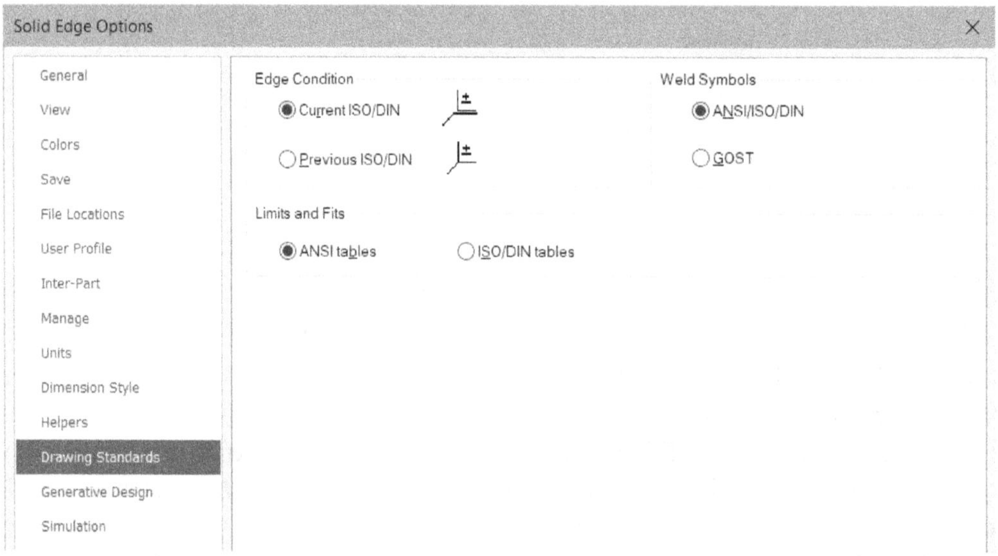

*Figure-53. Drawing Standards page of Solid Edge Options dialog box*

- Select the desired radio button from the **Edge Condition** area to define what standards should be used for edges. Similarly, you can set standard for weld symbols and Limits & Fits.

You will learn about other settings later in the book. Click on the **OK** button from the dialog box to apply settings.

## Add-Ins

**Add-Ins** are used to add extra functions in Solid Edge like eAssistant. Before using Add-Ins, you need to download and install Add-Ins from Solid Edge certified developers. The procedure is discuss next.

- Click on the **Add-Ins** tool from the **Settings** cascading menu. The **Add-In Manger** dialog box will be displayed; refer to Figure-54.
- Select the desired check box from the dialog box to activate the Add-In and click on the **OK** button. The related tools will be added in the **Ribbon**.

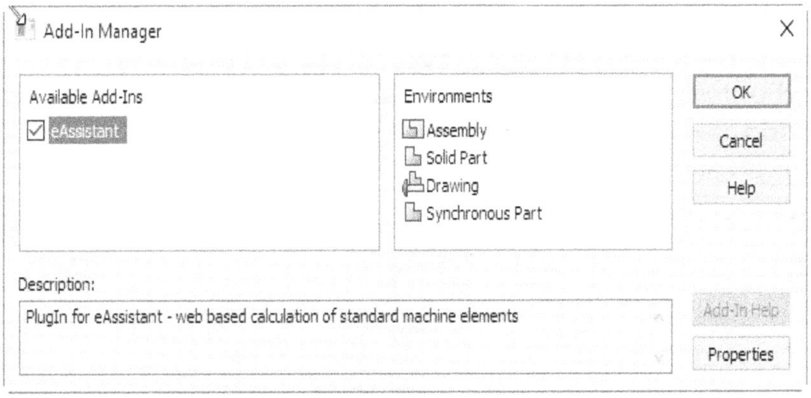

*Figure-54. Add In Manager dialog box*

## Customize

The option in **Customize** dialog box are used to customize interface of Solid Edge. The procedure to customize is discussed next.

- Click on **Customize** tool from the **Settings** cascading menu of the **Application** menu. The **Customize** dialog box will be displayed; refer to Figure-55.

Various options for customizing are discussed next.

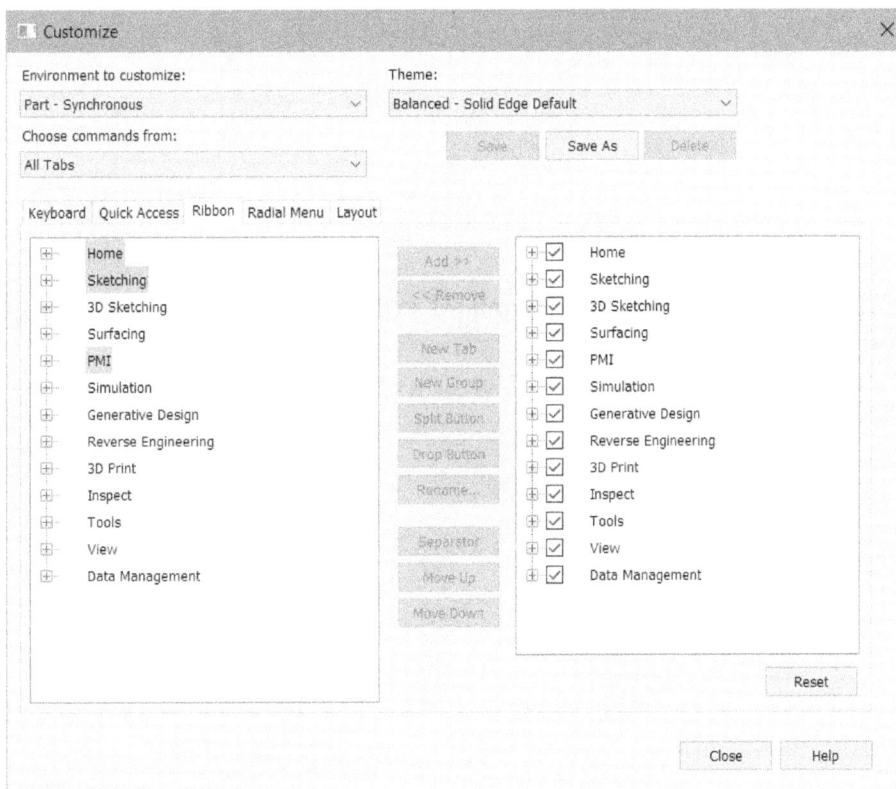

*Figure-55. Customize dialog box*

### Keyboard

The options in **Keyboard** tab are used to set keyboard shortcuts for various tools and options. The procedure to create keyboard shortcut is given next.

- Select desired environment of Solid Edge in which you want to create custom shortcut key from **Environment to customize** drop-down.
- Select desired option from **Choose commands from** drop-down to define the scope from which commands can be selected. The **Customize** dialog box will be displayed as shown in Figure-56.

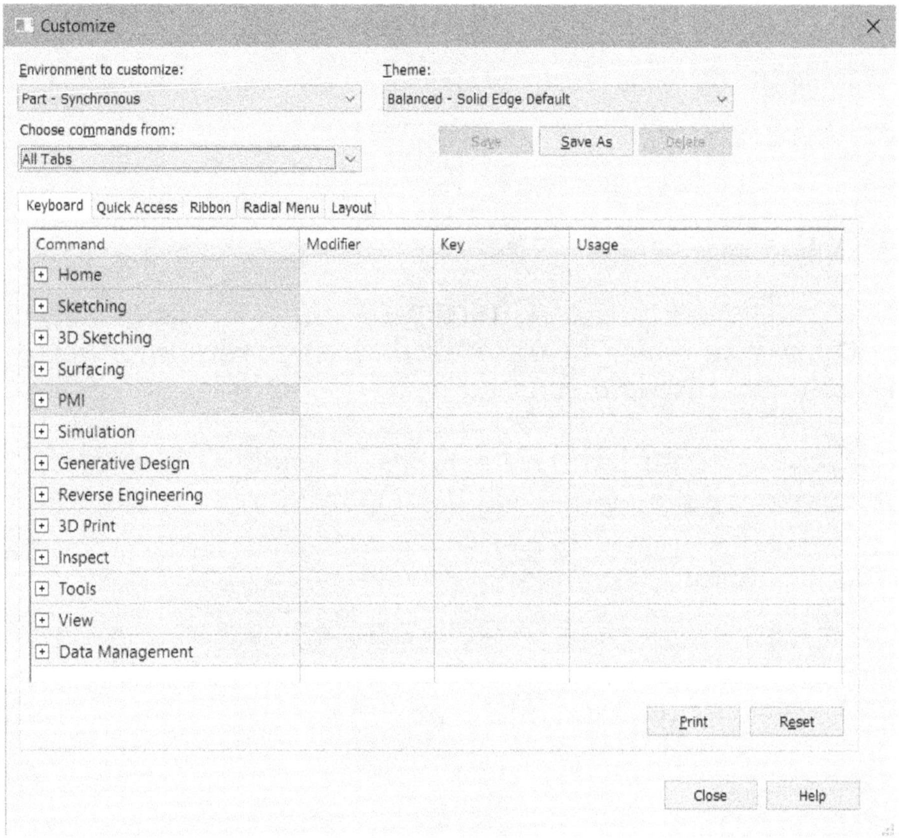

*Figure-56. Customize Keyboard*

- Select the desired tool from the **Command** column. Click on the **+** sign to expand the command groups.
- Set the desired modifier and key in the respective columns for selected command. Refer to Figure-57.

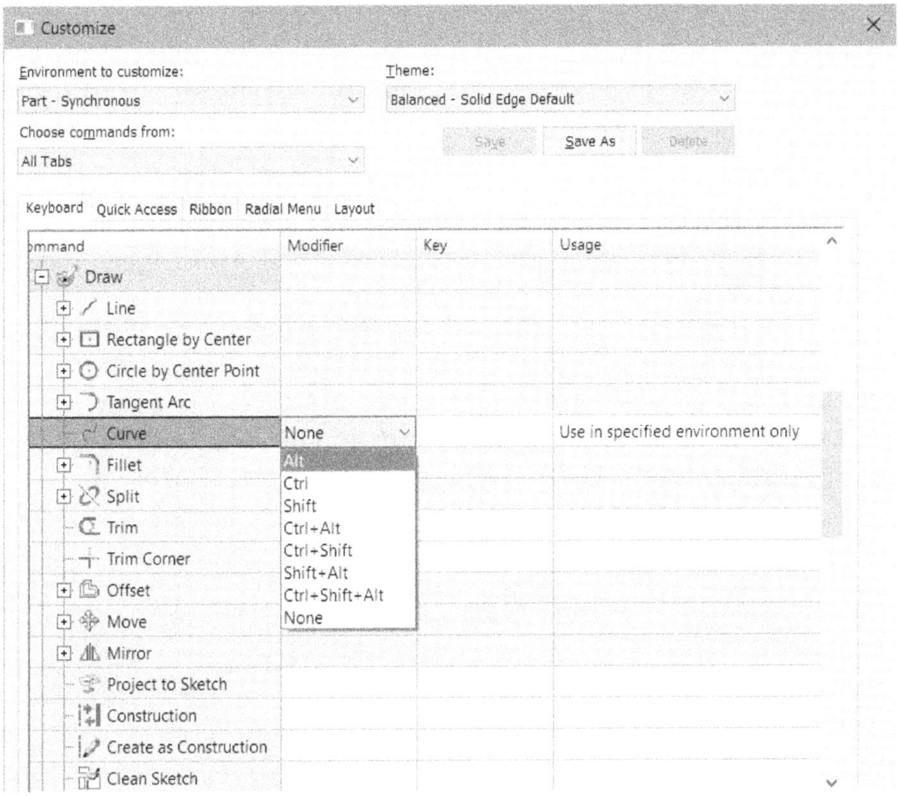

*Figure-57. Selecting button for keyboard shortcut*

## Quick Access

Quick Access->Environment to customize or Choose commands->select commands->Add

- Click on **Quick Access** tab in the dialog box to customize Quick Access Toolbar. The options in the dialog box will be displayed as shown in Figure-58.

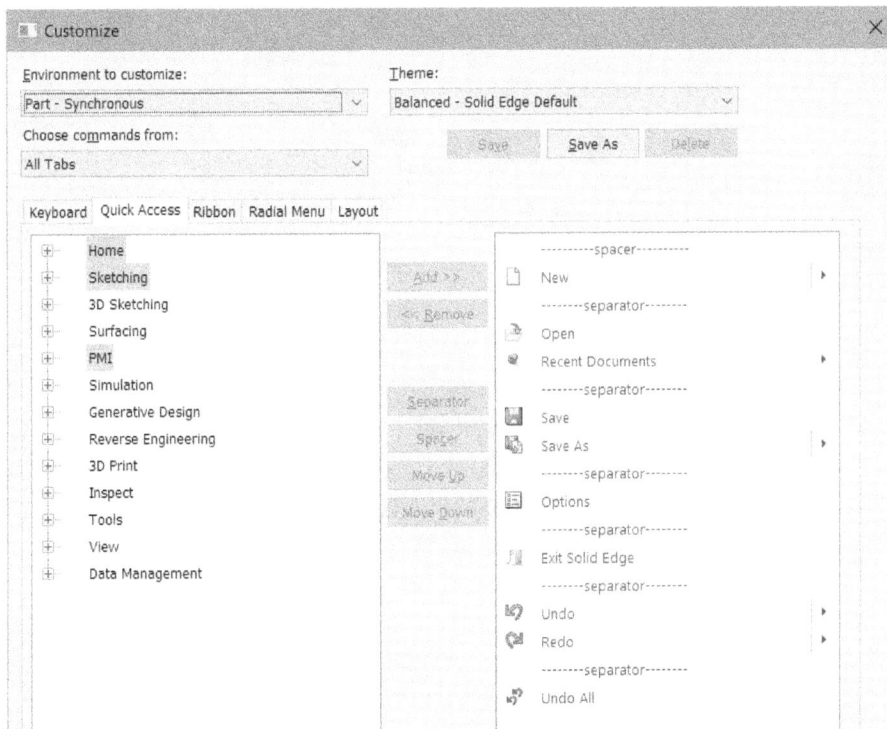

*Figure-58. Quick Access tab in Customize dialog box*

- Select the desired tool to be added in **Quick Access Toolbar** from the left area and click on the **Add** button. The tool will be added at the end of toolbar.
- Using the **Move Up** or **Move Down** tool to position the new tool in toolbar.

## Ribbon

The process of customizing **Ribbon** has already been discussed earlier in this chapter.

## Radial Menu

Radial menu->Environment to customise->Choose commands from->select commands and drag it

- **Radial Menu** is displayed on right-clicking in the drawing area. To customize the Radial Menu, click on the **Radial Menu** tab from **Customize** dialog box. The options to customize Radial Menu will be displayed; refer to Figure-59.

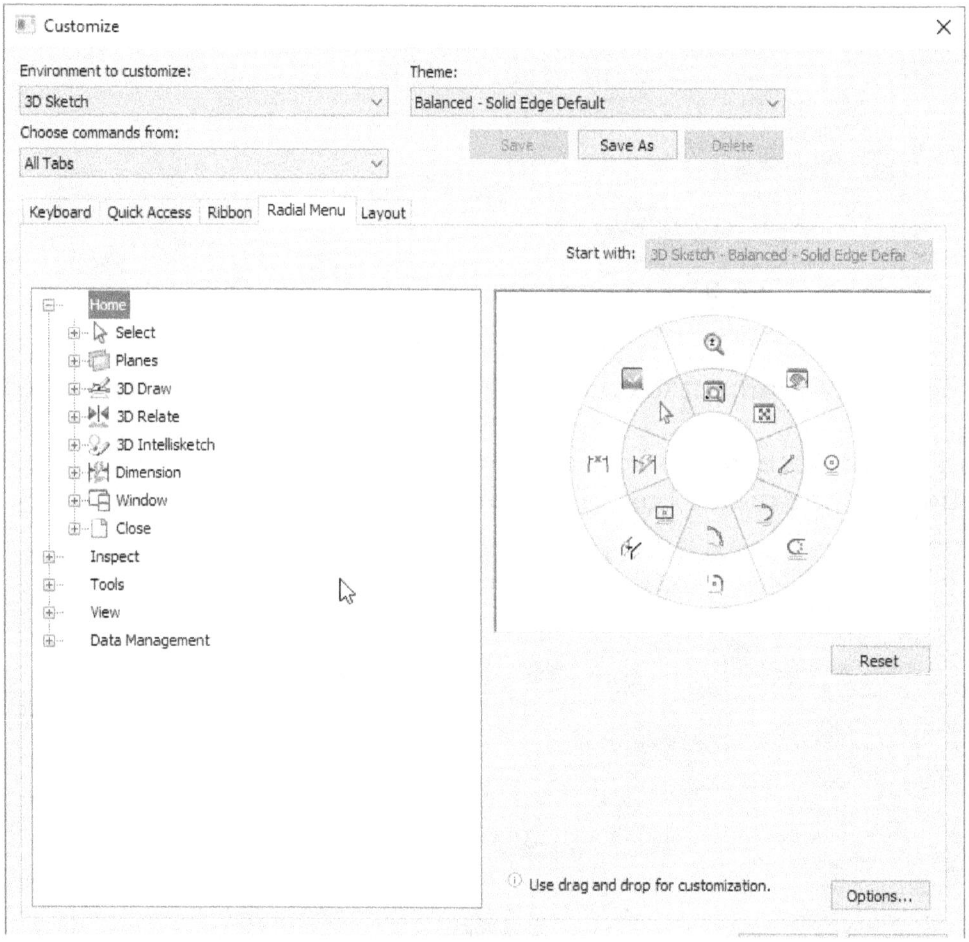

*Figure-59. Radial Menu tab in Customize dialog box*

- Expand the desired node from which you want to include tool in Radial menu.
- Select the desire tool and drag & drop to the Radial menu for replacing the tool; refer to Figure-60.

*Figure-60. Radial menu*

## Layout

- Click on **Layout** tab in the dialog box to modify interface layout of software like location of CommandBar, tooltips, and so on. The options in the dialog box will be displayed as shown in Figure-61. Set desired options for repositioning the interface elements. After setting desired parameters, click on the **Close** button to exit the dialog box. You will be asked whether to save new scheme. Save the theme as desired.

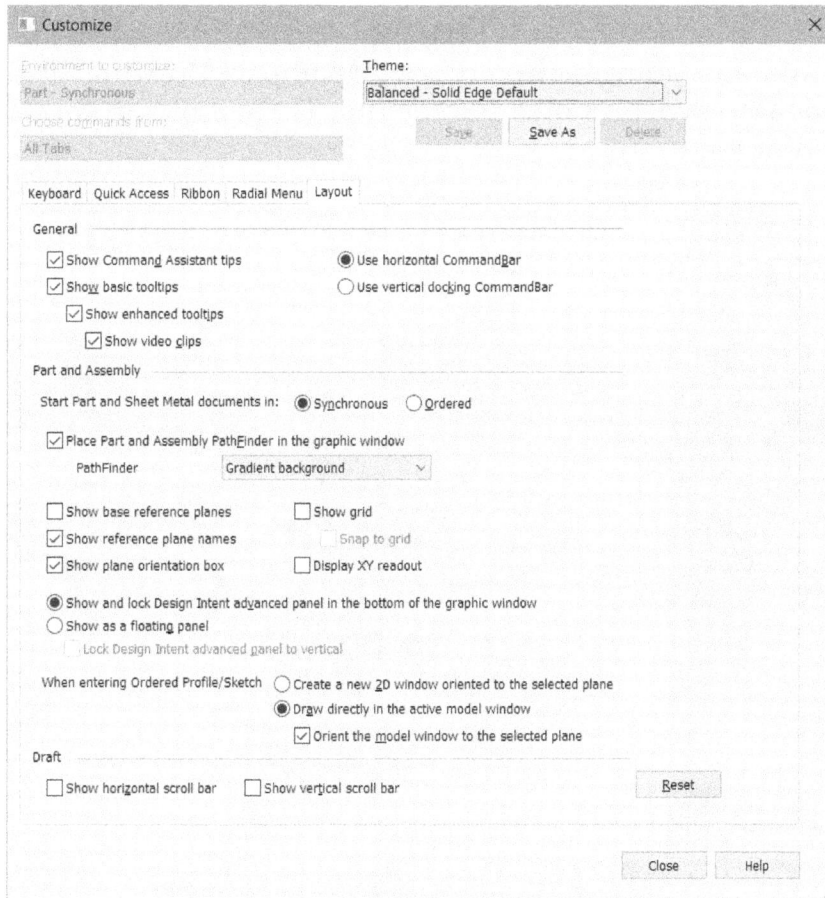

*Figure-61. Layout tab dialog box*

## Themes

Using the **Themes** tool, user can define the level of assistance provided by software related to tools and commands like displaying tips, help menus, and so on.

- To select a theme, click on the **Themes** tool from the **Settings** cascading menu of **Application** menu. The **Themes** menu will be displayed; refer to Figure-62.

- Select the desired theme option from the menu.

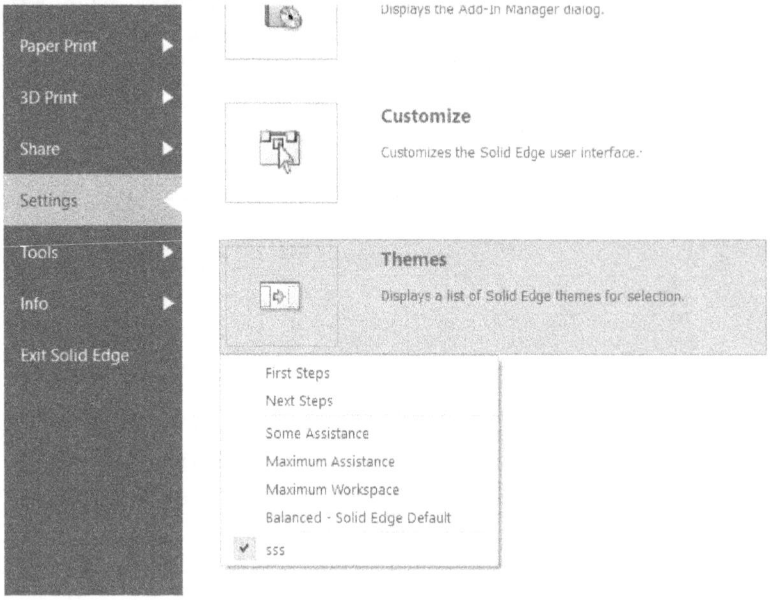

*Figure-62. Themes menu*

## Tools

The option in **Tools** cascading menu are use to compare drawing file and convert the file from traditional to synchronous documents. The options in the **Tools** cascading menu are shown in Figure-63.

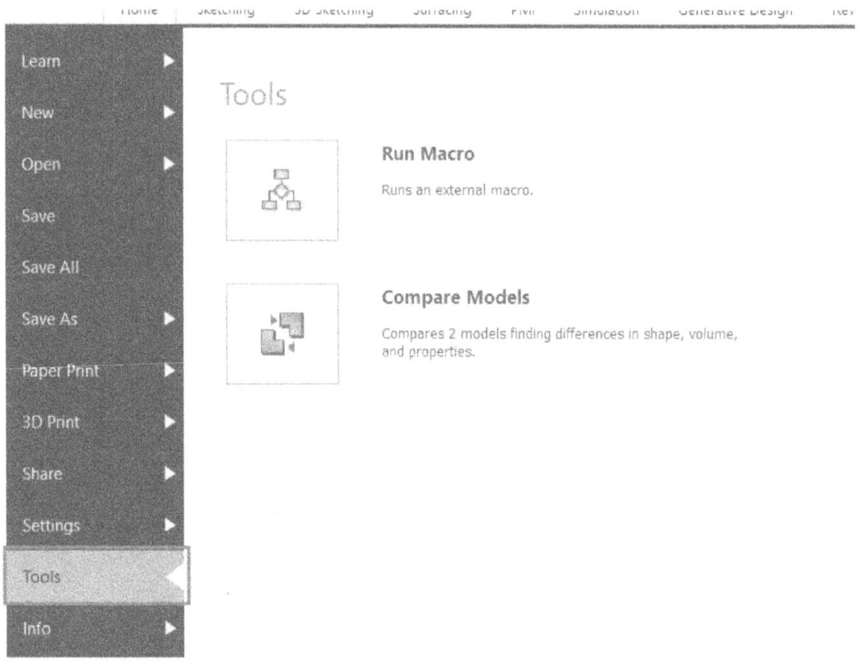

*Figure-63. Tools cascading menu*

## Compare Models

The **Compare Models** option is used to compare the drawing files. The procedure is given next.

- Click on the **Compare Models** tool from **Tools** cascading menu. The **Compare Models** dialog box will be displayed; refer to Figure-64.

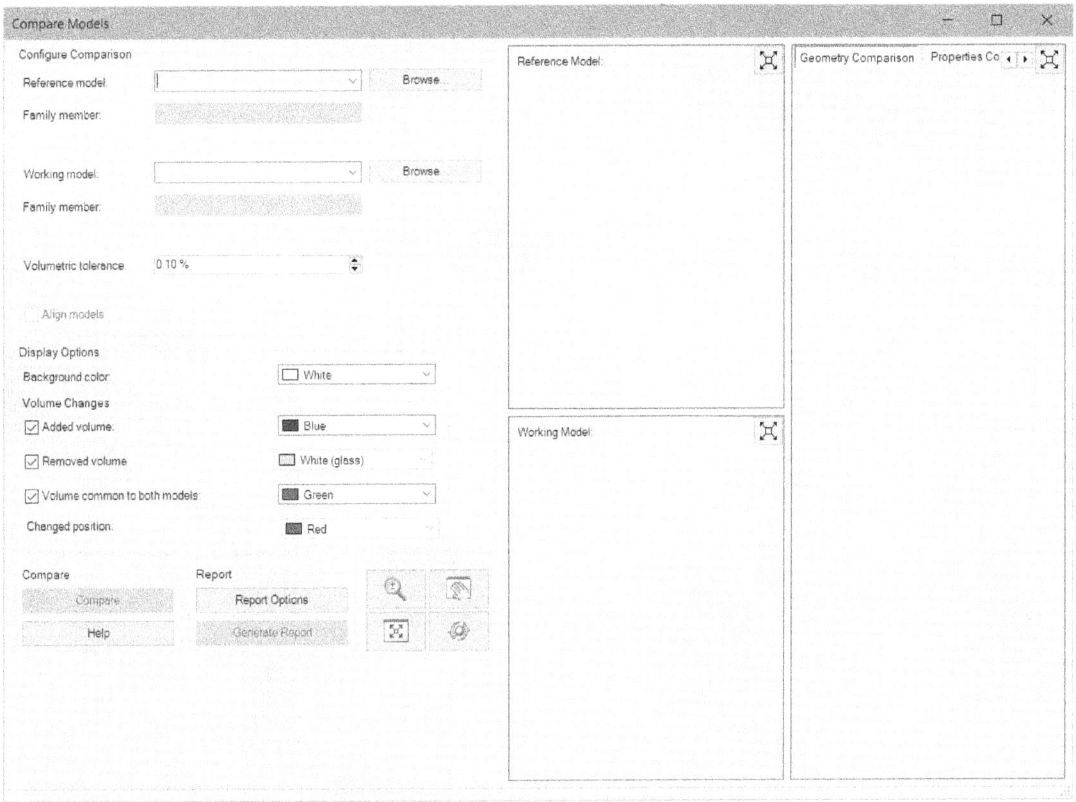

*Figure-64. Compare Models dialog box*

- Click on the **Browse** button for **Reference** edit box and select the desired file from **Open File** dialog box displayed; refer to Figure-65. Similarly, select the other file in **Working model** edit box.
- Now, click on **Compare** button from **Compare** area of the dialog box. The differences between two files will be displayed; refer to Figure-66.

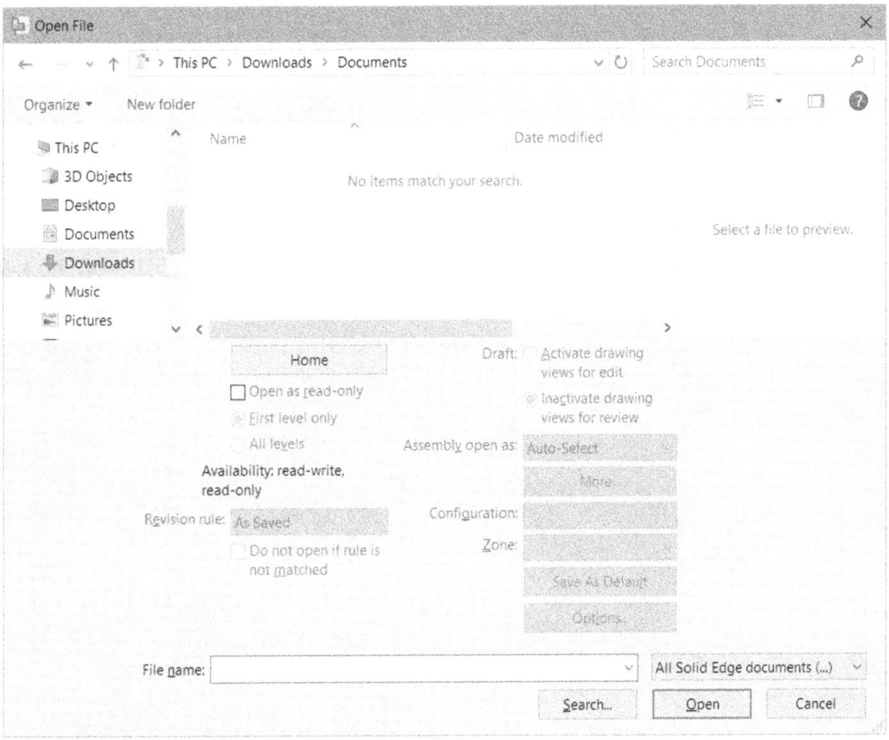

*Figure-65. Open file dialog box*

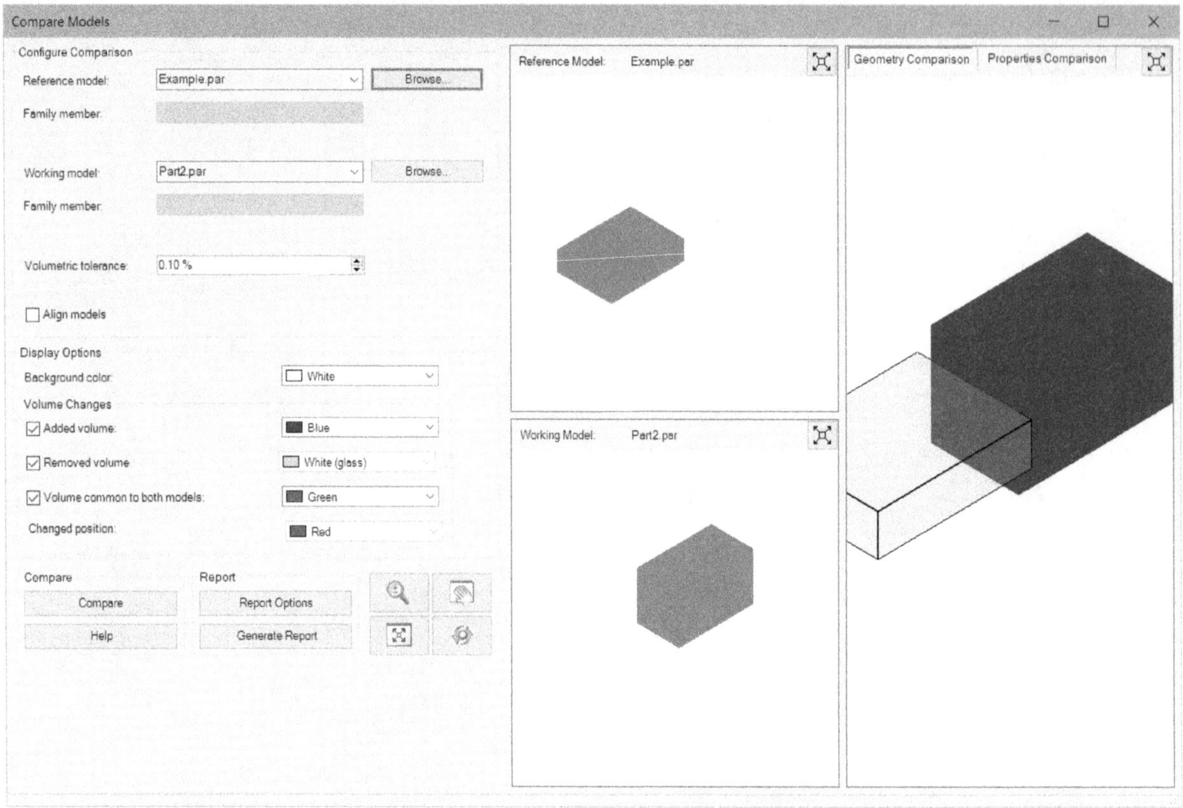

*Figure-66. Comparing drawings*

- Click on the **Report Options** button from the **Report** area to specify parameters related to report and then click on the **Generate Report** button. The **Save Comparison Report** dialog box will be displayed.
- Select the desired format for report from the **Report type** drop-down and specify the path where you want to save file in the **Report folder path** edit box. Click on the **Save Report** button to save the file.
- Close the **Compare Models** dialog box using **Close** button at top-right corner.

## Running External Macro

Macros are scripts that automate tasks in software. The procedure to run a macro is given next.

- Click on the **Run Macro** tool from the **Tools** cascading menu of the **Application** menu. The **Run Macro** dialog box will be displayed; refer to Figure-67.
- Select the desired macro file and click on the **Open** button. The related macro dialog box will be displayed; refer to Figure-68. Use the macro as per related documentation.

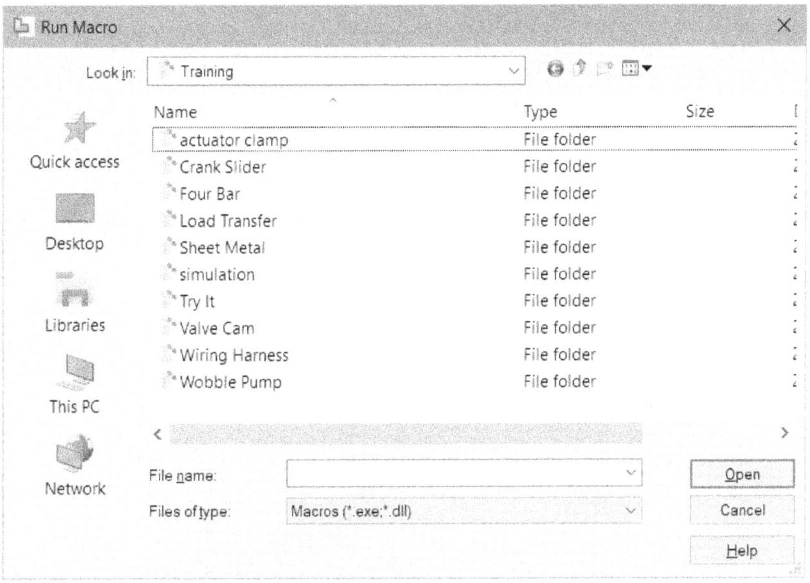

*Figure-67. Run Macro dialog box*

*Figure-68. Mouse Events dialog box*

# Info

The options in the **Info** cascading menu of the **Application** menu are used to modify document properties. The option in the **Info** cascading menu are shown in Figure-69 & Figure-70.

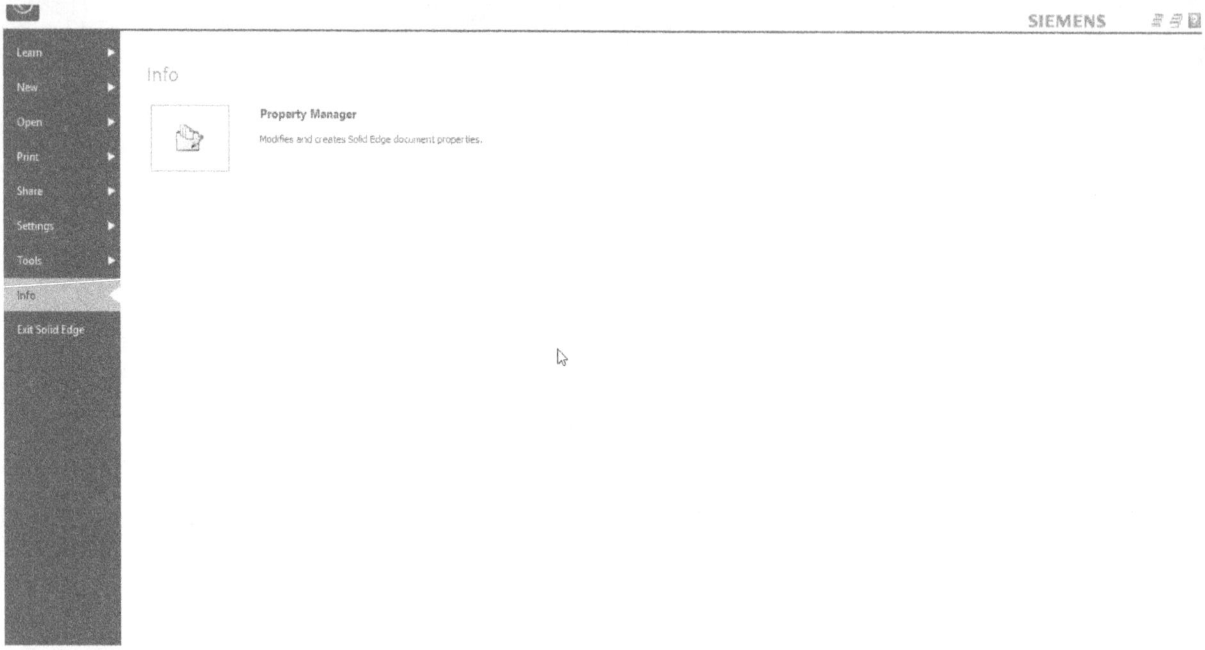

*Figure-69. Info cascading menu when no documents is open*

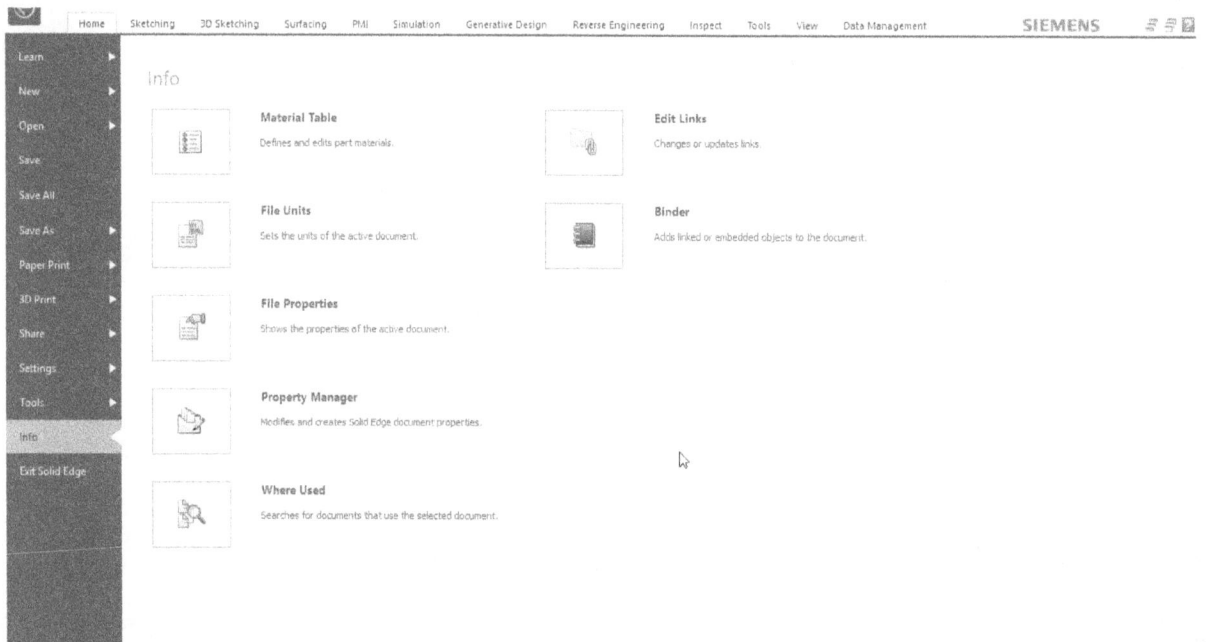

*Figure-70. Info cascading menu after opening new document*

## Material Table

The **Material Table** tool in the **Info** cascading menu is used to define and edit materials.

* Click on the **Material Table** option from the **Info** cascading menu. The **Material Table** dialog box will be displayed; refer to Figure-71.

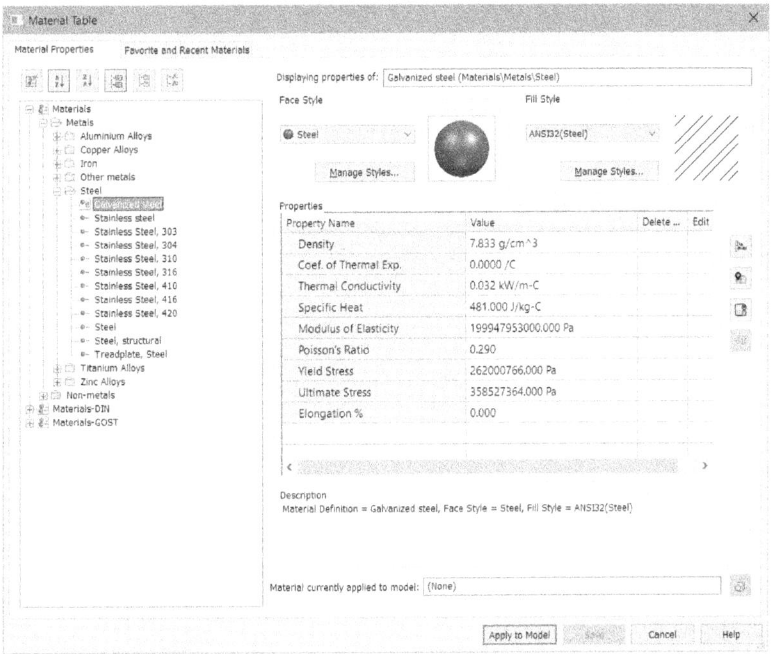

*Figure-71. Material Table dialog box*

- Select the material from the left list box, the preview and specifications of material are shown in the right window of dialog box.
- Modify the properties as desired and click on **Apply to Model** button.

## File Units

The **File Units** is used to set the unit of active documents. It is same as we discuss earlier in **Options** section.

## File Properties

The **File Properties** tool is used to show the properties of active document.

- Click on **File Properties** tool from **Info** cascading menu. The **File Properties** dialog box will be displayed; refer to Figure-72.

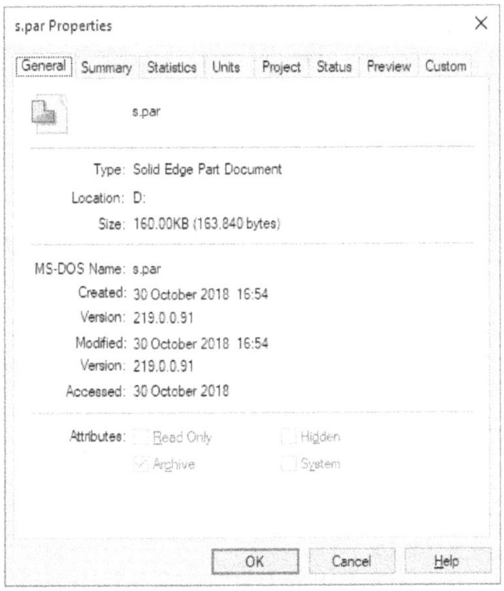

*Figure-72. Properties dialog box*

You can check and modify various general properties of the model in the dialog box.

## Property Manager

The **Property Manager** is used to modify Solid Edge document properties.

*   Click on the **Property Manager** tool from **Info** cascading menu. The **Property Manager** dialog box will be displayed; refer to Figure-73. Select the desire file to edit from the dialog box. You can edit any field for desired file by double-clicking.

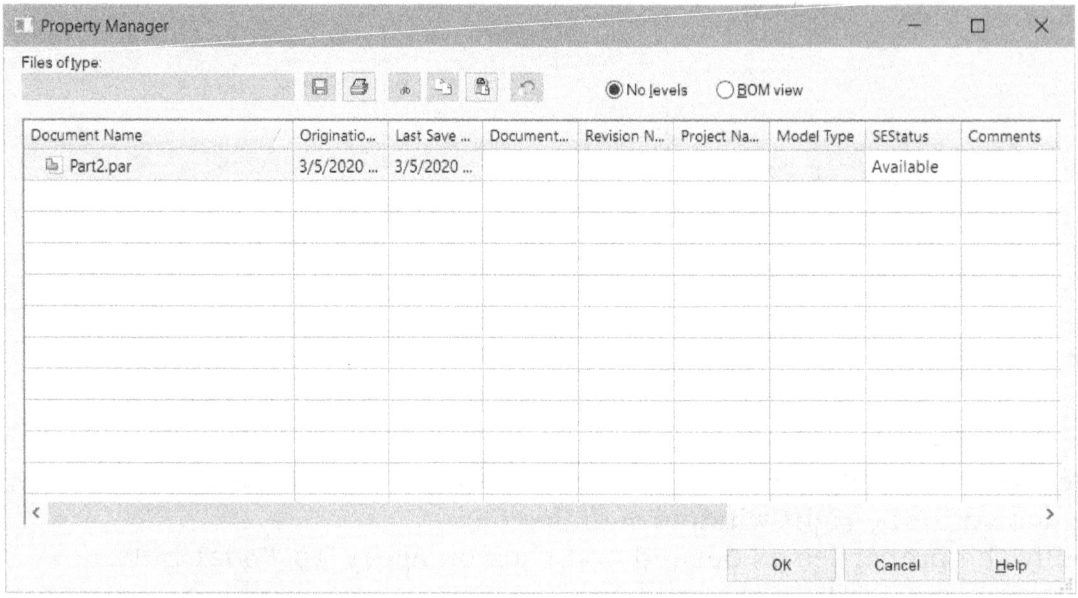

*Figure-73. Property Manager dialog box*

*   Click on the **OK** button from the dialog box to apply changes.

## Where Used

The **Where Used** option in **Info** cascading menu is used to search for documents that use the selected documents.

*   Click on the **Where Used** option from **Info** cascading menu, the **Where Used** dialog box will be displayed; refer to Figure-74.

*Figure-74. Where Used dialog box*

*   Select the desired folder from left list box and click on **Add** button. The selected folders will be added in the list where software will search for use of current file.

- Click on the **Next** button from the dialog box. The **Where Used Results** dialog box will be displayed with list of files where the current document was used.

## Edit Links

The **Edit Links** option in **Info** cascading menu is used to change and update links of particular document. The **Edit Links** option is active only when an assembly file with component is open.

- Click on **Edit Links** from the **Info** cascading menu. The **Links** dialog box will be displayed; refer to Figure-75.

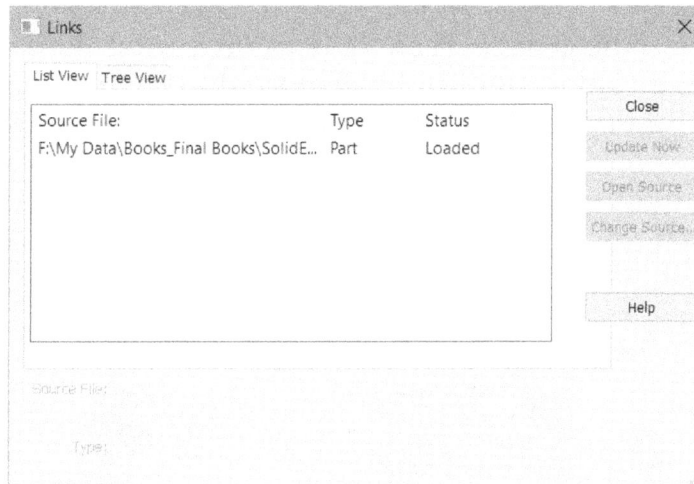

*Figure-75. Links dialog box*

- Select the file from the dialog box and select the desire operation to be performed.

## Binder

The **Binder** option in **Info** cascading menu is used to add linked or embedded objects to the document.

- Click on the **Binder** option from the **Info** cascading menu. The **Binder** dialog box will be displayed; refer to Figure-76.

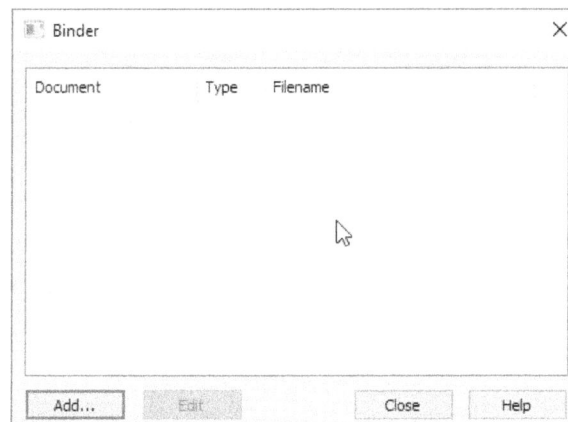

*Figure-76. Binder dialog box*

- Click on the **Add** button to bind a new document. The **Insert Object** dialog box will be displayed; refer to Figure-77. Set the desired parameters and click on the **OK** button.

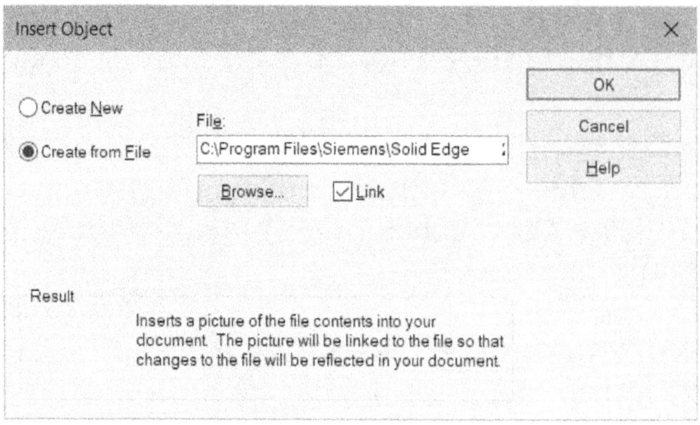

*Figure-77. Insert Object dialog box*

- If you have selected the **Create from File** radio button then click on the **Browse** button. The **Browse** dialog box will be displayed; refer to Figure-78.

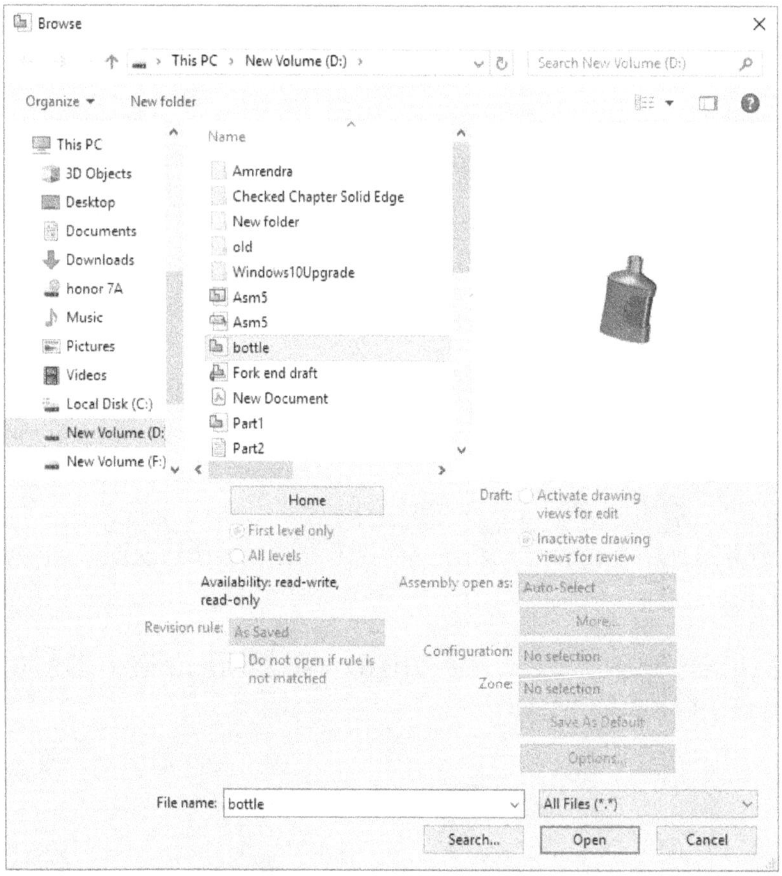

*Figure-78. Browse dialog box*

- Select the desired document from the dialog box and click on **Open** button.
- After that click on **OK** button from the **Insert Object** dialog box.
- If you want to create a new document then select the **Create New** radio button. The dialog box will be displayed as shown in Figure-79.

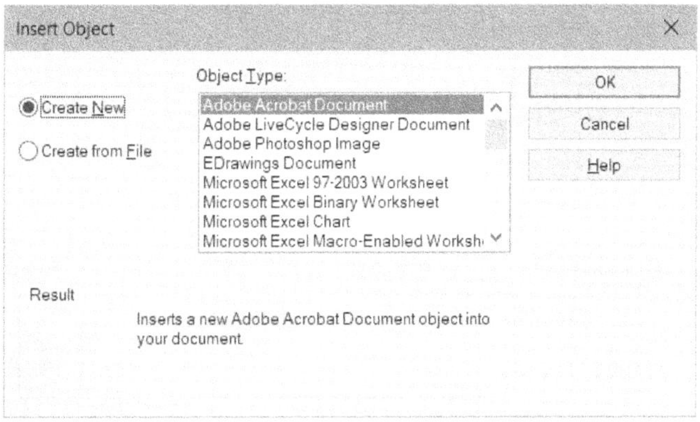

*Figure-79. Create New option in Insert Object dialog box*

- Select the desired option from the **Object Type** list and click on the **OK** button. A new document will be displayed in respective application.
- Similarly, you can add more documents. Click on the **Close** button from the dialog box.

## Exit Solid Edge

Click on the **Exit Solid Edge** tool from the **Application** menu to exit the Solid Edge software.

## MOUSE BUTTON FUNCTION

**Rotate view**
- To rotate the model in viewport: Press the Middle mouse button and drag.

**Pan**
- Hold down **SHIFT** key from keyboard and drag with middle mouse button.

**Zoom In/Out**
- For Zoom In/out scroll the middle mouse button up/down or hold down **CTRL** key and drag up/down with middle mouse button pressed.

## WORKFLOW IN SOLID EDGE

The first step in Solid Edge is to create a sketch. After creating sketch of the desired feature, create solid or surface model from that sketch. After doing the desired operations on the solid/surface model, we go for assembly or analysis. After, we are satisfied with the assembly/analyses, we create the engineering drawings from the model to perform manufacturing of the model into a real world object.

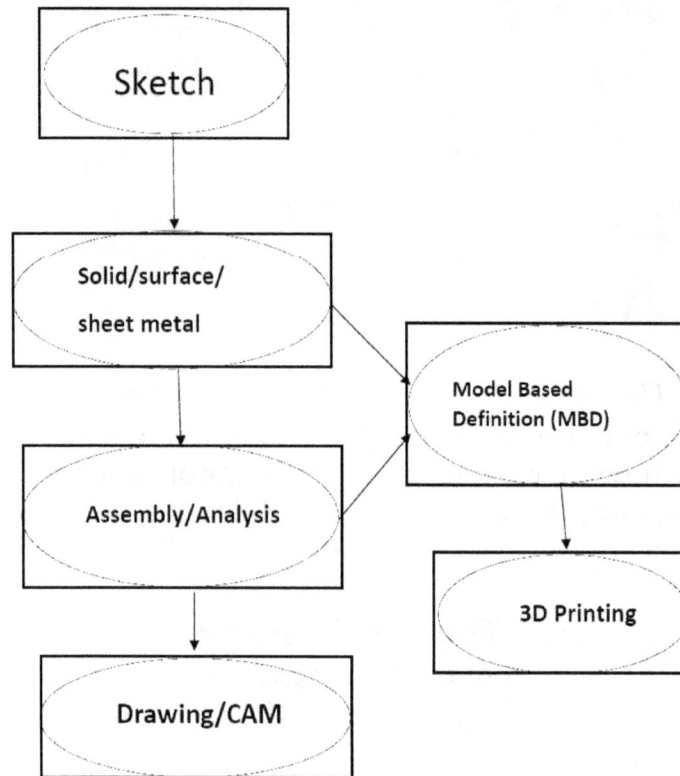

# SELF ASSESSMENT

1. The templates whose names end with part.par are used for creating assembly files. (T/F)

2. The templates whose names end with sheetmetal.psm are used to start sheetmetal design. (T/F)

3. The options in **Open** cascading menu are used to open CAD files in Solid Edge. (T/F)

4. The options in **Tools** cascading menu are used to compare drawings. (T/F)

5. To create new document, which of the following option can be used?
   (a) **New** tool
   (b) CTRL + N
   (c) SHIFT + N
   (d) Both a & b

6. Which of the following templates is used to start new drawing file?
   (a) draft.dft
   (b) part.par
   (c) assembly.asm
   (d) weldment.asm

7. Hold down **SHIFT** and drag with middle mouse button to
   (a) Rotate view
   (b) Pan
   (c) Zoom In/Out
   (d) Fit

8. The ....................... button display the application menu, which provide access to all documents level functions, templates, and standards.

9. The **PromptBar** shows the .............................................. you have selected.

10. What is the purpose of Add-Ins in Solid Edge?

11. Write down the steps to create new documents?

FOR STUDENT NOTES

# Chapter 2

# Sketching

## Topics Covered

The major topics covered in this chapter are:

- *Basic for Sketching*
- *Synchronous Sketching Environment*
- *Ordered Sketching Environment*
- *View Toolbar*
- *Sketch Creation Tools*
- *Relate Tools*
- *Dimensions*
- *Practical 1*
- *Practical 2*
- *Practice*

## BASICS FOR SKETCHING

In Engineering, sketching do not mean sketches of birds or animals. It means sketches that are based on real dimensions of real-world objects. Now, we are going to work with geometric entities like, Line, Arc, Circle, Rectangle, Ellipse, and so on which you have used in your school and engineering. But this time, you will be using the software in place of pencil, scale, and other geometry tools. Note that sketching is the base for most of 3D Models so you should be proficient in sketching.

To start with sketching, we must have a good understanding of planes in **Solid Edge**.

## PLANES

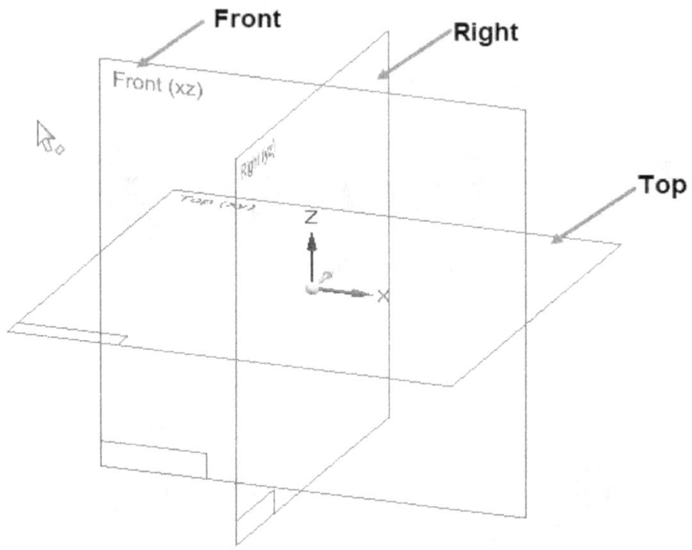

*Figure-1. Plane*

In Solid Edge, the planes are displayed in the same orientation as shown in Figure-1. To check the planes of Solid Edge, click on the planes (Front plane, Top plane, Right plane)in **Pathfinder**. You can also select the **Base Reference Planes** check box to display all the planes; refer to Figure-2.

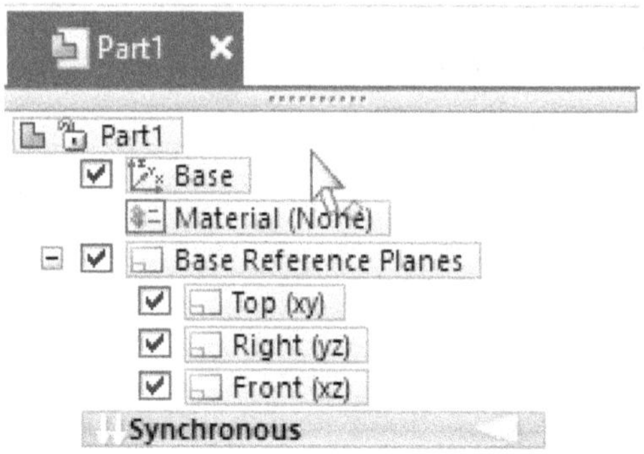

*Figure-2. Plane in Pathfinder*

- Select the plane on which you want to create sketch.

## Relation between sketch, plane and 3D model

Sketch has a direct relationship with plane and the outcome which is generally a **3D** model; refer to Figure-3. In this figure, rectangle is created on the XY plane which is also called **Top** plane. The circle is created on the YZ plane which is also called **Right** plane. The polygon is created on the XZ plane which is also called **Front** plane. In a 3D model, the geometry seen from the Top view should be drawn on the **TOP** plane. Similarly, geometry seen from the Right view should be drawn on **Right** plane and geometry seen from the Front view should be drawn on the **Front** plane.

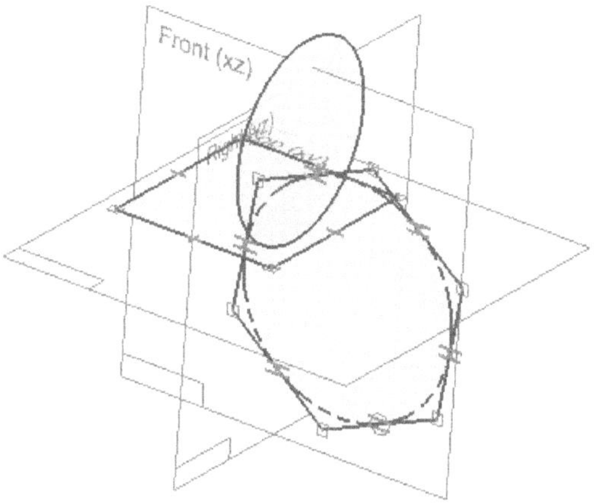

*Figure-3. Sketches created on different planes*

## STARTING SKETCH

In Solid Edge, Sketching environment can be opened as **Synchronous** and **Ordered**. To change synchronous to ordered, right-click in the empty area of drawing area. A shortcut menu will be displayed; refer to Figure-4.

* Select **Transition to Ordered** option from the shortcut menu to activate ordered sketching.

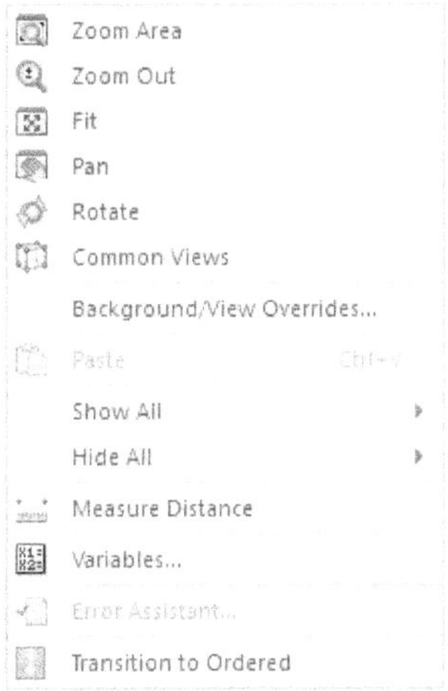

*Figure-4. Right click shortcut menu*

## Sketching Environment and Tools

In this section, we will discuss different types of sketching tools which are used in sketch creation.

## Synchronous Sketching Environment

The tools in Synchronous environment are available in **Home** tab of **Ribbon**; refer to Figure-5. You can directly use these tools and later select the plane where you want to create sketch entities.

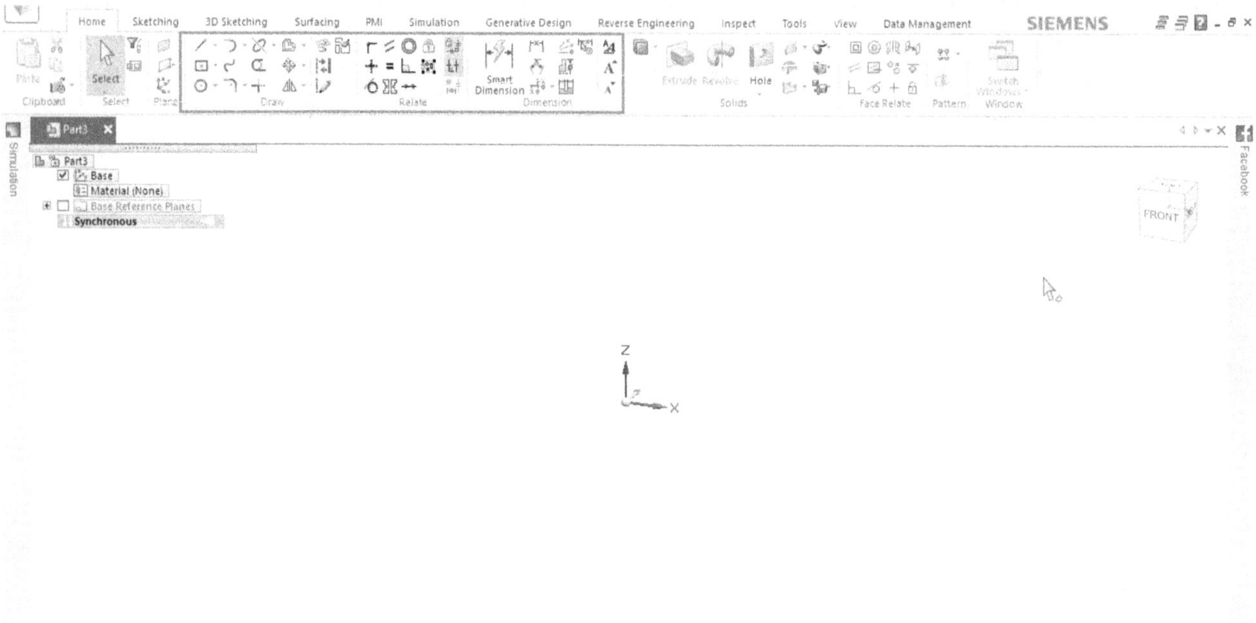

*Figure-5. Synchronous Sketch Environment*

## Ordered Sketching Environment

You need to switch to Ordered environment first before using the sketching tools of ordered environment. You can switch the environment as discussed earlier. The Ordered environment will be displayed as shown in Figure-6.

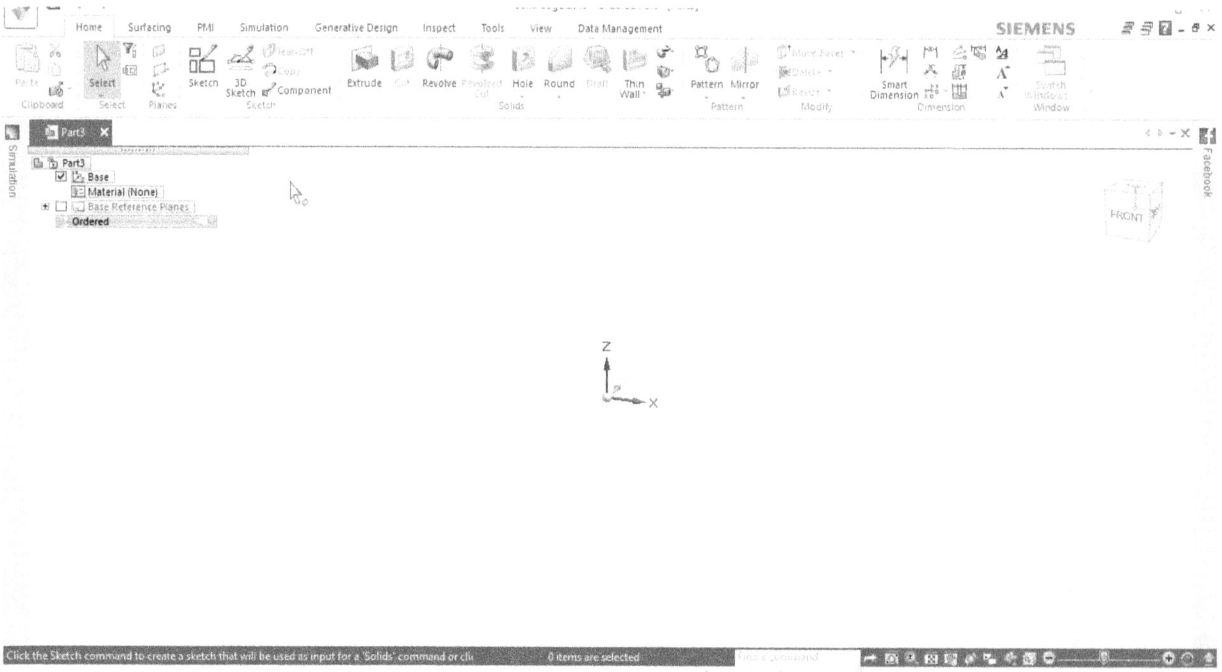

*Figure-6. Ordered Sketching Environment*

The procedure to create sketch in Ordered environment is given next.

- Click on the **Sketch** tool from the **Sketch** panel in the **Home** tab of **Ribbon**. You will be asked to select a sketching plane and related toolbar will be displayed; refer to Figure-7.

*Figure-7. Toolbar for sketch plane selection*

- Select the desired plane from the drawing area or **PathFinder**.

The tools to create sketch will be displayed.

# VIEW TOOLBAR

The **View Toolbar** contains tools to change the view and orientation of the model; refer to Figure-8.

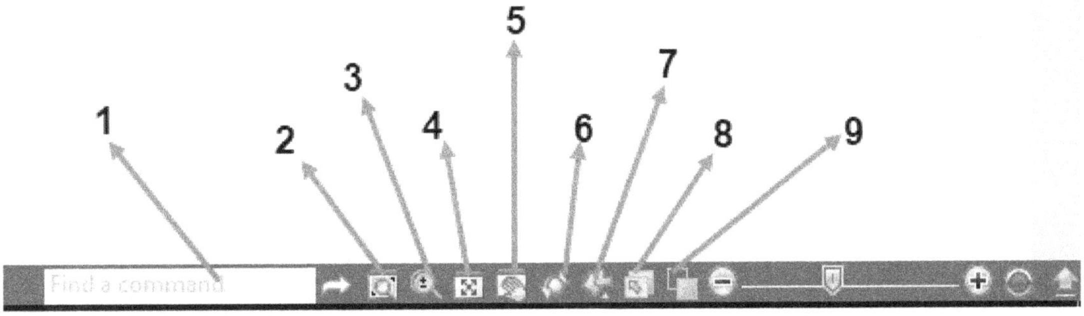

*Figure-8. View Toolbar*

1. **Command Finder** : The **Command Finder** tool is used to search tools in Solid Edge. For example, if you can not find **Line** tool then type **Line** in **Command Finder** and click on **Go** button ⬛. The tools with Line in their name will be displayed; refer to Figure-9.

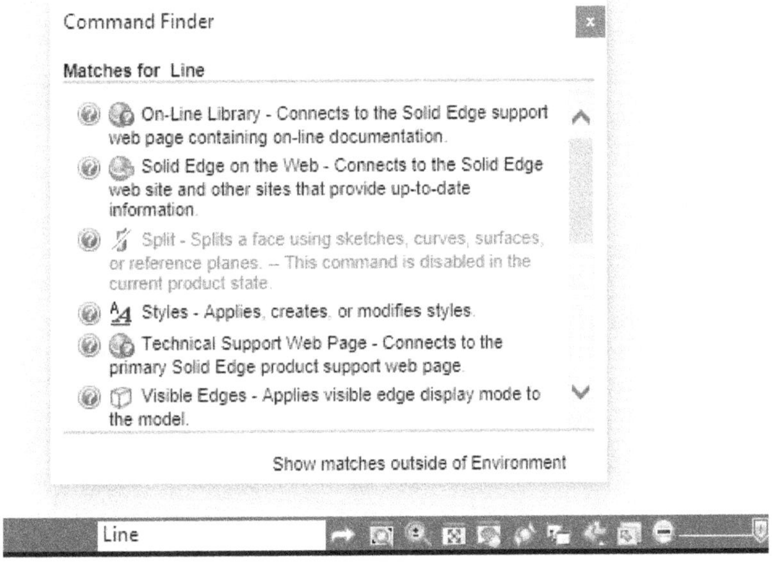

*Figure-9. Command Finder dialog box*

2. **Zoom Area** : The **Zoom Area** tool is used to display a specific area in the viewport zoomed to the full extent. To use this tool, click on it. The cursor will change to a zoom box selection cursor and you will be asked to create a boundary box surrounding the entities which you want to zoom in. Click to specify the starting point of the zoom box and then drag to the point till where you want to complete the zoom box. The area in the box will zoom automatically. Shortcut key for Zoom Area is **Alt + MMB**.

3. **Zoom** : The **Zoom** tool is used to Zoom-in and zoom-out the drawing area. After clicking this tool, drag cursor upward to zoom-out and drag cursor downward to zoom-in. Shortcut key for Zoom is **Ctrl + MMB**.

4. **Fit** : The **Fit** tool is used to display all the objects created in the viewport. To use this tool, click on the tool once. The objects will automatically fit in the current viewport.

5. **Pan** : The **Pan** tool is used to changes the viewed area by dragging the entities. Shortcut key is **Shift + MMB**.

6. **Rotate** : The **Rotate** tool is used to rotate the entities 360 degree. After selecting this tool, drag the cursor in desired direction to rotate. Shortcut key is **MMB**.

7. **Sketch View** : The **Sketch View** tool is used to make the sketching plane parallel to screen. Shortcut key is **Ctrl + H**.

8. **View Orientation** : The **View Orientation** tool is used to change the view orientation of model. When you click on this tool, a toolbox will be displayed as shown in Figure-10. These buttons orient the model as per their shapes and annotations.

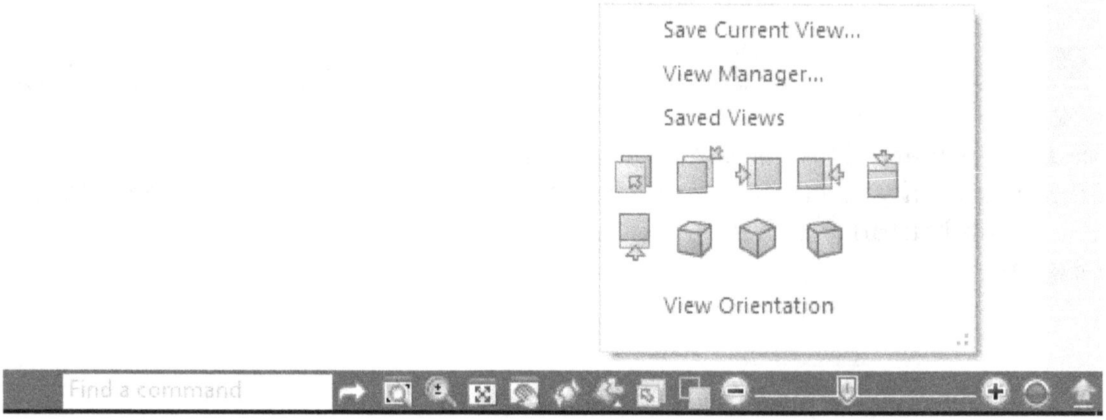

*Figure-10. View orientation toolbox*

9. **View Style** : These tools are used to display the model in different styles like shaded, hidden, no hidden, and so on.

# SKETCH CREATION TOOLS

The tools to create sketch are available in **Home** tab of **Ribbon**. These tools are discussed next.

## Line Tools

There are four tools in the **Line** drop-down for Synchronous as shown in Figure-11 and three line tools for Ordered as shown in Figure-12.

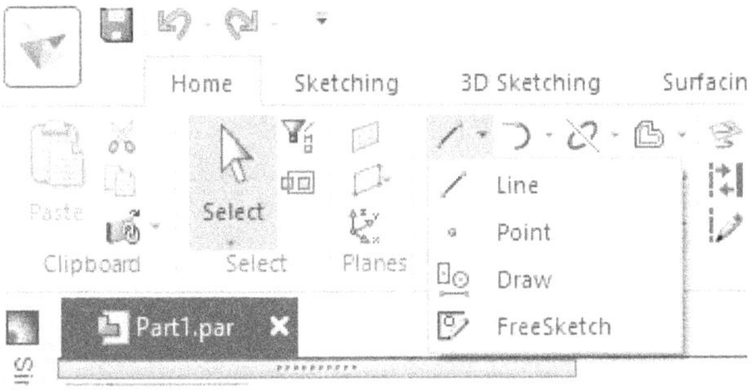

*Figure-12. Line tools in Ordered Environment*

*Figure-11. Line tools in Synchronous Environment*

## Line

The **Line** tool is used to create each type of lines in creating sketch. The procedure to create line is discussed next.

- Click on the **Line** tool from the **Line** drop-down. The **Line Command Bar** will be displayed as shown in Figure-13. Specify the origin point of line and drag the mouse. The preview is shown in Figure-14 & Figure-15.

*Figure-13. Line command bar*

- Click on **Line Color** button to change the color of line.
- Click on **Line Type** button to select the line type.
- Click on **Line Width** button ☰ to select the line width.
- Click on the **Line - Line** button to activate line creation mode.
- Click on **Line Arc** button to activate arc creation from **Line Command Bar**.
- Click on **Projection Line** button to draw projection line.
- Click **Projection Line Type** button to select the type of projection of line.
- After setting desired options in the **Command Bar**, click at desired location to specify the start point. You will be asked to specify end point of the line.
- If you move the cursor horizontally or vertically then respective constrain will be applied. Note that if there is another sketch entity then cursor will automatically snap to center, intersection, midpoint, or endpoints when moving cursor over the entity. Also, if you hold **SHIFT** key while moving the cursor then it move in increment of 15 degree.

*Figure-14. Line Created*                *Figure-15. Line Arc*

- Click at desired location to specify end point of line. You can also specify the length and angle in dynamic edit boxes displayed with line to specify end point. Press **TAB** to switch between dynamic edit boxes.
- After specifying end point of first line, you will be asked to specify end point of next line. Specify the end point parameters or press **ESC** to exit the tool.

## Point

Points are generally used as references for creating sketches entities. To create point, click on **Point** tool from the **Line** drop-down in **Draw** panel of **Home** tab in **Ribbon**. You will be asked to specify the location of point and **Point Command Bar** will be displayed; refer to Figure-16. Click at desired location to place point. You can also specify coordinates in the **X** and **Y** edits boxes and press **Enter** to create points.

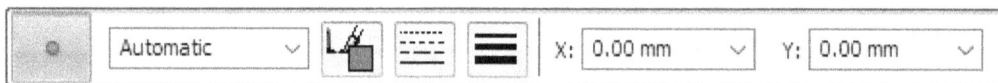

*Figure-16. Point Command Bar*

## Draw

The **Draw** tool is used to create precise geometry by sketching freehand with the help of pen, finger, or mouse. This tool is available in synchronous sketching environment only. On clicking the **Draw** tool from the **Line** drop-down in the **Home** tab of **Ribbon**, the **Draw Command Bar** will be displayed as shown in Figure-17.

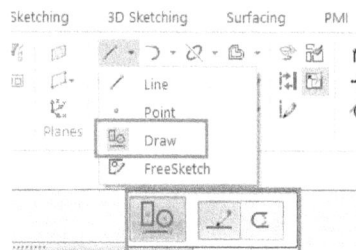

*Figure-17. Draw Command Bar*

### Creating Geometry

Select the **Create Geometry** button to dynamically create geometries. Create geometry by dragging your pen (if using digitizer), fingertip (if using touch screen device) or mouse.

Trim Geometry ⌒

Select the **Trim Geometry** button to trim geometry while dragging the cursor. Trim or delete by dragging your pen, finger, or mouse as discussed earlier.

## Freesketch

When you select **FreeSketch** from **Line** drop-down, then you can draw rough sketch of basic entities like arcs, lines, circles, rectangles, and so on. Based on drag motion, the respective sketch entities will be created. This tool is available in both Synchronous and Ordered Sketching Environment. On selecting **FreeSketch** tool, the **FreeSketch Command Bar** will be displayed as shown in Figure-18.

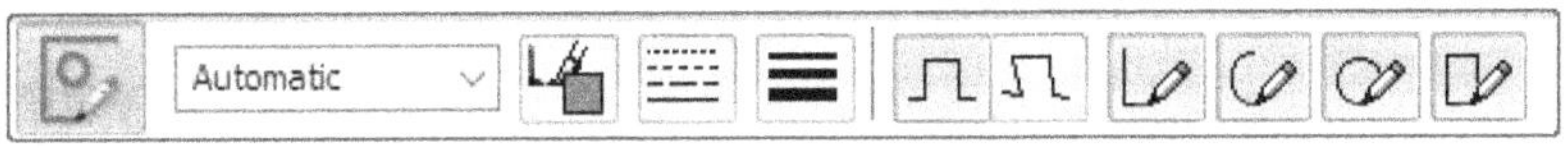

*Figure-18. FreeSketch CommandBar*

- **Adjust On** ⌐ : **Adjust On** button is active by default. It means when you draw any entities roughly. The software recognise the orientation of entities and convert it into the suitable entities.

- **Adjust Off** ⌐ : When you select this button, the software do not recognise the orientation of entities and keep the orientation the same as you draw them.
- You should try to draw entities with Adjust On and Adjust Off buttons to find out which suits better for creating an entity.
- Press **ESC** to exit the tool.

## Rectangle Tools

In **Rectangle** drop-down, there are four tools out of which three are used to draw rectangles and one of them is used for creating polygon. The tools of **Rectangle** drop-down are shown in Figure-19. These tools are discussed next.

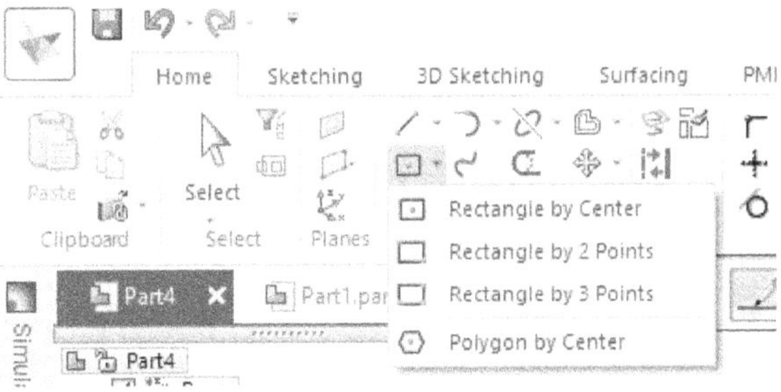

*Figure-19. Rectangle drop-down*

## Rectangle by Center

The **Rectangle by Center** tool is used to draw rectangle by specifying centre point and a diagonal corner point. The procedure is discussed next.

- Click on the **Rectangle by Centre** tool from **Rectangle** drop-down. The **Rectangle Command Bar** will be displayed; refer to Figure-20.
- Specify the centre point and diagonal point of rectangle. Refer to Figure-21.

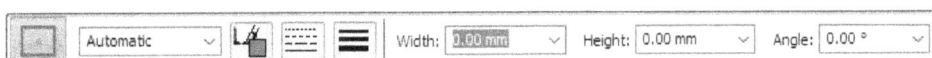

*Figure-20. Rectangle Command Bar*

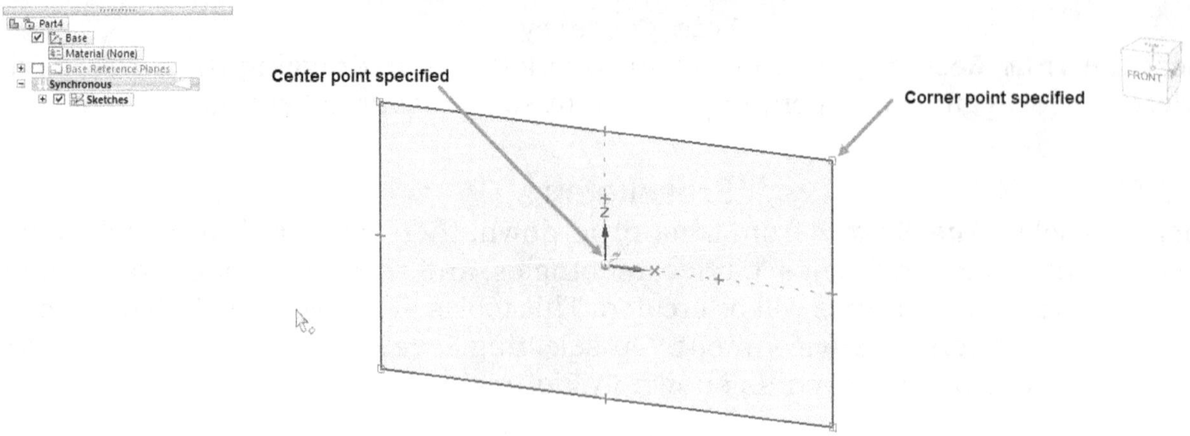

Center point specified

Corner point specified

*Figure-21. Rectangle by center*

- Press **ESC** to exit the tool.

## Rectangle by 2 Points

The **Rectangle by 2 Points** tool is used to draw rectangle by specifying two diagonal corner points. The procedure is discuss next.

- Click on the **Rectangle by 2 Points** tool from **Rectangle** drop-down. The **Rectangle Command Bar** will be displayed similar to Figure-20.
- Specify the first and second point of rectangle. Refer to Figure-22.

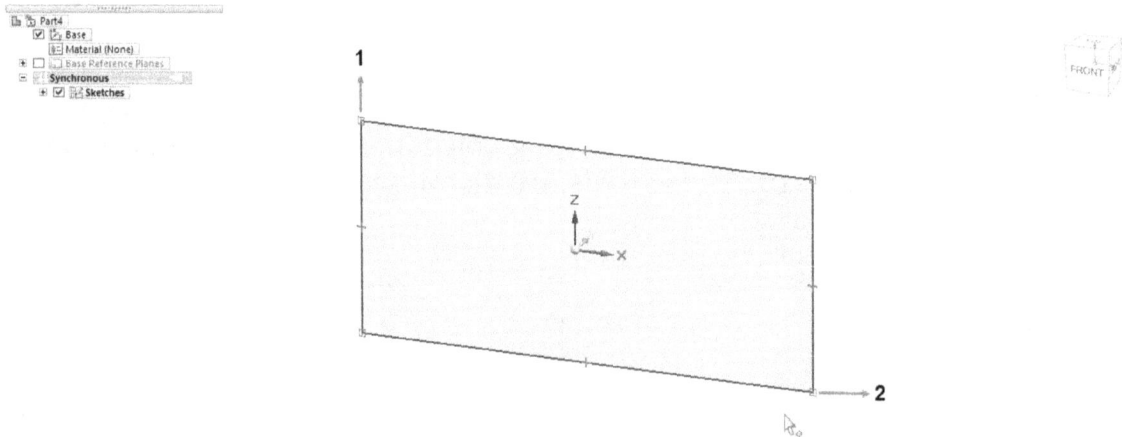

*Figure-22. Rectangle by 2 points*

## Rectangle by 3 Points

The **Rectangle by 3 Points** tool is used to draw rectangle by three corner points. The procedure is discuss next.

- Click on the **Rectangle by 3 Points** from **Rectangle** drop-down. The **Rectangle Command Bar** will be displayed similar to Figure-20.
- Specify the first, second, and third point of rectangle. Refer to Figure-23.

*Figure-23. Rectangle by three Points*

## Polygon by Center ⬡

The **Polygon by Center** tool is used to create regular polygon by defining number of sides and size of polygon. In Solid Edge, you can create polygon with number of sides ranging from 3 to 200.

- Click on the **Polygon by Center** tool from **Rectangle** drop-down. The **Polygon Command Bar** will be displayed; refer to Figure-24.

*Figure-24. Polygon Command Bar*

- **By Vertex** ⌒ : Select the **By Vertex** button from the **Command Bar** to create polygon using vertices; refer to Figure-25.

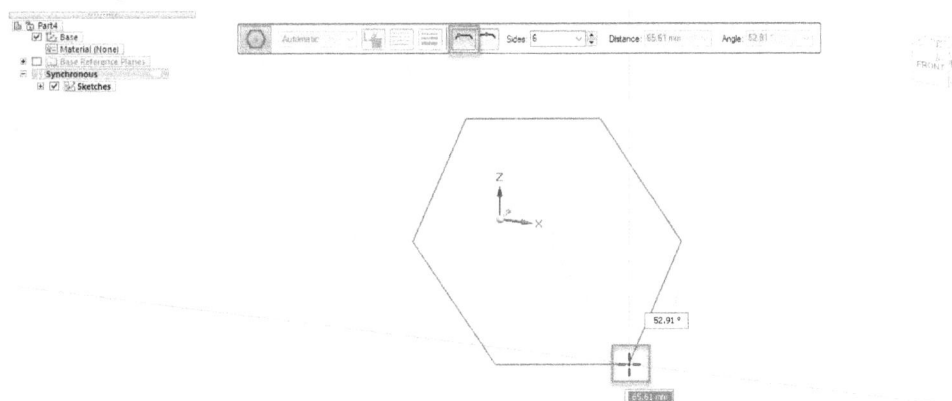

*Figure-25. Polygon by vertex*

- **By Midpoint** ⌒ : Select the **By Midpoint** button from the **Command Bar** to create polygon using mid points of sides of polygon; refer to Figure-26.
- Click at desired location to specify the center point of polygon and then specify the mid point or vertex point of polygon side.
- After creating polygon, press **ESC** to exit the tool.

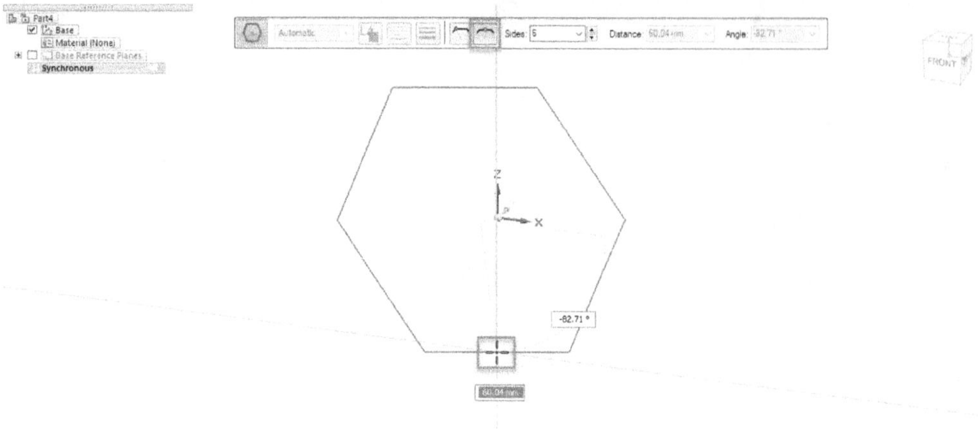

*Figure-26. Polygon by midpoint*

## Circle Tools ⊙

The **Circle** tool is used to draw circles. In **Circle** drop-down, there are five tools. Three tools are used to create circles and two tools are used to create ellipses; refer to Figure-27.

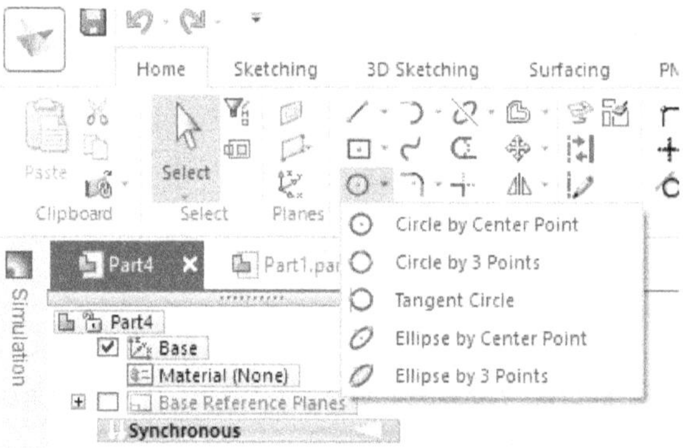

*Figure-27. Circle drop down*

## Circle by Center Point ⊙

The **Circle by Center Point** tool is the widely used tool to draw circle. The procedure to use this tool is given next.

• Click on the **Circle by Center Point** tool from the **Circle** drop-down. The **Circle Command Bar** will be displayed; refer to Figure-28.
• Specify diameter/radius of circle in the edit boxes of **Command Bar** and click at desired location to place the circle; refer to Figure-29.

*Figure-28. Circle Command Bar*

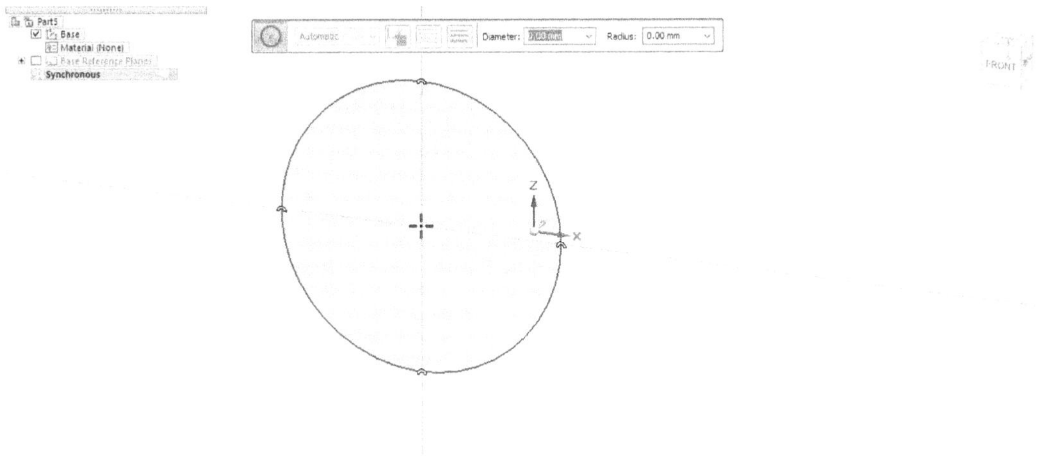

*Figure-29. Circle by center point*

## Circle by 3 Points ○

The **Circle by 3 Points** tool is used to draw circle but it is different from using **Circle by Center Point** tool. As the name suggest, you need to specify three points to draw circle. After specifying first and second point of circle, the cursor gets attached to third point of circle. Click at desired location to specify third point of circle; refer to Figure-30.

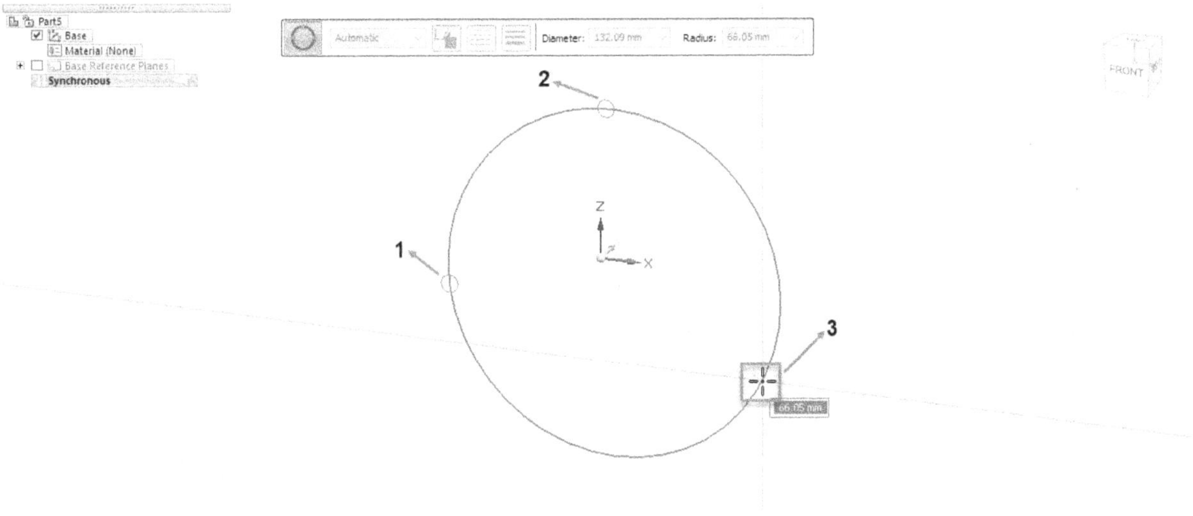

*Figure-30. Circle by 3points*

## Tangent Circle ○

The **Tangent Circle** tool is used to draw circle tangent to the existing one or two entities. To draw this type of circle, select the **Tangent Circle** tool from the **Circle** drop-down. To draw tangent circle, click on the two entities to which circle should be tangent or click on one entity and specify radius/diameter; refer to Figure-31.

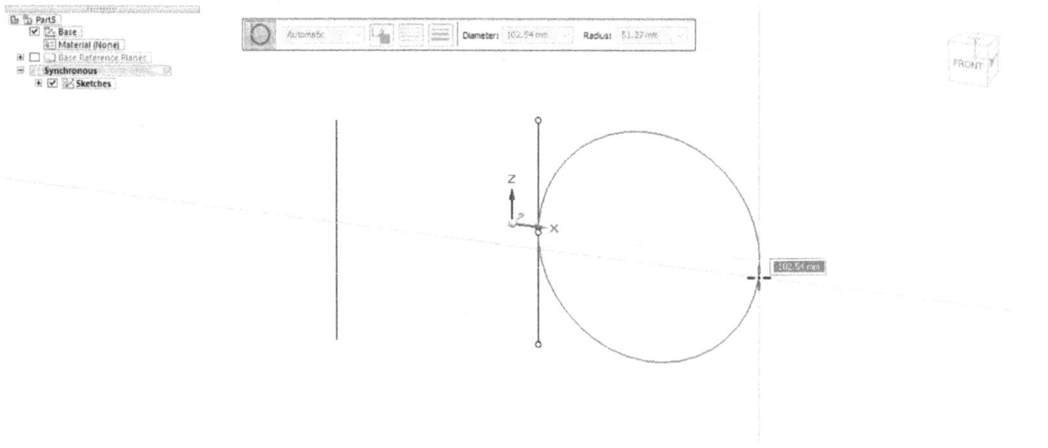

*Figure-31. Tangent circle*

## Ellipse Tools

There are two tools in **Circle** drop-down to create ellipses which are:-

*   Ellipse by Center Point
*   Ellipse by 3 Point

## Ellipse by Center Point

The **Ellipse by Center Point** tool is used to draw ellipse by specifying center point, major axis, and minor axis. The procedure is given next.

*   Click on the **Ellipse by Center Point** tool from **Circle** drop-down. The **Ellipse Command Bar** will be displayed; refer to Figure-32. You will be asked to specify the centre of the ellipse.
*   Click to specify the center. You will be asked to specify the radius along the major axis.
*   Click to specify the radius of major and minor axes or specify the values in the **Primary** and **Secondary** edit boxes of the **Command Bar**. You can also specify angle in the **Angle** edit box of **Command Bar**. The ellipse will be created ; refer to Figure-33.

*Figure-32. Ellipse Command Bar*

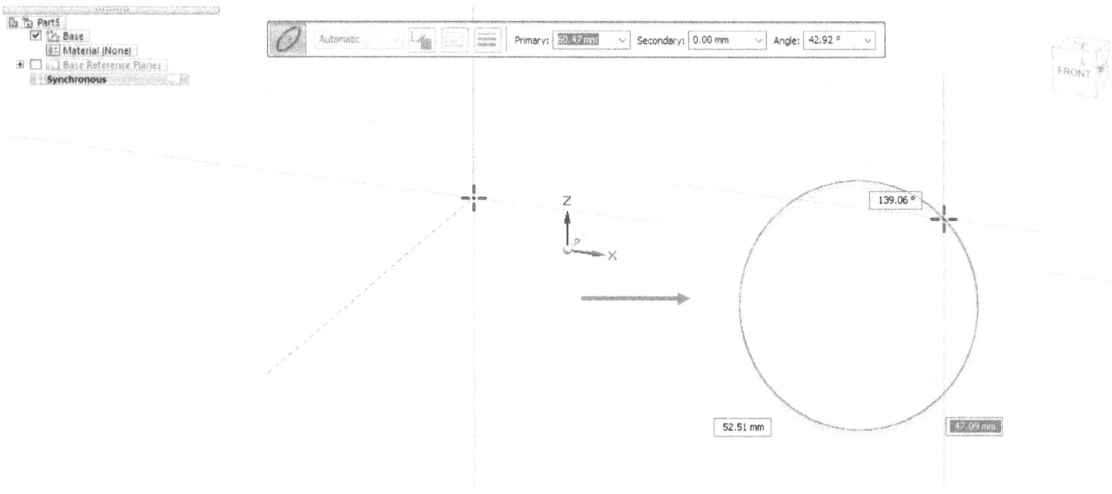

*Figure-33. Ellipse by center point*

## Ellipse by 3 Point

The **Ellipse by 3 Point** tool is used to create ellipse by three points. As the name suggest, you need to specify three points of ellipse. The procedure is given next.

*   Click on the **Ellipse by 3 Point** tool from **Circle** drop-down. The **Ellipse by 3 Point Command Bar** will be displayed similar to Figure-32.
*   Specify the first, second, and third point of ellipse. When you specify the first and second point, the annotations are displayed on the ellipse. Select the desired location to place third point of ellipse or specify parameters in the **Command Bar**. You can also create ellipse by providing dimensions of primary and secondary axes in **Command Bar** of ellipse. The preview is shown in Figure-34.

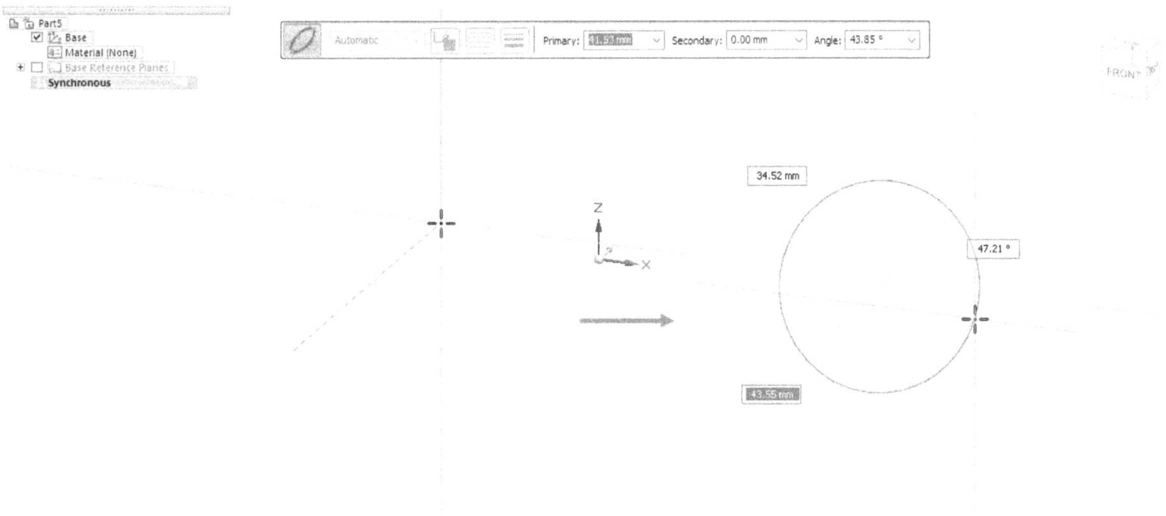

*Figure-34. Ellipse by 3 point*

# Tangent Arc Tools

There are three tools available in **Tangent Arc** drop-down of **Draw** panel in **Ribbon**; refer to Figure-35. These tools are discussed next.

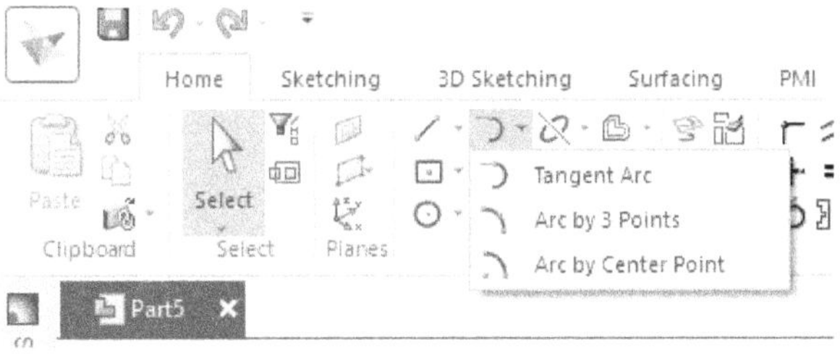

*Figure-35. Arc drop down*

## Tangent Arc

*   Click on the **Tangent Arc** tool from the **Arc** drop-down. The **Tangent Arc Command Bar** will be displayed; refer to Figure-36.

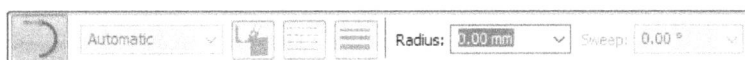

*Figure-36. Arc Command Bar*

*   Select the end point of the entity to which the arc should be tangent.
*   Click to specify the end point of the arc. Note that the tangent constrained is automatically applied at the arc. The preview of tangent arc is shown in Figure-37.

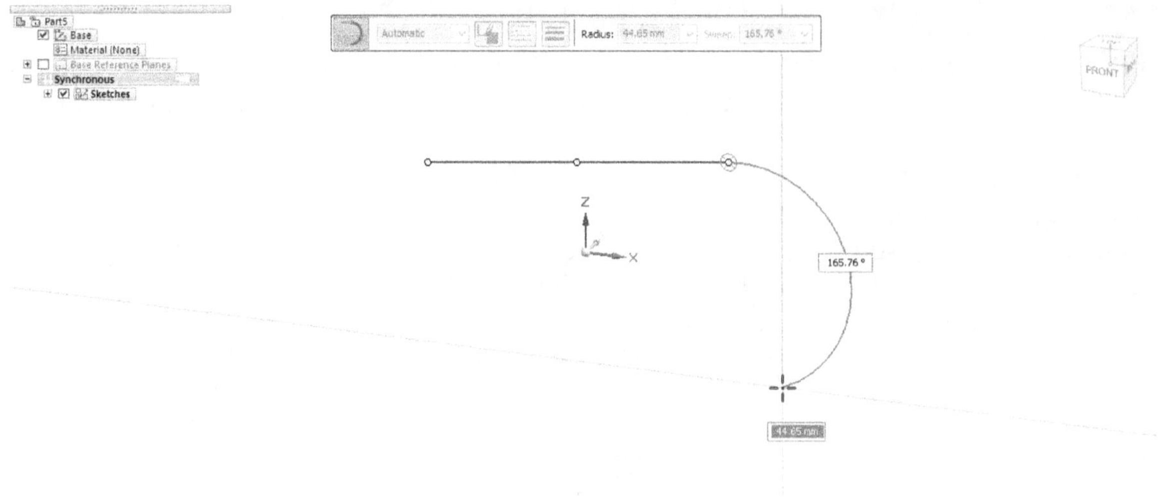

*Figure-37. Tangent arc*

## Arc by 3 Points

- Click on the **Arc by 3 Points** tool from **Arc** drop-down. The **Arc Command Bar** will be displayed as earlier.
- Click to specify the start point of the arc.
- Click to specify the end point of the arc.
- Click to specify a point on the arc to set the radius of the arc. The preview of arc by 3 points is shown in Figure-38.

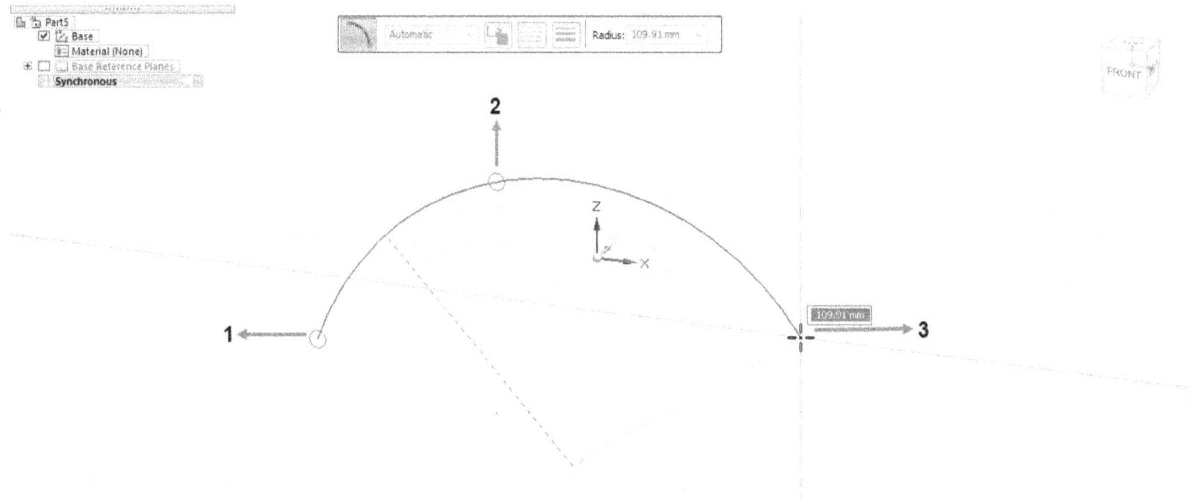

*Figure-38. Arc by 3 points*

## Arc by Center Point

- Click on the **Arc by Center Point** tool from **Arc** drop-down. The **Arc Command Bar** will be displayed as earlier.
- Click to specify the center point of the arc.
- Click to specify the start point of the arc.
- Click to specify the end point of the arc. The preview of arc by center point is shown in Figure-39. You can also specify the radius and sweep angles in **Command Bar** after specifying start point.

*Figure-39. Arc by Center Point*

# Curve Tool

The **Curve** tool is used to draw spline curve. The procedure is given next.

- Click on the **Curve** tool from **Draw** panel. The **Curve Command Bar** will be displayed; refer to Figure-40. You are asked to specify the points through which the curve should pass.
- One by one click in the viewport to specify the points of the curve.
- Press **ESC** key from the keyboard to exit the tool. The curve will be created; refer to Figure-41. Select the recently created curve to activate all the buttons in the **Command Bar**. Some of the important buttons in the **Command Bar** are discussed next.

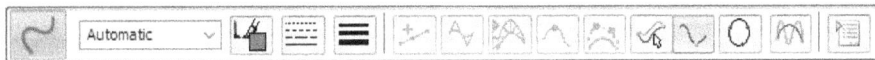

*Figure-40. Curve Command Bar*

*Figure-41. Curve created*

- **Close Curve** ⬭ : Select **Close Curve** button from **Command Bar** to draw close curve; refer to Figure-42.

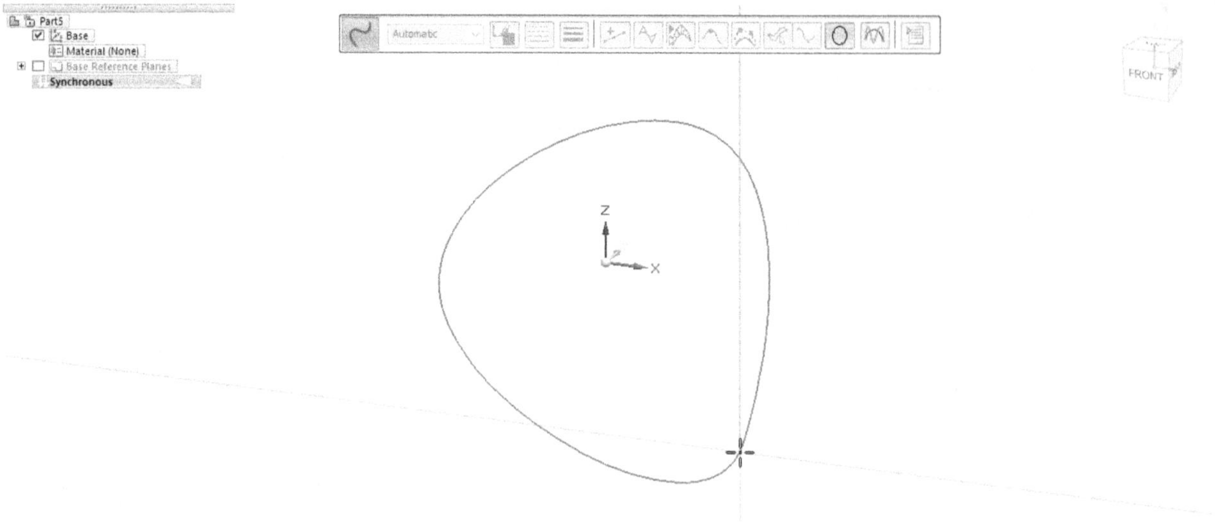

*Figure-42. Close curve*

**Simplify Curve** [icon]: The **Simplify Curve** button is used to reduce the number of points in the curve.

**Shape Edit** [icon]: Toggle the **Shape Edit** button from **Command Bar** to modify shape of full curve by dragging a point of curve.

**Local Edit** [icon]: Toggle the **Local Edit** button from the **Command Bar** to modify the section of curve by dragging respective point of curve.

Select the **Show Edit Points** and **Show Control Vertex Points** buttons to display respective points.

Similarly, you can display other parameters of curve. Click on the **Curve Options** button in the **Command Bar** to modify curve options. The **Curve Options** dialog box will be displayed; refer to Figure-43. Select the desired option from the drop-down to define smoothness of curve.

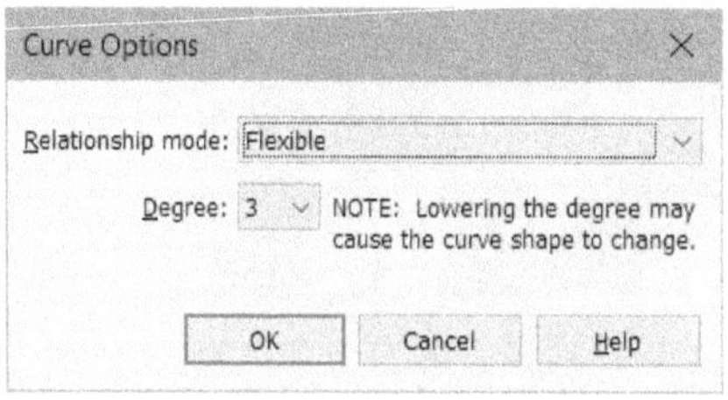

*Figure-43. Curve Options dialog box*

## Fillet Tools [icon]

There are two tools in **Fillet** drop-down to create fillets and chamfers; refer to Figure-44.

*Figure-44. Fillet drop-down*

## Fillet

The **Fillet** tool is used to create fillet/round at the corners created by intersection of two entities. Fillet sometimes also referred as round. Generally it is not advised to use this tool first if there are some major changes going to occur later in the sketch. The procedure to create fillet is given next.

*   Click on the **Fillet** tool from **Fillet** drop-down in **Draw** panel of **Home** tab in the **Ribbon**. The **Fillet Command Bar** will be displayed; refer to Figure-45.

*Figure-45. Fillet Command bar*

*   **Fillet- No Trim** ⌐ **Off**:- When this option in fillet command bar is **"OFF"**. It means trim takes place while performing fillet. Refer to Figure-46.

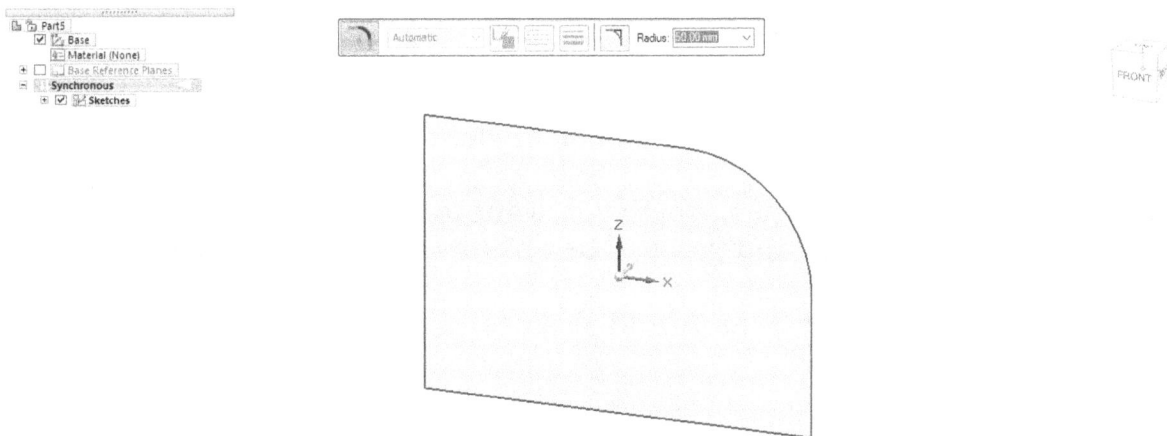

*Figure-46. Fillet trim*

*   **Fillet- No trim** ⌐ **On**:- After turning **"ON"** this option from command bar,the trim does not take place while performing fillet. Refer to Figure-47.

*Figure-47. Fillet no trim*

## Chamfer

The **Chamfer** tool is used to provide gradient line to the object. You can create chamfer by specifying chamfer angle, distance from first side, and distance from second side. The procedure to create chamfer is given next.

*   Click on the **Chamfer** tool from the **Fillet** drop down in the **Home** tab of **Ribbon**. The **Chamfer Command Bar** will be displayed; refer to Figure-48.

*Figure-48. Sketch Chamfer Command Bar*

*   Specify desired parameters like chamfer angle, distance A and distance B in respective edit boxes.
*   Click on two intersecting lines, where you want to apply chamfer. The preview of chamfer will be displayed and you will be able to switch A and B sides by cursor movement.
*   Click at the desired side to fix A and B sides of chamfer. The chamfer will be created; refer to Figure-49.

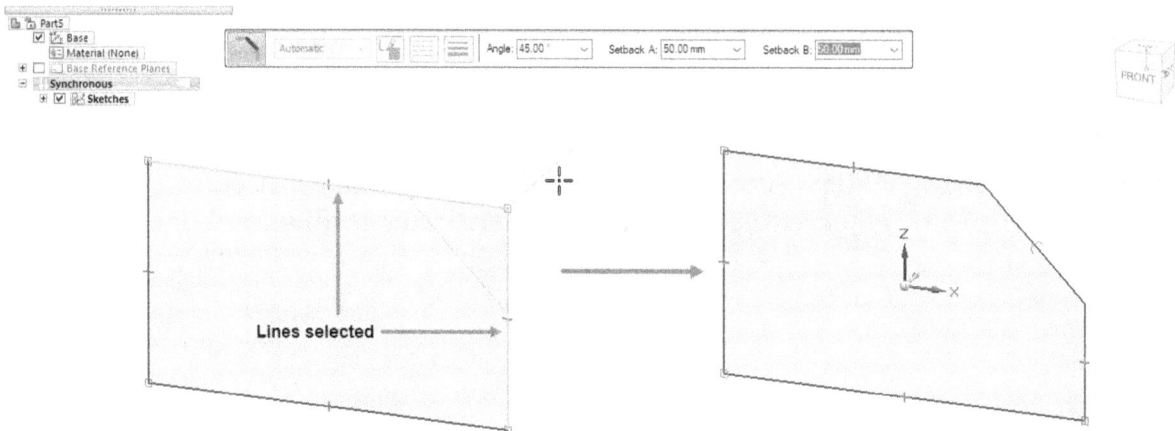

*Figure-49. Chamfer applied to sketch*

## Split Drop-down

In **Split** drop-down, there are two tools available **Split** and **Extend to Next**; refer to Figure-50. These tools are discussed next.

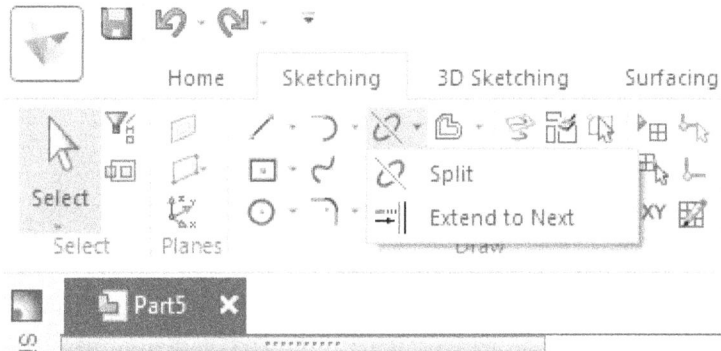

*Figure-50. Split drop-down*

## Split

The **Split** tool is used to divide a sketch entity into two or more segments. The procedure to use this tool is given next.

*   Click on the **Split** tool from the **Split** drop-down in the **Draw** panel of **Home** tab in the **Ribbon**. You will be asked to select the sketch entity to be split.
*   Select the object which you want to split. An arrow will be displayed showing the direction of split.
*   Click at desired location on sketch entity where you want to split the entity. A round circle is created on the object that indicates that the object has split; refer to Figure-51. The number of circles show number of splits created. Press **ESC** to exit the tool.

*Figure-51. Splitting line*

## Extend To Next

The **Extend to Next** tool is used to extend the line to the next object. If you have drawn a line but there is certain gap between it then you can use this tool to fill the gap. The procedure to use this tool is given next.

*   Click on the **Extend to Next** tool from the **Split** drop-down in the **Draw** panel of **Home** tab in the **Ribbon**. You will be asked to select entity to be extended.
*   Hover the cursor at desired side of sketch entity. Preview of extended sketch will be displayed; refer to Figure-52. If the sketch extension is as desired then click on the entity to create extended entity. Press **ESC** to exit the tool.

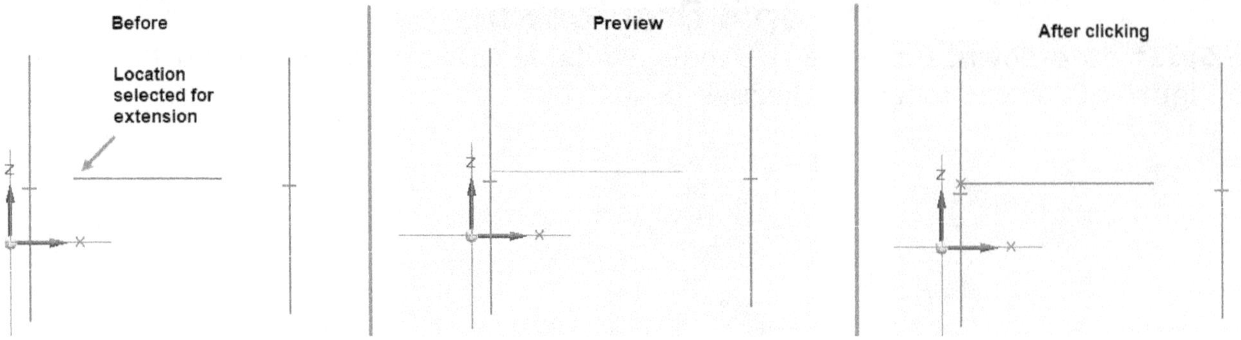

Before                              Preview                              After clicking

Location
selected for
extension

*Figure-52. Extend to next*

# Trim Tool

The **Trim** tool is used to delete unwanted segment of an entity in sketch. Suppose you have create a sketch and small segment of an entity is extending out of the sketch then you can use this tool to remove that extra segment. The procedure to use this tool is given next.

*   Click on the **Trim** tool from the **Draw** panel in the **Home** tab of **Ribbon**. You will be asked to select the entity to be trimmed.
*   Click at desired side of entity to be removed by trimming or you can create a freehand curve to trim desired side of curve; refer to Figure-53.

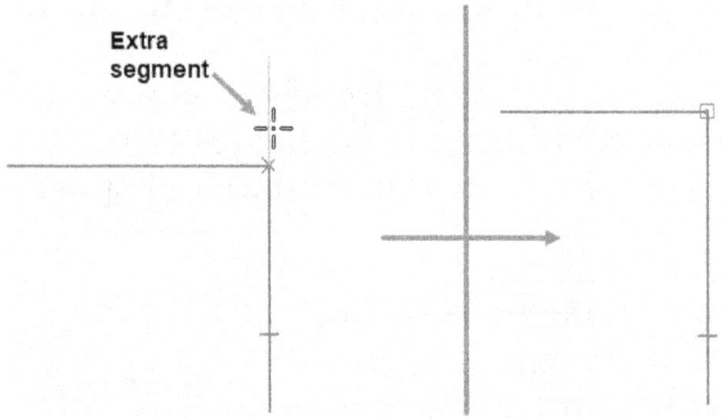

Extra
segment

*Figure-53. Trimming extra segment*

# Trim Corner Tool

The **Trim Corner** tool is used to trim corner or extend the corner, simultaneously. Suppose, you have created a sketch and you want their corners to be extend to next corner as shown in Figure-54. The procedure to use this tool is given next.

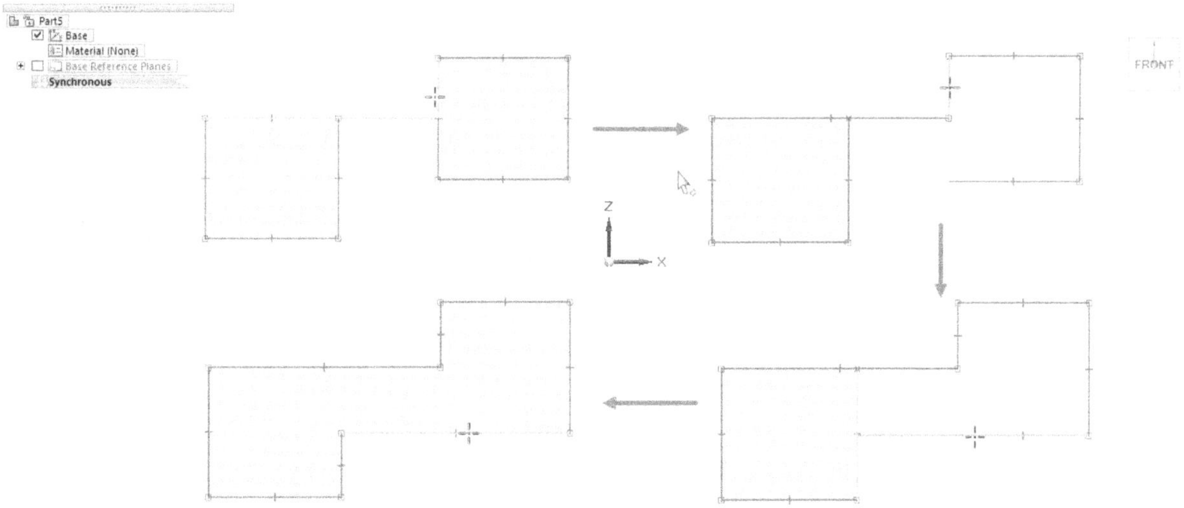

*Figure-54. Trim corner*

- Click on the **Trim Corner** tool from the **Draw** panel in the **Home** tab of **Ribbon**. You will be asked to select the first entity.
- Select the first entity and then hover the cursor on second entity. Preview of trim will be displayed; refer to Figure-55.

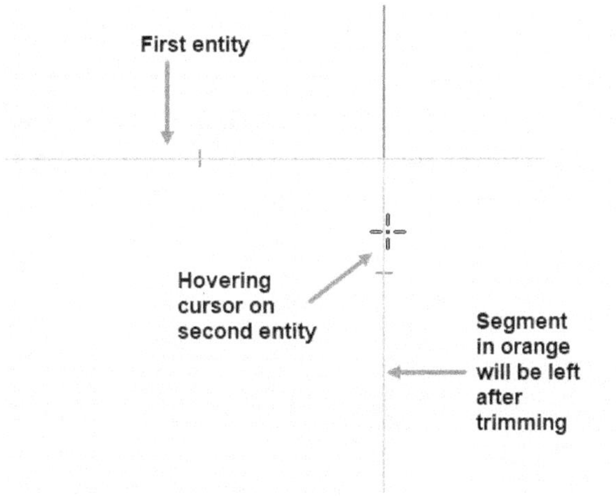

*Figure-55. Trimming corners*

- If the preview is as desired then click on the entity. The corner will be created by trimming.

## Offset Tools drop-down

There are two tools available in the **Offset** drop-down, **Offset** and **Symmetric Offset**; refer to Figure-56.

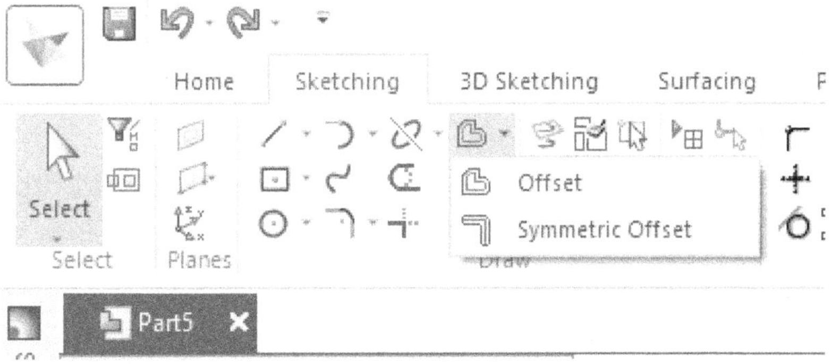

*Figure-56. Offset drop-down*

## Offset

The **Offset** tool is used to create copy of the selected entities at specified distance from them. If you are user of **Autocad** then this tool is the common tool being used while creating layout. The procedure to use this tool is given next.

*   Click on the **Offset** tool from the **Offset** drop-down in the **Draw** panel of **Home** tab in the **Ribbon**. The **Offset Command Bar** will be displayed; refer to Figure-57.

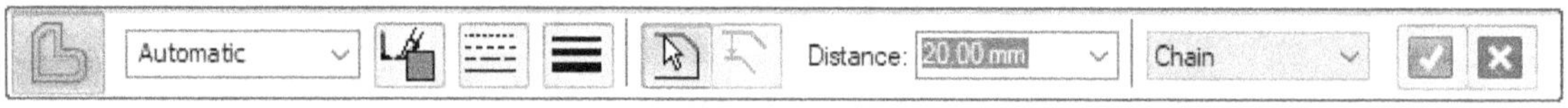

*Figure-57. Offset Command Bar*

*   Select the desired option from the **Offset-Select** drop-down to define whether you want to select single curve for offset or a chain of curves. Select the **Single** option from the drop-down to select single curves and select the **Chain** option from the drop-down to select chain of connected curves.
*   Specify the distance by which you want to offset curves in the **Distance** edit box of the **Command Bar**.
*   You can modify the line color, width, and type using the respective buttons in the **Command Bar** as discussed earlier.
*   Select the curves to be offset from the drawing area and click on the **OK** button from the **Command Bar** or right-click in the empty area. Preview of offset will be displayed with arrow showing direction of offset; refer to Figure-58.
*   If you want to modify the offset distance then you can do so by specifying value in the **Distance** edit box of **Command Bar**. Click on the desired side of original curves to create offset copy.

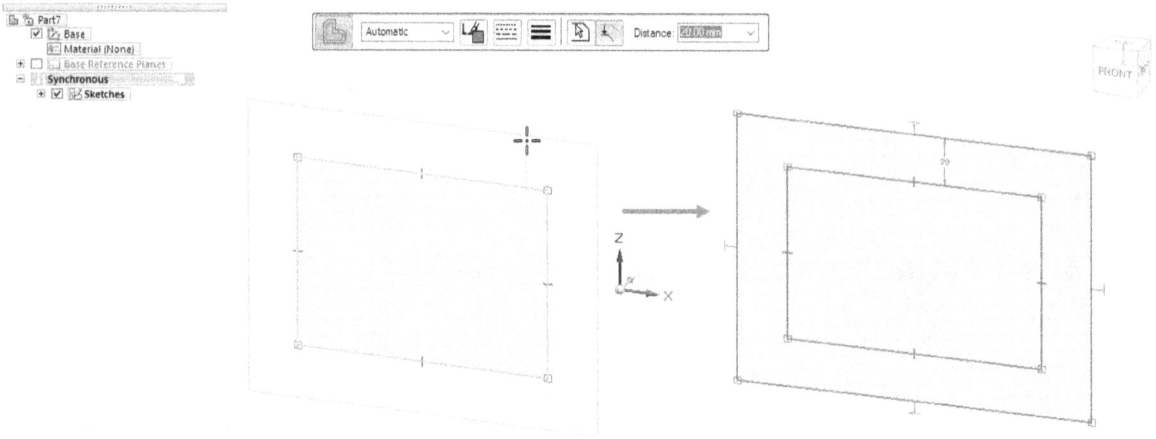

*Figure-58. Offset*

*   Press **ESC** to exit the tool.

## Symmetric Offset

The **Symmetric Offset** tool is used to create symmetric offset on both sides of selected entities. The procedure to use this tool is given next.

*   Click on the **Symmetric Offset** tool from the **Offset** drop-down in the **Draw** panel of **Home** tab in the **Ribbon**. The **Symmetric Offset Options** dialog box will be displayed; refer to Figure-59.

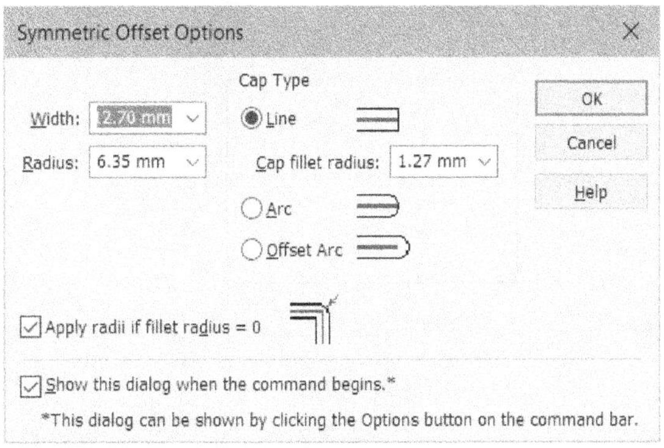

*Figure-59. Symmetric Offset Options dialog box*

- Specify the desired value of distance between two symmetric offset copies in the **Width** edit box.
- Select the **Apply radii if fillet radius = 0** check box to apply fillet at the corner even if not radius is applied in original.
- Specify the desired value of fillet radius in the **Radius** edit box.
- By default, **Line** radio button is selected in the dialog box and so a line connects the open end points of offset copies. You can also specify fillet at the capped corners by specifying value in the **Cap fillet radius** edit box.
- If you want to create a semi circular cap then select the **Arc** radio button. Similarly, select the **Offset Arc** radio button to create semi-circular cap at offset distance equal to radius of arc.
- After setting desired parameters, click on the **OK** button from the dialog box. The **Symmetric Offset Command Bar** will be displayed asking you to select the entities to be offset.
- Set the desired parameters in the **Command Bar** as discussed earlier and select the entities.
- Click on the **OK** button from the **Command Bar** or right-click to create symmetric offset copy; refer to Figure-60.
- Press **ESC** to exit the tool.

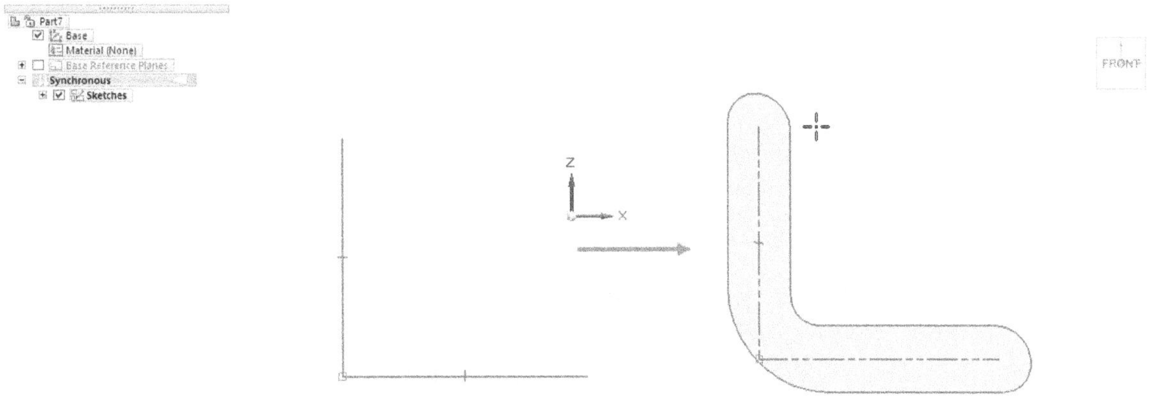

*Figure-60. Symmetric offset copy created*

## Move Tools

The tools in the **Move** drop-down are used to move and rotate selected sketch entities; refer to Figure-61. These tools are discussed next.

*Figure-61. Move drop down*

## Move ✛

The **Move** tool is used to move the entities from one position to another position. You can move the entities either by specifying coordinates or by dragging the entities. The procedure to use this tool is given next.

• Click on the **Move** tool from the **Move** drop-down in the **Draw** panel of **Home** tab in the **Ribbon**. The **Move Command Bar** will be displayed; refer to Figure-62.

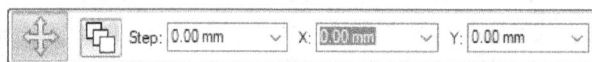

*Figure-62. Move Command Bar*

• **Move-Copy** 🔲 **Off** :- When **Move copy** button is off, then the entities can be moved by selecting and dragging the element from one location to another. In this case, the original entities will move. Refer to Figure-63.

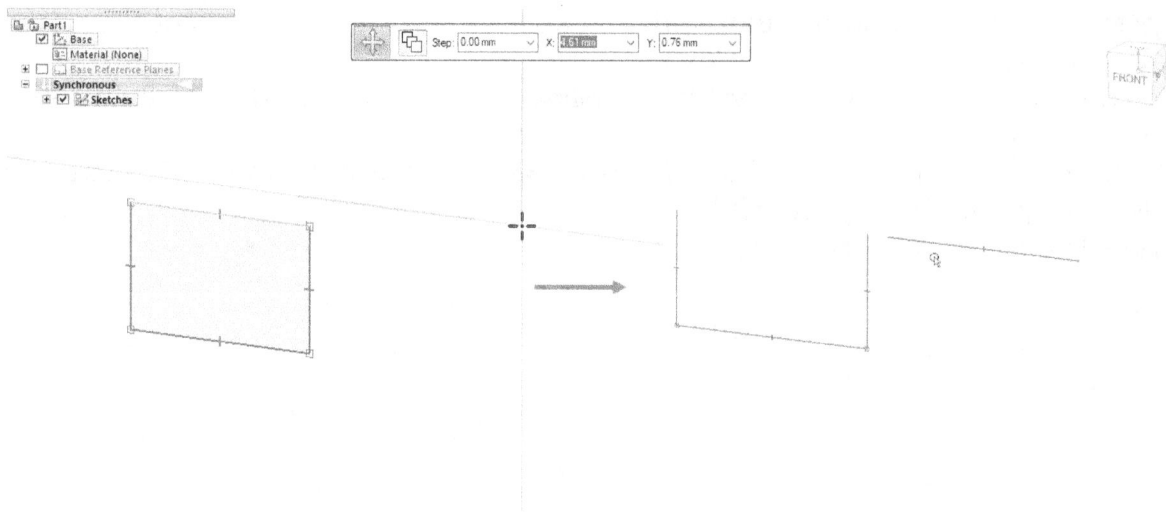

*Figure-63. Move copy off*

• **Move-Copy** 🔲 **On** :- When the **Move-copy** button is **On**, then the original entities remain at their location and copies of entities are moved from one location to another by selecting and dragging the entities. Refer to Figure-64.

*Figure-64. Move copy on*

## Rotate

The **Rotate** tool is available in the **Move** drop-down. This tool is used to rotate the entities by specified angle or by dragging.

- Click on the **Rotate** tool from the **Move** drop-down. The **Rotate Command Bar** will be displayed as shown in Figure-65.

*Figure-65. Rotate command bar*

- Select the entity that you want to rotate. You will asked to specify center point for rotation.
- Click at desired location/point to specify center point. You will be asked to specify start point for rotation.
- Specify the angle and the value at which the object will rotate.
- **Rotate-Copy** 🔲 **Off** :- When the **Rotate-copy** button is **Off**, then the entities can be moved by selecting and rotating the element from one location to another. Refer to Figure-66.

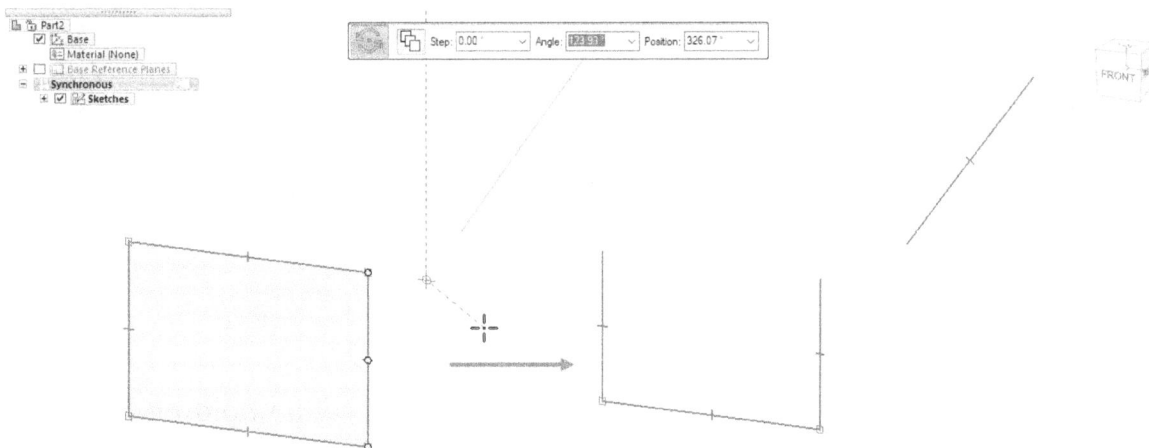

*Figure-66. Rotate copy off*

- **Rotate-Copy** 🗗 **On** :- When the **Rotate-copy** button is **On**, then the entities are moved and copied from one location to another by selecting and rotating the entity. Refer to Figure-67.

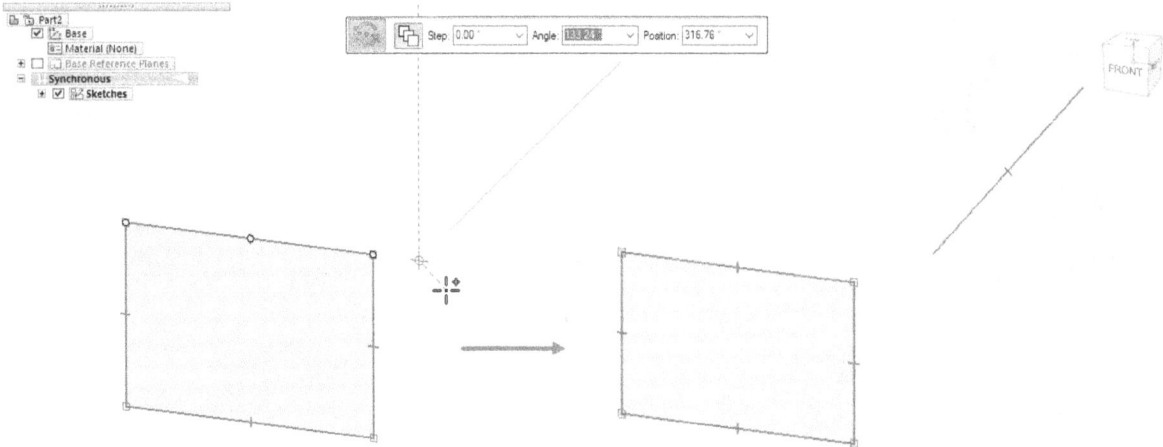

*Figure-67. Rotate copy on*

## Mirror Drop-down ⚐ ·

In **Mirror** drop-down, there are three tools; **Mirror**, **Scale**, and **Stretch**; refer to Figure-68.

*Figure-68. Mirror drop down*

## Mirror ⚐

The **Mirror** tool is used to create mirror copy of the selected entities with respect to a reference called mirror line. The procedure to create mirror entities is given next.

- Click on the **Mirror** tool from **Mirror** drop-down in **Draw** panel of the **Home** tab in the **Ribbon**. The **Mirror Command Bar** will be displayed;refer to Figure-69.

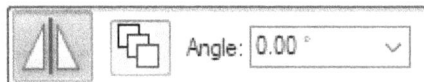

*Figure-69. Mirror Command bar*

- Select the entity/entities that you want to mirror copy while holding **CTRL** key (if there are multiple entities).
- Click on the reference line about which you want to mirror the entities. You can create multiple mirror copies by selecting different reference lines one by one.

- **Mirror-Copy** ⧉ **Off** :- When the **Mirror-copy** button is **Off**, then the entities can be mirror without being copied by selecting the entities and reference line. The base entity will be deleted after mirroring. Refer to Figure-70.

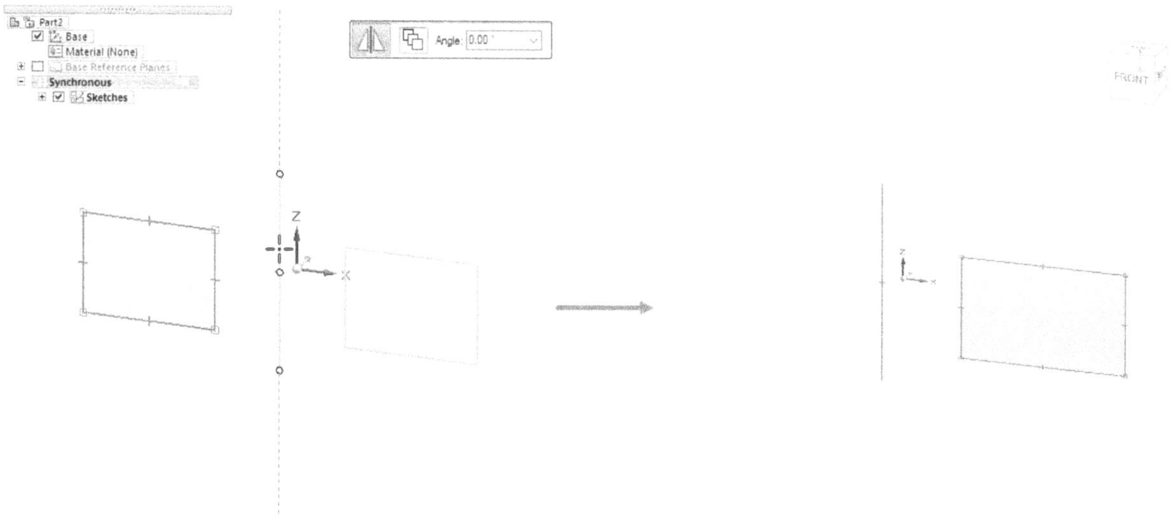

*Figure-70. Mirror copy off*

- **Mirror-Copy** ⧉ **On** :- When the **Mirror-copy** button is **On**, then the entities can be mirror copied to other location by selecting the entities and reference line. Refer to Figure-71.

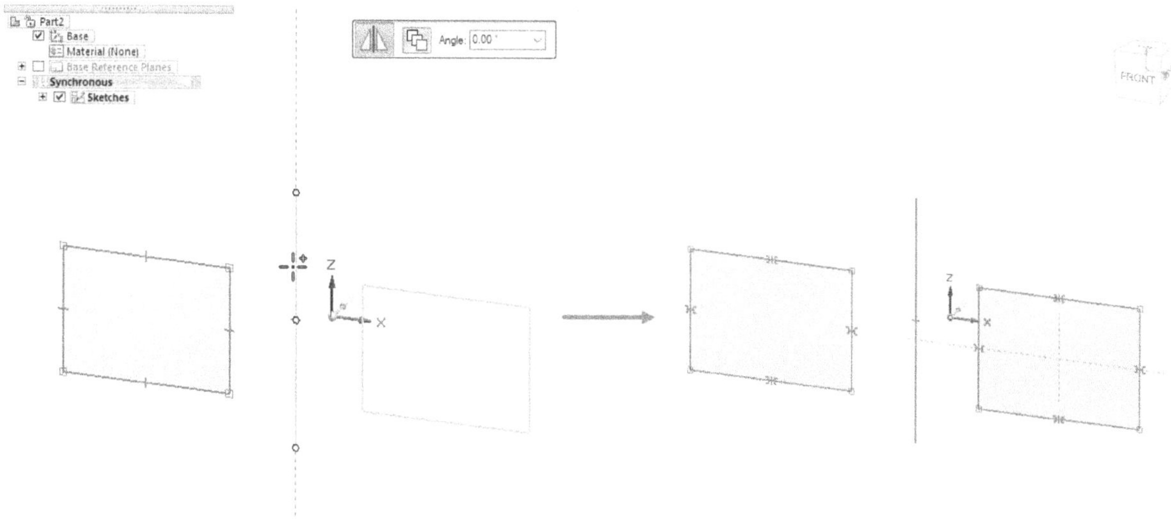

*Figure-71. Mirror Copy on*

## Scale

The **Scale** tool is available in the **Mirror** drop-down. This tool is used to increase or decrease the size of an entity by specifying scale value. The procedure to use this tool is given next.

- Click on the **Scale** tool from **Mirror** drop-down. The **Scale Command Bar** will be displayed as shown in Figure-72.

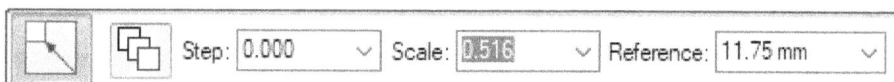

*Figure-72. Scale command bar*

- Select the entities that you want to scale up or scale down. You can select multiple entities by pressing and holding the **CTRL** key during selection. After selecting entities, you will be asked to specify center point for scaling.
- Select the desired base point about which you want to scale the entities.
- Specify desired scale value in the **Scale** edit box of **Command Bar** or use cursor movement for scaling.
- **Scale-Copy** 🔲 **Off** :- When the **Scale-Copy** button is **Off**, then the entities are scaled without being copied. In that case, the original entities will be deleted. Refer to Figure-73.

*Figure-73. Scale copy off*

- **Scale-Copy** 🔲 **On** :- When the **Scale-Copy** button is **On**, then the entities are scaled and copied. In that case, the selected entities will remain unchanged. Refer to Figure-74.

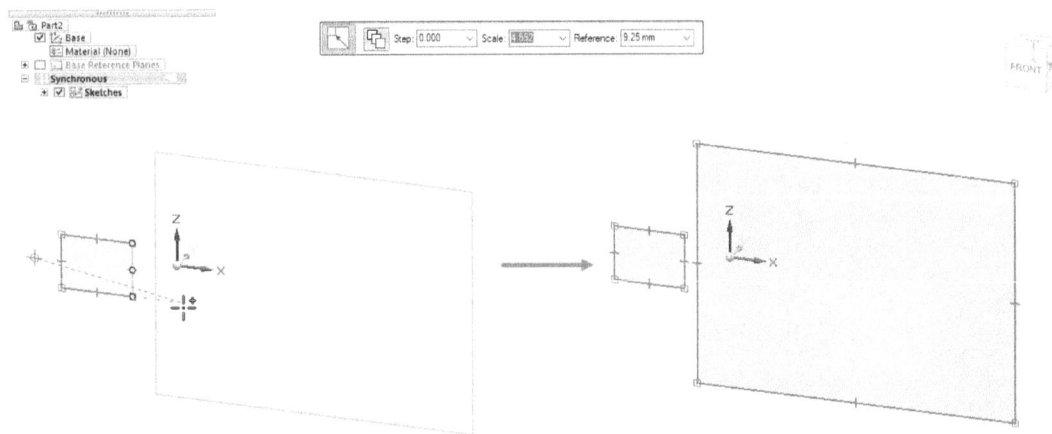

*Figure-74. Scale copy on*

## Stretch ▦

The **Stretch** tool is available in the **Mirror** drop-down. This tool is used to stretch any sketched entity which is constrained at one side. The procedure to use this tool is given next.

- Click on the **Stretch** tool from **Mirror** drop-down in **Draw** panel of **Home** tab in the **Ribbon**. The **Stretch Command Bar** will be displayed as shown in Figure-75.

*Figure-75. Stretch Command Bar*

- By using the cross-rectangle selection, select the portion of entities that you want to stretch. You will be asked to select point for moving entities.
- Select the base point on the sketch. The selected portion of sketch will get attached to cursor.
- Move the cursor to the desired position and click to stretch the entities.
- You can also stretch the entity by specifying desired values in the **X** and **Y** edit boxes of **Command Bar** for selected point; refer to Figure-76.

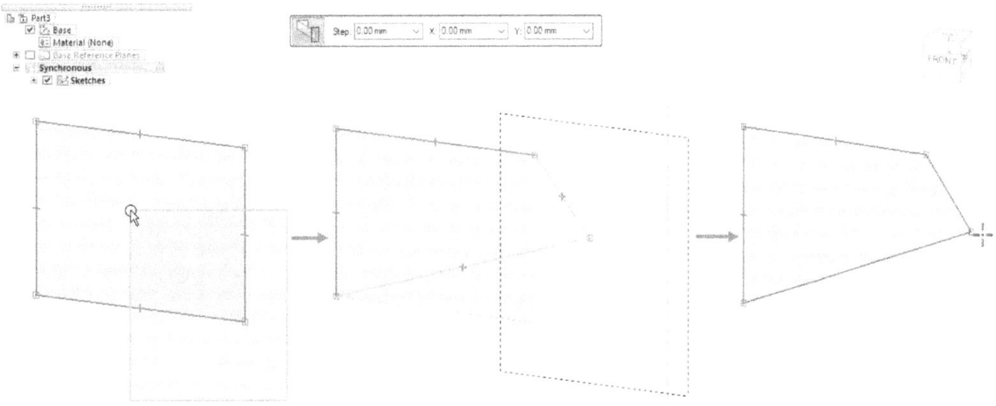

*Figure-76. Stretching entities*

## Construction Tool

The **Construction** tool is used to switch the entities between construction or profile. If you have drawn any sketch and you want to convert it in construction geometry then select the sketch entities and click on the **Construction** button. The procedure is discussed next.

- Click on the **Construction** tool from the **Draw** panel. The **Construction Command Bar** will be displayed as shown in Figure-77.

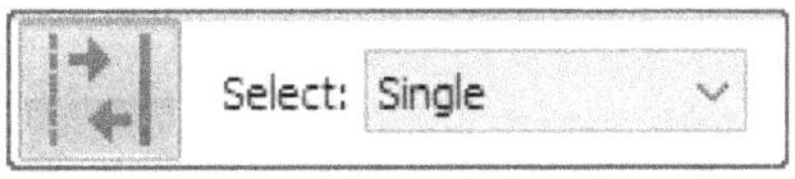

*Figure-77. Construction command bar*

- Select the **Single** option from the **Select** drop-down to select single curve at a time for conversion. Select the **Chain** option to select consecutively connected curves as a chain.
- Select the entities from drawing area which you want to convert in construction geometry; refer to Figure-78 and Figure-79.

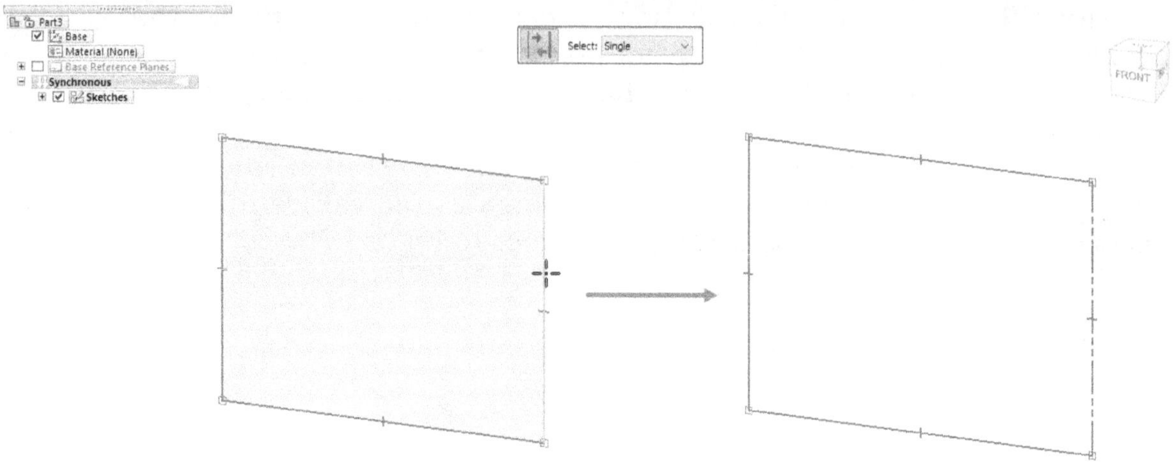

Figure-78. Single construction

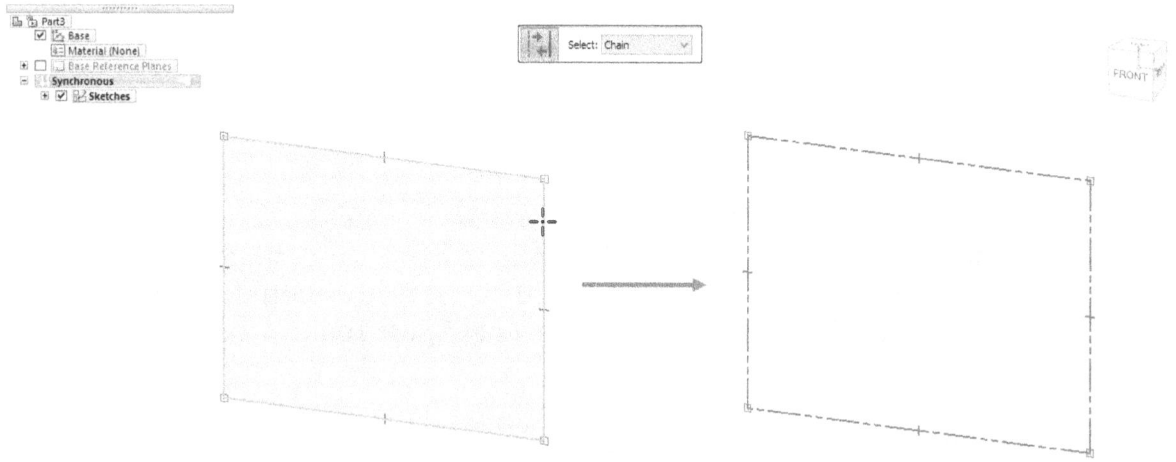

Figure-79. Chain construction

## Create As Construction Tool

The **Create As Construction** tool is used to draw construction entities. If you want to draw construction entities, you need to turn **"ON"** **Create As Construction** tool to create construction entities. If it is **"OFF"** then profile entities will be drawn. The procedure is discussed next.

- Click on the **Create As Construction** tool from the **Draw** panel in the **Home** tab of **Ribbon** to turn on the construction mode.
- Select the desired sketch creation tool from the **Draw** panel and draw the entities. Refer to Figure-80.

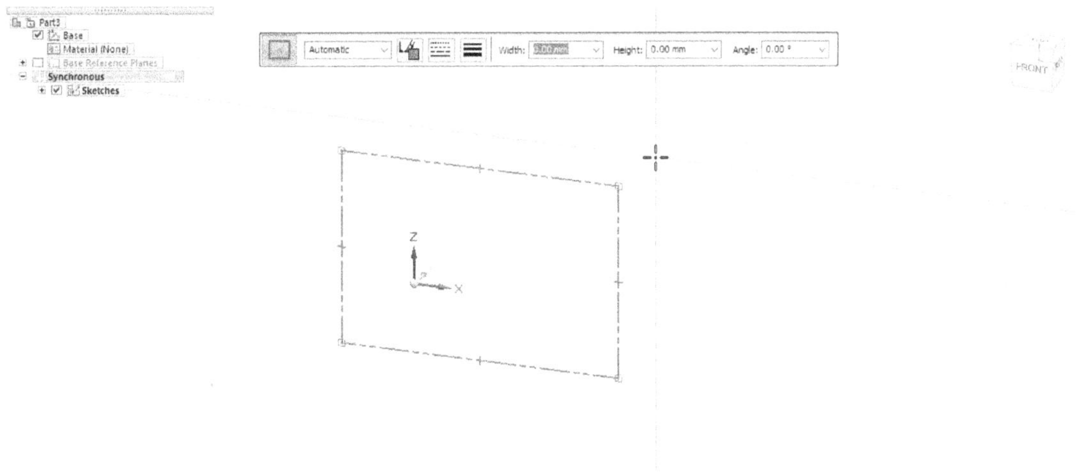

*Figure-80. Created as construction*

The other tools in this panel will be discussed later.

# RELATE TOOLS

The tools in the **Relate** panel are used to apply different types of relations between entities; refer to Figure-81. Relations are used to constrain the sketch entities geometrically. These constraints are used to constrain the shape/position of sketch entities with respect to other entities. The constrains available in **Relate** panel are discussed next.

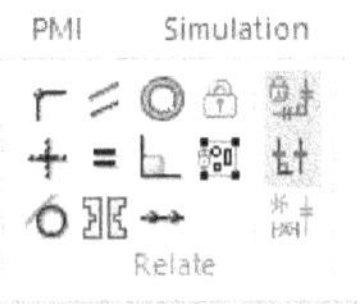

*Figure-81. Relate tools*

# Connects ⌐

The **Connect** tool is used to connect the two entities at specified points. To invoke this tool, select **Connect** from **Relate** panel in the **Home** tab of **Ribbon**. Select the entities on their snap points where they will be connected; refer to Figure-82.

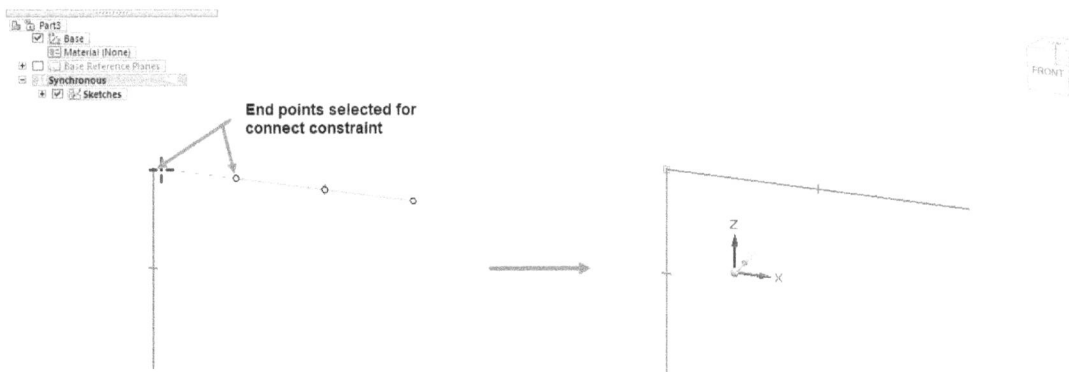

*Figure-82. Connecting points*

## Horizontal / Vertical ✛

**Horizontal** - The **Horizontal** constraint makes one or more selected lines or center line to becomes horizontal. You can select external entity such as an edge, plane, axis or sketch curve on external sketch that will act as a reference to apply this constraint. You can also make two or more points horizontally aligned using the Horizontal constraint. A point can be a sketch point, a center point, an endpoint, a control point of a spline, or an external entity such as origin, vertex, axis, or point in the external sketch. To apply this constraint, select the **Horizontal/Vertical** tools from **Relate** panel and select the desired entities.

**Vertical** - The vertical constraint makes one or more selected lines or center lines to become vertical. You can force two or more point to become vertically aligned by using Vertical Constraint. To apply this constraint, select the **Horizontal/Vertical** tools from the relate groups; refer to Figure-83.

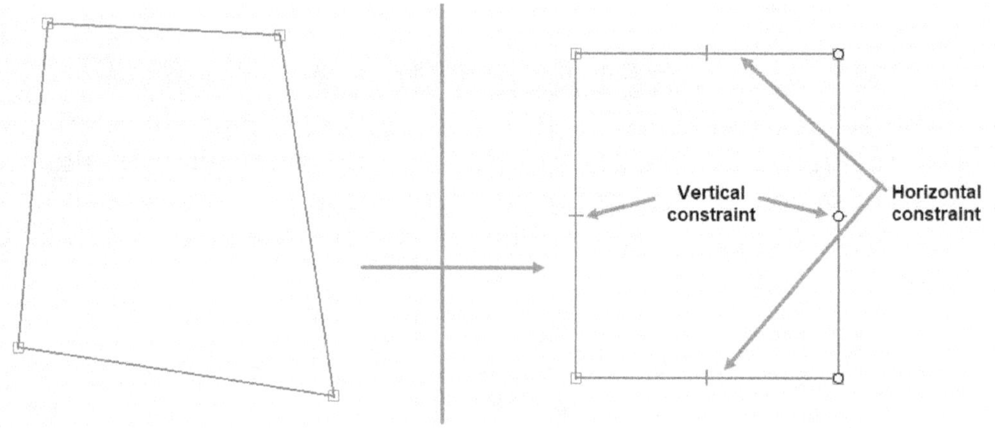

*Figure-83. Horizontal Vertical*

## Tangent ○

The **Tangent** constraint makes selected arc, circle, spline, or ellipse to become tangent to other arc, circle, spline, ellipse, line or edge. The procedure to apply constraint is given next.

*   Click on the **Tangent** button from the **Relate** panel in the **Home** tab of **Ribbon**. You will be asked to select the first entity.
*   Select the first entity and then select second entity. You can change the position of entities by selecting and dragging them if the sketch is not fully constrained; refer to Figure-84.

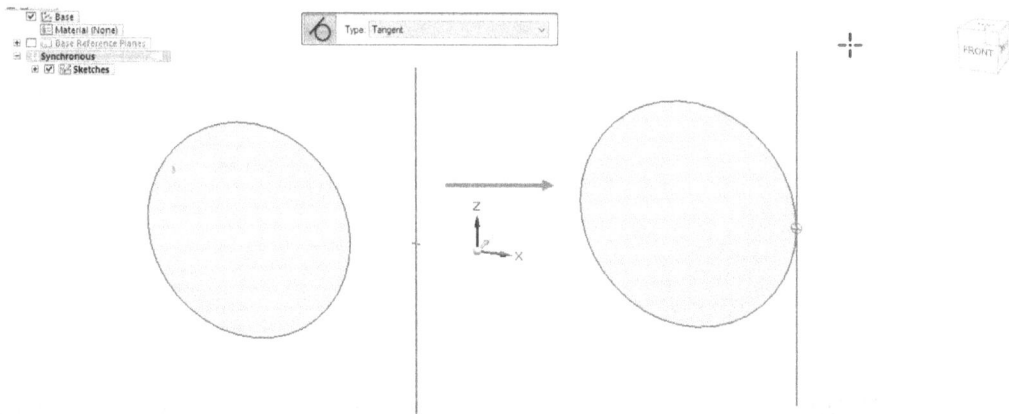

*Figure-84. Tangent constraint*

## Parallel

The **Parallel** constraint makes the selected lines parallel to each other. The procedure to apply this constraint is given next.

• Click on the **Parallel** tool from the **Relate** panel in the **Home** tab of **Ribbon**. You will be asked to select the entities.

• Select first line and then select second line. The constraint will be applied; refer to Figure-85.

*Figure-85. Parallel constraint*

## Equal =

The **Equal** constraint makes the selected lines to have equal length and the selected arcs, circles, or arc and circle to have equal radii. The procedure to apply this constraint is given next.

• Click on the **Equal** tool from the **Relate** panel in the **Home** tab of **Ribbon**. You will be asked to select the entities.

• Select first entity and then select second entity. The constraint will be applied; refer to Figure-86.

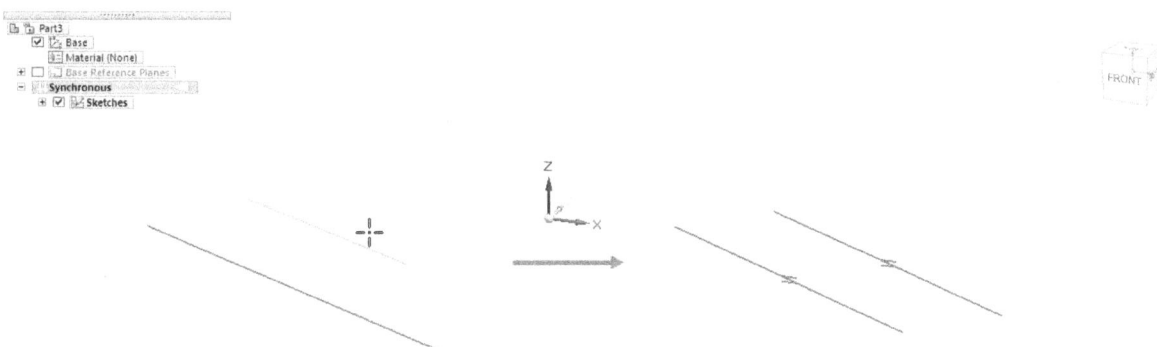

*Figure-86. Equal constraint*

## Symmetric ⚃

The **Symmetric** constraint makes two selected entities mirror copy of each other about a selected center line. The procedure is given next.

*   Click on **Symmetric** tool from the **Relate** panel in the **Home** tab of **Ribbon**. You will be asked to select the reference line.
*   Select the reference line. You will be asked to select the first entity.
*   Select first entity and then select second entity. The constraint will be applied; refer to Figure-87.

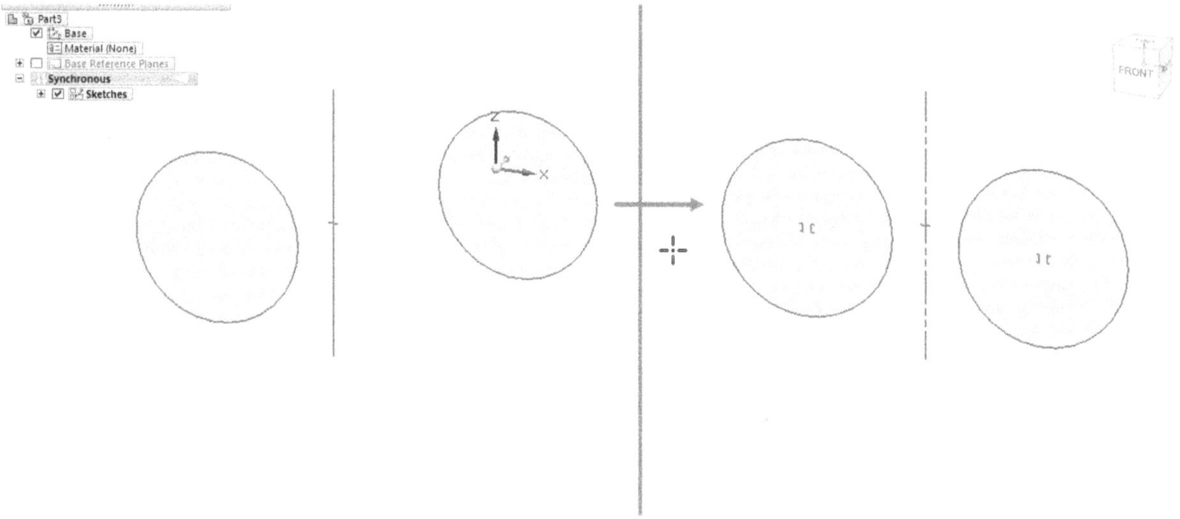

*Figure-87. Symmetric constraint*

## Concentric ○

The **Concentric** constraint makes selected arc or circle to share the same center point with other arc, circle, point, vertex or circular edge. The procedure to use this constraint is given next.

*   Click on the **Concentric** tool from the **Relate** panel in the **Home** tab of **Ribbon**.
*   Select the first entity and then select the second entity to be made concentric. The constraint will be applied; refer to Figure-88.

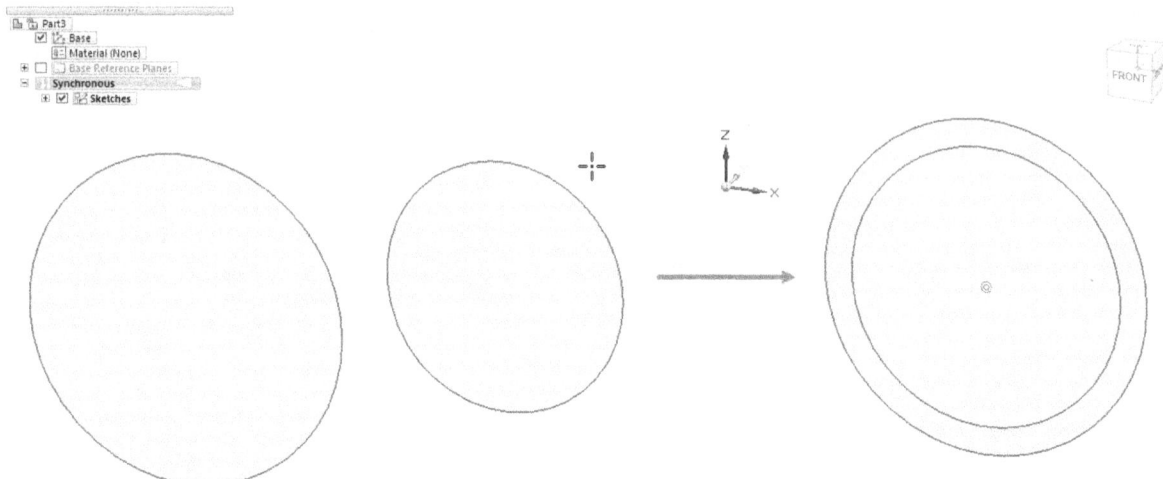

*Figure-88. Concentric constraint*

## Perpendicular ⌐

The **Perpendicular** constraint is used to make one entity perpendicular to another. It can make line perpendicular to circle, ellipse, arc or other entities. The procedure is given next.

- Click on the **Perpendicular** tool from **Relate** panel in the **Home** tab of **Ribbon**. You will be asked to select the entities.
- Select first entity which must be a line.
- Select second entity (line, arc, circle, ellipse, or other curve). The constraint will be applied; refer to Figure-89.

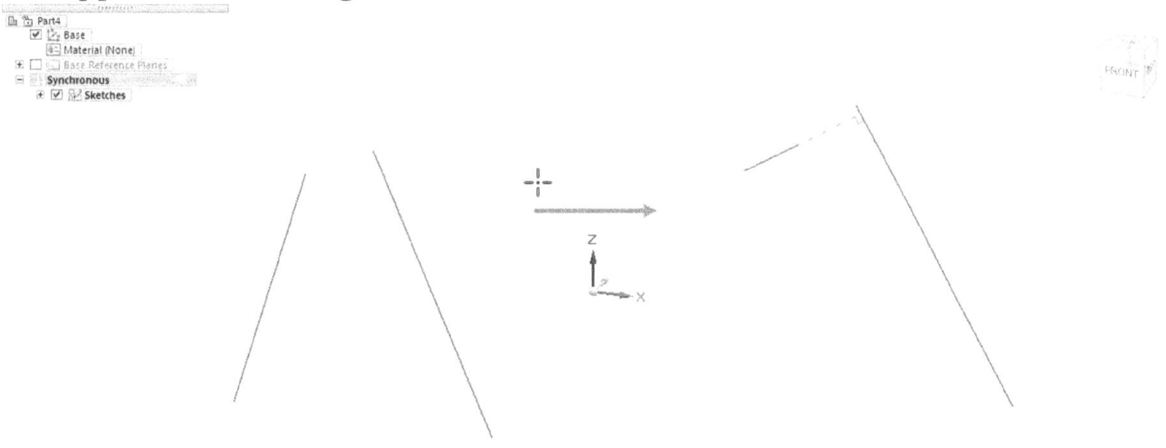

*Figure-89. Perpendicular constraint*

## Collinear ↔

The **Collinear** constraint makes the selected lines to lie on the same imaginary infinite line. The procedure is given next.

- Click on the **Collinear** from **Relate** panel in the **Home** tab of **Ribbon**. You will be asked to select the entities.
- Select first entity and then select the second entity. The constraint will be applied; refer to Figure-90.

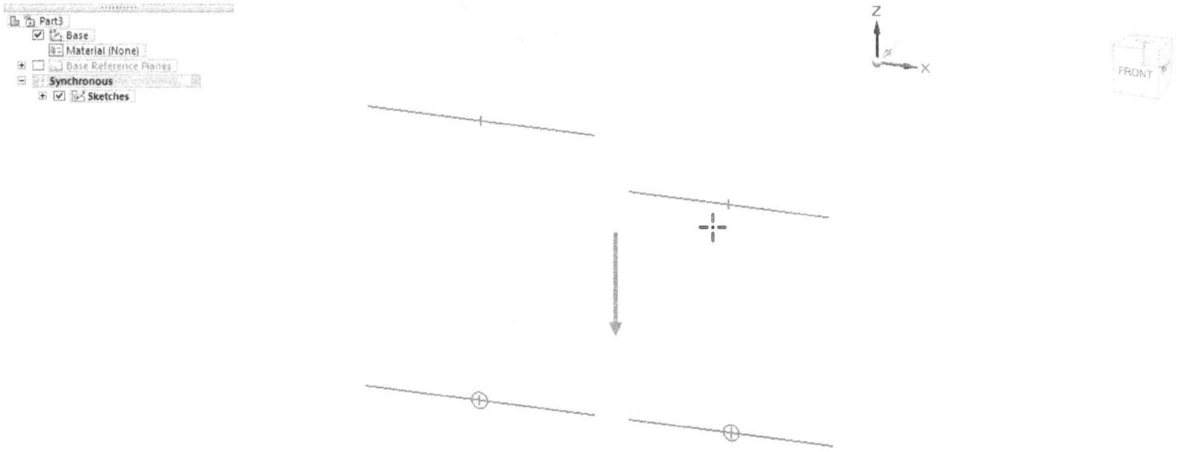

*Figure-90. Collinear constraint*

## Lock 🔒

The **Lock** tool is used to fix movement of entities and key points so that they are not modified accidently. Once a key point is locked, it cannot modified without unlocking it. A locked key point/entity can be moved by manipulation commands like Move/ Rotate and will be fixed at the new location after manipulation. You can apply multiple lock command at the same time.

## Rigid Set ⊡

The **Rigid Set** tool is available in **Relate** panel. This tool is used to create rigid sketch which means you can add multiple sketch entities to a rigid set and they will behave as single object. To procedure to apply this tool is given next.

- Select the **Rigid Set** tool from the **Relate** group in the **Home** tab of **Ribbon**. You will be asked to select the entities.
- Select multiple entities that you want to add in a single set and press **ENTER**. The entities will be added to the set; refer to Figure-91. Now, if you drag one element of the set then the complete set will move.

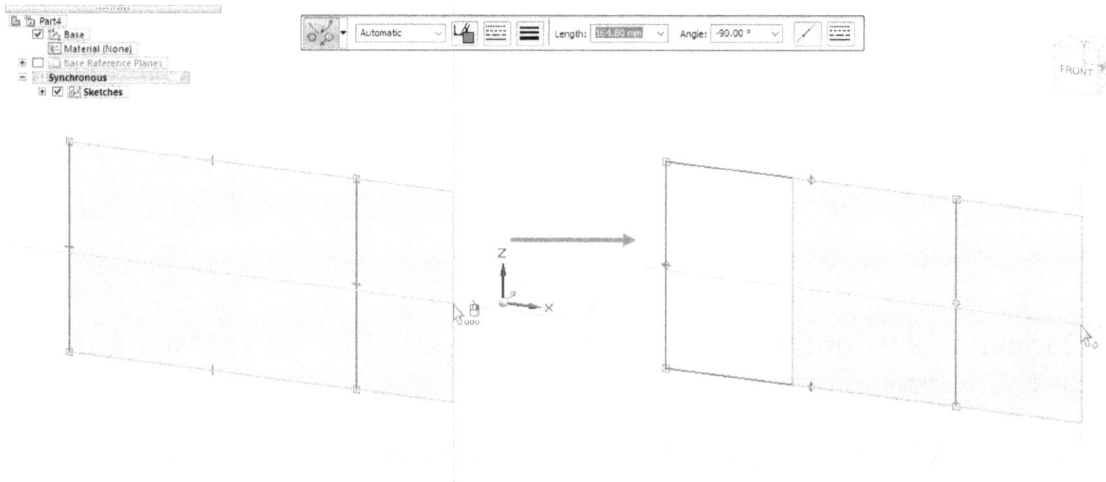

*Figure-91. Rigid set*

## Symmetry Axis ⫼

The **Symmetry Axis** button is used to convert selected line into symmetry axis. After converting a line to symmetry axis if you use **Symmetric** button to apply the constraint then this converted line will be automatically used as symmetry axis.

## Maintain Relationships ⊕

The **Maintain Relationships** tool is used to activate or deactivate the automatic recognition of geometric constraints while creating the sketch. If this tool is turn **"ON"** then the application or software automatically detects the geometric constraint. It depend upon the user to turn **"ON"** or turn **"OFF"** this option. You can delete constraint by selecting it and pressing **DELETE** key any time.

## Relationships Handle ⫴

The **Relationships Handles** tool is available in the **Relate** panel. This tool is used to display or hide constraints applied on the model. Note that hiding a constraint will not remove the constraint. If this tool is turn **"ON"**, it show the geometric constraints which are applied to the sketch. Refer to Figure-92.

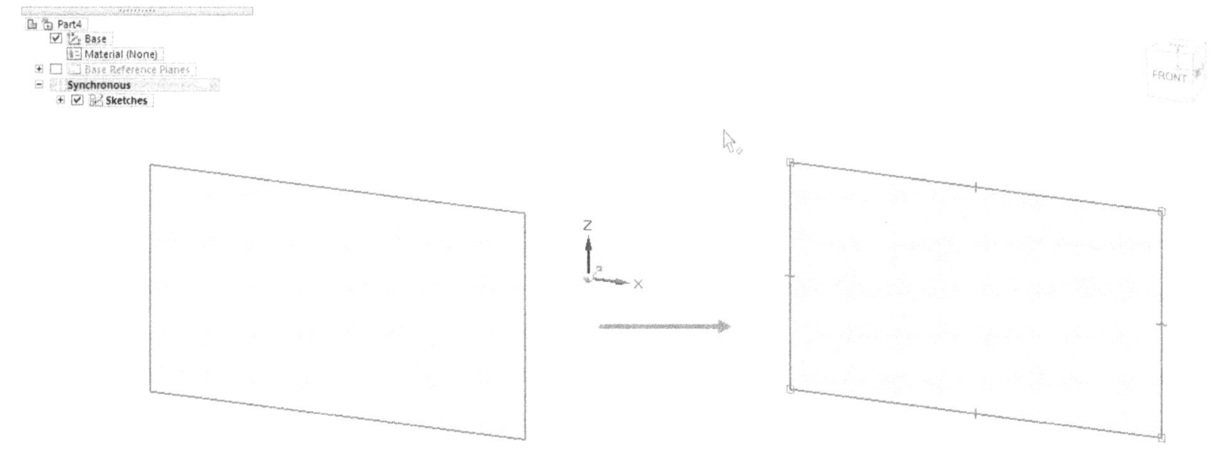

*Figure-92. Relationship handle*

# Relationship Assistant

The **Relationship Assistant** tool is used to apply dimensions to selected entities with respect to selected horizontal and vertical references. The procedure to use this tool is given next.

- Click on the **Relationship Assistant** tool from the **Relate** panel in the **Home** tab of **Ribbon**. You will be asked to select the entities to be dimensioned and **Relationship Assistant Command Bar** will be displayed; refer to Figure-93.

*Figure-93. Relationship Assistant Command Bar*

- One by one click on the elements that you want to be dimensioned and press **ENTER**. You will be asked to select horizontal dimension reference.
- Select the desired line or edge to define horizontal reference. You will be asked to specify vertical reference.
- Select the desired line to define vertical reference. The dimensions will be created automatically; refer to Figure-94.

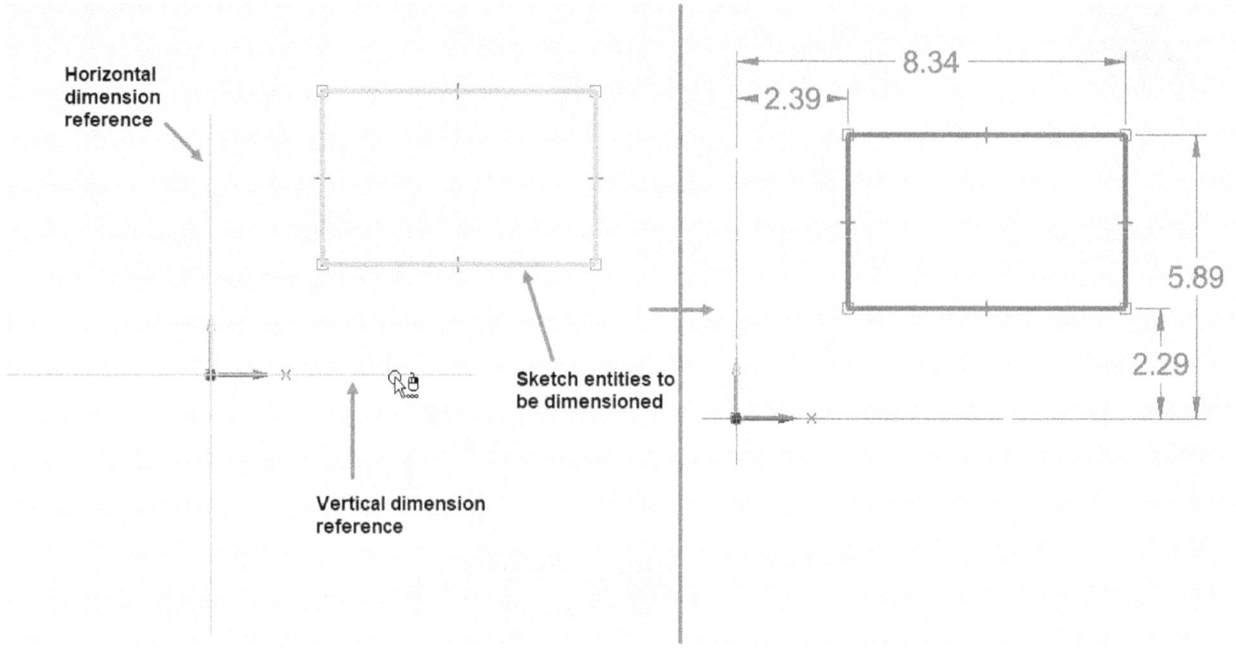

*Figure-94. Sketch dimensioned by relationship assistant*

* Press **ESC** to exit the tool.

# DIMENSIONS

**Dimensions** are used to specify size of the sketch entities. For example; specifying length of the line, specifying diameter of the circle, and so on. The tools to dimension sketch entities are available in the **Dimension** panel. These tools are discussed next.

## Smart Dimension

The **Smart Dimension** tool is used to dimension various entities like line, arc, circle, and so on in the sketch. The procedure to use **Smart Dimension** tool is discussed next.

* Click on the **Smart Dimension** tool from the **Dimension** panel in the **Home** tab of **Ribbon**. The **Smart Dimension Command Bar** will be displayed as shown in Figure-95.

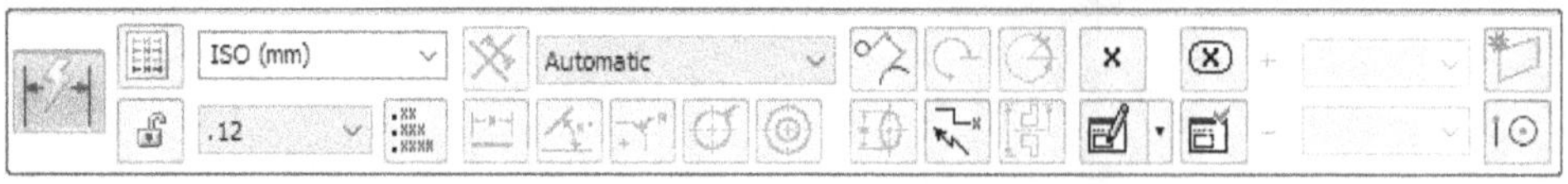

*Figure-95. Smart Dimension Command Bar*

The procedure to dimension various entities using the **Smart Dimension** tool is given next.

### Dimensioning Lines

* After selecting **Smart Dimension** tool, select the line from the middle. The dimension will get attached to cursor; refer to Figure-96.

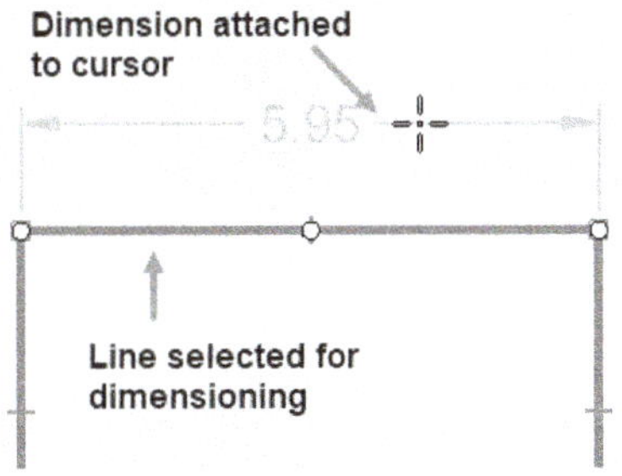

*Figure-96. Dimensioning a line*

- Click at desired location to place the dimension. A dynamic input box will be displayed for specifying value of dimension.
- Enter the desired value in input box if you want to change the length of line or click in the empty area to lock the displayed dimension.

## Dimensioning Inclined Lines

- After selecting **Smart Dimension** tool, select the line from the middle. An inclined dimension will get attached to cursor.
- Click at desired location to place the dimension. A dynamic input box will be displayed for modifying the length of line; refer to Figure-97.

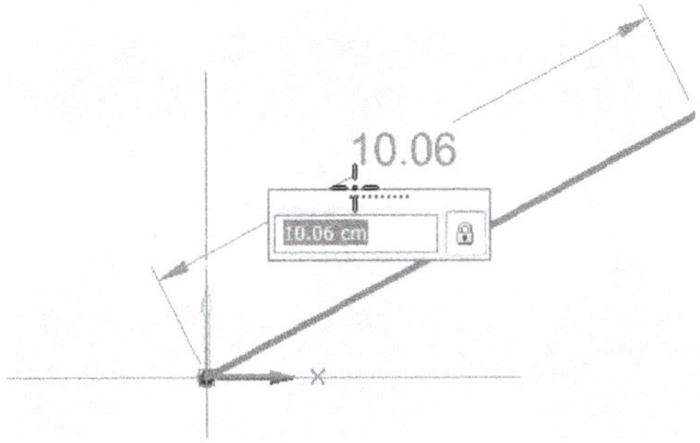

*Figure-97. Dynamic input box for dimensioning*

- Enter the desired value or click in the empty area to lock the dimension.

Now, we need to define angle dimension for inclined line.

- Make sure the **Smart Dimension** tool is still active. Click on the inclined line and then reference line/edge from which you want to dimension angle; refer to Figure-98. You will be asked to select a key point for specifying 3 point angle.

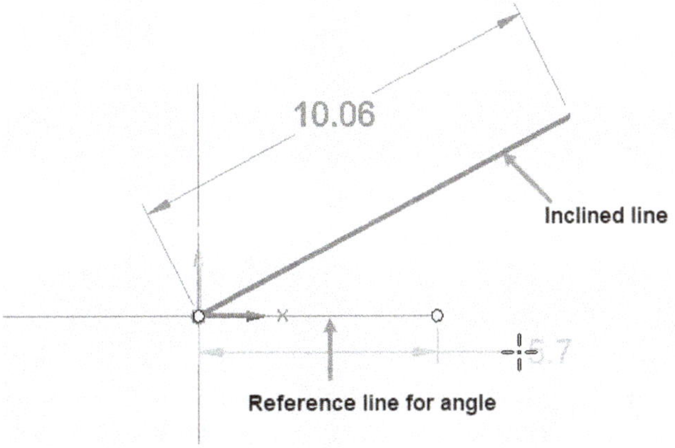

*Figure-98. Selecting lines for angle dimension*

- Click at the point where reference line and inclined line intersects. The angle dimension will get attached to cursor; refer to Figure-99.

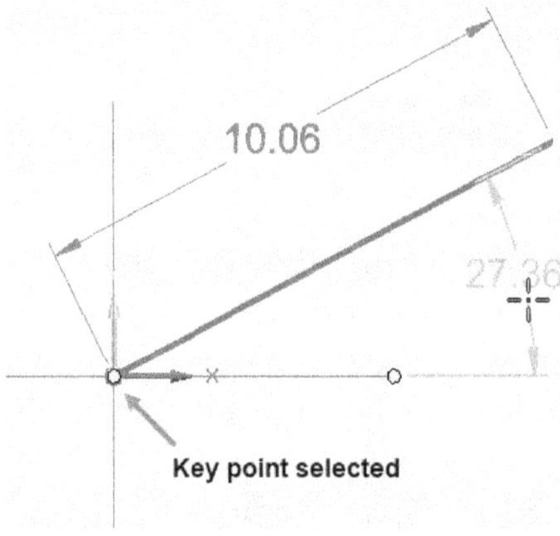

*Figure-99. Placing inclined dimension*

- Click at desired location to place dimension. A dynamic input box will be displayed for modifying the dimension.
- Enter the desired value in the input box. The dimension will be applied.

## Dimensioning a Circle

- Click on the **Smart Dimension** tool from the **Dimension** panel in the **Home** tab of **Ribbon** if tool is not activated yet.
- Click on the circumference of circle. The diameter dimension will get attached to cursor. Note that depending on the location of cursor, the single arrow or double arrow line will be displayed for diameter dimensioning; refer to Figure-100.

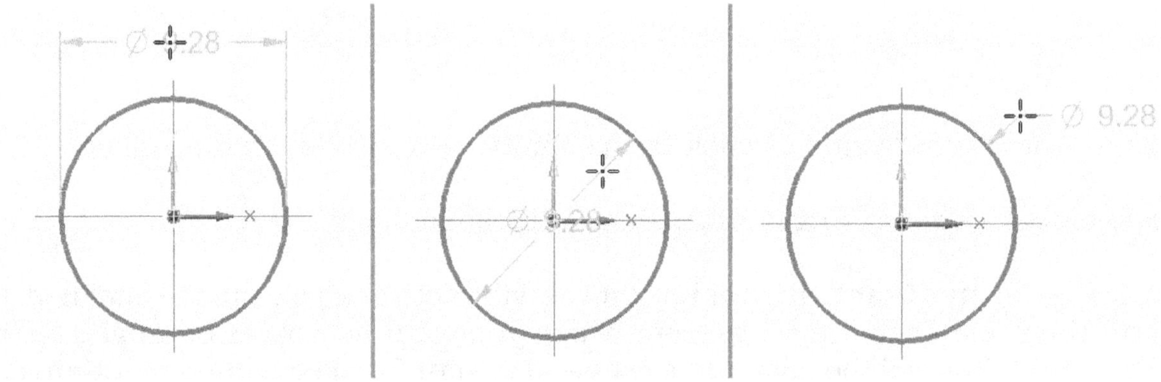

*Figure-100. Dimensioning a circle*

- Click at desired location to place the dimension and enter the desired value in the dynamic input box.

## Dimensioning an Arc

- After activating the **Smart Dimension** tool, click on the circumference of arc. The radius dimension will get attached to cursor.
- Click at desired location to place the dimension and enter the value of radius in dynamic input box.
- To specify arc angle, click on the circumference of arc while **Smart Dimension** tool is active and then press **A** from keyboard. The arc angle dimension will be displayed; refer to Figure-101. If you again press **A** from keyboard then arc length

dimension will be displayed; refer to Figure-102. You can also switch between various arc dimensioning styles by selecting the respective button from **Smart Dimension Command Bar**; refer to Figure-103.

*Figure-101. Angular arc dimension*

*Figure-102. Arc length dimension*

*Figure-103. Buttons for changing dimension style*

- Click at desired location to place the dimension and specify dimension value in the dynamic input box.

## Dimensioning a Chamfer

- After activating the **Smart Dimension** tool, select the inclined line of chamfer and place horizontal or vertical dimension to specify length of chamfer by holding **SHIFT** key from keyboard while placing the dimension; refer to Figure-104.

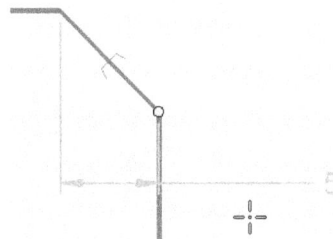

*Figure-104. Placing horizontal dimension*

- Enter the desired value in the dynamic input box to specify length of chamfer.
- To specify angle, select the inclined line again and click on the **Smart Dimension - Angle** button from the **Command Bar** or press **A** key. The angle dimension will get attached to cursor; refer to Figure-105.

*Figure-105. Angle dimension*

- Place the dimension at desired location and enter the angle value in dynamic input box.

## Creating Distance Between Dimension

The **Distance Between** tool is used to give linear distance between the elements or key points. This tool is available in the **Dimension** panel. Select the edge and corresponding edges to give dimension and click at desired location to place the dimension; refer to Figure-106. You can specify the dimension value in the same way as discussed earlier.

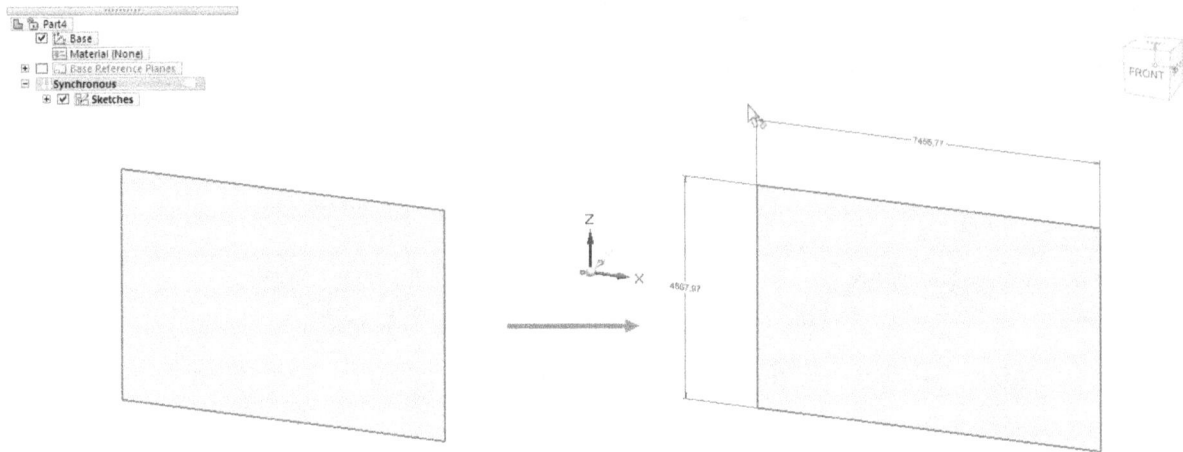

*Figure-106. Distance between*

## Dimensioning an Ellipse

*   After activating **Distance Between** tool, one by one select two opposite quadrant points of the ellipse. The dimension will get attached to cursor; refer to Figure-107.

*Figure-107. Selecting points of ellipse*

*   Click at desired location to place the dimension and enter the value in dynamic input box.
*   Similarly, you can dimension other side of ellipse.

## Creating Angle Between Dimension

The **Angle Between** tool is used to dimension angle between elements or key points. When you select the elements or key points, the dimension is placed in one of four quadrants. As you move the cursor, the dimension dynamically changes to another quadrant. You can place angular dimensions in stacked or chained dimension groups; refer to Figure-108.

*Figure-108. Angle between*

## Creating Coordinate Dimension

The **Coordinate Dimension** tool is used to dimension entities by their coordinates relative to selected reference origin. The procedure to create coordinate dimensions is given next.

- Click on the **Coordinate Dimension** tool from the **Dimension** panel in the **Home** tab of **Ribbon**. You will be asked to select reference origin for dimensioning.
- Click at desired point for specifying origin. The 0 reference dimension will get attached to cursor; refer to Figure-109.

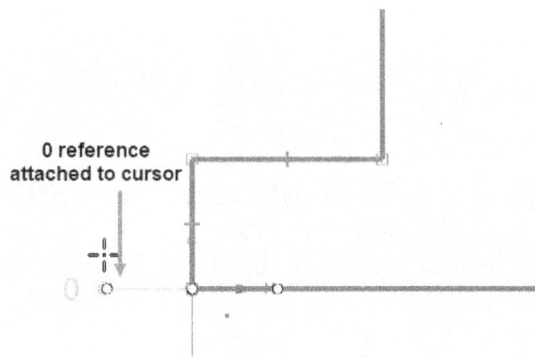

*Figure-109. Reference dimension attached to cursor*

- Click at desired location to place zero reference. You can create horizontal reference or vertical reference based on defining the location.
- One by one click on the lines which you want to dimension. The coordinate dimensions aligned to selected reference dimension will be displayed; refer to Figure-110. You can specify the value of dimension as discussed earlier. If you want to change the location of dimension then press **ESC**, select the dimension and drag it using the key point displayed on dimension. Note that by default all the dimensions will move by dragging. To move a single dimension, press **ALT** key while dragging.

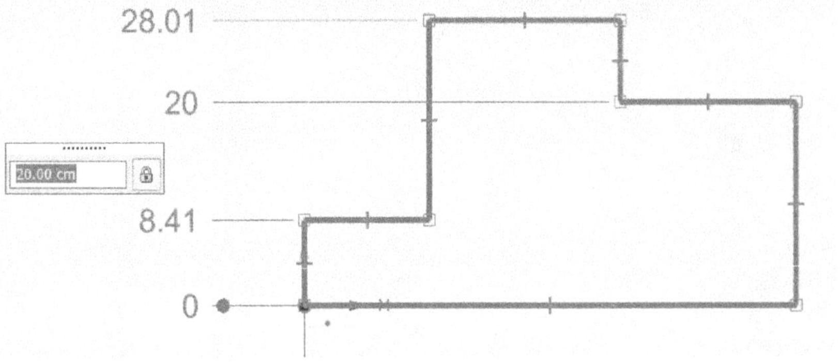

*Figure-110. Creating coordinate dimensions*

- After you have specified the dimensions in one direction. Press **ESC** key and then restart the tool.
- Now, select the next zero references and repeat the procedure. Figure-111 shows a sketch dimensioned by coordinates.

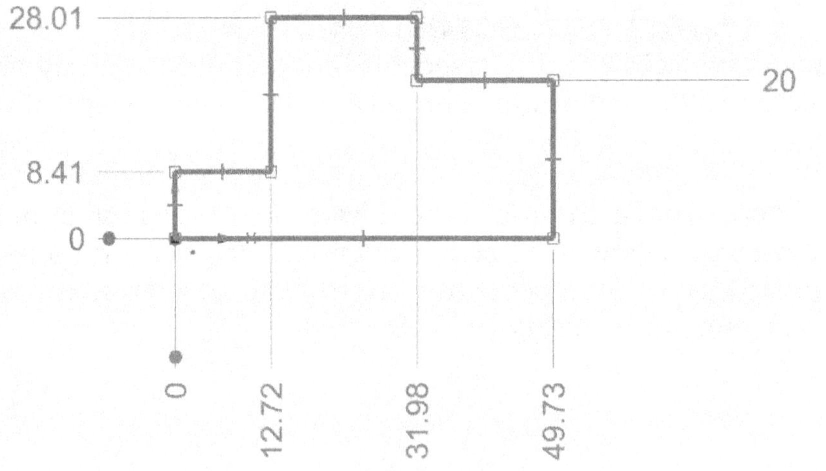

*Figure-111. Coordinate dimension created*

## Angular Coordinate Dimension

The **Angular Coordinate Dimension** tool is used to dimension the angle with coordinate values. The procedure is same as discussed earlier for coordinate dimensioning. Refer to Figure-112.

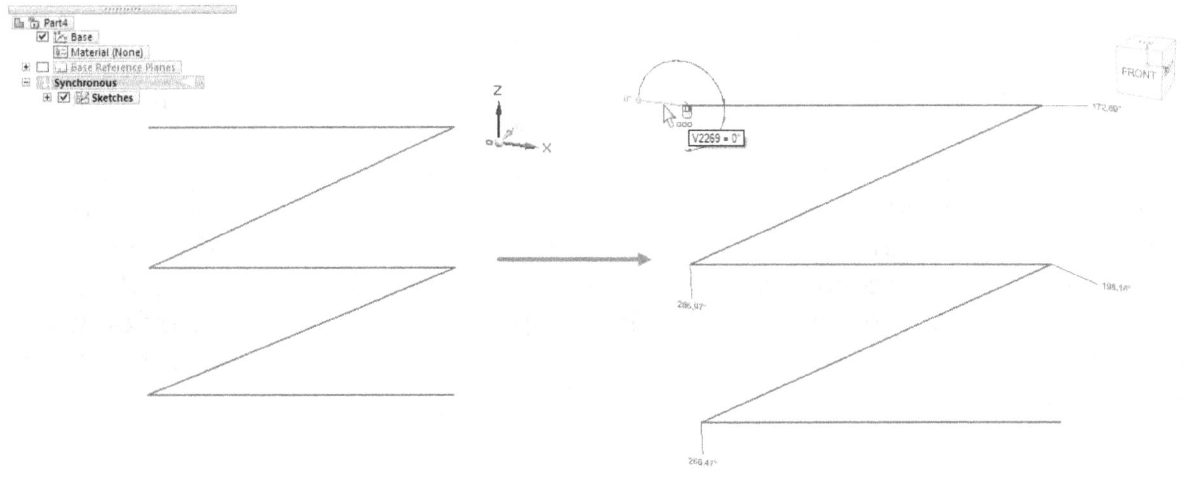

*Figure-112. Angular Coordinate Dimension*

## Symmetric Diameter

The **Symmetric Diameter** tool is used to specify symmetric diameter between two element or key points from the center line. The procedure of this tool is discussed next.

- Click on the **Symmetric Diameter** tool from the **Dimension** panel in the **Home** tab of **Ribbon**. You will be asked to select the reference origin and the **Symmetric Diameter Command Bar** will be displayed.
- Select the reference point or center line. You will be asked to select entities to be dimensioned.
- Select the entities to be dimensioned. The dimensions will be created; refer to Figure-113.
- Press **ESC** to exit the tool.

*Figure-113. Symmetric diameter dimensions created*

## Creating and Using Dimension Axis

The **Dimension Axis** tool is used to set reference for other dimensions. The procedure to use this tool is given next.

- Click on the **Dimension Axis** tool from the **Dimension** panel in the **Home** tab of **Ribbon**. You will be asked to select key points or line for defining dimension axis.
- Select the desired line or two key points. A green line will be displayed for an instance showing that reference dimension axis has been created.
- Now, click on the **Smart Dimension** or **Distance Between** tool to create dimensions. The respective **Command Bar** will be displayed.
- Select the **Use Dimension Axis** tool from the **Distance Between - Orientation** drop-down of **Command Bar**. Preview of dimension axis will be displayed; refer to Figure-114.

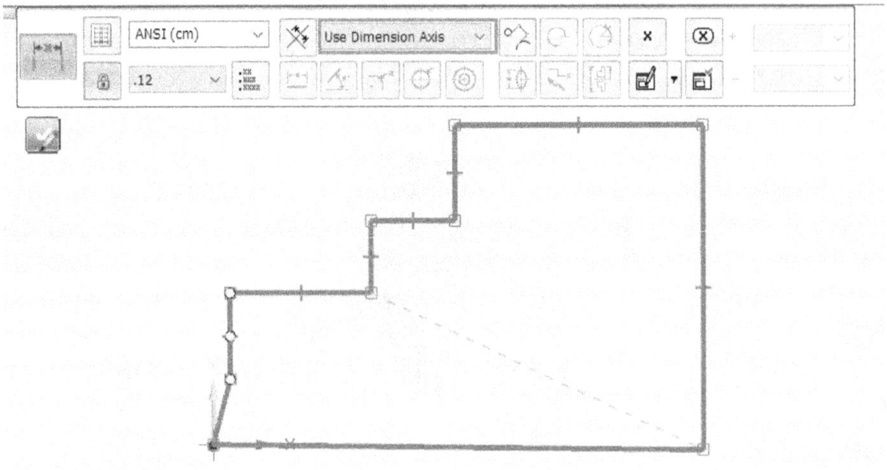

*Figure-114. Use Dimension Axis option*

- One by one select the entities to be dimensioned and place the dimensions at desired locations; refer to Figure-115.

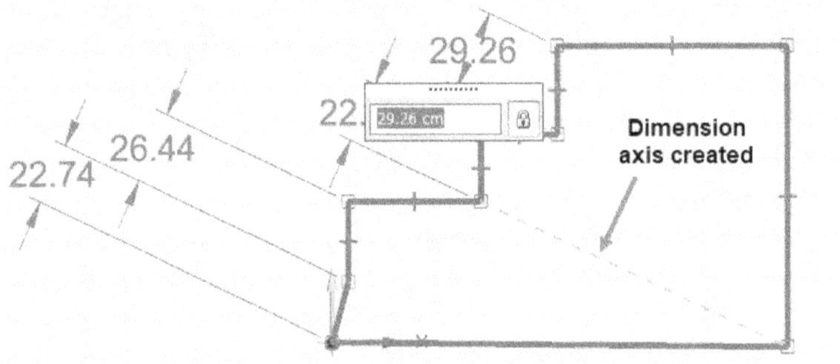

*Figure-115. Dimensions aligned to dimension axis*

## Activating Auto Dimensioning

The **Auto-Dimension** tool is used to toggle auto-dimensioning scheme in sketch. When this button is toggle **ON** then dimensions will be automatically created on specifying value while creating sketch entities.

You will learn about other dimension toggle buttons later in this book.

## ORDERED TOOLS

When we transition from **Synchronous** to **Ordered** then some more tools are displayed for sketching in the **Ribbon**. These tools are discussed next.

The procedure to transition from **Synchronous** to **Ordered** has been discussed earlier. After changing from **Synchronous** to **Ordered** modeling, the **Ribbon** will be displayed as shown in Figure-116. In ordered to start a sketch, we have two type of option:-

1. Sketch ( For 2D sketch)
2. 3D Sketch.

In this topic, we will discuss about 2D sketch. 3D sketch will be discussed in next chapter.

*Figure-116. Ordered Ribbon*

# Starting 2D Sketch in Ordered Environment

In **Synchronous** environment, you can selected the sketching tools and directly draw on a plane. In **Ordered** environment, you need to select a sketching plane first then you can use the sketching tools to draw sketch. The procedure to start a 2D sketch is given next.

- Click on the **Sketch** tool from the **Sketch** panel in the **Home** tab of **Ribbon**. The **Sketch Command Bar** will be displayed; refer to Figure-117 and you will be asked to select a sketching plane.
- Select the desired plane or face of solid. The selected sketching plane will become parallel to screen. The interface to draw 2D sketch will be activated and the **Ribbon** will be displayed as shown in Figure-118.

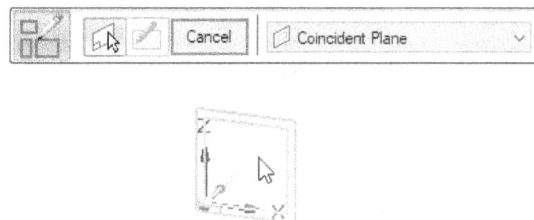

*Figure-117. Sketch Command Bar*

*Figure-118. 2D Ribbon*

After creating 2D sketch, click on the **Close Sketch** button from **Close** panel in the **Home** tab of **Ribbon** to exit sketching mode.

# Tools In Draw Group

Most of the tools in **Draw** panel have been discussed earlier like Line, Rectangle, Circle, Arc, Curve, Fillet, and so on. Here, we will discuss rest of the tools available in the **Draw** panel like Convert to Curve, Fill, and so on.

## Convert To Curve

The **Convert to Curve** tool is used to convert analytic geometry into a B-spline curves while using analytic creation tools. Suppose, you have drawn a line in a model but you want a curve in place of line for modifications. Instead of deleting the line and creating the curve using respective tool, you can turn line into curve by invoking **Convert to Curve** tool. The procedure is discussed next.

- Select the entities to be converted and click on the **Convert to Curve** tool from the **Draw** panel in the **Home** tab of **Ribbon**. The selected entities will be converted to curves.
- Press **ESC** key to exit the tool.
- Now, if you select the entity then drag points will be displayed on entities and **Command Bar** will be displayed as shown in Figure-119.

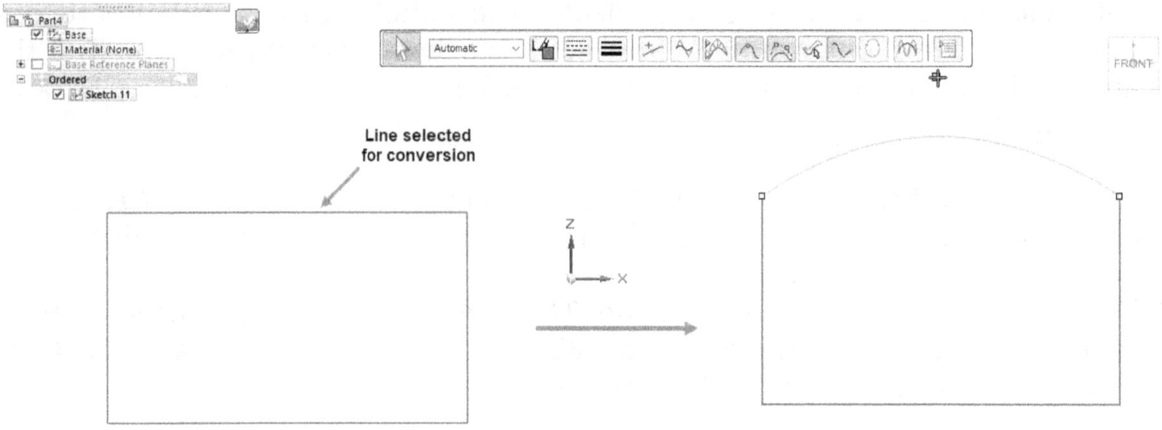

*Figure-119. Convert to Curve*

## Fill

The **Fill** tool is used to hatch the closed region. The procedure is discussed next.

- Click on the **Fill** tool from the **Draw** panel in the **Home** tab of **Ribbon**. The **Fill Command Bar** will be displayed; refer to Figure-120.

*Figure-120. Fill Command Bar*

You can change color, angle, and spacing of grid lines using the **Command Bar**.

- Click in the closed region where you want to apply hatching. The fill will be created; refer to Figure-121.

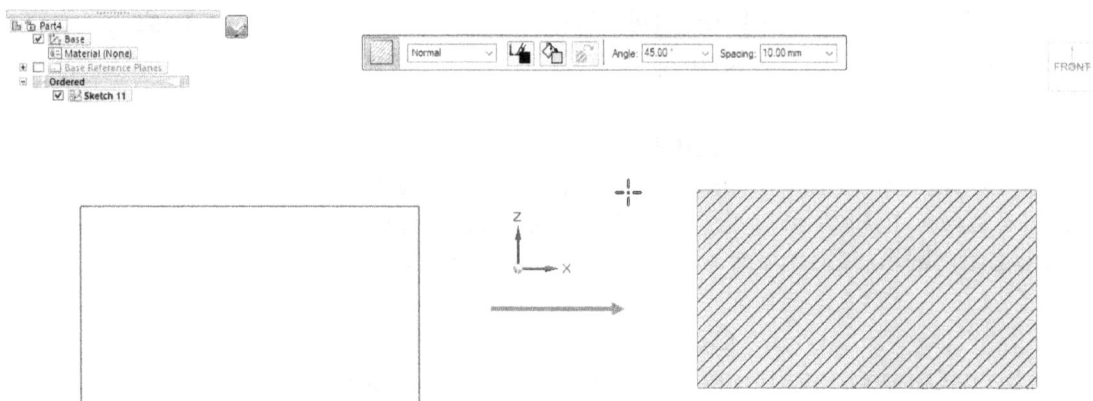

*Figure-121. Hatch fill created*

## Show Grid

The **Show Grid** tool is used to display grids in sketching environment; refer to Figure-122. These grid lines can be used as reference for creating entities.

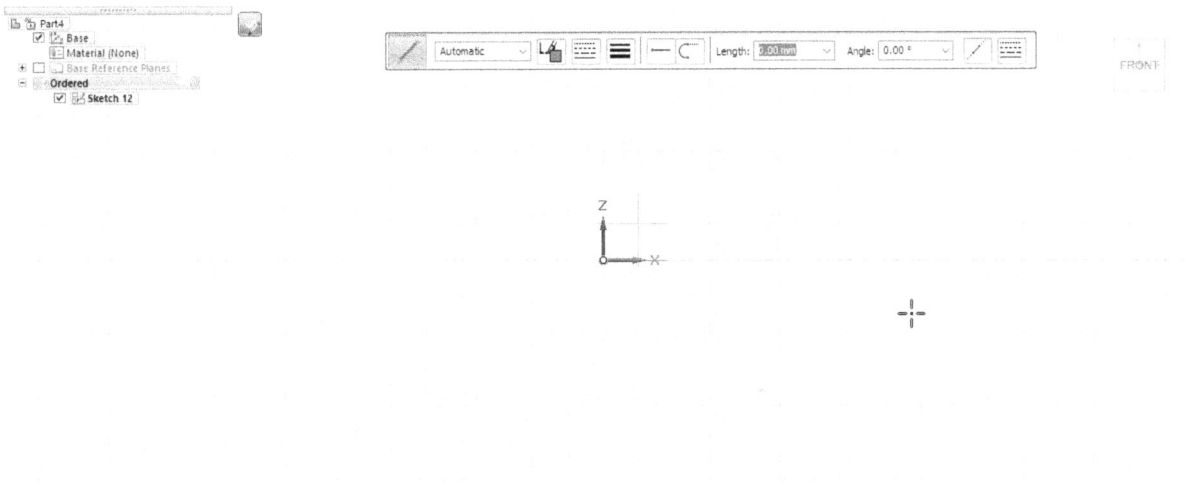

Figure-122. Show Grid

## Snap to Grid

The **Snap to Grid** tool is used to snap cursor to the grid. To invoke this tool, click on the **Snap to Grid** tool from the **Draw** panel in the **Home** tab of **Ribbon**; refer to Figure-123.

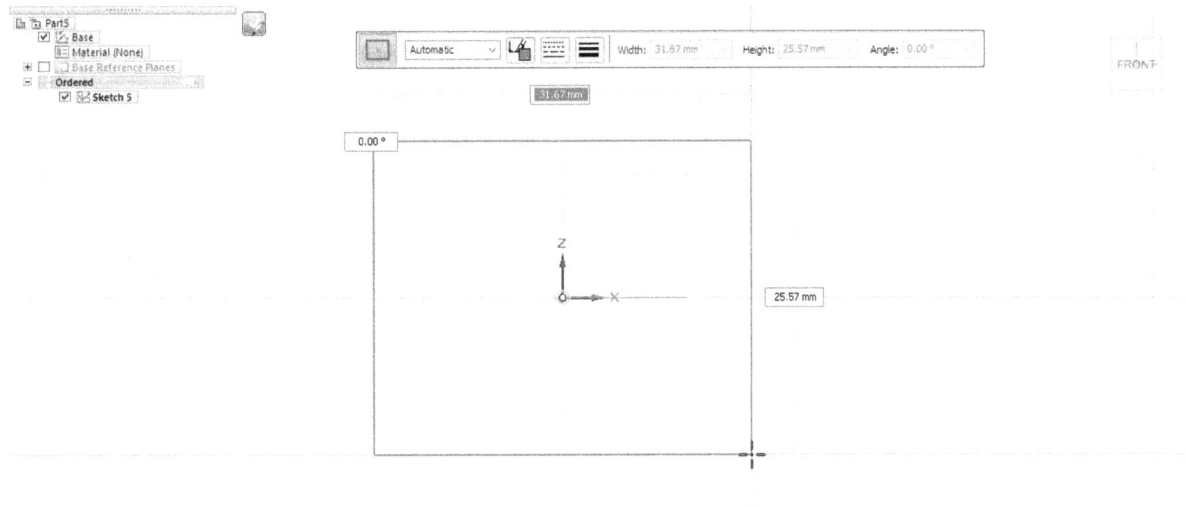

Figure-123. Snap to Grid

## Grid Option

The **Grid Option** tool is used to set the grid appearance. On selecting this tool, the **Grid Options** dialog box will be displayed; refer to Figure-124. The options of this dialog box are discussed next.

• Click on the **Grid Options** tool from the **Draw** panel in the **Home** tab of **Ribbon**. The **Grid Options** dialog box will be displayed; refer to Figure-124.

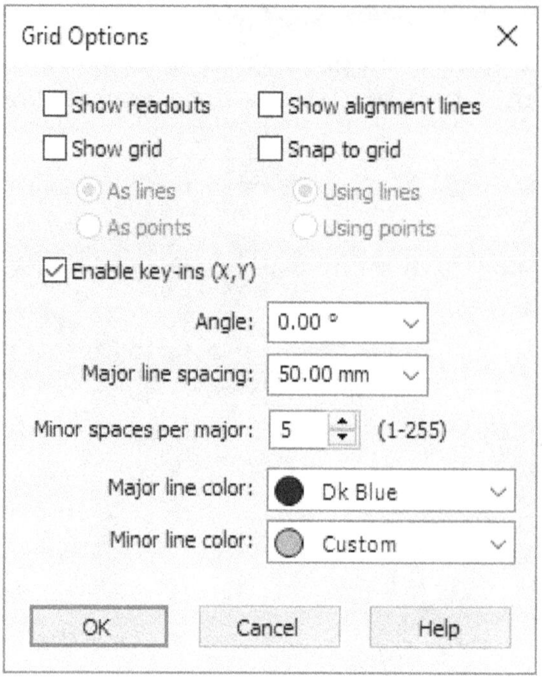

*Figure-124. Grid Option*

- Select the **Show readouts** check box to display coordinates while creating sketch entities.
- Select the **Show Grid** check box and then select the desired radio button below the check box to display lines or points of grid.
- Select the **Show alignment lines** check box to display horizontal and vertical lines on the cursor for alignment while creating sketch entities.
- Select the **Snap to grid** check box to automatically snap cursor to grid and then select the desired radio button below the check box to define whether you want to snap cursor to grid lines or grid points.
- Select the **Enable key-ins (X,Y)** check box to enter the X and Y coordinates in the dynamic input box displayed while creating sketch entities; refer to Figure-125.

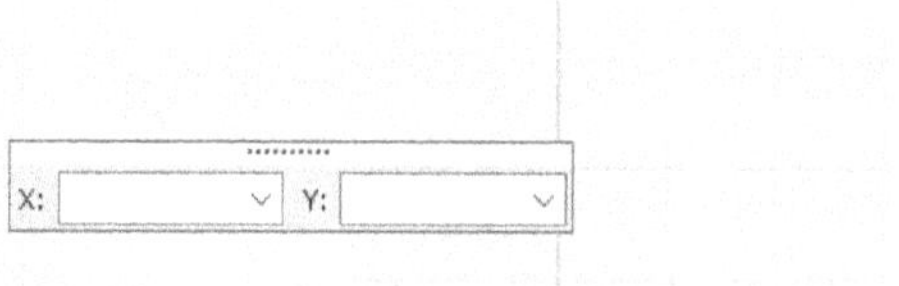

*Figure-125. Dynamic input box for coordinates*

- Specify the desired value in **Angle** edit box to rotate grid lines to specified angle.
- Specify the desired value in **Major line spacing** edit box to define gap between two consecutive major grid lines.
- Specify the desired value in **Minor spaces per major** edit box to define how many minor grid lines are to be created between two consecutive major lines. You can specify value from minimum 1 to maximum 255 in this edit box.
- Select the desired colors for grid lines in the **Major line color** and **Minor line color** drop-downs.
- After setting desired parameters, click on the **OK** button to apply settings.

## Repositioning Origin

The **Reposition Origin** tool is used to change position of grid origin at specified location. The procedure to use this tool is given next.

*   Click on the **Reposition Origin** tool from the **Draw** panel in the **Home** tab of **Ribbon**. You will be asked to specify location for new grid origin.
*   Click at desired location to reposition the grid origin. The grids will move accordingly.

## IntelliSketch Tools

In this panel, you will find various check boxes used to apply constraints in sketch intelligently while creating the sketch; refer to Figure-126. Note that using Intellisketch options, you can snap cursor to key points of other sketches and solid model edges as well. The options in this panel are discussed next.

*Figure-126. Intellisketch tools*

## End Point

Select the **End Point** check box to enable display and snapping of end points of sketch entities while creating sketch; refer to Figure-127.

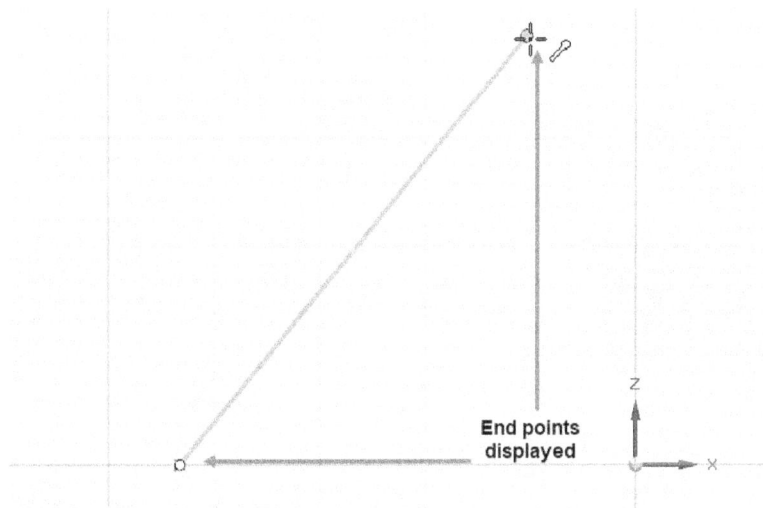

*Figure-127. Endpoint displayed*

## Midpoint

Select the **Midpoint** check box to enable display and snapping of mid points of sketch during sketch creation; refer to Figure-128.

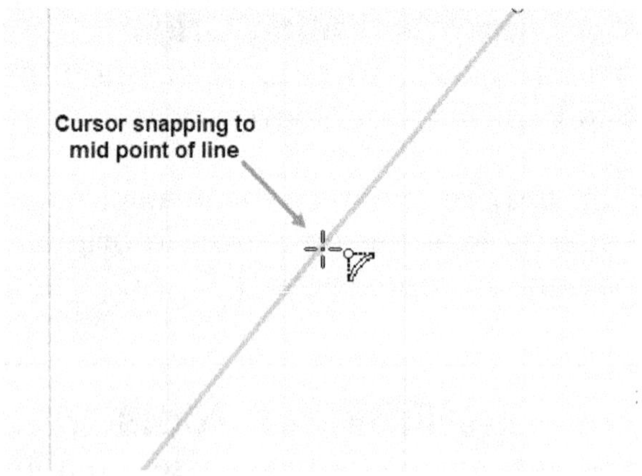

*Figure-128. Midpoint displayed*

## On Element

Select the **On Element** check box to display and snap to points on sketch entities or edges of other solids; refer to Figure-129.

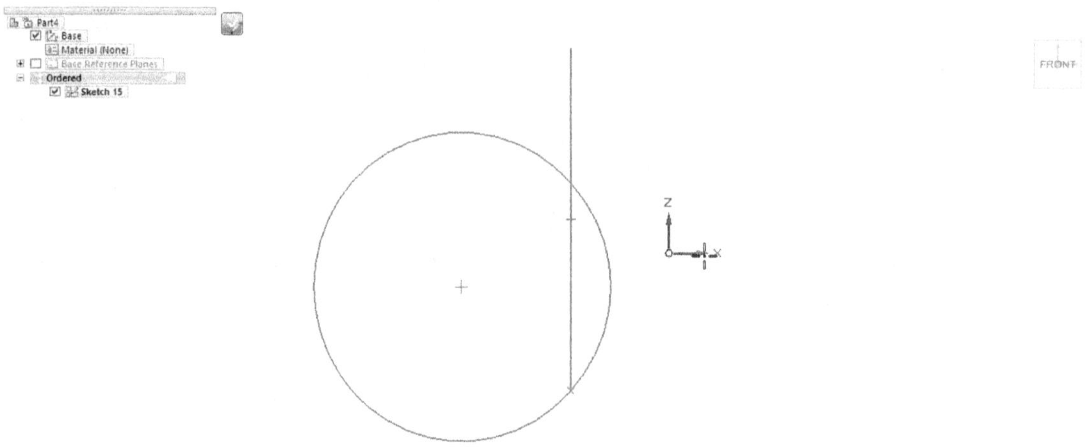

*Figure-129. On element*

## Center

The **Center** check box is used to display and snap to center of the circle, arc, and other round entities; refer to Figure-130. Note that center point is displayed when you hover cursor over the curve.

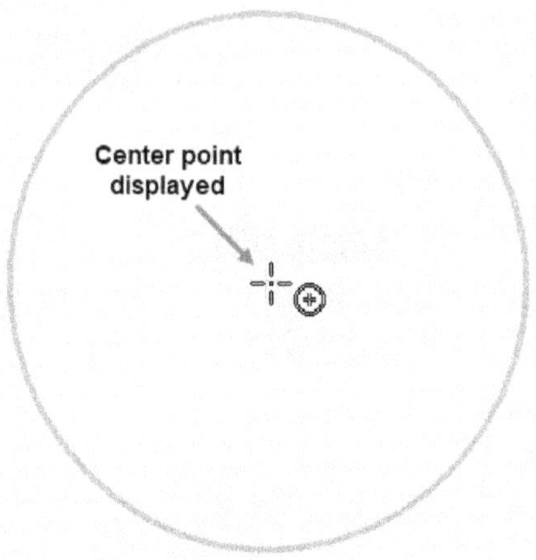

*Figure-130. Center*

## Intersection

The **Intersection** check box is used to display and snap to intersection point of two entities; refer to Figure-131.

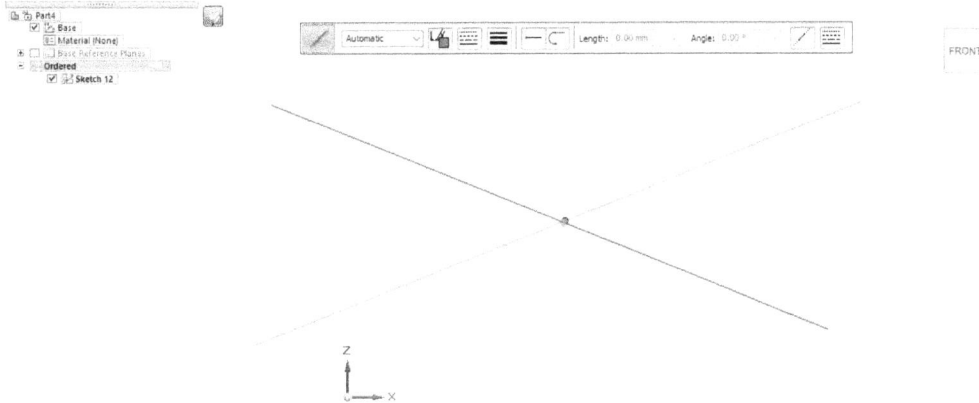

*Figure-131. Intersection*

## Silhouette

The **Silhouette** check box is used to display and snap to silhouette points on the arc, circle, ellipse, and other curves; refer to Figure-132.

*Figure-132. Silhouette*

## Edit Points

This check box is used to display and snap to the dynamic edit point of the curves while creating other sketch entities. Refer to Figure-112.

Edit points

*Figure-133. Edit points displayed on curve*

## Curve Control Vertex

The **Curve Control Vertex** tool is used to display and snap to the curve control vertices of curves while creating sketch; refer to Figure-134.

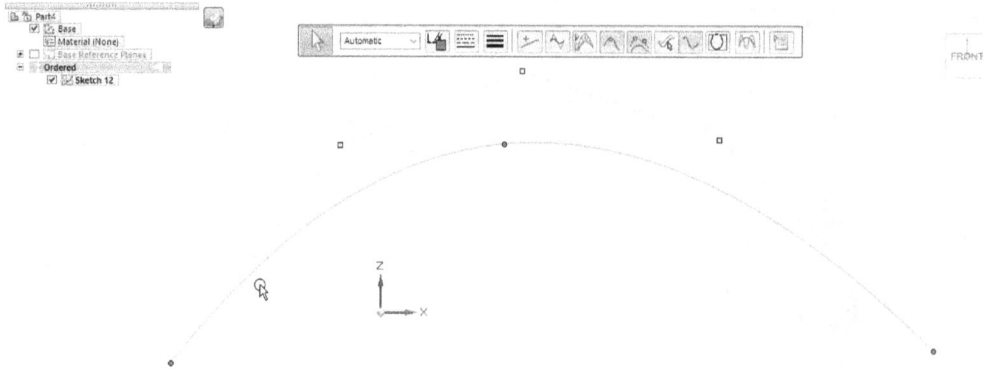

*Figure-134. Curve control vertex*

## Parallel

The **Parallel** tool is used to automatically recognize when a line is parallel to earlier created line while creating sketch; refer to Figure-135.

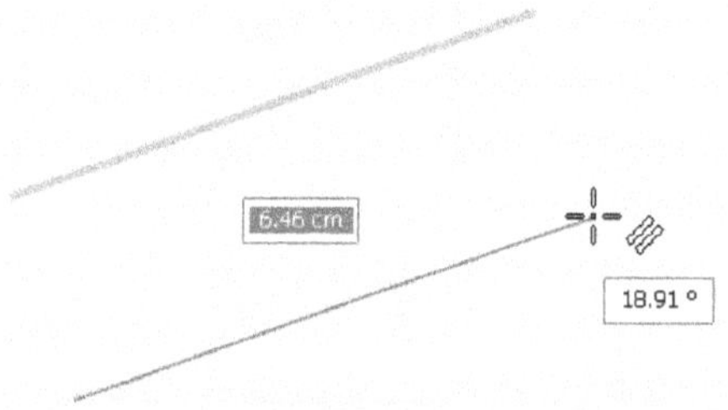

*Figure-135. Parallel constraint recognition*

## Horizontal or Vertical

The **Horizontal or Vertical** check box is used to recognize when a line is either Horizontal or Vertical with respect to X axis of sketch. Refer to Figure-136.

*Figure-136. Horizontal vertical line recognition*

## Perpendicular

The **Perpendicular** tool is used to recognize when a line or curve is perpendicular to another line/curve. Refer to Figure-116.

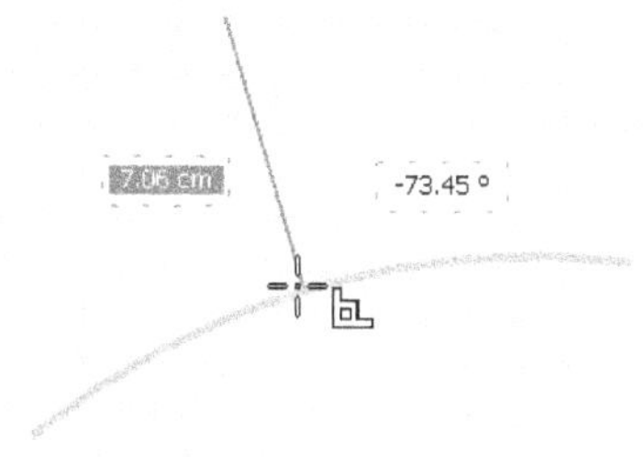

Figure-137. Perpendicular

## Tangent

The **Tangent** check box is used to recognize when a line/curve is tangent to another curve while creating; refer to Figure-138.

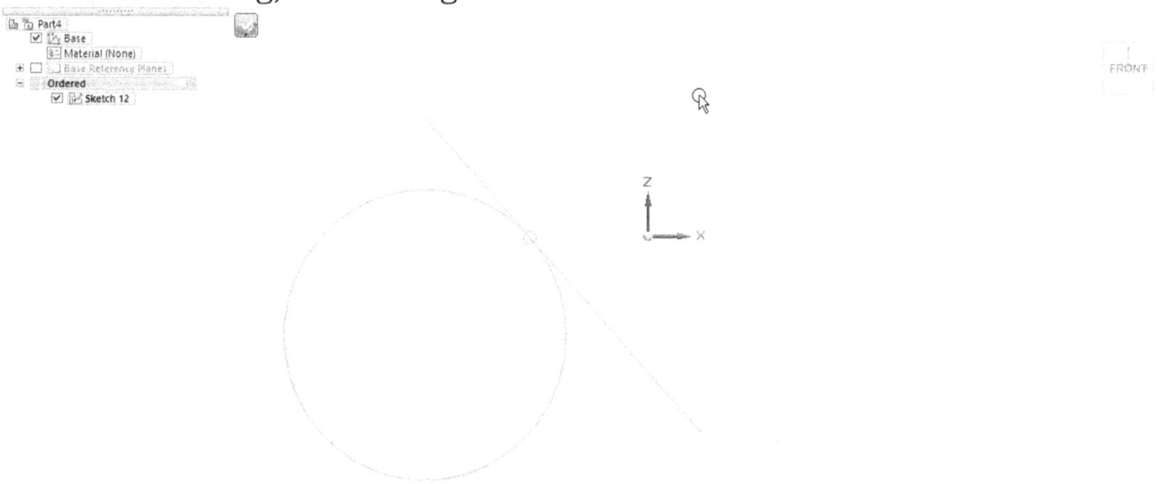

Figure-138. Tangent

## Intellisketch Options

The **IntelliSketch Options** tool is used to set parameters related to intellisketch interface. The procedure to use this tool is given next.

• Click on the **IntelliSketch Options** tool from the **IntelliSketch** panel in the **Home** tab of **Ribbon**. The **IntelliSketch** dialog box will be displayed; refer to Figure-139.
• Select the check boxes from the **Relationships** tab to activate respective snaps as discussed earlier.
• Click on the **Auto-Dimension** tab to specify parameters related to auto dimensioning; refer to Figure-140. The options in this tab have been discussed earlier.
• Click on the **Cursor** tab in the dialog box to set size of cursor and snap points in the drawing; refer to Figure-141.
• After setting desired parameters in the dialog box, click on the **OK** button.

*Figure-139. Intellisketch Options*

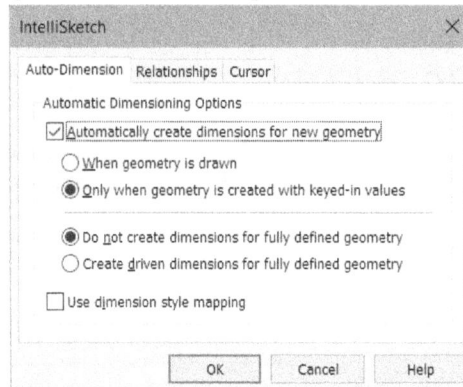

*Figure-140. Auto-Dimension tab*

*Figure-141. Cursor tab*

## Relationship Colors

The **Relationship Colors** tool is used to display sketch element in colors that indicate relationship status like sketch defined and so on. The colors can be modified in the **Color** page of the **Solid Edge Options** dialog box. To show color indications for relationships, you need to turn **"ON"** this toggle button.

## Alignment Indicator

The **Alignment Indicator** tool is used to indicate horizontal and vertical relationship while creating sketch entities.

## Styles

The **Styles** tool is used to manage the styles of text, dimensions, hatching, and so on. The procedure to use this tool is discussed next.

- Click on the **Styles** tool from the **Dimension** panel in the **Home** tab of **Ribbon**. The **Style** dialog box will be displayed as shown in Figure-142.

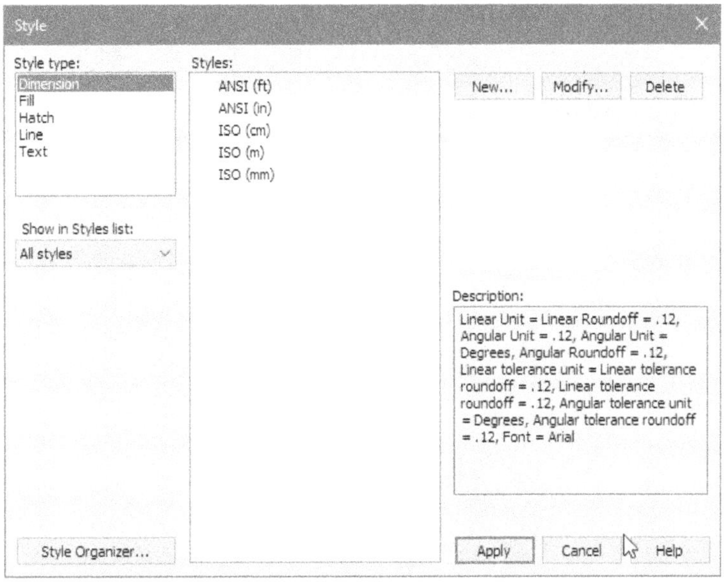

*Figure-142. Styles*

- Select the desired object from the **Style type** list box whose properties are to be modified. The list of default available styles available for the object will be displayed in the **Styles** section of dialog box and details of style will be displayed in the **Descriptions** area of dialog box.
- To modify properties of selected style, click on the **Modify** button from the dialog box. The related modify dialog box will be displayed. The procedures to modify styles of different objects are given next.

## Modifying Style of Dimension

- After selecting a dimension style, click on the **Modify** button from the dialog box. The **Modify Dimension Style** dialog box will be displayed; refer to Figure-143.

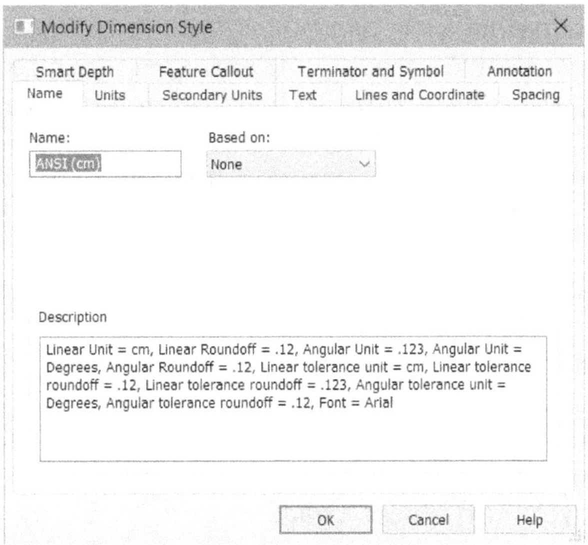

*Figure-143. Modify Dimension Style dialog box*

- Specify desired name of style in the **Name** edit box. If the style is based on any other style then you can select it from the **Based on** drop-down. The properties of style selected in drop-down will be copied.

### Units tab

- Click on the **Units** tab to modify unit parameters of the property. The options will be displayed as shown in Figure-144.

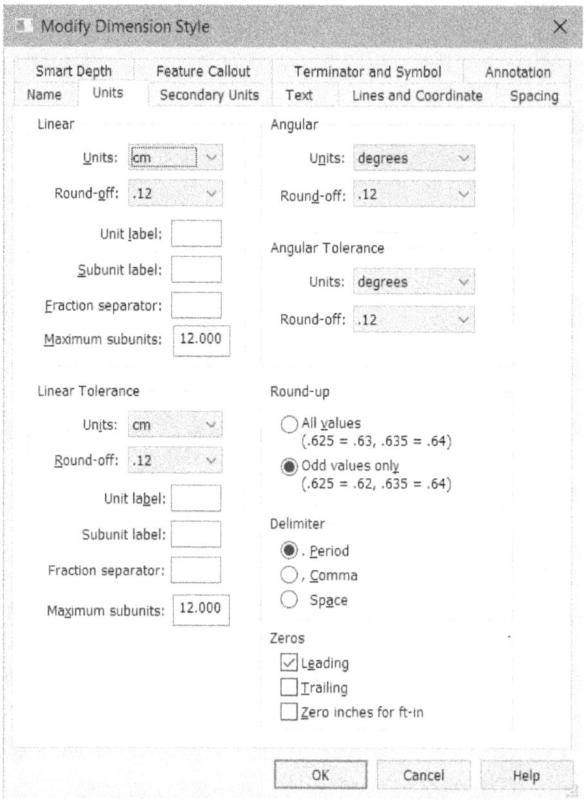

*Figure-144. Units tab*

- Select the desired option from the **Units** drop-down to define unit for dimensions.
- In the **Round-off** drop-down, you need to specify upto what accuracy the dimension will be created. You can create a dimension upto 7th place after decimal.
- In the **Unit label** edit box, you can specify characters to be used as unit label.
- In the **Subunit label** edit box, you can specify the characters to be used for subunit label of dimension.
- In the **Fraction separator** edit box, you can specify upto 5 characters by which whole value will be separated from the fraction value in dimension. If no value is specified in this edit box then fraction will be separated by single space like 15 2/3.
- In the **Maximum subunits** edit box, you can specify number of subunits available in one primary unit. For example, if primary unit is Feet then you can specify 12 as subunit which are Inches.
- You can set the above discussed parameters in the **Linear**, **Linear Tolerance**, **Angular**, and **Angular Tolerance** areas of the dialog box.
- Select the desired radio button from the **Round-up** area to define how dimensions will be rounded to nearest accuracy level.
- Select the desired radio button from the **Delimiter** area to define decimal delimiter. You can select period, comma, or space as delimiter.
- In the **Zeroes** area, you can select **Leading** and **Trailing** check boxes to specify whether you want to add zeros before decimal and after decimal if there is no number before and after decimal, respectively.
- Select the **Zero inches for ft-in** check box from the **Zeroes** area to display zero if there is no inch value in the dimension.

### Secondary Units tab

The options in the **Secondary Units** tab are used to specify properties of secondary unit if applied in dimension. The options in this tab are same as discussed for Units tab.

### Text tab

The options in the **Text** tab are used to modify text font and other related parameters.

*   Click on the **Text** tab from the **Modify Dimension Style** dialog box. The options will be displayed as shown in Figure-145.

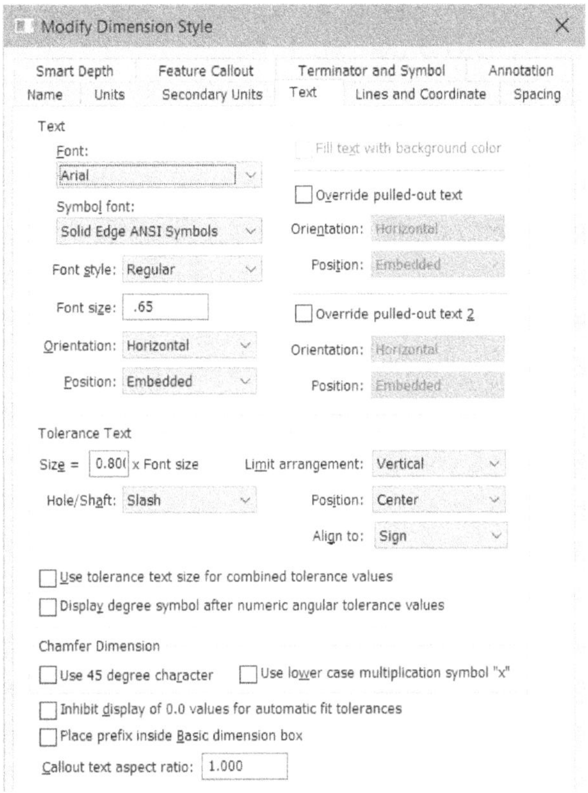

*Figure-145. Text tab in Modify Dimension Style dialog box*

*   Select the desired font from the **Font** drop-down. The dimensions will be displayed in selected font.
*   Select the desired option from the **Symbol Font** drop-down to define font for symbols used in dimension.
*   Select the desired option from the **Font style** drop-down to define the style of font.
*   In the **Font size** edit box, you can specify the size of dimension text.
*   Select the desired option from the **Orientation** drop-down to define whether text will be written horizontal, vertical, parallel to dimension line, or perpendicular to dimension line.
*   Select the desired option from the **Position** drop-down to define where to place the dimension.
*   Select the **Override pulled-out text** check box if you want to set different parameters for pulled-out text of dimension.
*   Specify the desired value in the **Size** edit box to define reduction in size for tolerance text.
*   Select the desired option from the **Limit Arrangement** drop-down to define how limits will be placed for tolerance in dimension.

- Set the desired position and alignment parameters for tolerance in the **Position** and **Align to** drop-downs.
- Select the **Use tolerance text size for combined tolerance values** check box to display upper and lower tolerance values as combined if they are same.
- Select the **Display degree symbol after numeric angular tolerance values** check box to display degree symbol after angular tolerance.
- Select the **Use 45 degree character** check box to display 45 degree value for chamfer.
- Select the **Use lower case multiplication symbol "x"** check box to add "x" between chamfer value and angle value.
- Select the **Inhibit display of 0.0 values for automatic fit tolerances** check box to not display 0.0 value in tolerance when fit tolerance is generated automatically.
- Similarly, set other parameters in the dialog box.

### Lines and Coordinate tab

The options in the **Lines and Coordinate** tab are used to modify parameters related to lines and coordinates; refer to Figure-146.

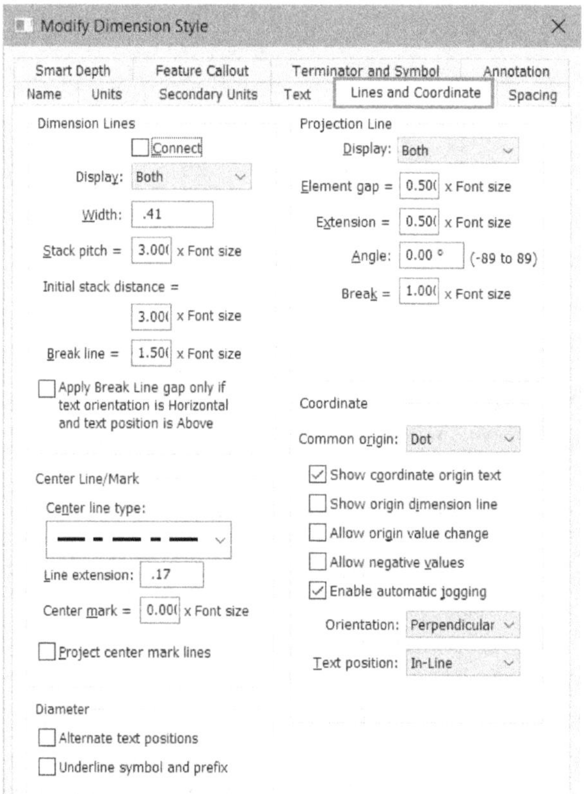

*Figure-146. Lines and Coordinate tab*

- Select the **Connect** check box if you want dimensions lines to automatically extend upto terminators of dimension.
- Set the desired option in the **Display** drop-down to define which side of dimension lines will be displayed for a given dimension.
- In the **Width** edit box, you can specify the width of dimension line.
- Set the desired value in the **Stack pitch** edit box to define gap between two stacked dimensions.
- Set the desired value in the **Initial stack distance** edit box to define gap between two innermost stacked dimensions.

- Set the desired value in **Break Line** edit box to define length of break line in case of stack dimensions.
- Select the **Apply Break Line gap** check box if stacked text is oriented horizontally and you want to apply gap between stacked dimensions.
- Select the desired option from the **Center line type** drop-down to define what type of line will be used for showing centerline in the drawing/sketch.
- Similarly, you can specify the line extension, center mark size and other parameters in **Center Line/Mark** area.
- Select the **Alternate text positions** check box in **Diameter** area to place dimension text alternatively left and right when dimensions are stacked.
- Select the **Underline symbol and prefix** check box to underline symbols and prefixes in the dimension.
- The options in the **Projection Line** area are used to modify projection lines created for projected views in the drawing.
- The options in the **Coordinate** area are used to set parameters for displaying coordinate system in the drawing.

### Spacing tab

The options in **Spacing** tab are used to modify various gaps in dimensions like text clearance gap from dimension line, vertical gap between dual dimensions, and so on; refer to Figure-147. Set the desired values in the edit boxes to modify gaps.

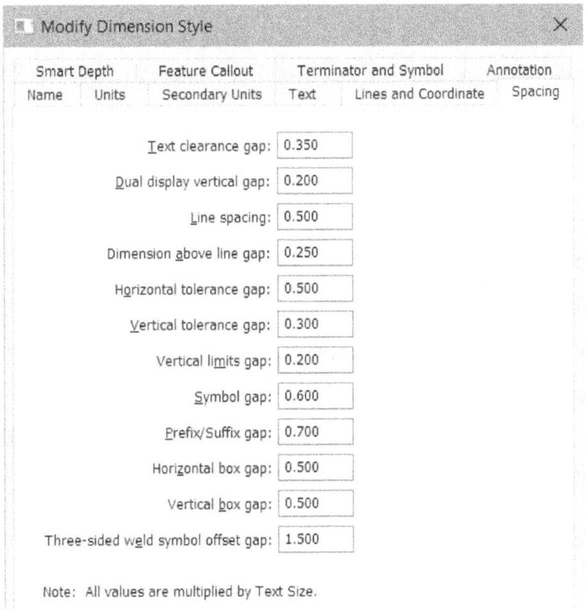

*Figure-147. Spacing tab*

The other options in this dialog box will be discussed later in chapter related to drafting.

## Increase PMI Font

The **Increase PMI Font** tool is used to increase the font size of all the text and dimensions in the drawing.

## Decrease PMI Font

The **Decrease PMI Font** tool is used to decrease the font size of all the text and dimensions in the drawing.

# PRACTICAL - 1

Create the sketch as shown in Figure-148. Also, dimension the sketch as per the figure.

## Steps to be performed

• Start a new part file and switch to Ordered environment of designing.
• Select a sketching plane and activate sketching mode.
• Create the sketch using **Line** tool.
• Apply the dimensions using the **Smart Dimension** tool.
• Save the file.

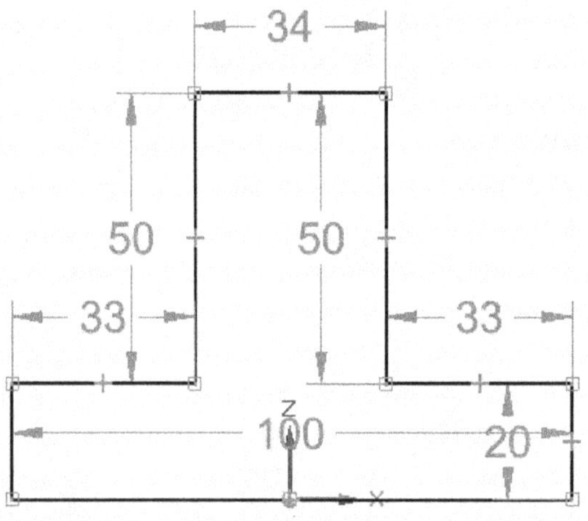

*Figure-148. Practical 1*

## Starting Sketching Environment

• Start Solid Edge if not started already.
• Click on the **New** button from the **Application** menu and click on the **New** tool from the cascading menu. The **New** dialog box will be displayed; refer to Figure-149.

*Figure-149. New Solid Edge document dialog box*

- Select the **ANSI Metric** option from the Standard Templates list at the left in the dialog box and double-click on the **ansi metric part.par** template at the right in the dialog box. The part environment of Solid Edge will be displayed; refer to Figure-150. You can draw sketch in **Synchronous** or **Ordered** sketch environment as discussed earlier. In this tutorial, we will create the sketch in ordered environment.
- Right-click in the empty area of application window and select the **Transition to Ordered** option from the shortcut menu displayed. The Ordered environment will become active.
- Click on the **Sketch** tool from the **Sketch** panel in the **Home** tab of the **Ribbon**. You will be asked to select a sketching plane.
- Expand the **Base Reference Planes** node from the **Path Finder** at the left in the application window and select the **Front (xz)** plane. You can also check the default planes by hovering cursor on the coordinate system. On selecting plane, various tools to create sketch will be displayed in the **Ribbon**.

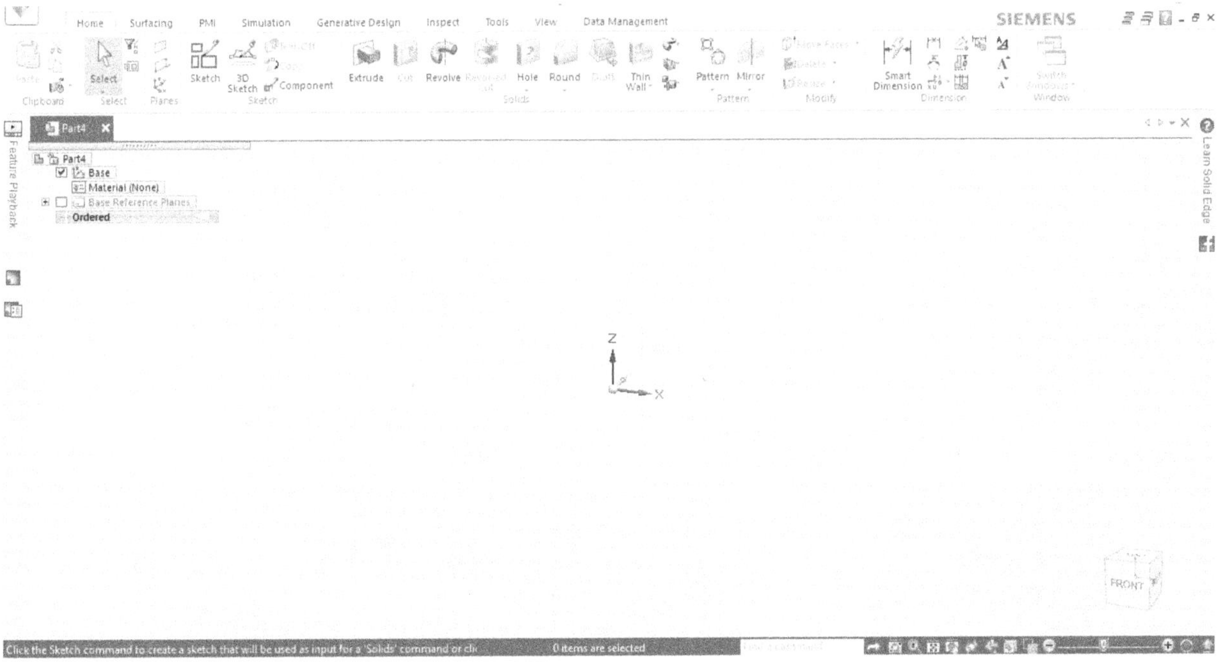

*Figure-150. Solid Edge Application window*

- Click on the **Line** tool from the **Draw** panel in the **Home** tab of the **Ribbon**. The **Line** tool will become active and you will be asked to specify the start point of line.
- Specify the start point of line and draw a line of length **100** at **0** degree angle. Refer to Figure-151.

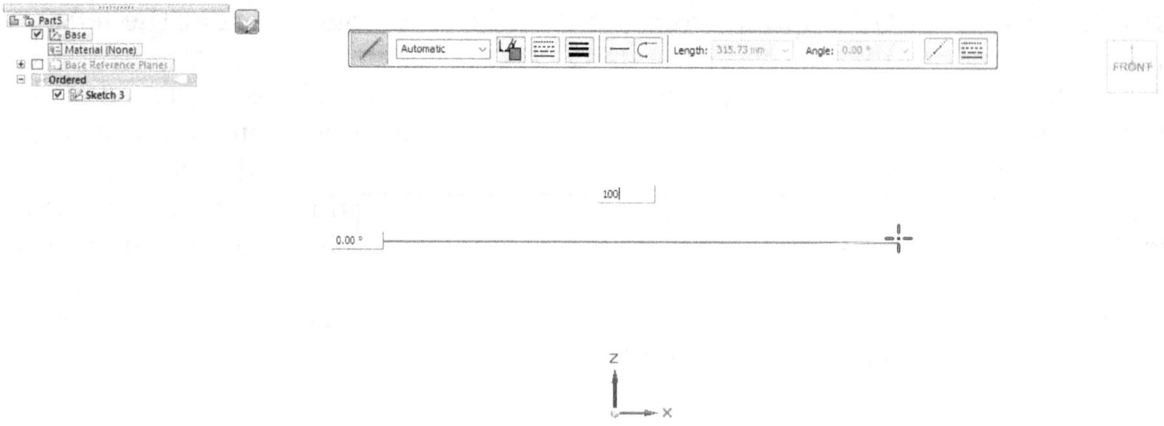

*Figure-151. Starting Creation Line*

- Move the cursor upwards and draw a line of **20** with **90** degree angle.
- Move the cursor left and draw a line of **33** with **180** degree angle.
- Move the cursor upwards and draw a line of **50** with **90** degree angle.
- Again, move the cursor left and draw a line of **34** with **180** degree angle.
- Move the cursor downwards and draw a line of **50** with **-90** degree angle or make equal to adjacent line.
- Move the cursor left and draw a line of **33** with **180** degree angle.
- Move the cursor downwards and draw a line of **20** with **-90** degree angle. After creating the sketch, press **ESC** to exit the tool.
- Note that sketch is not fully defined yet. To make sketch fully defined, we need to connect the middle point of line of 100 length with origin. Then the sketch will be fixed at its place and will become fully defined.
- Click on the **Connect** tool from the **Relate** panel in the **Home** tab of the **Ribbon**. You will be asked to select the entities to be connected.
- Select the middle point of the line and origin. The middle point of line will get connected with origin; refer to Figure-152.
- Press **CTRL+S** from keyboard and save the file at desired location.

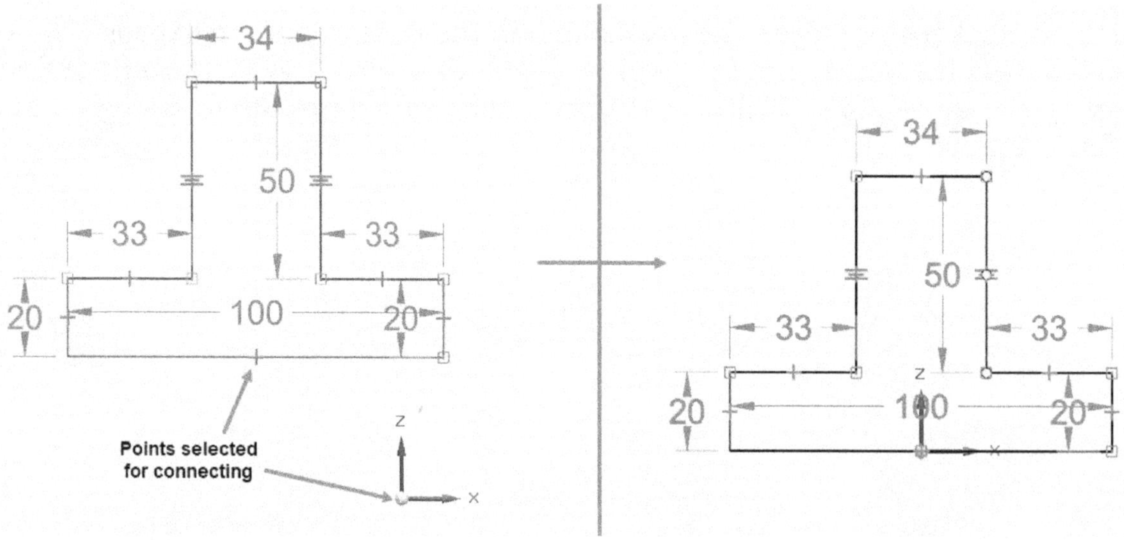

*Figure-152. Connecting mid point of line with origin*

# PRACTICAL - 2

In this practical, we will create the sketch as shown in Figure-153.

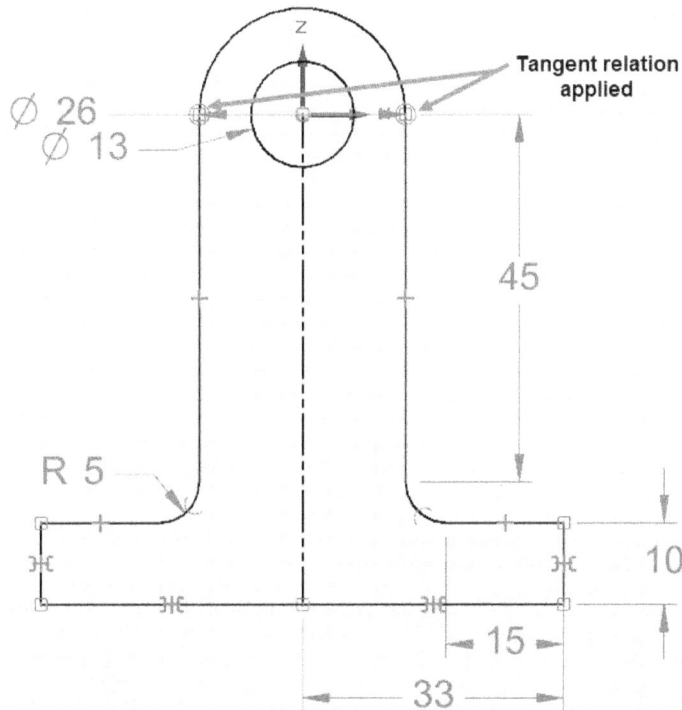

*Figure-153. Practical 2*

## Steps to be performed

- Start a new part file and switch to Synchronous environment of designing.
- Select a sketching plane and activate sketching mode.
- Create the sketch using **Line**, **Circle**, **Fillet**, and **Arc** tools.
- Apply the dimensions using the **Smart Dimension** tool.
- Save the file.

## Starting New Document

- Start Solid Edge if not started already.
- Click on the **New** button from the **Application** menu and click on the **New** tool from the cascading menu. The **New** dialog box will be displayed.
- Select the **ANSI Metric** option from the Standard Templates list at the left in the dialog box and double-click on the **ansi metric part.par** template at the right in the dialog box. The part environment of Solid Edge will be displayed. You can draw sketch in **Synchronous** or **Ordered** sketch environment as discussed earlier. In this tutorial, we will create the sketch in synchronous environment which is default selected when starting a new document.

## Creating circle

- Click on the **Circle by Center Point** tool from the **Circle** drop-down in the **Draw** panel of the **Sketching** tab in the **Ribbon**. You will be asked to specify the center point.
- Click on the origin to specify center point of circle. Specify the diameter of circle as **13**. Similarly, draw a circle with same center point of diameter **26**. Refer to Figure-154.

- Click on the **Sketch View** button from the **View** toolbar at the bottom right corner of the application window or press **CTRL + H** from keyboard to orient sketching plane parallel to screen.

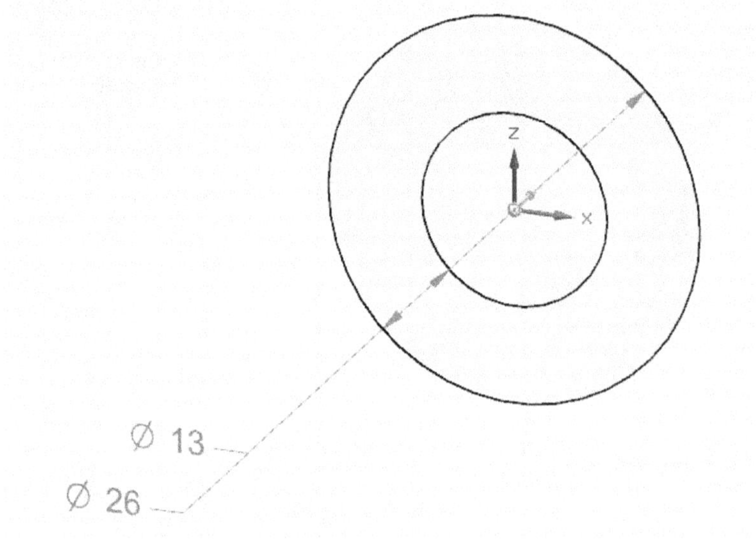

*Figure-154. Creating Circle*

## Creating Lines

- Click on the **Line** tool from the **Line** drop-down in the **Draw** panel of **Sketching** tab in the **Ribbon**. You will be asked to specify the start point of the line.
- Select the **silhouette** point of the circle as a start point; refer to Figure-155. Move the cursor downward and draw a line of **50** with **90** degree angle.
- Move cursor towards right and draw a line of **20** with **00** degree angle.
- Move cursor downwards and draw a line of **10** with **90** degree angle.
- Move cursor towards left and draw a line of **33** with **180** degree angle.
- Click on the **Create as Construction** tool from the **Draw** panel in the **Ribbon** and draw a construction line upto center point of circle; refer to Figure-155. Press **ESC** to exit the tool. Deselect the **Create as Construction** tool to exit construction mode.

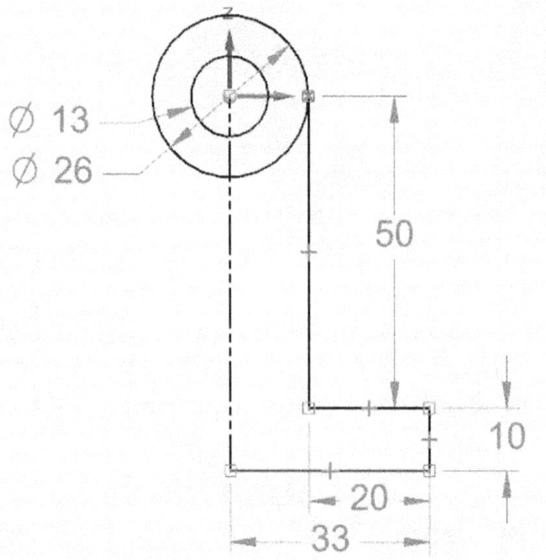

*Figure-155. Creating line*

## Creating Mirror

- Click on the **Mirror** tool from the **Mirror** drop-down in the **Draw** panel of the **Ribbon**. Select all the lines recently drawn except construction line while holding **CTRL** key. You will be asked to select line for mirror reference.
- Select the construction line as reference line for mirror. The mirror copy will be created; refer to Figure-156. Press **ESC** to exit the tool.

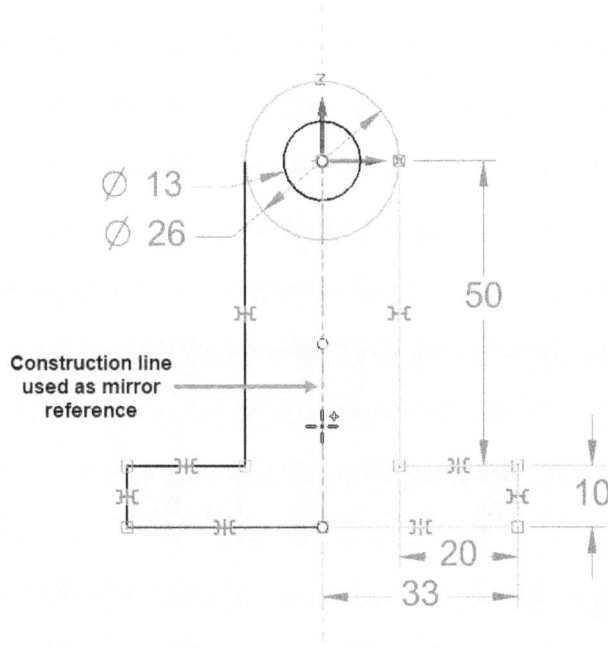

*Figure-156. Mirror copy generated*

## Applying Fillet

- Click on the **Fillet** tool from the **Fillet** drop-down in the **Draw** panel of the **Ribbon**. You will be asked to select two elements between which fillet will be applied and the **Fillet Command Bar** will be displayed.
- Specify the radius of fillet as **5** in the **Radius** edit box of the **Command Bar**. Select the lines between which fillets are to be created; refer to Figure-157. Press **ESC** to exit the tool.

*Figure-157. Fillet*

## Trim

- Click on the **Trim** tool from the **Draw** panel in the **Sketching** tab of **Ribbon** and trim the lower portion of circle with diameter **26**; refer to Figure-158.

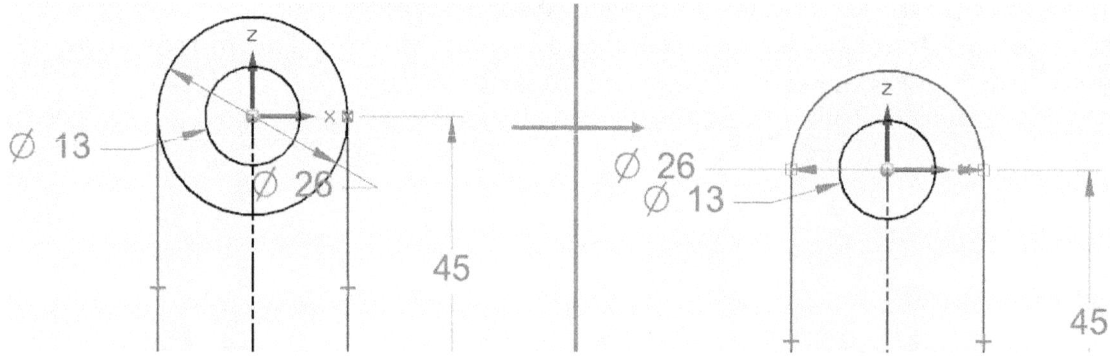

*Figure-158. Trim circle*

## Applying Constraints and Dimensions

- Click on the **Tangent** tool from the **Relate** panel in the **Sketching** tab of **Ribbon** to make the lines tangent to the trimmed circle so that the sketch should be fully defined.
- Click on the **Smart Dimension** tool from the **Dimension** panel in **Ribbon** and dimension all the entities as shown in Figure-159.

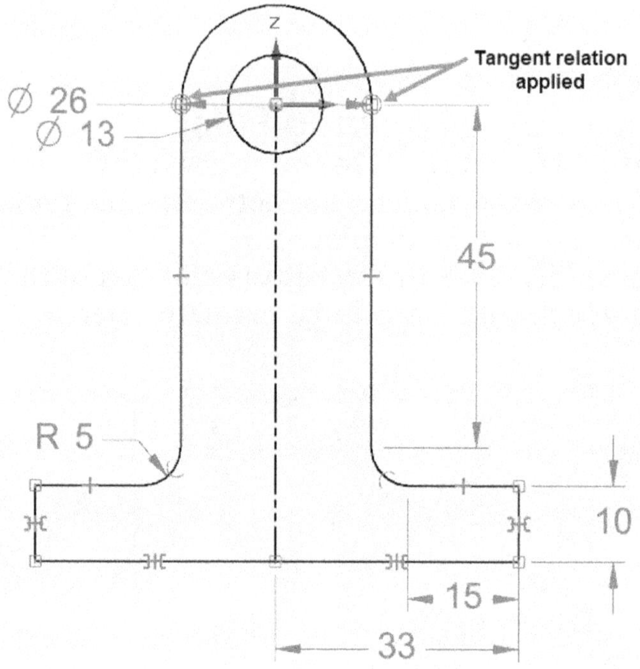

*Figure-159. Tangent relation and dimensions applied*

# PRACTICE 1

In this practice session, you will create a sketch for the drawing shown in Figure-160.

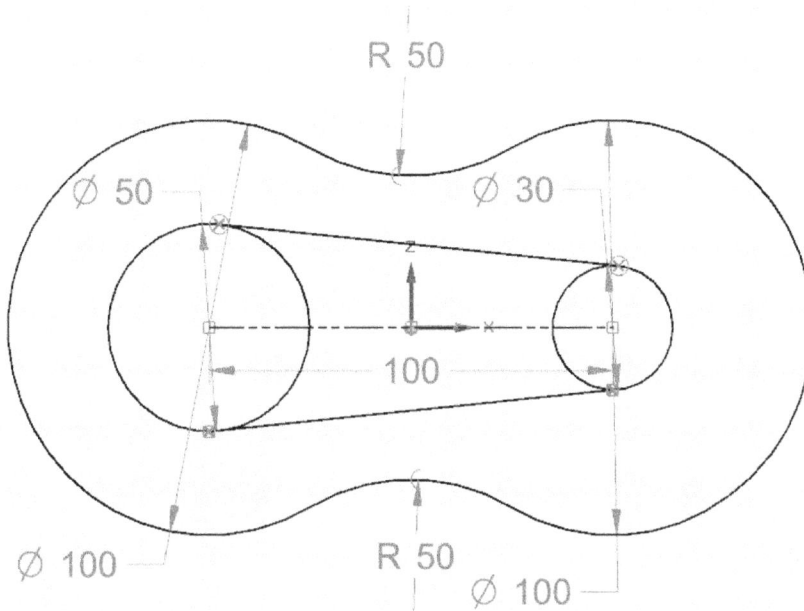

*Figure-160. Practice 1*

# PRACTICE - 2

In this practice, you will create a sketch for drawing shown in Figure-161.

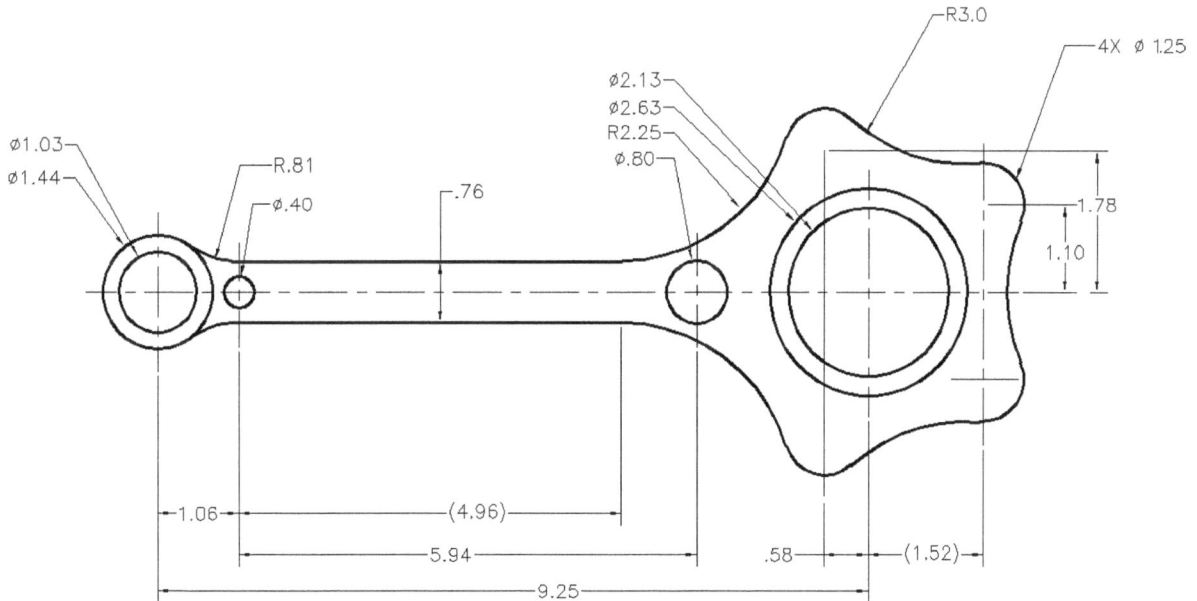

*Figure-161. Practice 2*

FOR STUDENT NOTES

# Chapter 3

# 3D Sketch and Solid Modeling

## Topics Covered

The major topics covered in this chapter are:

- *3D Sketch*
- *Solid Modeling tools*
- *Planes and Coordinate System*
- *Modifying tools*
- *Practical 1*
- *Practical 2*
- *Practice*

# INTRODUCTION

In previous chapter, we have discussed about the 2D sketch and related tools. In this chapter, we will discuss about 3D sketch and modeling tools.

# 3D SKETCH

The sketch creation tools are same in both **2D** and **3D** but in 3D sketch we are using all three coordinate **X,Y,Z** in simultaneously. The procedure to create a 3D sketch is discussed next.

- Click on the **New 3D Sketch** tool from the **New Sketch** panel in the **3D Sketching** tab of **Ribbon** in case of **Synchronous** environment or click on the **3D Sketch** tool from the **Sketch** panel in the **Home** tab of **Ribbon** in case of **Ordered** environment. If you are using the **Synchronous** environment then a new sketch will be added in the **Path Finder** and you need to right-click on the sketch and activate it for editing. Here, we will discuss the procedure for **Ordered** environment.
- After selecting **3D Sketch** tool from the **Ribbon**, the tools in **Ribbon** will be displayed as shown in Figure-1.

*Figure-1. 3D Sketch Ribbon*

- Select desired sketch creation tool from the **3D Draw** panel in the **Home** tab of **Ribbon** and draw sketch as desired. Note that the sketch entity will be automatically aligned to nearest axis while drawing sketch. You can use the **View Cube** to change orientation while drawing sketch.
- If you want to create sketch entity on a plane or along an axis then select it from the **OrientXpress** after activating the tool; refer to Figure-2.

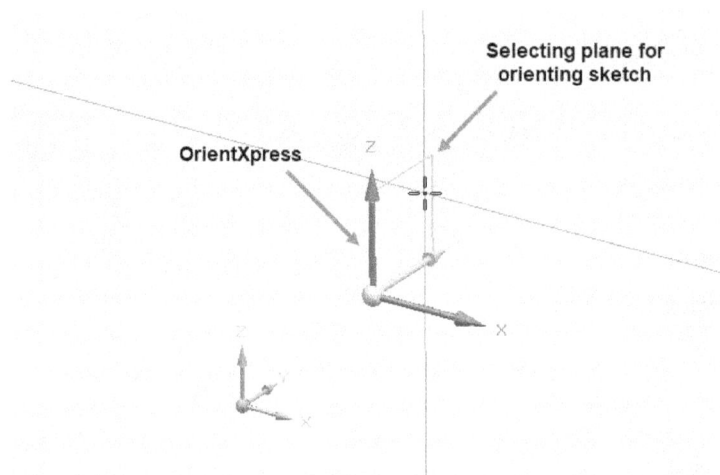

*Figure-2. OrientXpress*

- After selecting the desired orientation, create the sketch as desired; refer to Figure-3.

*Figure-3. 3D Sketch Creation*

- After creating sketch, click on the **Close 3D Sketch** button from the **Close** panel in the **Ribbon** to exit.

**Tips:**
- You can use 3D sketch tools to connect various edges of 3D model using lines, arcs, and so on.
- The relations available in the **3D Relate** panel are applicable on external sketches, edges, vertices, and faces of 3D models as well.
- Using the **Routing Path** tool, you can create guidelines for piping, electrical wiring, and other systems.

# EXTRUDE

The **Extrude** tool is used to create a solid volume by adding height to the selected sketch. In other words, this tool adds material in the direction perpendicular to the plane of sketch while using boundaries of sketch. There are two ways to create extrude feature in Solid Edge: In first method, select the **Extrude** tool and then make sketch by selecting a plane. In second method, sketch is created first and then **Extrude** tool is activated. After activating tool, you need to select the sketch for creating extrude feature. The procedure to create extrude features by second method is given next.

- Create a 2D sketch with non intersecting curves and forming one close loop chain; refer to Figure-4.

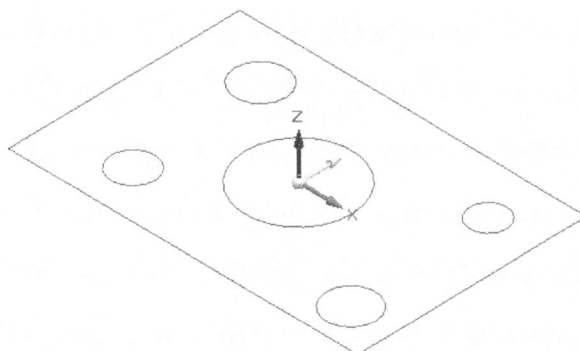

*Figure-4. Sketch created for extrude feature*

- Click on the **Extrude** tool from the **Solids** panel in the **Home** tab of **Ribbon**. You will be asked to select plane or planar face. Select the **Select from Sketch** option from the **Extrude - Create-From Options** drop-down in the **Extrude Command Bar** as shown in Figure-5.

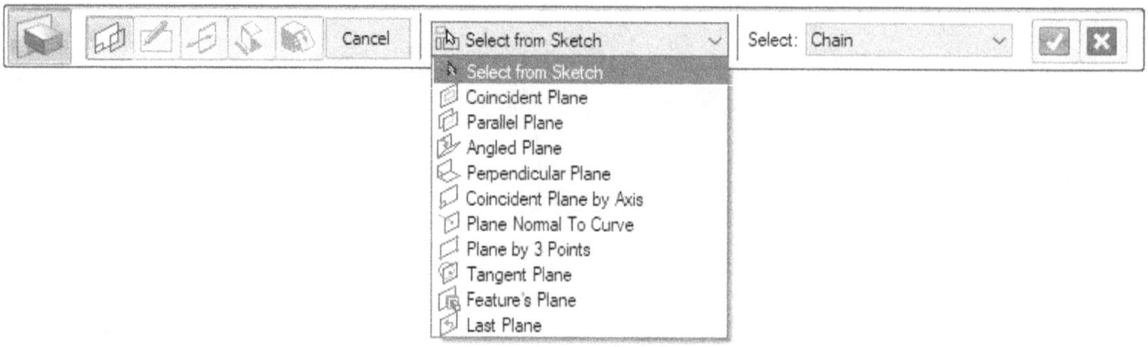

*Figure-5. Extrude - Create-From Options drop-down*

- Select the sketch from the drawing area. Note that you can select multiple loops in the sketch by selecting them one by one. After selecting sketch, right-click in the drawing area or click on the **OK** button from the **Command Bar**. The top face of extrude feature will get attached to cursor and the **Extrude Command Bar** will be displayed as shown in Figure-6.

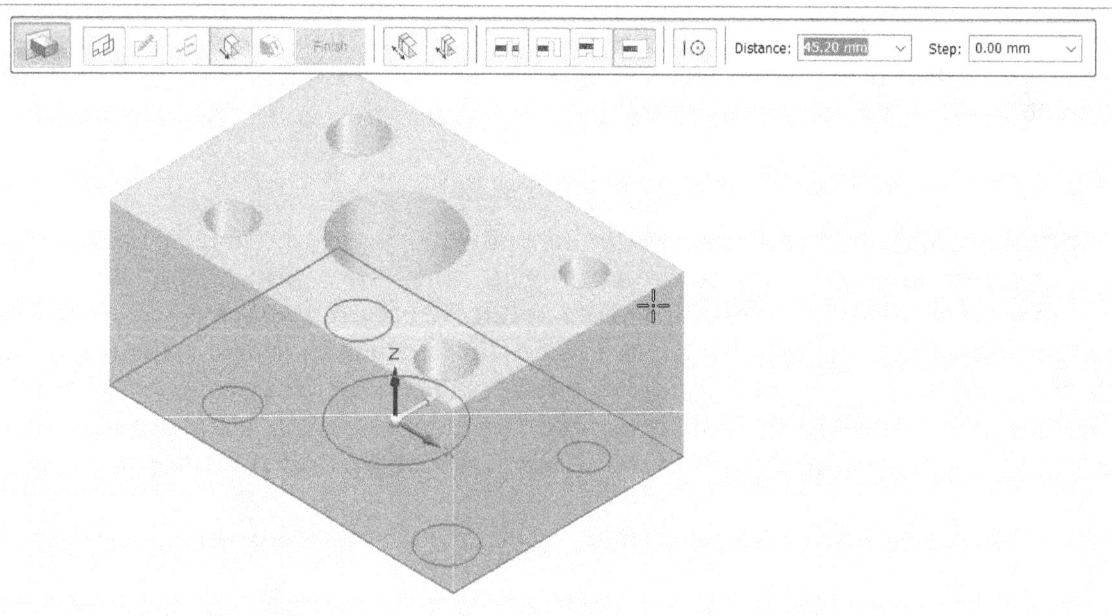

*Figure-6. Extrude Command Bar*

- Specify desired distance in the **Distance** edit box of **Command Bar** and left click to define side where you want to create the extrude feature with respect to sketching plane. The extrude feature will be created; refer to Figure-7. Note that if you want to modify any step of extrude creation then you can select respective button from the **Command Bar**. After setting parameters, click on the **Finish** button from the **Command Bar** to create the feature.

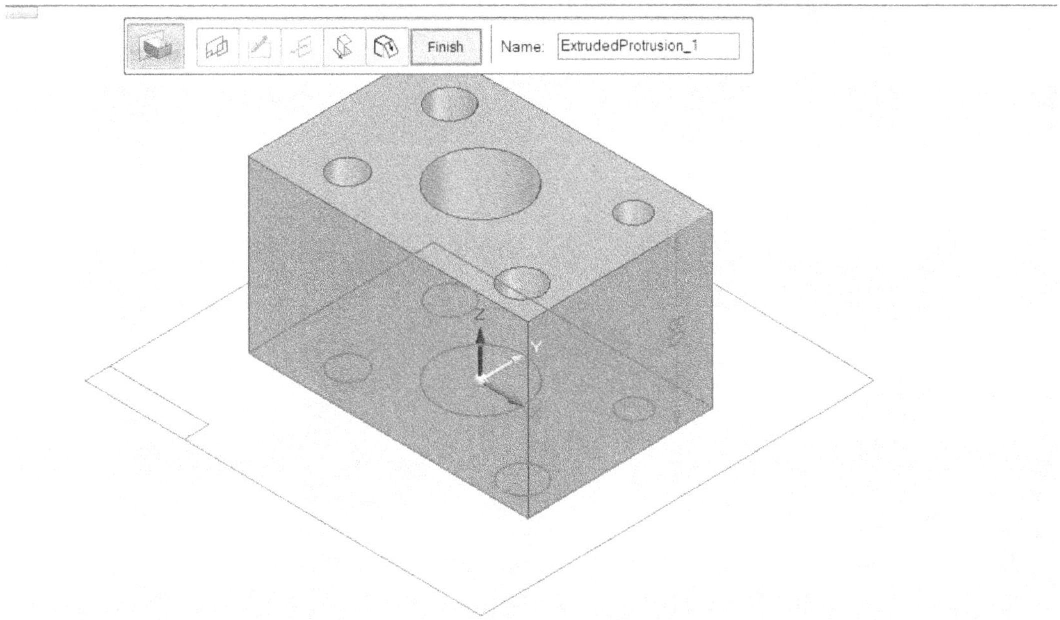

*Figure-7. Extrude Creation*

Various options in the **Extrude Command Bar** at **Extrude - Extent** step are discussed next.

**Extrude - Finite Extent** :- Select the **Finite Extent** button when you want to extrude the sketch up to the finite extent means user will define the limit of extrude; refer to Figure-8.

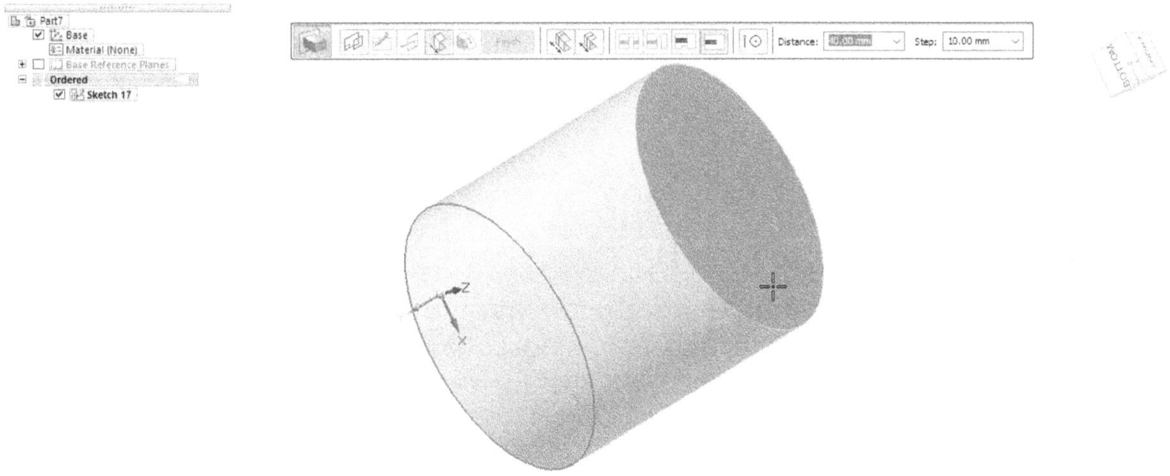

*Figure-8. Extrude Finite Extent*

• **Extrude - Through All** :- Select this button when you want to extrude the sketch through all of the solids earlier created. Note that if there is no face in perpendicular direction of sketching plane then this option will not generate any feature. Refer to Figure-9.

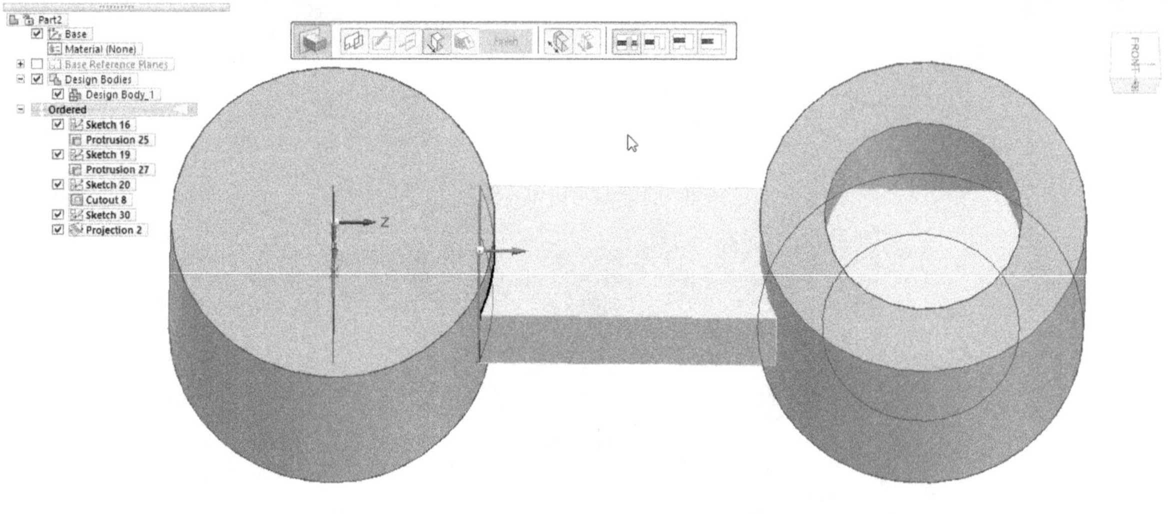

*Figure-9. Extrude Through All*

- **Extrude - Through Next** :- Select this button to extrude the sketch upto the next face/surface in direction perpendicular to sketching plane. Refer to Figure-10.

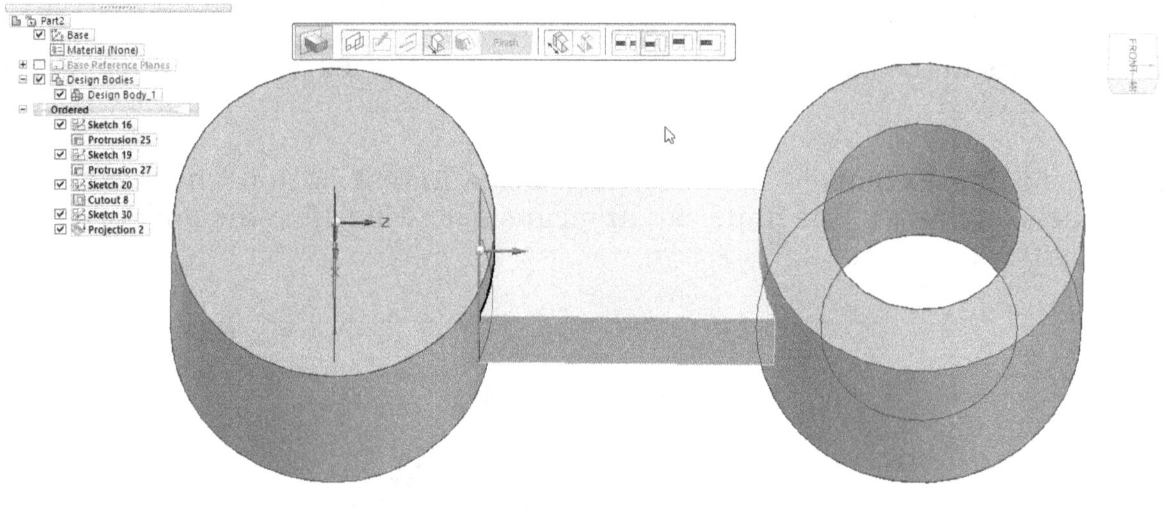

*Figure-10. Extrude Through Next*

- **Extrude - From/To Extent** :- This button is used to extrude the sketch from the selected reference location to another location in perpendicular direction. After selecting this button from **Command Bar**, select the face to define "From" point and then select another face to define "To" point for extrude feature; refer to Figure-11.

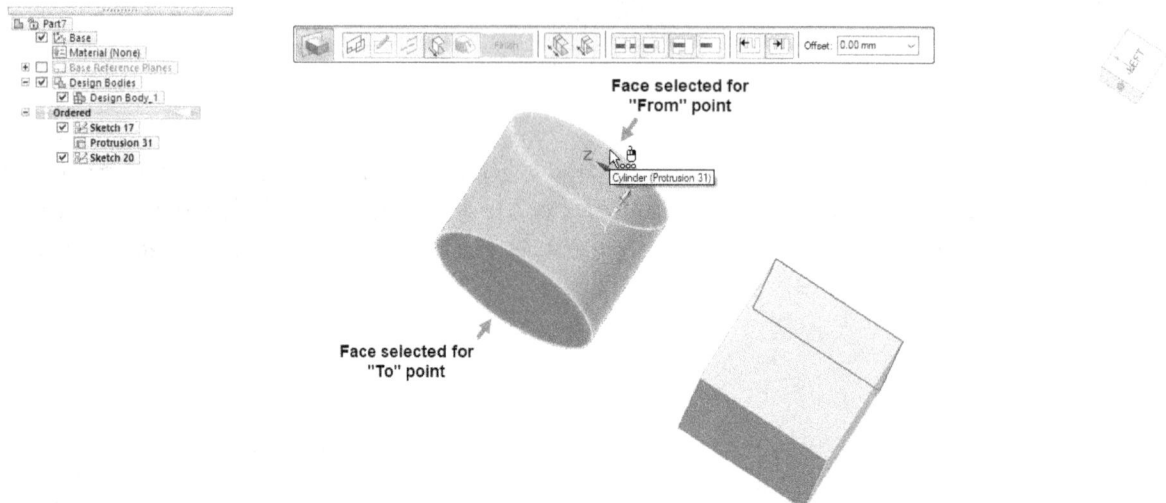

*Figure-11. Extrude From/To Extent*

- **Extrude - Symmetric Extent** 🔲 :- Select this button to extrude the sketch symmetrically on both sides of sketching plane; refer to Figure-12.

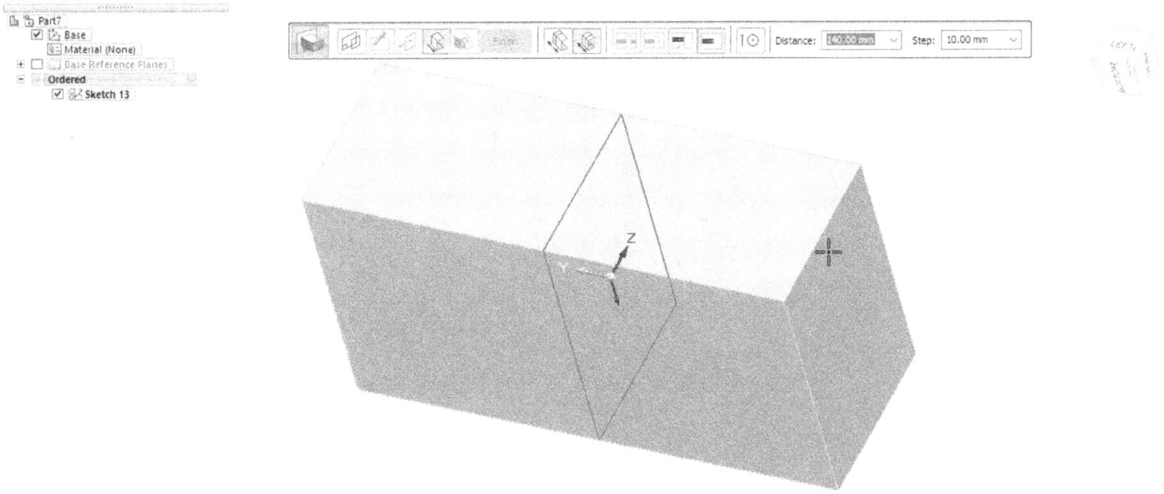

*Figure-12. Extrude Symmetric Extent*

- **Extrude - Non-Symmetric Extent** 🔲 :- The **Non-Symmetric Extent** button is used to extrude the sketch on both sides of sketching plane but you can specify different heights for both sides individually; refer to Figure-13.

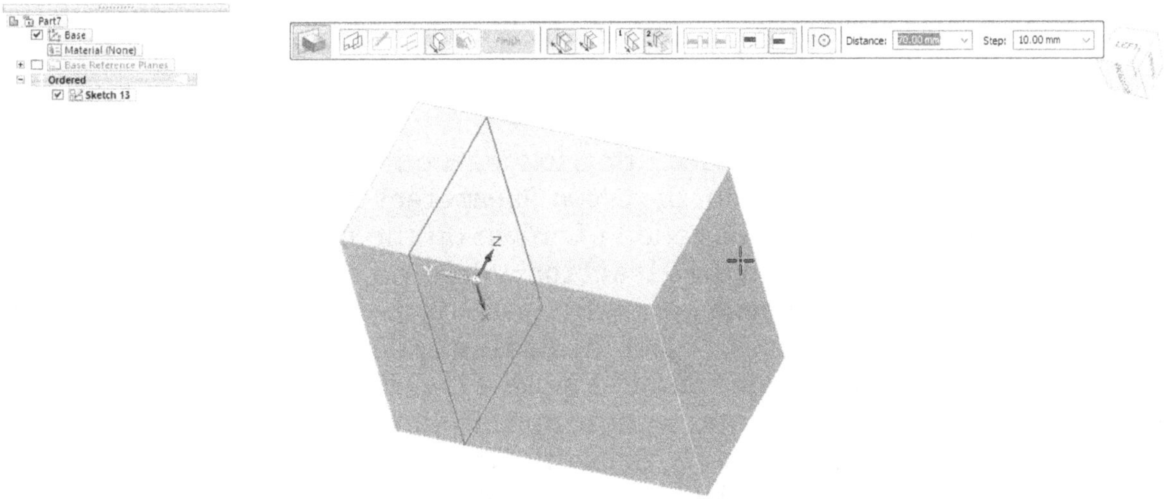

*Figure-13. Extrude Non Symmetric Extent*

After specifying extent of extrude feature, click on the **Extrude - Treatment Step** button from the **Command Bar**. The options in **Command Bar** will be displayed as shown in Figure-14.

*Figure-14. Extrude Command Bar with Treatment options*

Various options in the **Extrude Command Bar** at **Extrude - Treatment** step are discussed next.

- **Extrude - No Treatment** 🔲 : - Select this button if you do not want to apply any draft or crown modification to the extrude feature being created.
- **Extrude - Draft** 🔲 : - Select this button if you want to create taper side faces in the extrude feature; refer to Figure-15. After selecting this button, specify the

desired value of draft angle in the **Angle** edit box of the **Command Bar**. If you want to change the direction of draft then click on the **Flip** button at the end in the **Command Bar**.

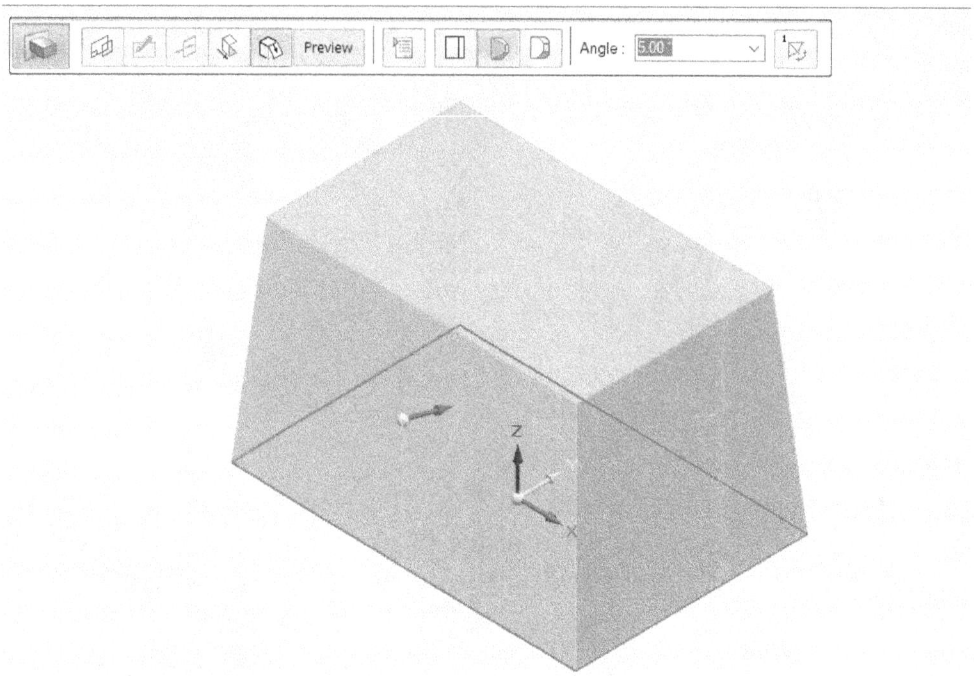

*Figure-15. Draft applied in extrude feature*

- **Extrude - Crown** 🗗 : - Select this button to create curves side faces of the extrude feature. On selecting this button, the **Crown Parameters** dialog box will be displayed along with arrow displaying direction of crown on the model; refer to Figure-16. Select the desired option from the **Direction 1** drop-down to define how crown will be created. Select the **No Crown** option from the drop-down if you do not want to create crown on the side faces. Select the **Radius** option if you want to specify the radius of crown on side faces; refer to Figure-17. Select the **Radius and take-off** option from the drop-down if you want to specify start angle along with radius for creating crown; refer to Figure-18. Select the **Offset** option from the drop-down if you want to specify the offset distance from side walls up to which crown walls will span. You can use the **Offset and take-off** option from the drop-down to specify take-off value as well. You can flip the direction and curvature of crown by using the **Flip Side** and **Flip Curvature** buttons respectively from the dialog box.

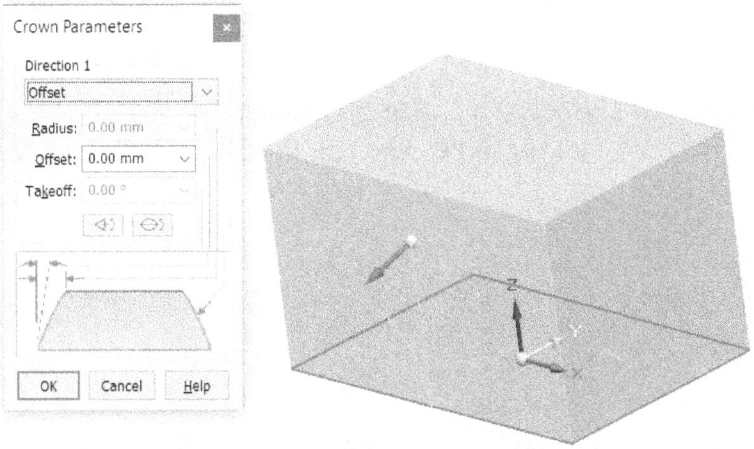

*Figure-16. Crown Parameters dialog box*

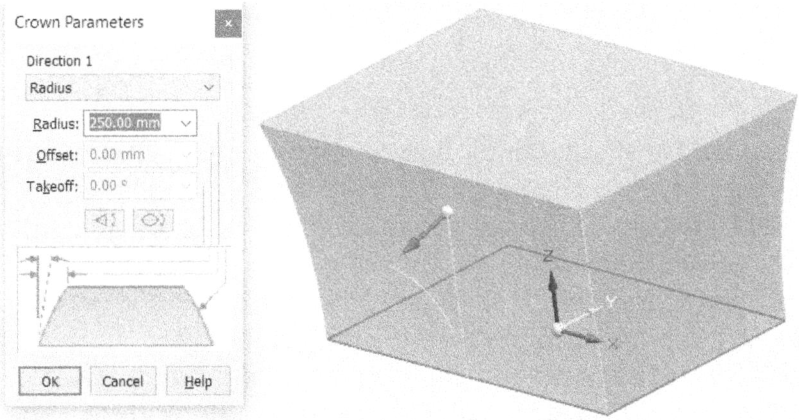

*Figure-17. Radius parameter specified for crown*

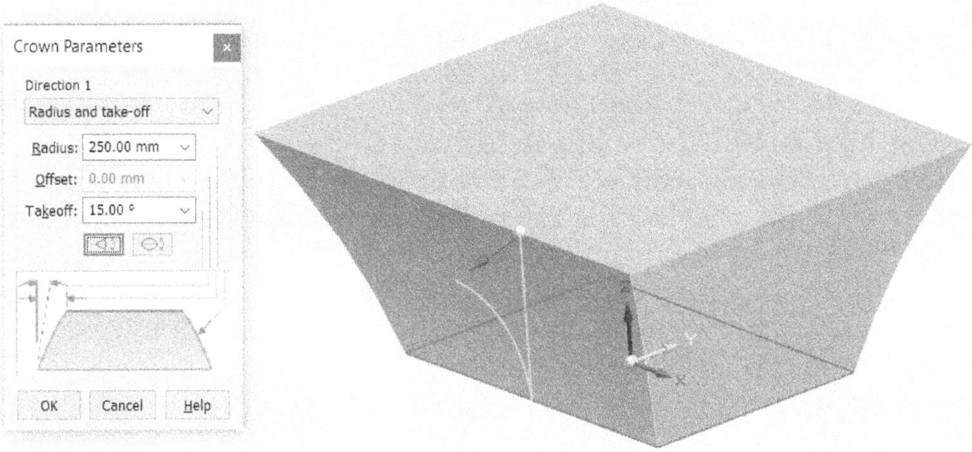

*Figure-18. Radius and takeoff parameters specified for crown*

- Right-click in the empty area to check the preview of extrude feature. If the feature is created as desired then right-click again to create the feature.

## Project To Sketch

The **Project To Sketch** tool is used when you need the projection of any face, edge, or sketch entity in another sketch. In this way, you can create the sketch entities from the projection of other features. The tool is available in the **Draw** group. The procedure is discussed next.

*   Click on **Project To Sketch** from the **Draw** panel in the **Home** tab of **Ribbon** after activating sketch. The **Project to Sketch Options** dialog box will be displayed; refer to Figure-19.

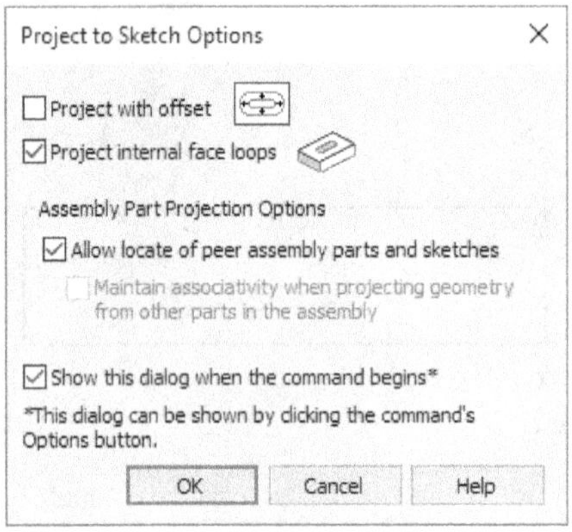

*Figure-19. Project To Sketch Options dialog box*

*   Select the **Project with offset** check box to specify offset distance after projecting edges/faces/other sketch entities.
*   Select the **Project internal face loops** check box to project internal loops as well when a face is selected for projection.
*   Select the **Allow location of peer assembly parts and sketches** check box to include entities of peer assemblies as well for projection.
*   Select the **Show this dialog when the command begins** check box to display the current dialog box when **Project to Sketch** tool is activated.
*   After setting desired parameters, click on the **OK** button. The **Project to Sketch Command Bar** will be displayed; refer to Figure-20.

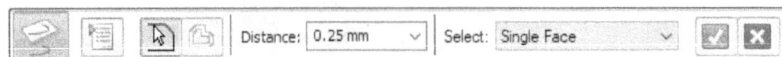

*Figure-20. Project To Sketch Command Bar*

*   Set the desired offset distance in the **Distance** edit box if needed.
*   Select the desired option from the **Select** drop-down to define selection filter. For example, if you want to select complete loops when clicking on one entity of loop then select the **Loop** option from the drop-down.
*   After setting desired parameters, select the desired edges/faces/curves external to current sketch. After selection, right-click or press **ENTER** to create projection; refer to Figure-21. Press **ESC** to exit the tool.

*Figure-21. Project To Sketch Creation*

## Copying or Moving Sketch Elements to New Sketch

The **Tear-Off** tool is used to copy/move sketch elements from one sketch to a new sketch. The procedure to use this tool is given next.

- Click on the **Tear-Off** tool from the **Sketch** panel in the **Home** tab of **Ribbon**. You will be asked to select the plane/face on which you want to copy/move the other sketch elements.
- Select the desired face/plane where you want to create new sketch. You will be asked to select sketch elements to be copies/moved and the **Tear-Off Sketch Command Bar** will be displayed; refer to Figure-22.

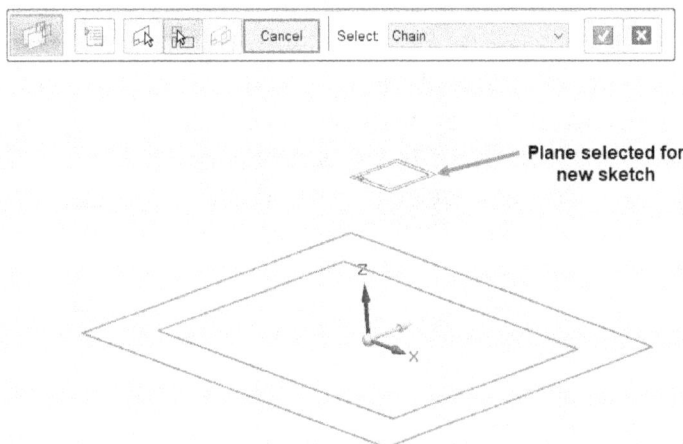

*Figure-22. Tear-Off Sketch Command Bar*

- Select the entities to be copied/moved. Preview of copying will be displayed.
- Click on the **Tear-Off Sketch Options** button from the **Command Bar**. The **Tear-Off Sketch Options** dialog box will be displayed; refer to Figure-23.

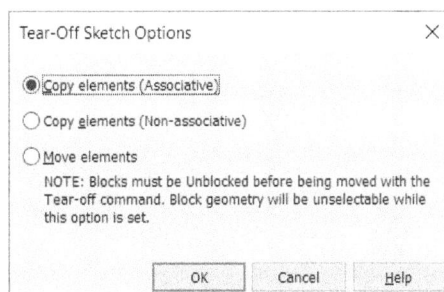

*Figure-23. Tear-Off Sketch Options dialog box*

- Select the **Copy elements (Associative)** radio button if you want the copied elements to be modified automatically if original sketch elements are modified. Select the **Copy elements (Non-associative)** radio butotn if you want the copied elements to be not linked with original elements after copying. Select the **Move elements** radio button if you want the original elements to move at new plane selected for sketch. Note that you need to break the blocks before copying/moving by this method. (You will learn about blocks later). After setting parameters, click on the **OK** button from the **Tear-Off Sketch Options** dialog box.
- Right-click in the empty area to perform copy/move of elements. A new sketch will be created and you will be asked to specify name of the sketch in **Command Bar**.
- Specify the desired name in the **Name** edit box and click on the **Finish** button from the **Command Bar**.
- Press **ESC** to exit the tool.

## CUT

The **Cut** tool is used to remove material from model. It can create simple hole and other cuts using the close loop sketch sections. This tool is available in the **Solids** panel of **Ribbon**. The procedure to create a cut is discussed next.

- Click on the **Cut** tool from the **Solids** panel in the **Home** tab of the **Ribbon**. The **Cut Command Bar** will be displayed; refer to Figure-24.

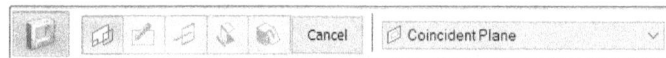

*Figure-24. Cut Command Bar*

- Select plane or features face, create a sketch which you want to use for removing material. Note that if you are using plane to create sketch then the plane should be parallel to the face through which you want to cut the material.
- After creating sketch, click on the **Close Sketch** button from the **Close** panel in the **Home** tab of **Ribbon**. Preview of cut feature will be displayed with extent of feature attached to cursor and the **Cut Command Bar** will be displayed as shown in Figure-25.

*Figure-25. Cut Command Bar*

- Set the desired parameters in the **Command Bar**. The feature preview will be displayed; refer to Figure-26.

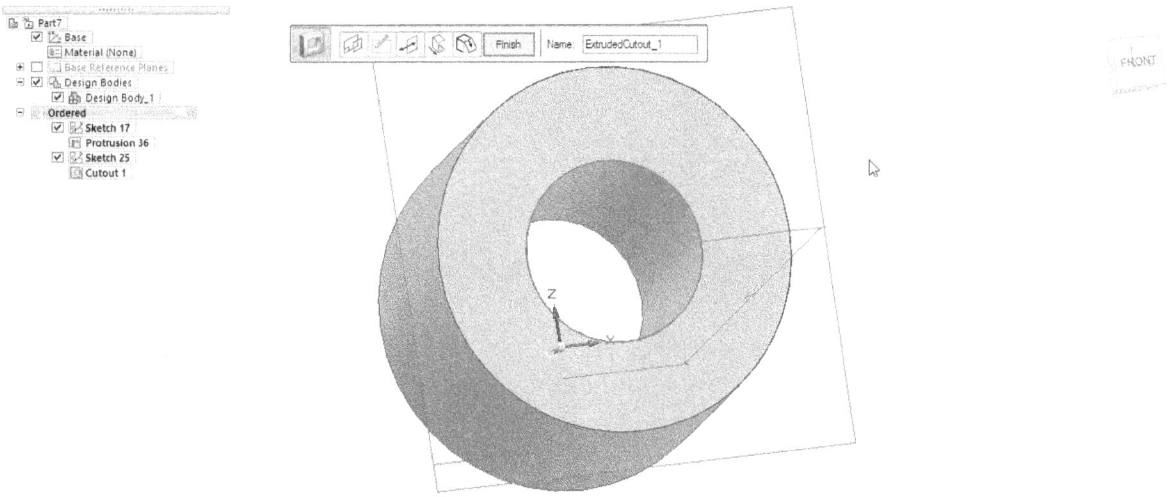

*Figure-26. Preview of cut*

Various option in the **Command Bar** for Extent are discussed next.

- **Through all Cut** 🔲 :- If this option is chosen then the cut is created through all part. Refer to Figure-27.

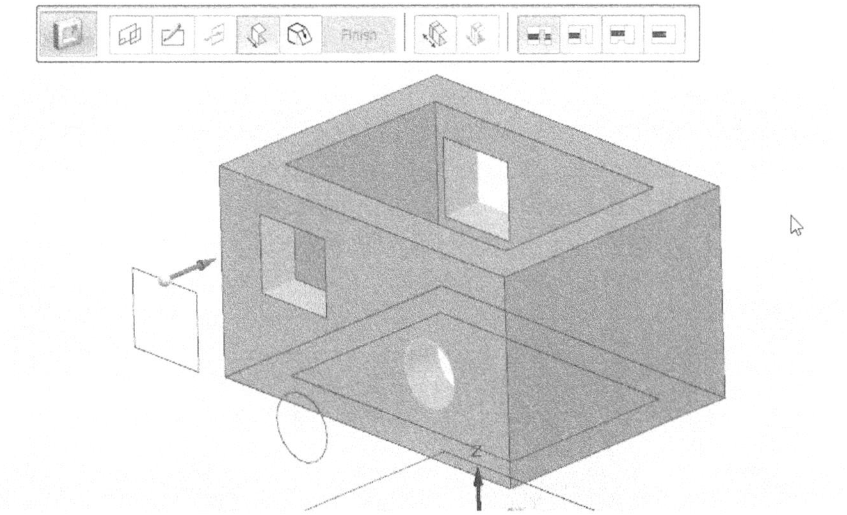

*Figure-27. Cut Through All*

- **Through Next** 🔲 :- If this option is selected then the cut is made through next face of the model; refer to Figure-28.

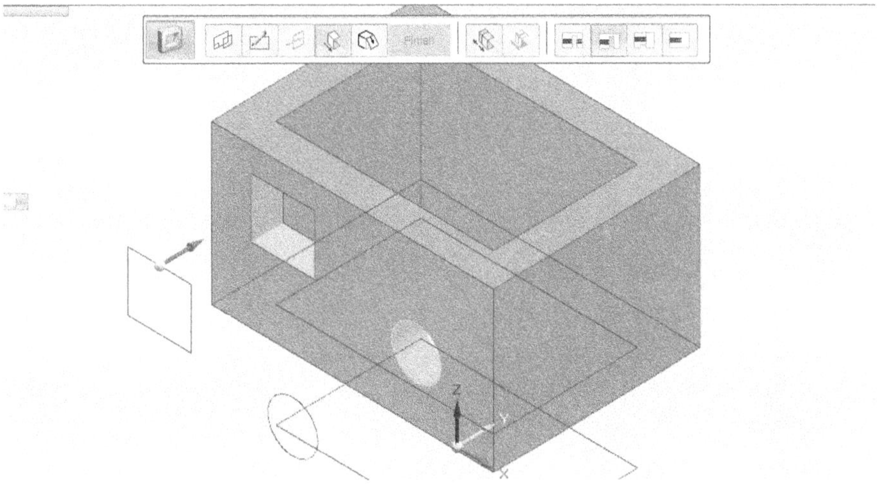

*Figure-28. Cut Through Next*

- **Cut - Finite Extent** ▣ :- If this option is selected then you need to define the distance upto which the cut will be created in the **Distance** edit box; refer to Figure-29.

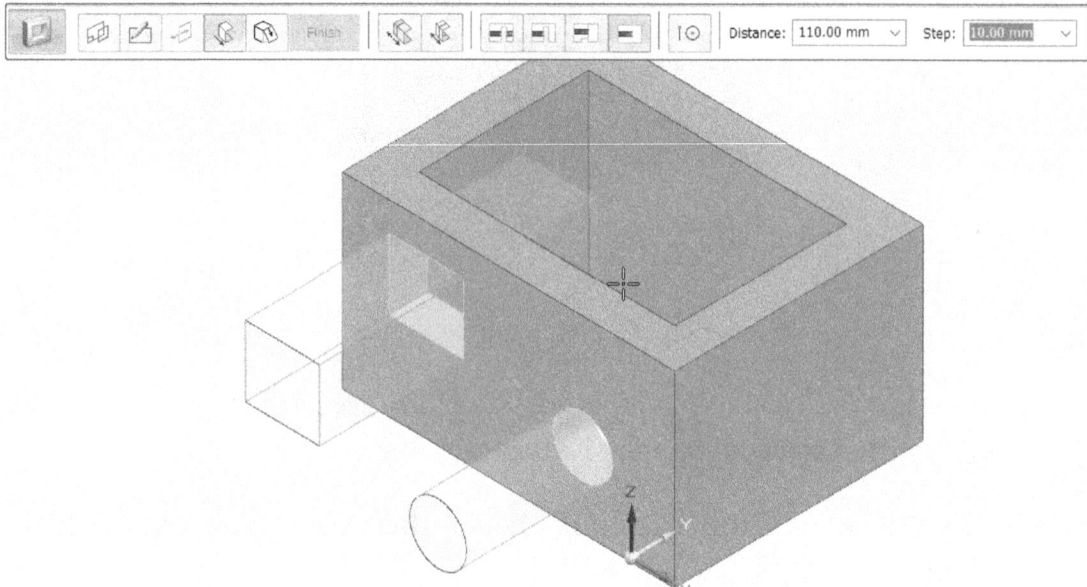

*Figure-29. Cut Finite Extent*

- **Cut - From/To Extent** ▣ :- Select this option if you want to define the start point and end point of cut feature by selecting faces, planes, or other references. Note that the selected references should be in parallel planes for defining extents. After selecting the **From/To Extent** button, select the start reference and then end reference. The preview of feature will be displayed; refer to Figure-30.

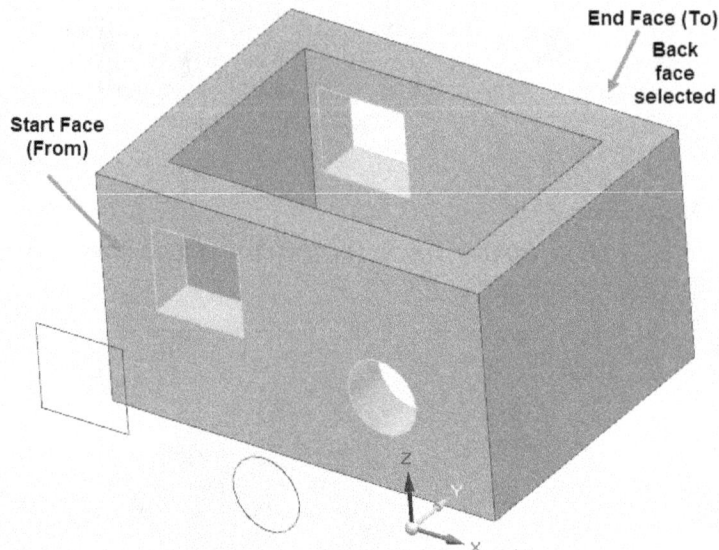

*Figure-30. Cut From/To Extent*

The other options in the **Command Manager** have been discussed earlier.

Before we further discuss the 3D modeling tools, it is important to understand the use of reference planes and Coordinate System in 3D Modeling.

# CREATING PLANES AND COORDINATE SYSTEM

The tools in **Planes** panel are used to create planes and coordinate system; refer to Figure-31. Various tools in this panel are discussed next.

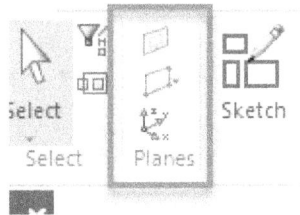

*Figure-31. Planes tool*

## Creating Coincident Plane

The **Coincident Plane** tool is used to create plane coincident with selected face or reference plane. The procedure to use this tool is given next.

• Click on the **Coincident Plane** tool from the **Planes** panel of **Home** tab in the **Ribbon**. You will be asked to select face/reference plane at which new plane will be created.
• Select the desired face/plane. The plane will be created and move handles will be displayed; refer to Figure-32.

*Figure-32. Creating coincident plane*

• Using the handles, you can translate or rotate the plane as need. After changing orientation and position of plane, click in the empty area of screen. The plane will be created.

## Creating Parallel Plane

The **Parallel** tool in **More Planes** drop-down is used to create plane parallel to selected face or plane at specified distance. The procedure to use this tool is given next.

• Click on the **Parallel** tool from the **More Planes** drop-down in the **Planes** panel of **Home** tab in the **Ribbon** in **Ordered** environment; refer to Figure-33. You will be asked to select a planar face or reference plane.

- Select the desired planar face/plane. Preview of plane will be displayed with **Parallel Command Bar**; refer to Figure-34.

*Figure-33. More Planes drop-down*

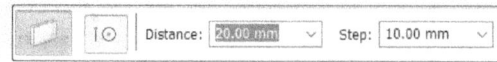

*Figure-34. Preview of parallel plane*

- Specify the desired value of distance in the **Distance** edit box of **Command Bar** and press **ENTER**. You will be asked to define side on which plane will be created with respect to selected plane/face.
- Click on the desired side to create the plane.

## Creating Plane at Angle

The **Angled** tool in **More Planes** drop-down is used to create plane at an angle to selected face/reference plane. The procedure to use this tool is given next.

- Click on the **Angled** tool from the **More Planes** drop-down in the **Planes** panel of **Home** tab in the **Ribbon**. You will be asked to select a reference plane/face.
- Select the desired face/plane with respect to which angle will be specified for new plane. You will be asked to select face/edge/plane to used as base for new plane.
- Select the edge (recommended), face, or plane intersecting with previous selected face/plane. You will be asked to click near the end point of selected edge to define orientation of plane.
- Click at desired side of selected edge. The plane will get attached to cursor and you will be asked to specify the value of angle; refer to Figure-35.
- Specify the desired value in **Angle** edit box of **Command Bar** and press **ENTER**. You will be asked to specify the side at which you want to create the plane.
- Click on the desired side of selected face to create plane above or below the selected face.

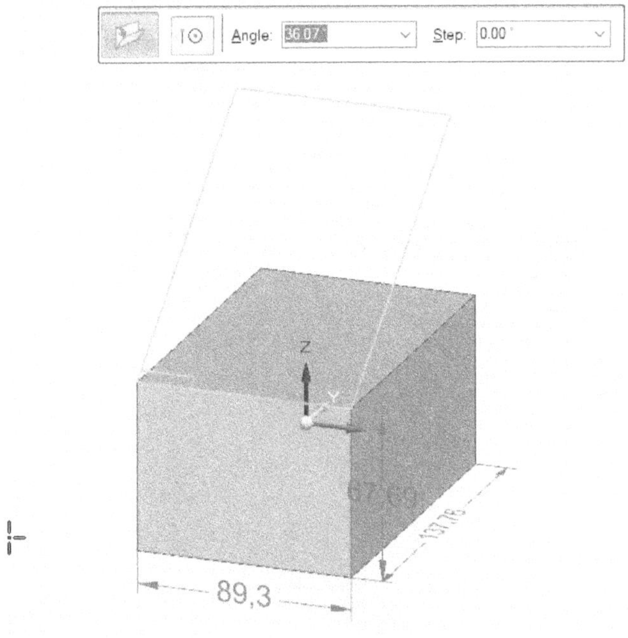

*Figure-35. Specifying angle for plane*

## Creating Perpendicular Plane

The **Perpendicular** tool in **More Planes** drop-down is used to create plane perpendicular to selected planar face/reference plane. The procedure to use this tool is given next.

- Click on the **Perpendicular** tool from the **More Planes** drop-down in the **Planes** panel of **Ribbon**. You will be asked to select a plane/planar face.
- Select the desired planar face/plane. You will be asked to select the base of plane which can be face, edge or plane.
- Select the desired plane/edge/face. Rest of the procedure is same as discussed for plane at angle.

Similarly, you can use the other tools in **More Planes** drop-down. The **Coincident by Axis** tool is used to create plane coincident to selected intersecting planes or plane and edge. The **Normal to Curve** tool in **More Planes** drop-down is create plane perpendicular to selected curve or edge. The **By 3 Points** tool is used to create a plane using points; refer to Figure-36. The **Tangent** tool in **More Planes** drop-down is used to create plane tangent to selected curved face.

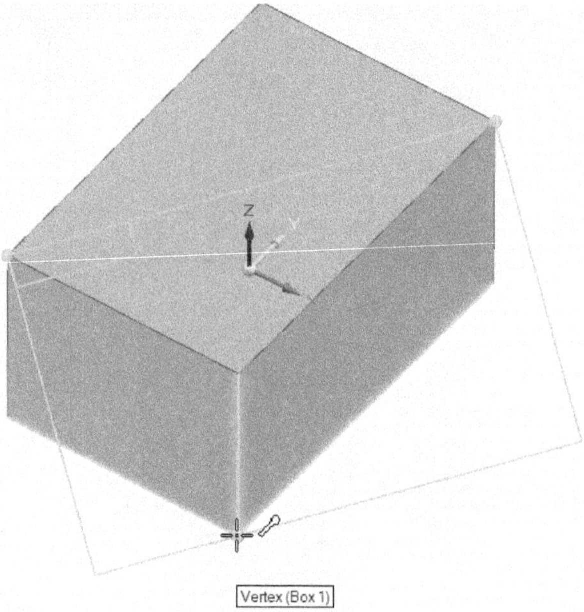

*Figure-36. Creating plane using 3 points*

## Creating Coordinate System (Ordered Environment)

The **Coordinate System** tool is used to create custom coordinate system at desired location. The procedure to use this tool is given next.

*   Click on the **Coordinate System** tool from the **Planes** panel in the **Home** tab of **Ribbon**. The **Coordinate System Options** dialog box will be displayed; refer to Figure-37.

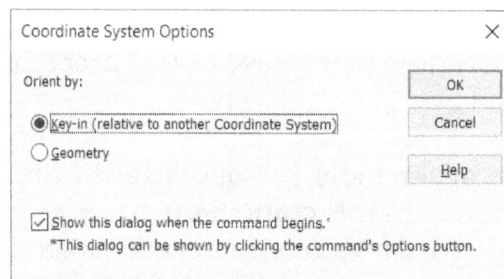

*Figure-37. Coordinate System Options dialog box*

*   Select the desired option from the dialog box to specify how the coordinate system's location and orientation will be defined. Select the **Key-in** radio button to create coordinate system in reference to another coordinate system. Select the **Geometry** radio button if you want to place the coordinate system at desired point and specify its orientation. We will discuss the **Geometry** option here, you can apply it to **Key-in** option as well.
*   Select the **Geometry** radio button from the dialog box and click on the **OK** button. The **Coordinate System Command Bar** will be displayed; refer to Figure-38 and you will be asked to specify the key point of placing coordinate system.

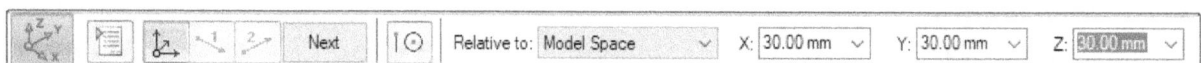

*Figure-38. Coordinate System Command Bar*

- Click at desired location or specify the coordinates in X, Y and Z edit boxes of **Command Bar** to place the coordinate system. You will be asked to define the direction references.
- Select the desired edge to define direction of X axis in coordinate system and press **ENTER**. An arrow will be displayed attached to cursor denoting the direction of X axis.
- Click on the desired side of selected location to define the X axis direction. You will be asked to specify the Y axis direction. Set the Y axis direction in the same way.
- On specifying the direction, preview of coordinate system will be displayed; refer to Figure-39.
- Click on the **Finish** button to create the coordinate system and press **ESC** to exit the tool.

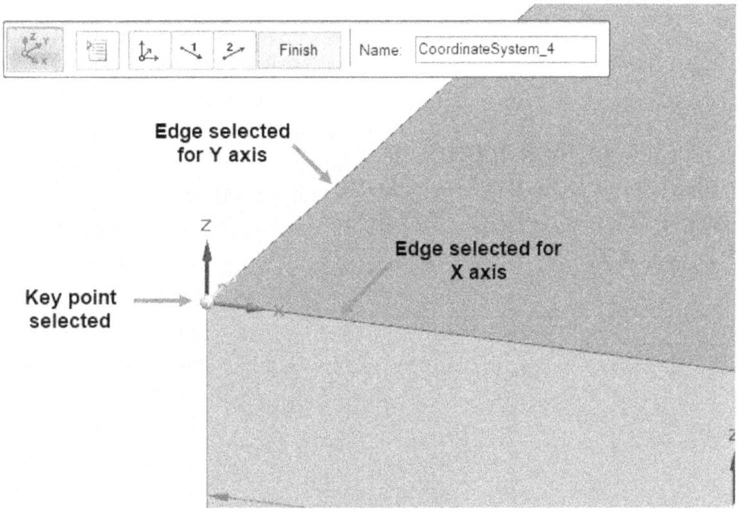

*Figure-39. Preview of Custom Coordinate System*

# REVOLVE

The **Revolve** tool is used to create revolve protrusion features. This tool is available in the **Solids** panel of **Home** tab in the **Ribbon**. The procedure to create a revolve is discussed next.

- Click on the **Revolve** tool from the **Solid** panel in the **Home** tab of **Ribbon**. The **Revolve Command Bar** will be displayed; refer to Figure-40 and you will be asked to select a plane for creating base sketch.

*Figure-40. Revolve Command Bar*

- If you have a sketch earlier created then you can use it as discussed earlier or you can select a plane and create the sketch; refer to Figure-41.

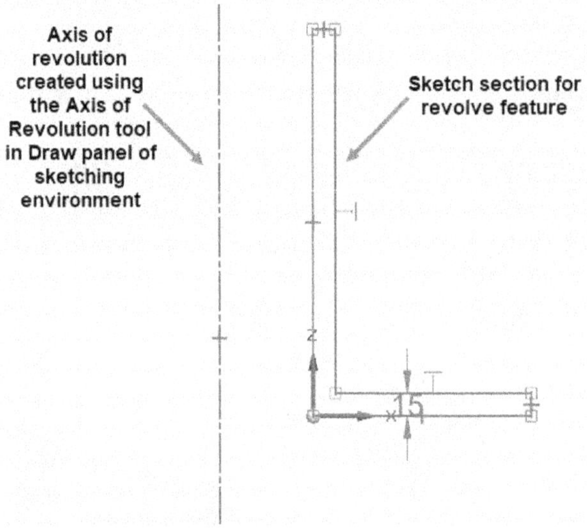

Axis of
revolution
created using
the Axis of
Revolution tool
in Draw panel of
sketching
environment

Sketch section for
revolve feature

*Figure-41. Sketch created for revolve feature*

- After creating sketch, click on the **Close Sketch** tool from the **Close** panel in the **Home** tab of **Ribbon**. The preview of feature will be displayed with **Revolve Command Bar**; refer to Figure-42.

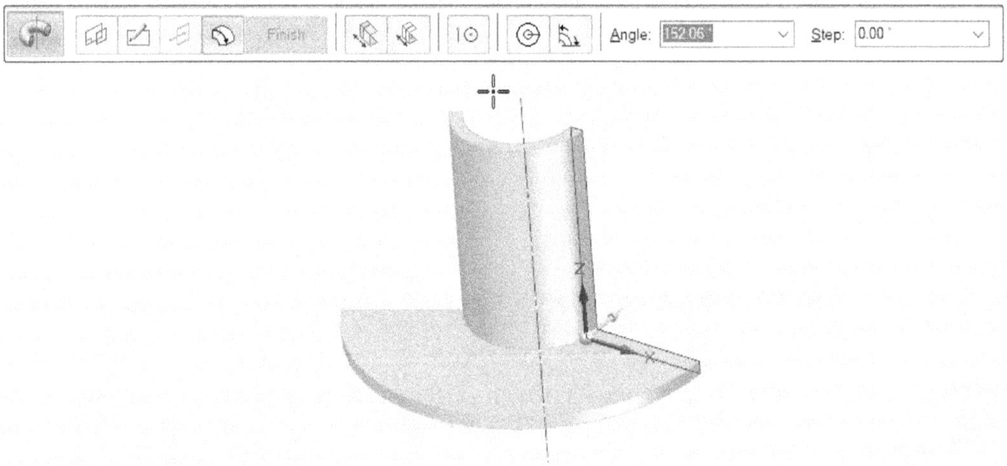

*Figure-42. Revolve Command Bar*

- If you want to create a full 360 revolve feature then select the **Revolve 360** button from the **Command Bar**. If you want to specify the angle span for feature then select the **Finite Extent** button and specify the value in **Angle** edit box of **Commmand Bar**; refer to Figure-43.

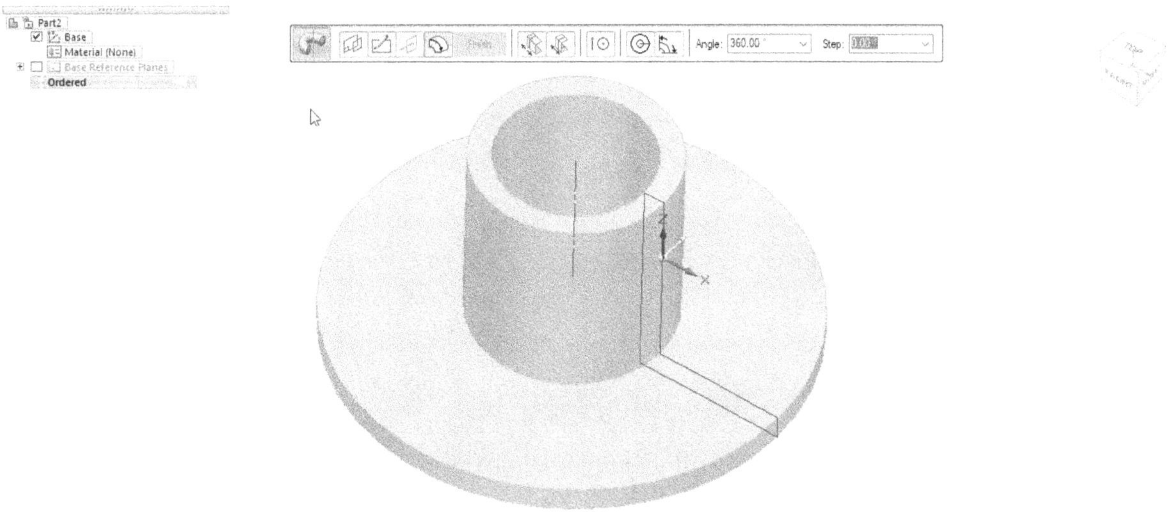

*Figure-43. Revolve Creation*

- After setting desired parameters, click on the **Finish** button from the **Command Bar** or right-click in empty area to create the feature. Press **ESC** to exit the tool.

## REVOLVE CUT

The **Revolve Cut** tool is used to remove material from the model by revolving close loop sketch section about an axis. This tool is available in the **Solids** panel of **Ribbon**. The procedure is same as we have discussed for the **Revolve** tool. The difference is that in case of revolve, we add material and in case of revolve cut, we remove material from the model. Refer to Figure-44.

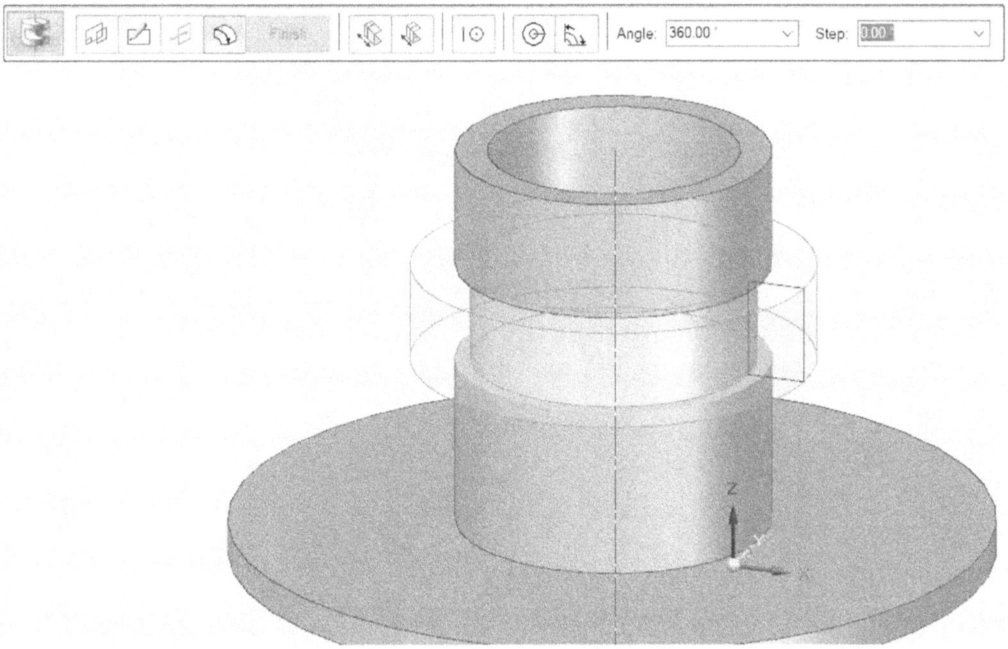

*Figure-44. Revolve Cut Creation*

## HOLE TOOLS

There are three tools available in the **Hole** drop-down: **Hole**, **Thread**, and **Slot**; refer to Figure-41. These tools are discussed next.

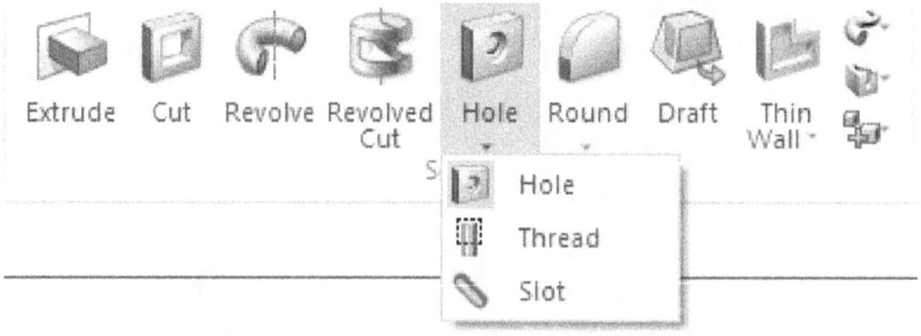

*Figure-45. Hole drop-down*

# HOLE

The **Hole** tool is used to create holes that comply with the real machining tools. Solid Edge has library of standard holes that can be created in the solid model. You can use this standard library or you can create a customized hole by using the tool. The procedure to use this tool is given next.

- Click on **Hole** tool from **Hole** drop-down in the **Solids** panel of **Home** tab in the **Ribbon**. The **Hole Command Bar** will be displayed; refer to Figure-46 and you will be asked to select plane/face for creating hole.

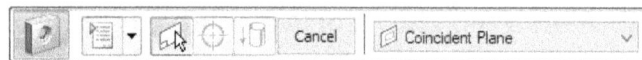

*Figure-46. Hole Command Bar*

- Select the desired face/plane on which you want to create the hole. The sketching environment will be displayed along with **Hole Circle Command Bar**; refer to Figure-47.

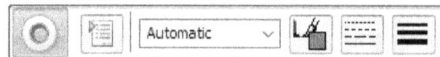

*Figure-47. Hole Circle Command Bar*

- **Hole Options** [icon] :- This button is used to select the type of hole. On clicking this button, the **Hole Options** dialog box will be displayed; refer to Figure-48. Using the options in this dialog box, you can define the shape and size of the holes. You can create five types of holes using the options in this dialog box. The methods to define properties of these holes are discussed next.

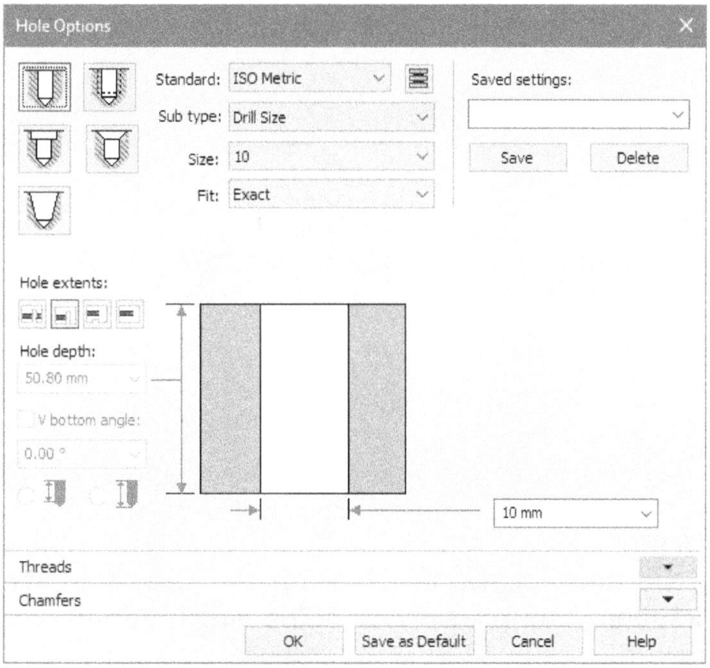

*Figure-48. Hole Option dialog box*

## Creating Simple Hole

- Select the **Simple Hole** button from the dialog box. The dialog box will be displayed; refer to Figure-48.
- Select the desired option from the **Standard** drop-down to define database standard of holes.
- Select the desired option from the **Sub type** drop-down to define how hole size will be selected. There are three options in this drop-down viz. **Drill Size**, **Dowel**, and **General Screw Clearance**.
- Select the desired option from the **Size** and **Fit** drop-downs to define size of holes.
- Select the desired button from the **Hole extents** area to define the depth of holes. If you have selected the **Finite Extent** button then you can create V type bottom of holes by selecting the **V bottom angle** check box. After selecting the check box, specify the desired angle value in the edit box below it.
- Specify the desired diameter and depth of hole in the edit boxes displayed.
- Expand the **Chamfers** section of the dialog box and select the desired check boxes to create start and end chamfers on the hole. Preview of chamfer will be displayed in the preview area of the dialog box.
- After setting desired parameters in the dialog box, click on the **OK** button from the dialog box. You will be asked to specify the location of centers of holes.
- Click at desired locations to specify the hole positions; refer to Figure-49.

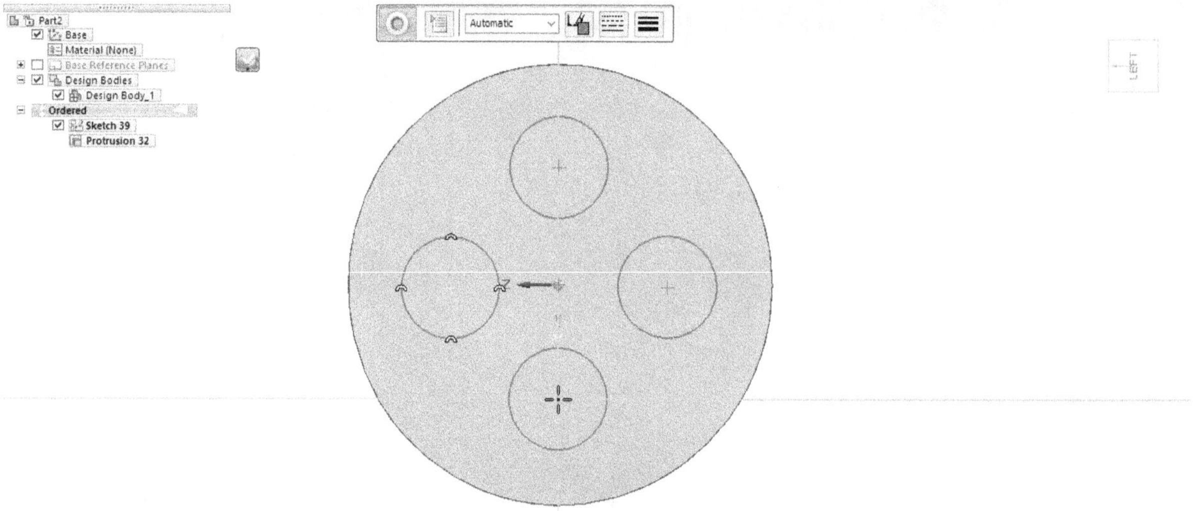

*Figure-49. Selecting Hole Location*

- After specifying the hole locations, click on the **Close Sketch** button from the **Close** panel of **Ribbon**. The preview of holes will be displayed with the **Command Bar**; refer to Figure-50.

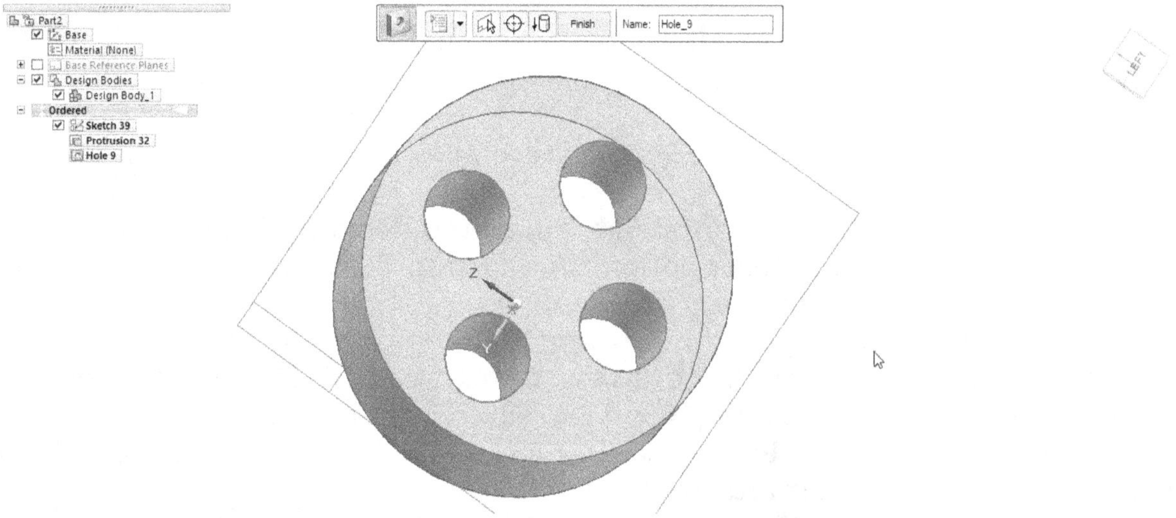

*Figure-50. Hole Creation*

## Creating Threaded Hole

- Select the **Threaded** button from the dialog box. The options in the dialog box will be displayed as shown in Figure-51.
- Set the desired standard and sub-type for defining thread standard and then select the size of hole with related pitch. Preview of hole will be displayed in the dialog box.
- Expand the **Threads** section of the dialog box and select the desired radio button to define the manufacturing parameter for creating hole on machine. Set the desired diameter value in the respective edit box.
- Specify the value of pitch and thread depth in respective edit boxes of the **Threads** section in the dialog box.
- Rest of the options in this dialog box are same as discussed for standard hole. Click on the **OK** button and specify location for theaded holes as discussed earlier.

## Creating Counterbore Holes

- Select the **Counterbore** button from the dialog box. The options in dialog box will be displayed as shown in Figure-52.

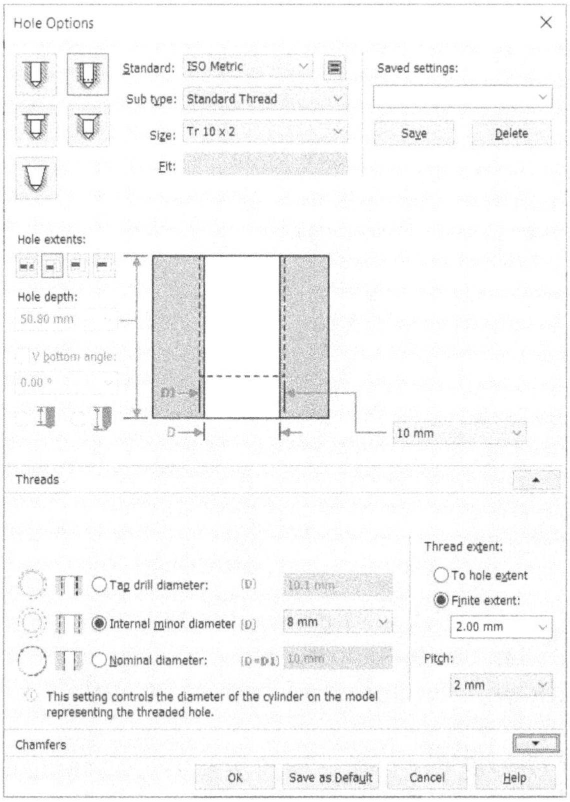

*Figure-51. Hole Options dialog box for threaded hole*

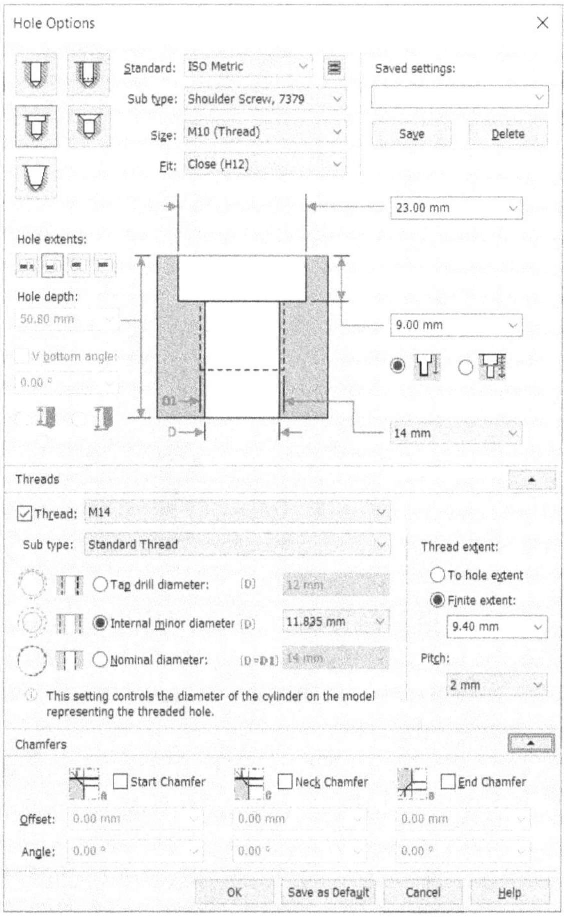

*Figure-52. Hole Options dialog box for counterbore hole*

- The options in this dialog box have been discussed earlier. After specifying parameters of hole, click on the **OK** button from the dialog box and place the hole centers at desired locations as discussed earlier.

Similarly, you can create Countersink and Taper holes.

## Thread

The **Thread** tool is used to create thread in existing hole or on a cylindrical surface. This tool is available in the **Hole** drop-down of **Solids** panel in the **Ribbon**. The procedure to create thread using this tool is discussed next.

- Select **Thread** from **Hole** drop-down in **Solids** panel of **Home** tab in the **Ribbon**. The **Thread Options** dialog box will be displayed; refer to Figure-53. Select the desired option from the **Standard** and **Type** drop-downs to specify type of threads.

*Figure-53. Thread Options dialog box*

- Select the desired radio button from the **Internal Thread (Holes)** area of the dialog box to specify diameter of internal threads. If you are creating external threads then nominal diameter of threads will be selected automatically based on shaft diameters.
- After setting desired parameters, click on the **OK** button. The **Thread Command Bar** will be displayed; refer to Figure-54 and you will be asked to select cylindrical surface.
- Select the cylindrical surface and then select the bottom edge of cylinder. The **Command Bar** will be displayed as shown in Figure-55. Specify the desired parameters in the **Command Bar** and click on the **Finish** button. The threads will be created; refer to Figure-56.

*Figure-54. Select Cylinder Step Command Bar*

*Figure-55. Thread Command Bar*

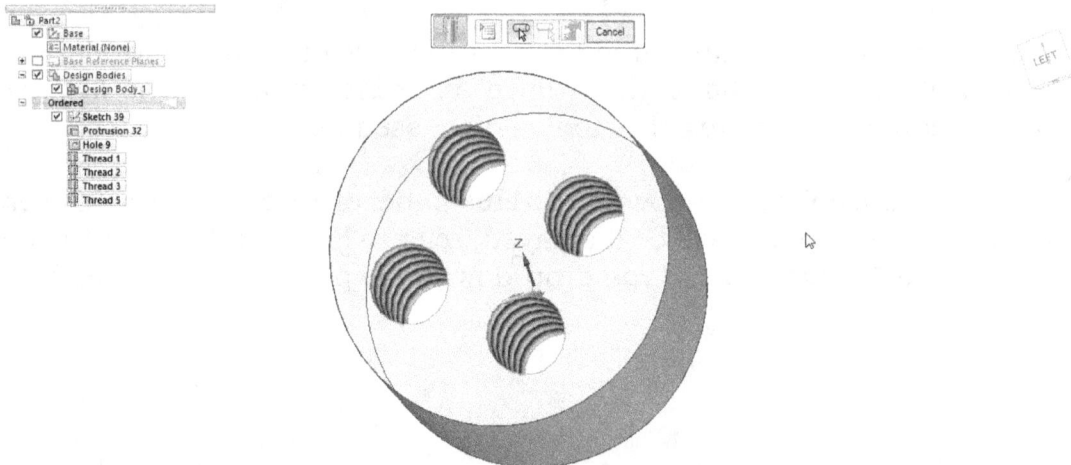

*Figure-56. Thread Creation*

- Press **ESC** to exit the tool.

# SLOT

The **Slot** tool is used to create slots using tangent continuous sketch elements. This tool is available in the **Hole** drop-down list of **Solids** panel in the **Ribbon**; refer to Figure-30. The procedure to use this tool is discussed next.

- Click on the **Slot** tool from the **Hole** drop-down in the **Solids** panel of the **Home** tab in the **Ribbon**. The **Slot Command Bar** will be displayed; refer to Figure-57.

*Figure-57. Slot Command Bar*

- Click on the **Slot Options** button from the **Slot Command Bar** to define slot width, path offset, depth offset, and so on. The **Slot Options** dialog box will be displayed; refer to Figure-58.
- Select the **Counterbore** check box if you want to create counterbore slot of specified parameters. You can create two types of counterbore slots; **Recessed** and **Raised** by selecting respective radio button below the selected check box.
- Specify the desired value of slot width in the **Slot width** edit box. If you have opted to create counterbore slot then specify the desired values in **Path offset** and **Depth offset** edit boxes.
- By default, the **Flat end** radio button is selected so rectangular shaped cuts are created in the slot. If you want to create slot with semi-circular shaped ends then select the **Arc end** radio button.
- After setting desired parameter in the dialog box, click on the **OK** button. You will be asked to select tangent continuous sketch elements.

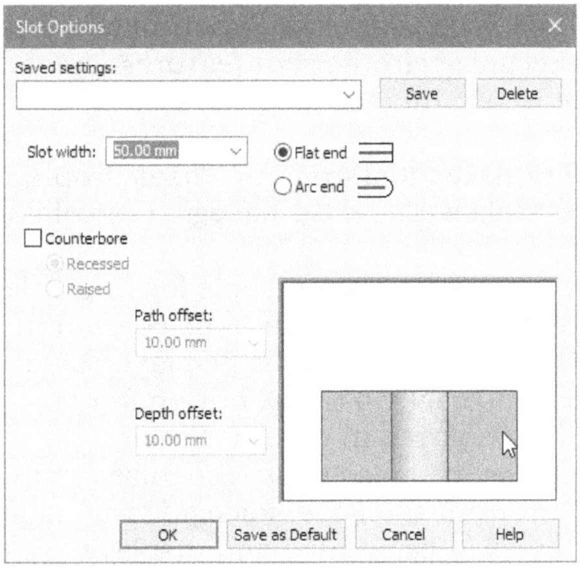

*Figure-58. Slot Options Dialog Box*

- Select the sketch elements if you want to use earlier created elements for slot creation or select the desired plane or face on which you want to create sketch for slot.
- After selecting plane/face, create a sketch on that plane and click on **Close Sketch** button. Preview of the slot will be displayed with **Command Bar** as shown in Figure-59.

- Specify the desired value of depth in the **Distance** edit box or set desired extent as discussed earlier. After specifying the value, click in the drawing area to create the slot.
- Click on the **Finish** button from the **Command Bar** to create the feature. Press **ESC** to exit the tool.

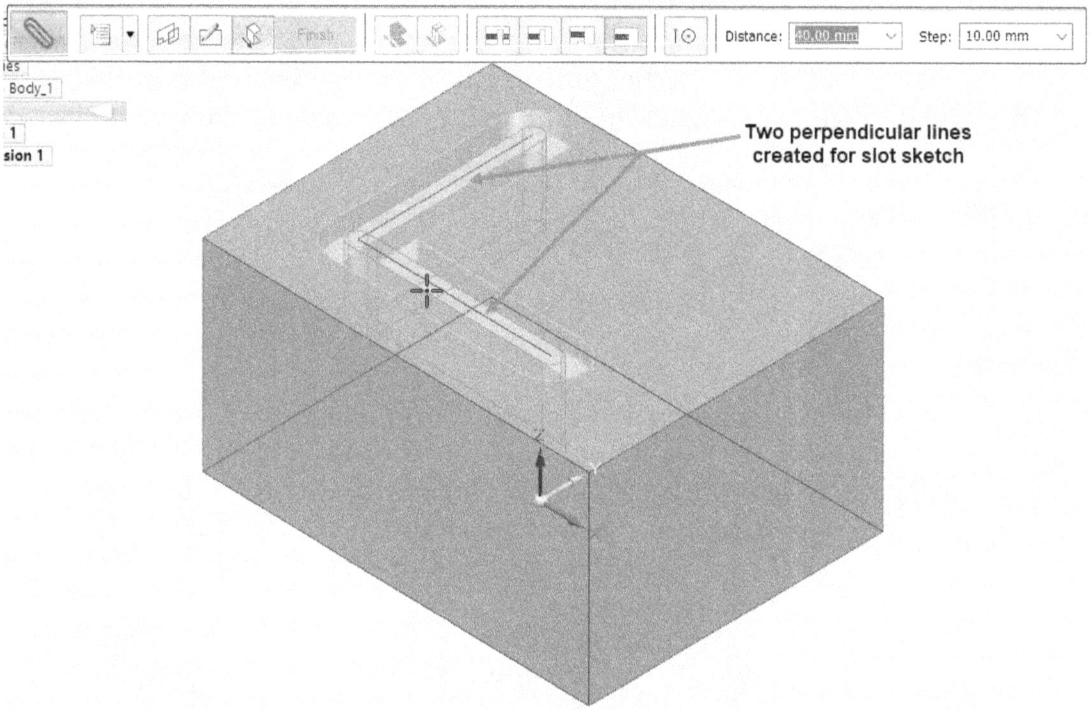

*Figure-59. Slot Creation*

## Recognizing Holes

The **Recognize Holes** tool is available in Synchronous environment. This tool is used to automatically recognize circular cut features (created by sketch) as holes if possible. The procedure to use this tool is given next.

- Click on the **Recognize Holes** tool from the **Holes** drop-down in the **Solids** panel of the **Home** tab in the **Ribbon**. The **Hole Recognition** dialog box will be displayed; refer to Figure-60.
- If you want to change parameter for any hole for your reference then click on the button from the **Select Alternate Hole Type** column in the table for respective hole. The **Hole Options** dialog box will be displayed as discussed earlier. Set the parameters as desired and click on the **OK** button from the dialog box. The preview of selected hole will be displayed in different color.
- Select the check boxes for the holes to be recognized from the **Recognize** column.
- Click on the **OK** button from the **Hole Recognition** dialog box to recognize the holes. The hole features will be created and displayed in the **Path Finder**.

Figure-60. Hole Recognition dialog box

# ROUND

The **Round** tool is used to apply fillets at sharp edges or face of solid model. **Round** tool is available in the **Solids** panel. The function of this tool is similar to fillet in sketch. The difference is that here we are creating round in solid model. The procedure to use this tool is discuss next.

- Click on the **Round** tool from the **Solids** panel in the **Home** tab of **Ribbon**. The **Round Command Bar** will be displayed; refer to Figure-61.

Figure-61. Round Command Bar

- Click on the **Round Options** button from the **Round Command Bar** to define the shape and type of fillet. The **Round Options** dialog box will be displayed; refer to Figure-62.

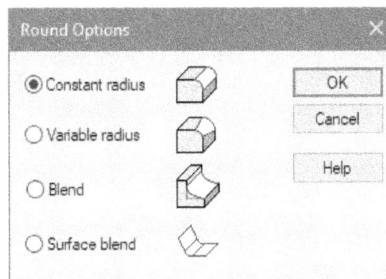

Figure-62. Round Options dialog box

## Creating Constant Radius Round

- Select the **Constant radius** radio button if you want to create fixed radius value fillet and click on the **OK** button from the dialog box.
- Click on the **Round Parameters** button from the **Round Command Bar** to define how round will be created on corners and tangent edges. The **Round Parameters** dialog box will be displayed; refer to Figure-63.

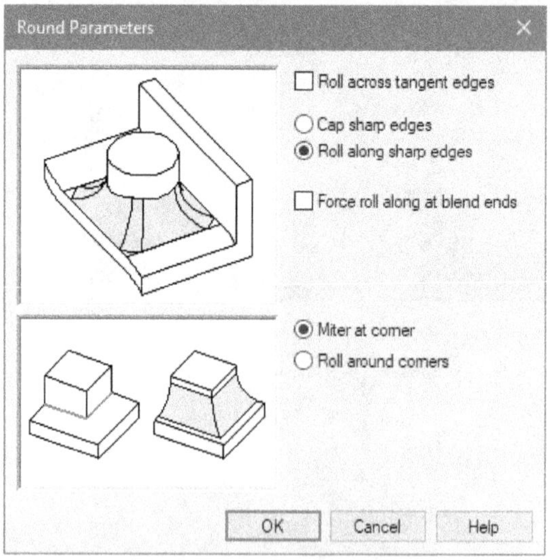

*Figure-63. Round Parameters dialog box*

- Select the **Roll across tangent edges** check box to create tangential edges of the fillet at the base.
- Select the **Cap sharp edges** check box to make fillets without sharp edges. Note that you should select **Curvature Continuous** option from **Command Bar** before selecting this check box to get best finish. Select the **Roll along sharp edges** check box to round sharp edges using a roller ball of size equal to round radius. Select the **Force roll along at blend edges** check box to round sharp edges of the blend ends.
- Select the **Miter at corner** radio button to remove extra fillet material at the corners while creating the feature. Select the **Roll around corners** radio button to create round base at the corner.
- Set the desired option in the dialog box and click on the **OK** button from the dialog box.
- Select the desired option from the **Shape** drop-down to define how the surface continuity will be followed. Select the **Tangent Continuous** option if you want to create rounds with edges tangent to connected surfaces. Select the **Curvature Continuous** option from the drop-down if you want the round to follow curvature of connected faces/surfaces.
- Select the desired option from the **Select** drop-down to set selection filter for selecting entitites to create round; refer to Figure-64.

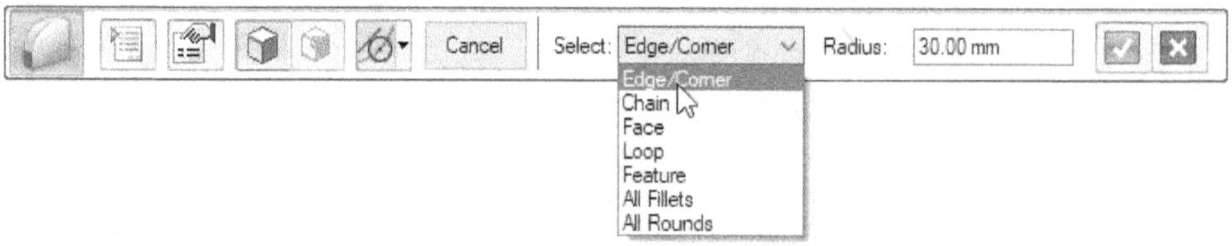

*Figure-64. Selection Type drop down*

- Select the desired edges/faces/features to be used for creating round. Preview of round will be displayed; refer to Figure-65.
- Specify the desired value of radius in the **Radius** edit box and click on the **OK** button from the **Command Bar**.

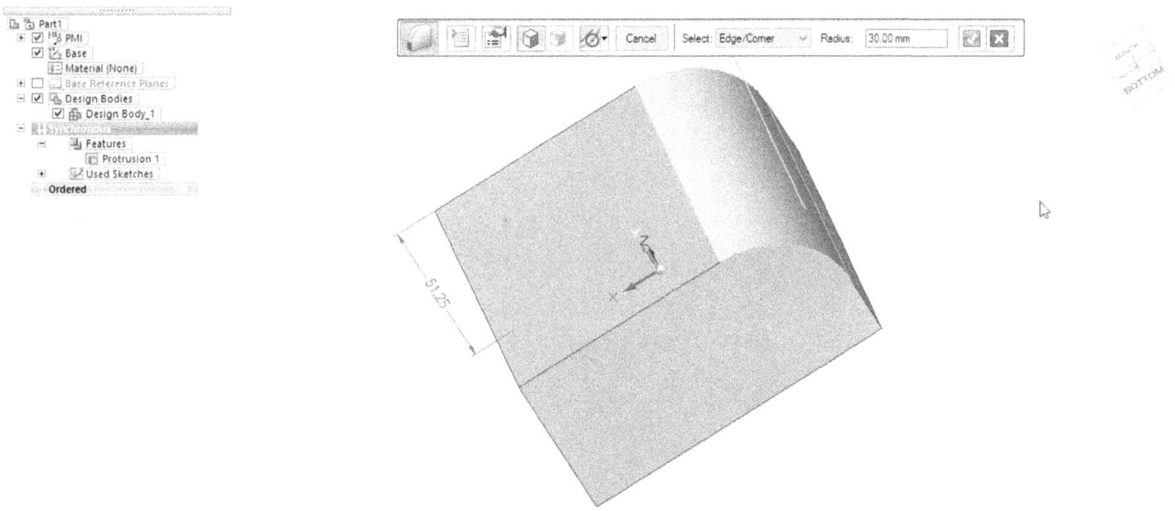

*Figure-65. Round Created*

## Creating Variable Radius Round

- Select the **Variable radius** radio button from the **Round Options** dialog box and click on the **OK** button. The **Round Command Bar** will be displayed as shown in Figure-66.

*Figure-66. Round Command Bar for variable radius*

- Select the desired edges on which you want to apply variable radius round and then click on the **OK** button from the **Command Bar**. You will be asked to select a vertex on which you want to specify radius for round.
- Select the desired vertex/point and specify the radius in the **Command Bar**; refer to Figure-67. Click on the **OK** button from the **Command Bar**. You will be asked to select next vertex for specifying radius.

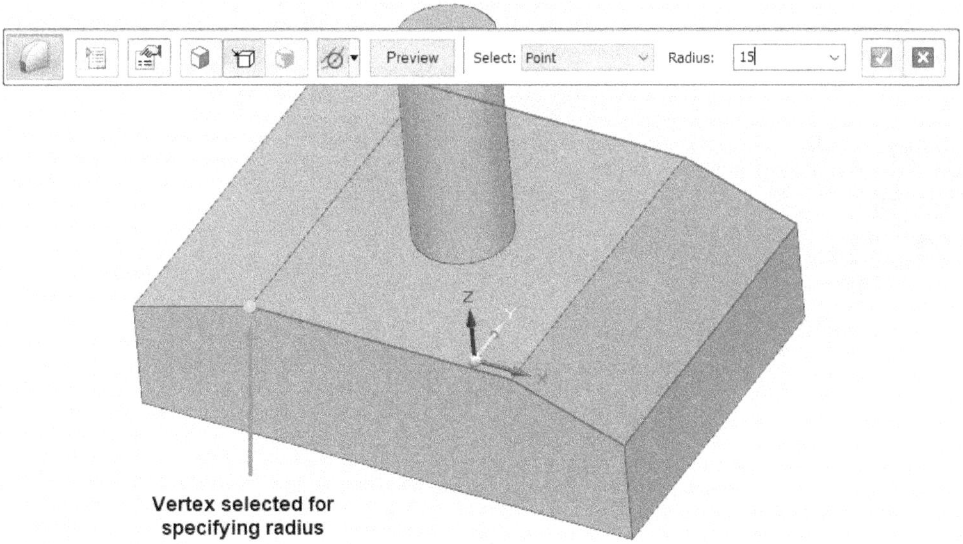

Vertex selected for
specifying radius

*Figure-67. Specifying radius at a vertex*

- Repeat the previous step until you have specified radius at desired vertices of selected edges; refer to Figure-68.
- Click on the **Preview** button to check the preview of round and click on the **Finish** button from the **Command Bar** to create the feature.

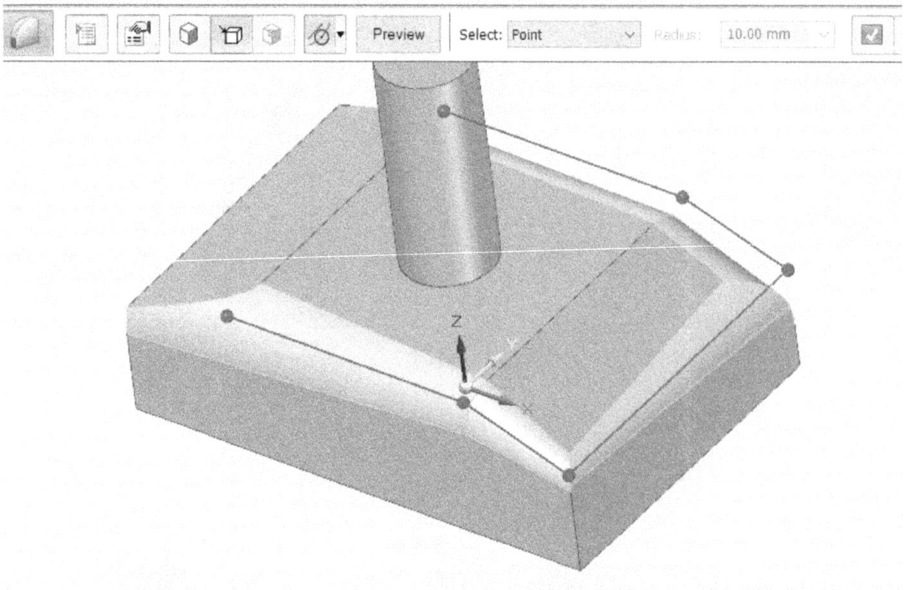

*Figure-68. Variable radius specified on vertices*

## Creating Blend Round

- Select the **Blend** radio button from the **Round Options** dialog box to blend the edge at intersection of two faces/surfaces. Click on the **OK** button from the dialog box. The **Round Command Bar** will be displayed as shown in Figure-69 and you will be asked to select faces/surfaces.

*Figure-69. Round Command Bar*

- Select the two faces which you want to blend using the round. Preview of round feature will be displayed; refer to Figure-70.

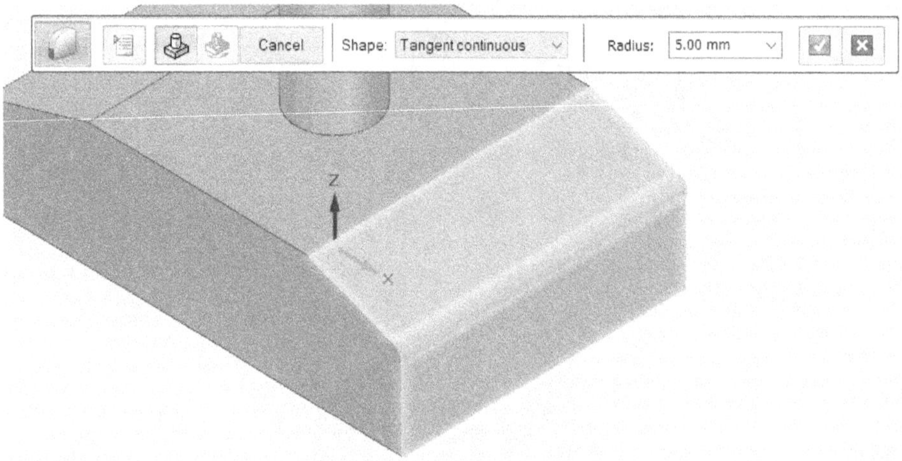

*Figure-70. Preview of blend round*

- Select the desired option from the **Shape** drop-down to define the shape of blend. The **Tangent continuous** and **Curvature continuous** options have been discussed earlier. Select the **Constant width** option to create round of specified width value. Select the **Chamfer** option from the drop-down to create chamfer at blending edge of specified setback. Note that chamfers will be created at 45 degree by default. Select the **Bevel** option from the drop-down to chamfer the edge with different setbacks for both sides. Note that the parameter specified in **Value** edit box defines the ratio of bevel length of both sides. If **1** is specified in the **Value** edit box then both sides will be equal. Select the **Conic** option from the drop-down to create conical shaped blend. The shapes of blend for different options are shown in Figure-71.

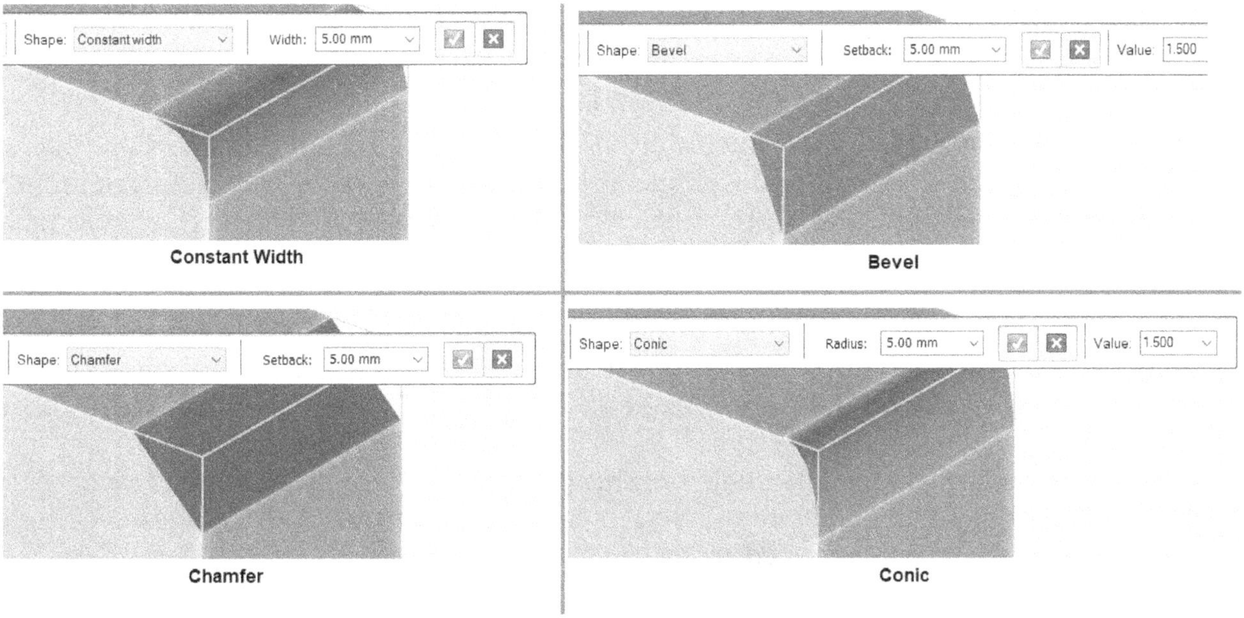

*Figure-71. Shapes for different types of rounds*

- After setting desired parameters, click on the **OK** button to create the round.

## Creating Surface Blend

- Select the **Surface Blend** radio button from the **Round Options** dialog box and click on the **OK** button. You will be asked to select two surfaces for creating surface blend.
- Select the desired surfaces and create the round as discussed earlier for blend rounds. Note that sometimes you need to correct the direction of blend for creating this feature.

## CHAMFER

The **Chamfer** tool is used to bevel the sharp edges of the solid model. This tool works in the same way as the sketch chamfer do. The procedure to use this tool is discussed next.

- Click on **Chamfer** tool from **Round** drop-down in the **Solids** panel of **Home** tab in the **Ribbon**. The **Chamfer Command Bar** will be displayed; refer to Figure-72.

*Figure-72. Chamfer Command Bar*

- Click on the **Chamfer Options** button from the **Command Bar**. The **Chamfer Options** dialog box will be displayed; refer to Figure-73.

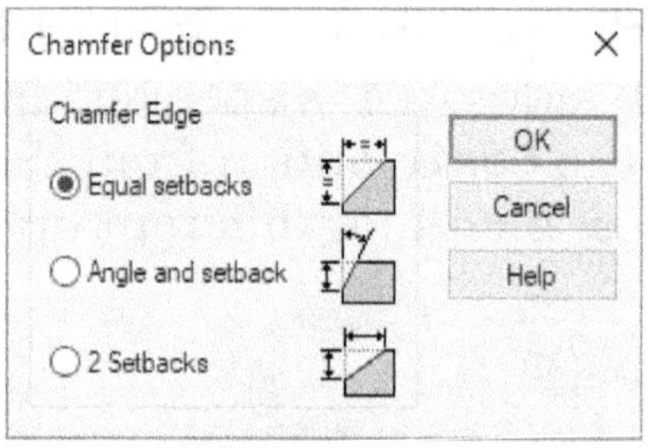

*Figure-73. Chamfer Options*

There are three ways in which you can create chamfer: By specifying equal setbacks on both sides, by specifying chamfer angle and length, or by specifying different length for both sides of chamfer. The procedure to create chamfer by all three options is almost similar. Here, we will use the **Angle and setback** option. You can apply the same procedure for other types.

* Select the **Angle and setback** radio button from the dialog box and click on the **OK** button. You will be asked to asked to select the face containing edges to be chamfered.
* Select the desired face and click on the **OK** button from the **Command Bar**. The **Chamfer Command Bar** will be displayed as shown in Figure-74. Note that the angle and distance will be measured from the selected face.

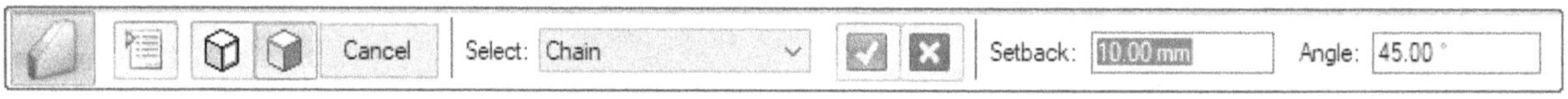

*Figure-74. Setback And Angle Selection*

* Select the edges of the face to be chamfered and specify the desired parameters in the **Command Bar**. Preview of chamfer will be displayed; refer to Figure-75.

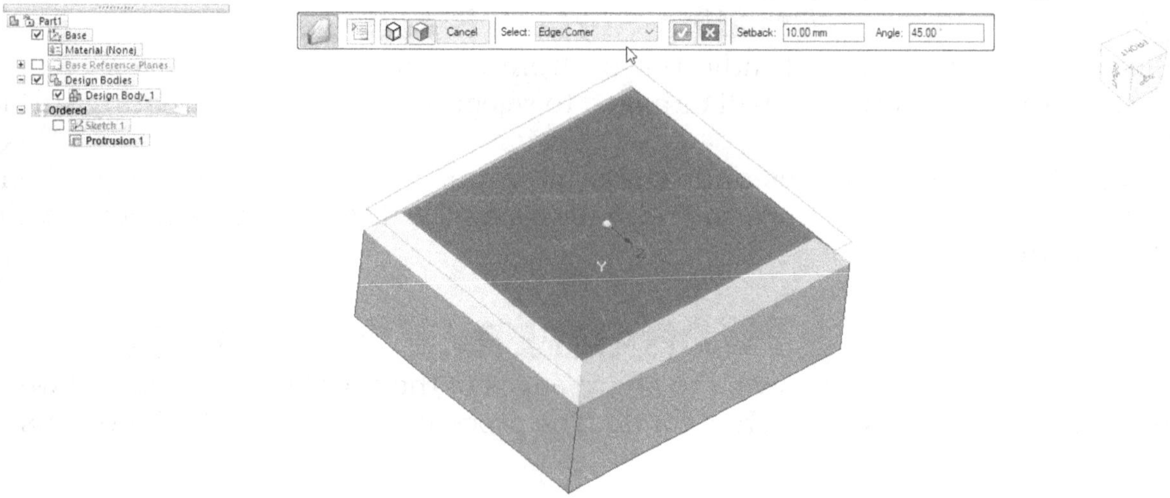

*Figure-75. Chamfer Created*

* Click on the **OK** button from the dialog box to create chamfers. Press **ESC** to exit the tool.

# DRAFT

The **Draft** tool is used to apply taper to the faces of a solid model. This tool is mainly useful when you are designing components for molding or casting. The taper applied using **Draft** tool allows easy and safe ejection of part from the dies. The procedure to use this tool is given next.

- Click on the **Draft** tool from the **Solids** panel in the **Home** tab of the **Ribbon**. The **Draft Command Bar** will be displayed; refer to Figure-76 and you will be asked to select a face/plane to be used reference for applying draft angle to other faces/ surfaces.

*Figure-76. Draft Command Bar*

- Select the desired face/plane. You will be asked to select side faces to be used for applying draft angle.
- Select the desired faces for applying draft angle. The **Command Bar** will be displayed as shown in Figure-77.

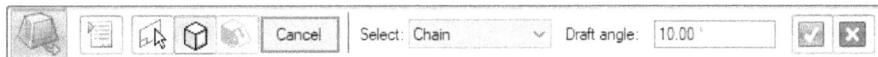

*Figure-77. Specify Draft Angle*

- Specify desired angle value for draft in the **Draft angle** edit box. After setting desired value, click on the **OK** button from the **Command Bar** or press **ENTER**.
- If you want to add more faces for draft then you can select them now. After that click on the **Next** button from the **Command Bar** as shown in Figure-78. You will be asked to specify the direction of draft angle.

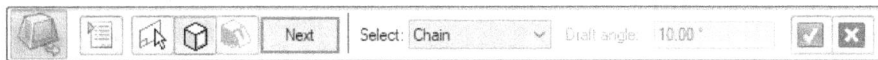

*Figure-78. Next Command Bar*

- Click on the desired side of draft faces to define direction; refer to Figure-79. Draft will be created and draft angle annotation will be displayed on the model; refer to Figure-80.

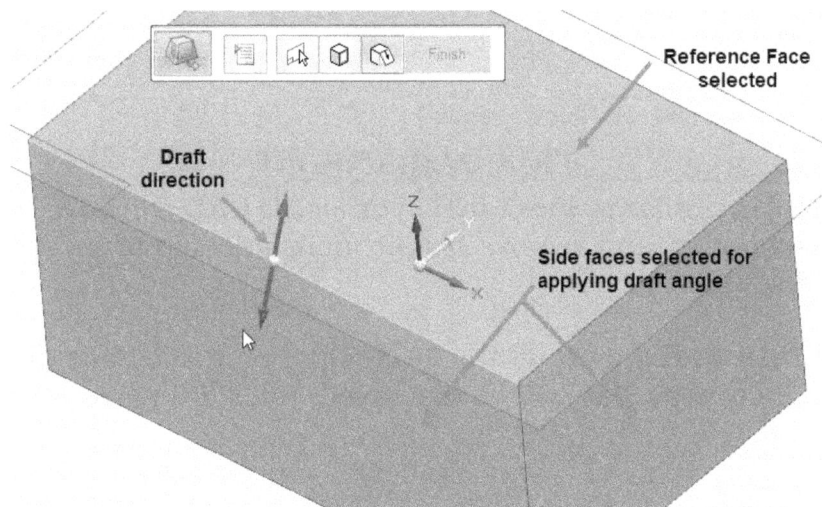

*Figure-79. Draft direction handle displayed*

- Click on the **Finish** button from the **Command Bar** to create feature and press **ESC** to exit the tool.

*Figure-80. Draft created*

# THIN WALL DROP-DOWN

The tools in **Thin Wall** drop-down (refer to Figure-81)are used to create various structure supporting features like rib, web network, and so on. The Thin Wall, Thin Region, and Vent tools are used to remove material from the solid.

Various tools available in this drop-down are discussed next.

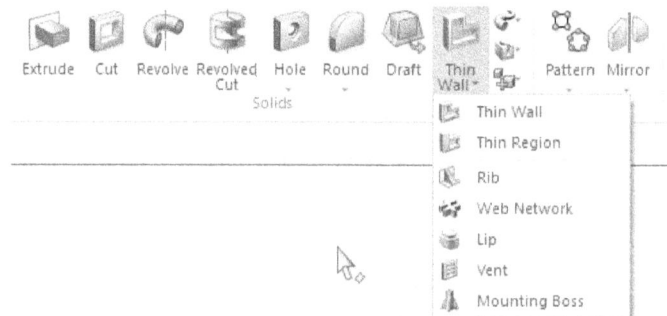

*Figure-81. Thin Wall drop down*

## Thin Wall (Shell)

The **Thin Wall** tool is similar to **Shell** tool in other 3D CAD software. This tool is used to make solid part hollow and remove one or more selected faces. The procedure to use this tool is discussed next.

- Click on the **Thin Wall** tool from the **Solids** panel in the **Home** tab of the **Ribbon**. The **Thin Wall Command Bar** will be displayed; refer to Figure-82.

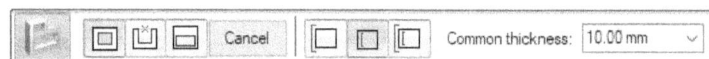

*Figure-82. Thin Wall Command Bar*

- Specify the desired value of thickness in the **Common thickness** edit box. The walls of solid will have this amount of thickness after applying thin wall feature.
- Select the desired button from the **Command Bar** to define which side the material will be offset for creating walls. Select the **Offset Outside** button to create thin wall of specified thickness outside the faces of current solid. Select the **Offset Inside** button to create wall inside the faces of current solid. Select the **Symmetrical** button to use faces of solid as midplane for creating thin wall.

- After setting desired parameters, press **ENTER**. You will be asked to select faces to be removed from shell body and the **Command Bar** will be displayed as shown in Figure-83.
- Select the desired faces to be removed; refer to Figure-84. Note that specified thickness will be applied to all the faces even if you do not remove any face of the model.

*Figure-83. Thin Wall Preview Command Bar*

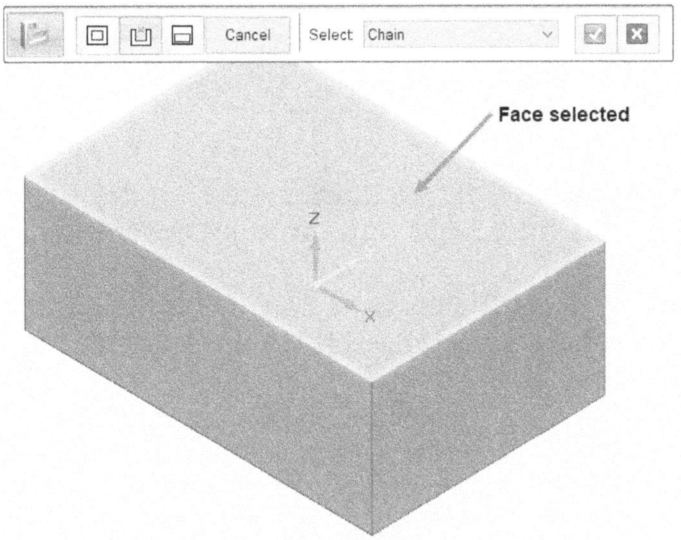

*Figure-84. Face selected for removing*

- After selected face(s), right-click in the empty area or press **ENTER** to confirm selection.
- If you want to specify unique thickness of any face then click on the **Unique Thickness** button from the **Command Bar**. You will be asked to select the faces.
- Select the desired face(s) and enter the thickness value in **Unique Thickness** edit box.
- After setting all the parameters, click on the **Preview** button to check the preview of feature; refer to Figure-85

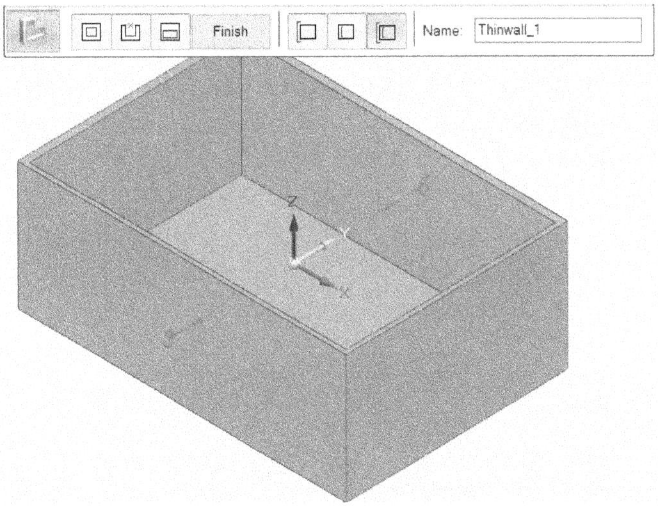

*Figure-85. Thin Wall Created*

- Click on the **Finish** button from the **Command Bar**. The thin wall feature will be created.

# THIN REGION

The **Thin Region** tool is used to make a closed portion of the part as thin feature. This tool is available in **Thin Wall** drop-down. The procedure to use this tool is discussed next.

- Click on the **Thin Region** tool from the **Thin Wall** drop-down of **Solids** panel in the **Home** tab of **Ribbon**. The **Thin Region Command Bar** will be displayed as shown in Figure-86.

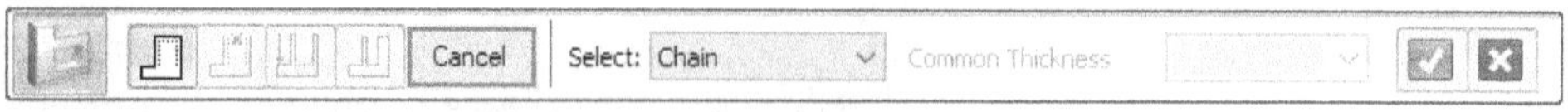

*Figure-86. Thin Region Command Bar*

- Select the desired enclosed face chain in the model to be converted to thin feature and specify the thickness value in the **Common Thickness** edit box; refer to Figure-87.
- Click on the **OK** button from the **Command Bar** after specifying parameters. You will be asked to select faces to be removed from the feature.

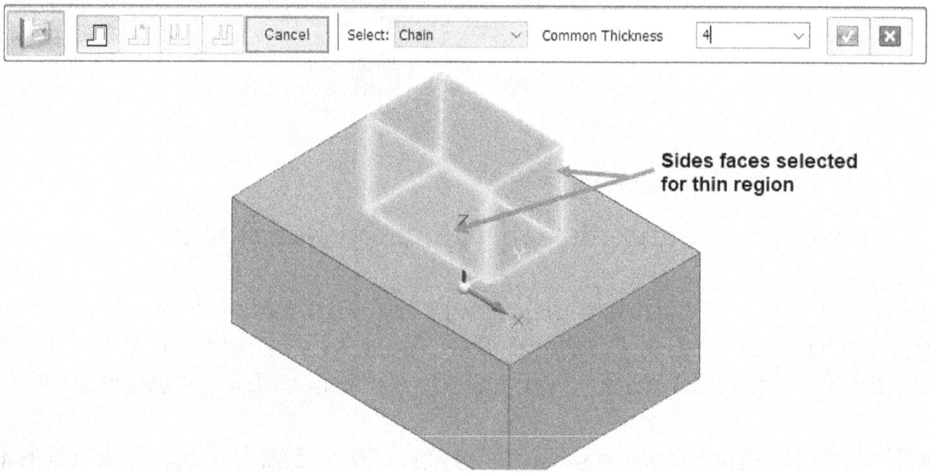

*Figure-87. Faces selected for thin region*

- Select the desired faces as discussed earlier. Click on the **Capping Faces** button from the **Command Bar** and select the faces to be used for creating cap of thin feature. Similarly, you can use the **Unique Thickness** button to specify different thickness for selected faces.
- After setting parameters, click on the **Preview** button to check the feature. Now, click on **Finish** button. Refer to Figure-88.

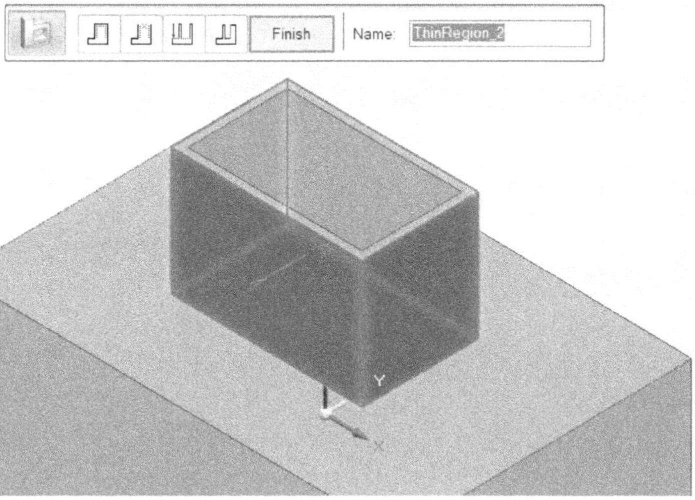

*Figure-88. Thin Region created*

- Press **ESC** to exit the tool.

# RIB

The **Rib** tool is used to create support in the structures to increase their strength. You can find use of rib in various fixtures that are fastened to the wall or in the building columns to support objects. The procedure to create rib feature is given next.

- Click on the **Rib** tool from the **Thin Wall** drop-down in the **Solids** panel of **Home** tab in the **Ribbon**. You will be asked to select a planar face or plane to be used as reference base for creating rib. Note that selected plane/face will be at the mid of rib feature.
- Select the sketch if there is an existing one for rib or select the plane/face and create an open sketch; refer to Figure-89.
- Close the sketch by clicking on the **Close Sketch** tool from the **Close** panel in the **Ribbon**. Preview of the rib feature will be displayed; refer to Figure-90.

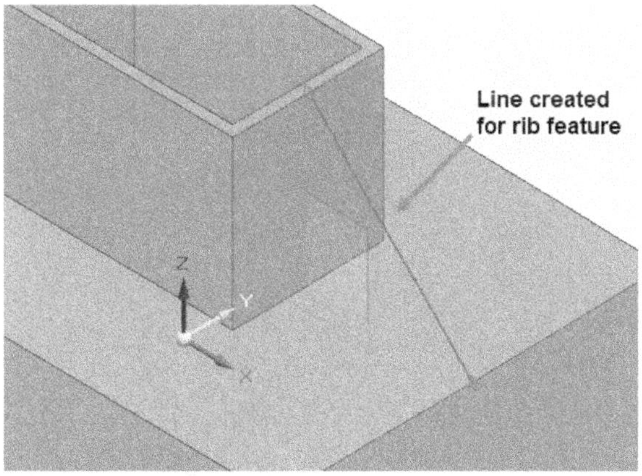

*Figure-89. Sketch created for rib feature*

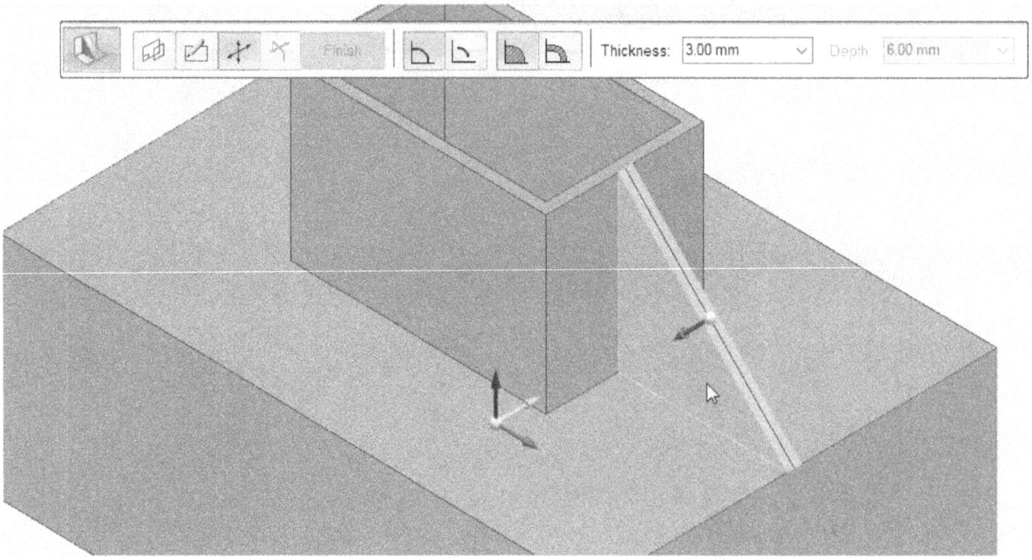

*Figure-90. Preview of Rib feature*

- The direction of rib feature is defined the position of cursor with respect to rib sketch. If you move cursor below the rib sketch then direction will be downward, if you move cursor to above the sketch then direction will be upward, similarly you can set left or right direction for rib feature.

- Select the **Extend Profile** button from the **Command Bar** if you want to extend the rib sketch profile automatically so that it intersects with nearby faces to form closed profile. Select the **No Extend** button if you do not want to extend profile automatically.

- Select the **Extend to Next** button from the **Command Bar** to automatically create rib feature up to next part surface in-line. If you want to specify the depth of rib feature then select the **Finite Depth** button from the **Command Bar** and specify the value in the **Depth** edit box.

- After specifying desired parameters, click at desired location to create the feature. Preview of feature will be displayed; refer to Figure-91.

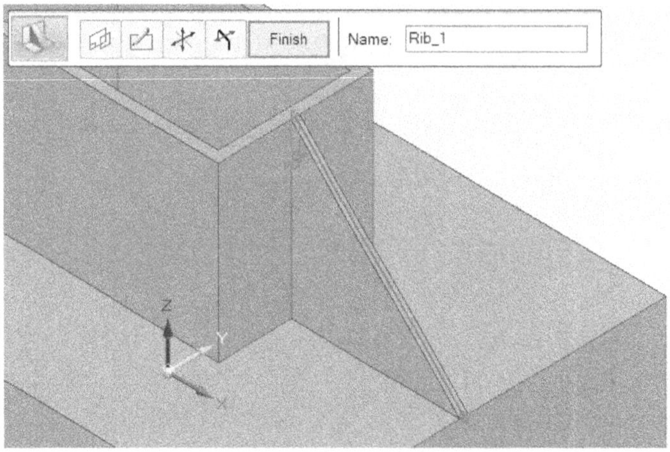

*Figure-91. Preview of Rib feature*

- Click on the **Side Step** button from the **Command Bar** if you want to change the side of rib feature with respect to sketch plane. You can place the rib feature on the left, right, or center of sketch plane.

- After setting desired parameters, click on the **Finish** button from the **Command Bar** to create the feature. Press **ESC** to exit the tool.

# WEB NETWORK

The **Web Network** tool is used to create web structure on entity using intersecting lines/curves. The procedure to use this tool is discuss next.

- Click on the **Web Network** tool from the **Thin Wall** drop-down. The **Web Network Command Bar** will be displayed; refer to Figure-92 and you will be asked to select sketch entities or create a sketch on desired plane.

*Figure-92. Web Network Command Bar*

- Select the desired plane to create sketch or select the **Select from Sketch** option from the **Create-From Options** drop-down and select the desired sketch entities; refer to Figure-93.

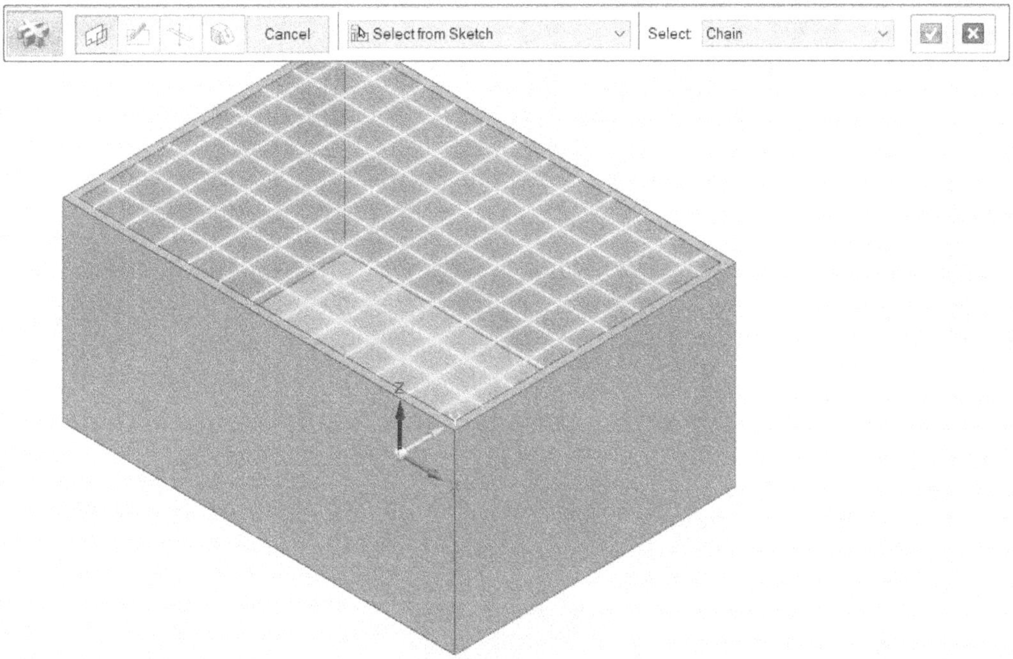

*Figure-93. Sketch entities selected for web network feature*

- After selecting sketch entities, click on the **OK** button from the **Command Bar**. The **Web Network Command Bar** will be displayed with preview of feature; refer to Figure-94 and you will be asked to define the direction of feature.

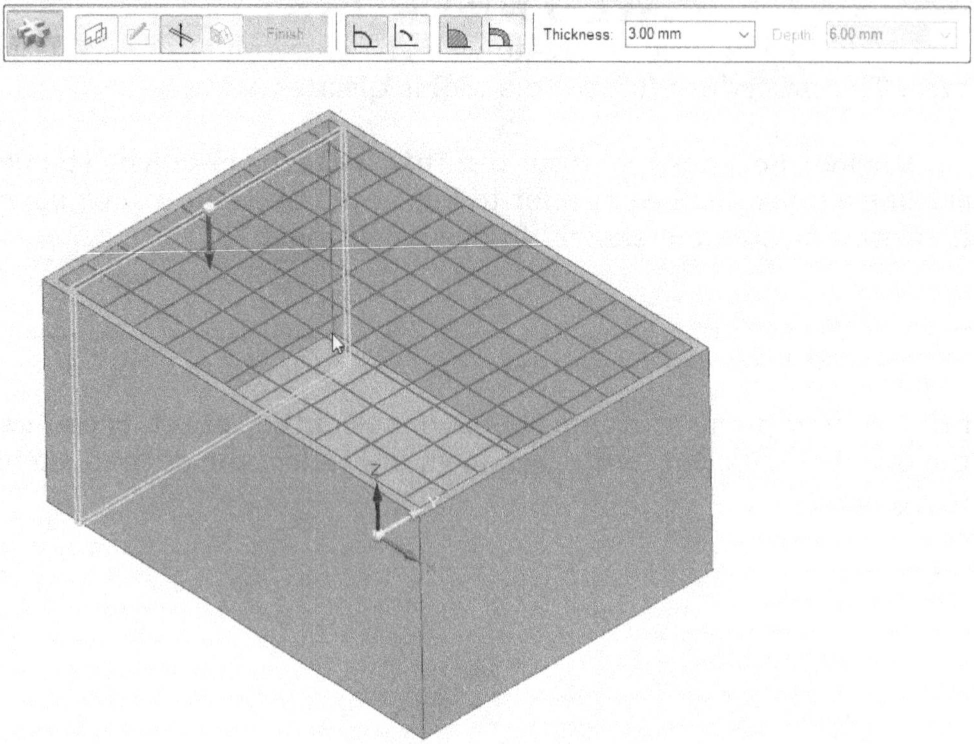

*Figure-94. Web Network Customization Command Bar*

- Specify Thickness and direction of web network. The feature will be created; refer to Figure-95. Note that the direction should be towards close region so that it can add material. The other options in the **Command Bar** are same as we discuss for rib feature.

- Click on **Finish** button from the **Command Bar** to complete feature creation. Press **ESC** to exit the tool.

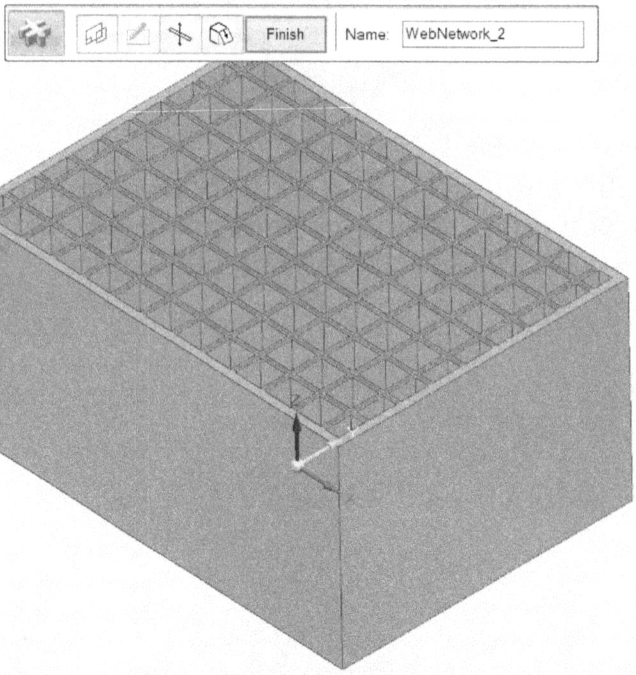

*Figure-95. Web Network Created*

# LIP

The **Lip** tool is used to create groove or lip on the surface of entity. Using this tool, you can add or remove material from the solid to form lip at the end face. The procedure of this tool is discussed next.

*   Click on **Lip** tool from the **Thin Wall** drop-down in the **Solids** panel of **Home** tab in the **Ribbon**. The **Lip Command Bar** will be displayed; refer to Figure-96 and you will be asked to select an edge of the model for creating lip feature.

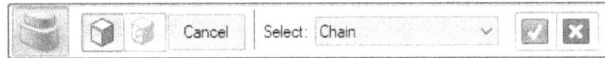

*Figure-96. Lip Command Bar*

*   Select the desired edge of model and click on the **OK** button from the **Command Bar**. The options to specify width and height of lip feature will be displayed in the **Command Bar**; refer to Figure-97 and you will be asked to specify direction of feature.

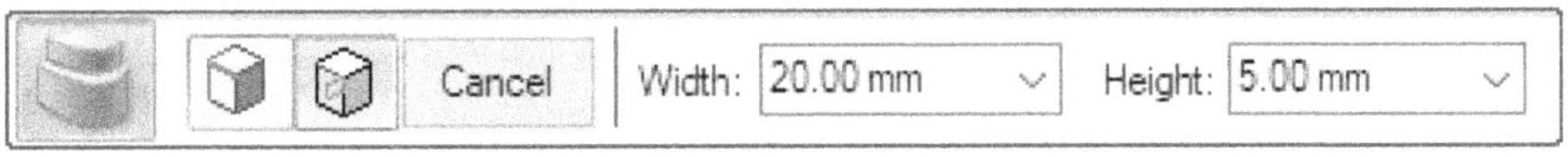

*Figure-97. Lip Customization Command Bar*

*   Specify the desired parameters in the **Command Bar** and then click at desired side of selected edge to define direction of feature. Note that based on selected direction, the material will be added or removed from the model.
*   After specifying direction, the preview of feature will be displayed; refer to Figure-98. Click on the **Finish** button from the **Command Bar** to create the feature. Press **ESC** to exit the tool.

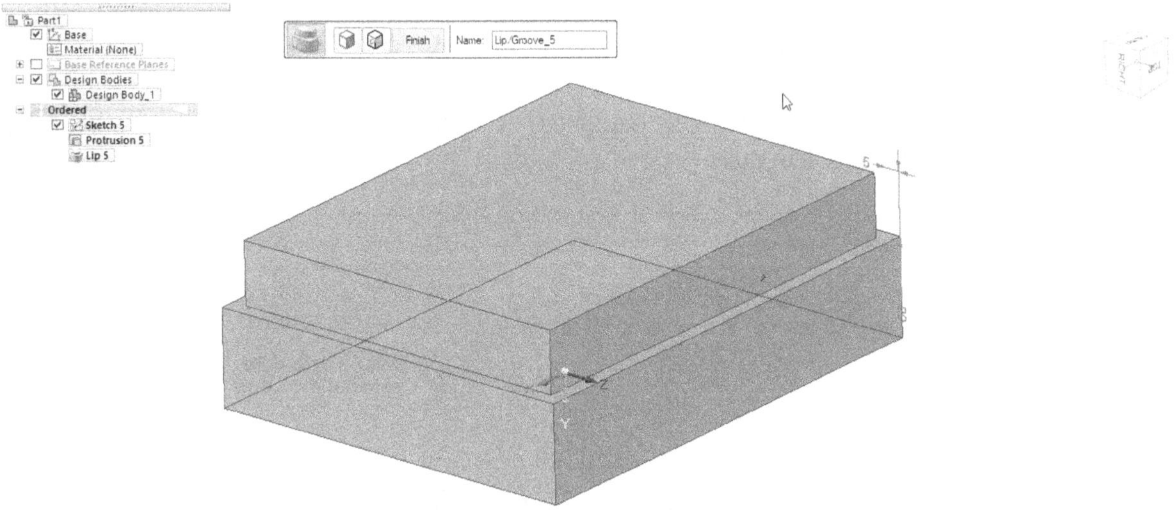

*Figure-98. Lip Created*

# VENT

The **Vent** tool is used create vent features. This feature is like web network but difference is that it removes material from solid face. To create vent, we need to create a sketch on the surface of solid. It must be closed sketch with intersecting curves. The procedure to use this tool is discuss next.

*   Click on the **Vent** tool from the **Thin Wall** drop-down in the **Ribbon**. The **Vent Options** dialog box will be displayed with **Command Bar**; refer to Figure-99 and Figure-100.

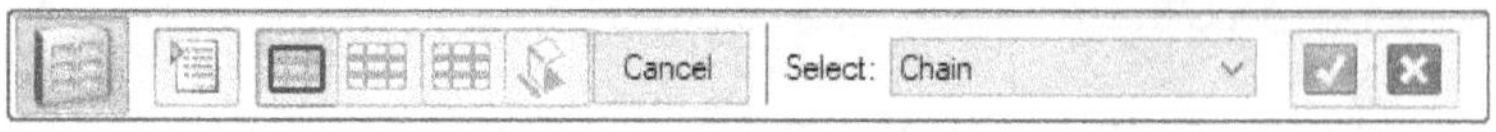

*Figure-99. Vent Command Bar*

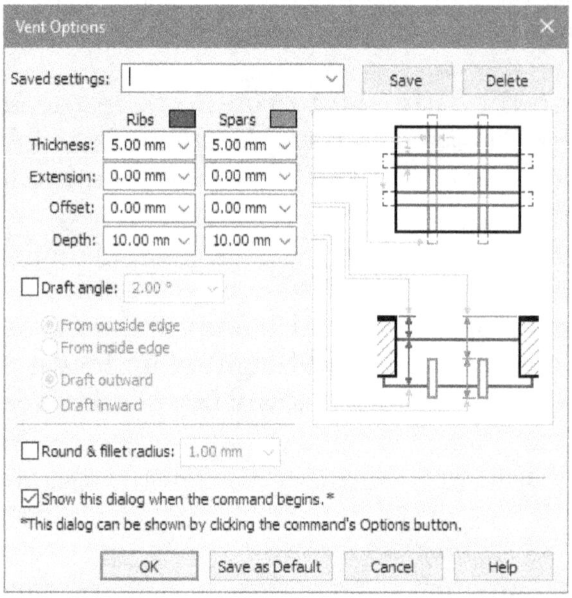

*Figure-100. Vent Options dialog box*

- Set the desired parameters for ribs (horizontal) and spars (vertical) in respective edit boxes in the dialog box. Select the **Draft angle** check box and specify the angle if you want to create taper walls of vents. Select the **Round & fillet radius** check box if you want to apply fillets and rounds to the members of vent. Set the desired values for selected parameters and click on the **OK** button from the dialog box. You will be asked to select the sketch boundary for vent.
- Select the closed loop existing sketch chain to define boundary of vent; refer to Figure-101 and click on the **OK** button from the **Command Bar**. You will be asked to select sketch entities for ribs.

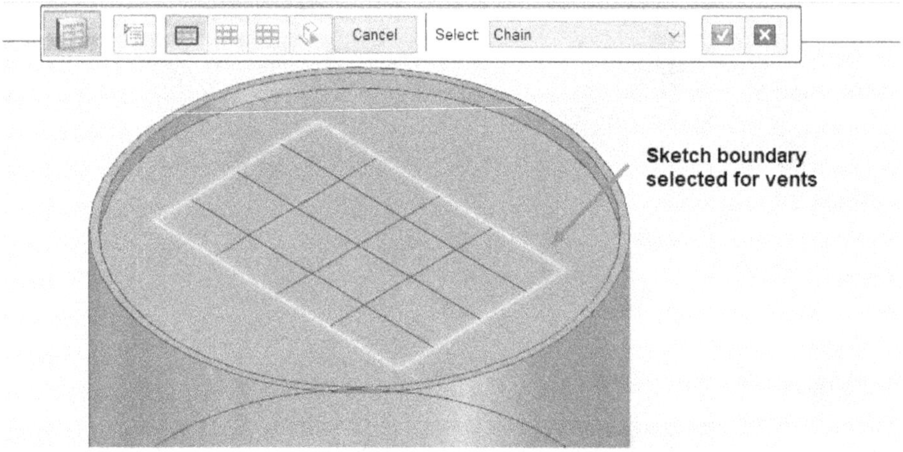

*Figure-101. Selecting sketch boundary of vents*

- Select the sketch entities for rib section of vent and press **ENTER**. You will be asked to select entities for spars.
- Select the desired sketch entities for spars; refer to Figure-102 and press **ENTER**. You will be asked to define direction of cut.

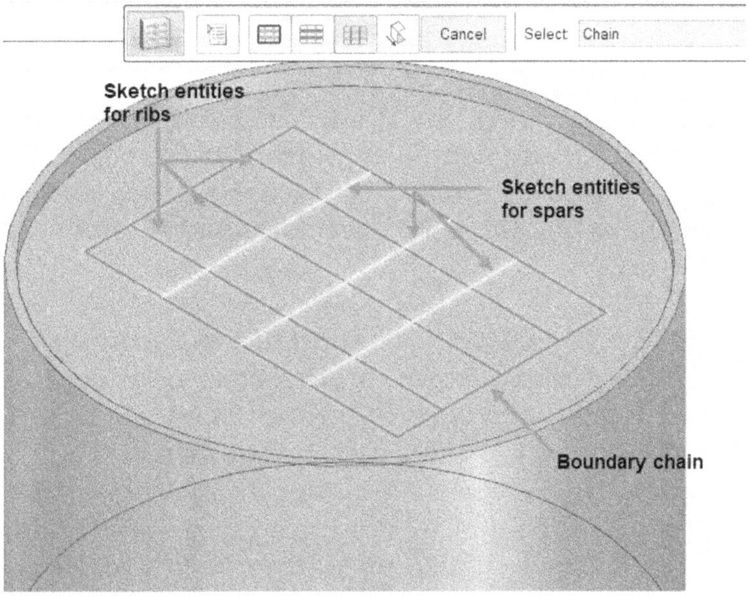

*Figure-102. Sketch selection for vent feature*

- Set the direction upward/downward as needed. Note that the direction must be towards the solid region on which vent should be created because vents remove material. Note that if you have created vent on solid filled model then it will automatically become hollow below the vents.
- If you want to manually specify the depth of vent then click on the **Extent Step** button from the **Command Bar**. The **Command Bar** will be displayed as shown in Figure-103. The options in this **Command Bar** as same as discussed earlier for other features.

*Figure-103. Vent Customization Command Bar*

- After setting direction and extents, preview of the feature will be displayed; refer to Figure-104.

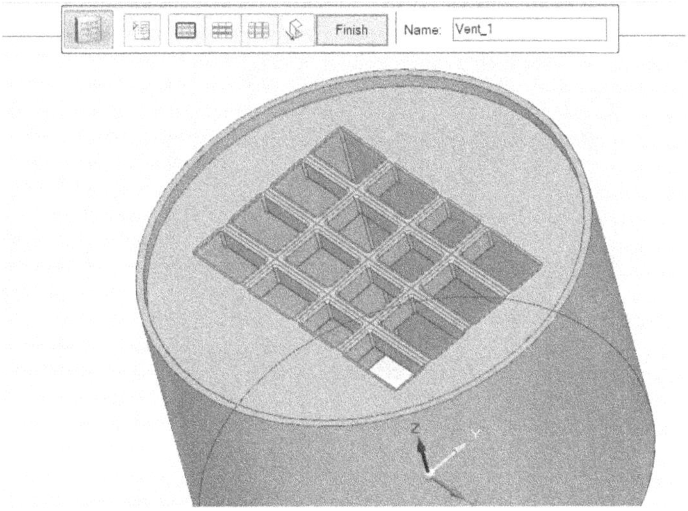

*Figure-104. Vent created*

- Click on the **Finish** button from the **Command Bar** to create the feature. Press **ESC** to exit the tool.

## MOUNTING BOSS

The **Mounting Boss** is used create simple cylindrical boss features with rib supports. You can add draft angle, round, and so on to the feature as discussed for vents. Mounting boss features are used to connect bodies with fasteners. The procedure to use this tool is given next.

- Click on the **Mounting Boss** tool from the **Thin Wall** drop-down in the **Solids** panel of **Home** tab in the **Ribbon**. The **Mounting Boss Command Bar** will be displayed asking you to select plane/face where you want to place the mounting boss.
- Set the desired option in the **Create-From Options** drop-down of the **Command Bar** and select the plane on which you want to mount top of the feature. Note that the feature height will be decided by distance of selected plane from the surface of existing solid. The base sketch of mounting boss feature will get attached to cursor and you will be asked to specify the positions of mounting boss features.

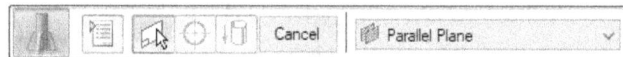

*Figure-105. Mounting Boss Command Bar*

- Click at desired locations to specify the positions of features. Note that you can also use the sketch dimensioning tools to locate sketch features. After specifying positions, close the sketching environment by using the **Close Sketch** button from the **Ribbon**. You will be asked to specify direction of feature.
- Click in the direction towards face of solid. Preview of feature will be displayed; refer to Figure-106.

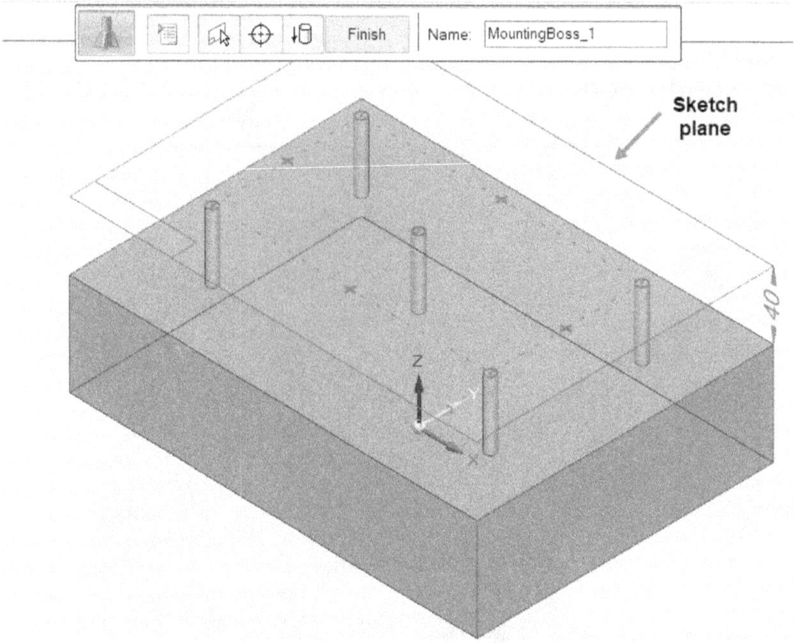

*Figure-106. Preview of mounting boss feature*

- Click on the **Mounting Boss Options** button from the **Command Bar**. The **Mounting Boss Options** dialog box will be displayed; refer to Figure-107.

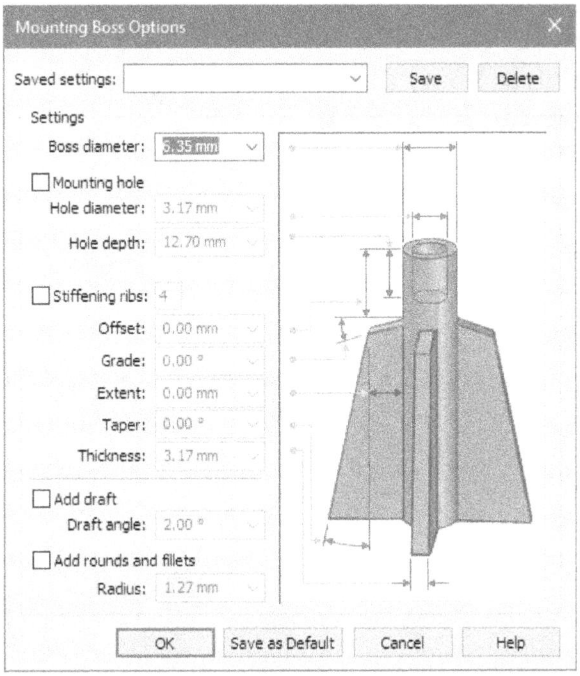

*Figure-107. Mounting Boss Options dialog box*

- Specify desired parameters in the dialog box and click on the **OK** button from the dialog box. The mounting boss features will be modified accordingly; refer to Figure-108.

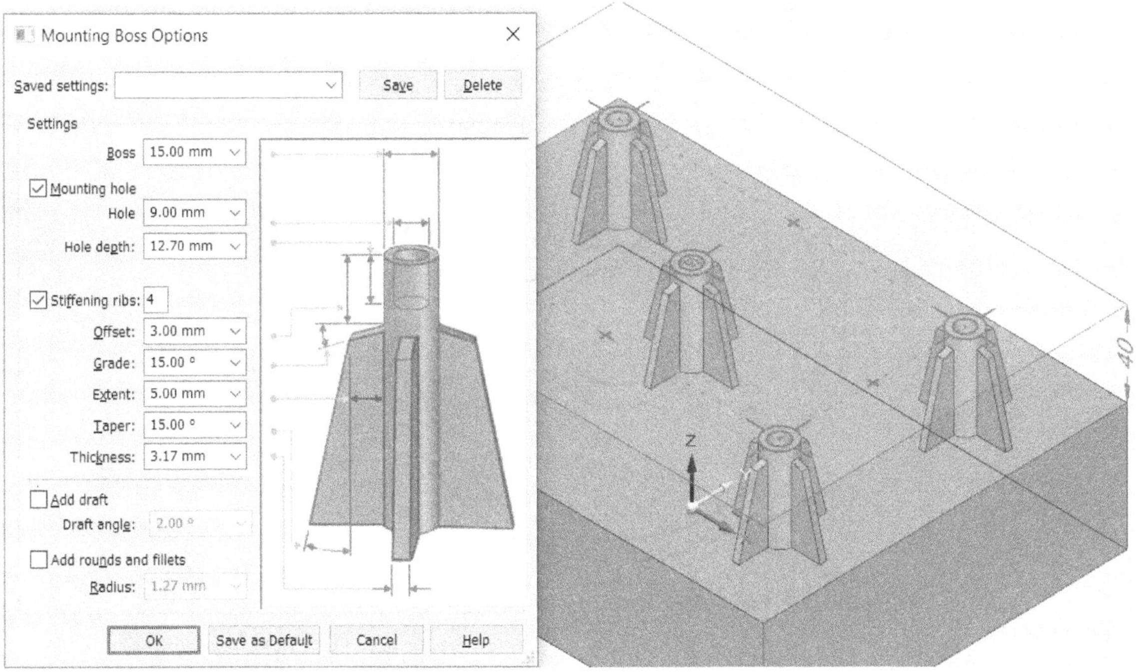

*Figure-108. Mounting boss features after modification*

- Click on the **Finish** button from the **Command Bar** to create the feature. Press **ESC** to exit the tool.

# PRACTICAL - 1

Create the model (isometric view) as shown in Figure-109. The views of the model with dimensions are given in Figure-110.

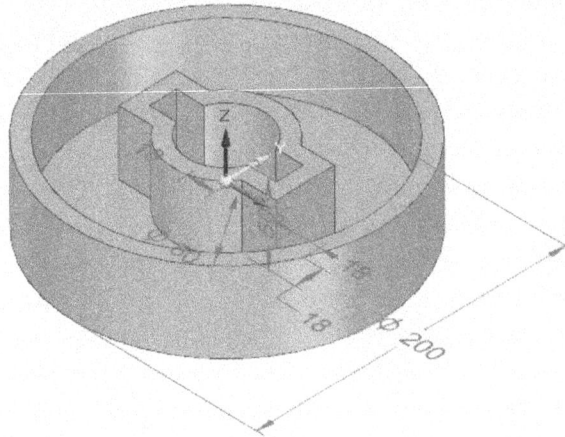

*Figure-109. Practical Model 1*

*Figure-110. Views for practical 1*

Before we start working on the practical, it is important to understand two terms; first angle projection and third angle projection. These are the standards of placing views in the engineering drawing. The views placed in the above figure are using third angle projection. In first angle projection, the top view of model is placed below the front view and right side view is placed at left of the front view. You will learn more about projection in chapter related to drafting.

## Starting Solid Edge Modeling environment and creating Extrude feature

- Double-click on the Solid Edge icon from desktop if you have not started Solid Edge.
- Click on the **New** tool from the **Menu bar** or press **CTRL+N**. The New dialog box will be displayed. Select **iso metric part.par** template from the dialog box and click on the **OK** button.
- You can create this part in **Synchronous** as well as **Ordered** environment. Here, we will create this part in **Synchronous** environment.
- Select the **Circle by Center Point** tool from the **Circle** drop-down in the **Draw** panel of **Home** tab in the **Ribbon** and select the **Top** plane from the **Base Reference Planes** node in the **Path Finder**.
- Draw a circle of diameter **200** at the centre of coordinate system.
- Again, draw a circle of diameter **50** on the center of coordinate system. Select the **Top** face of **ViewCube** to make the sketch plane parallel to screen.
- Click on the **Rectangle by Center** tool from the **Draw** panel in the **Home** tab of **Ribbon** and draw the rectangle as shown in Figure-111.
- Click on the **Trim** tool from the **Draw** panel in the **Home** tab of **Ribbon** and trim extra portion of the sketch; refer to Figure-112.

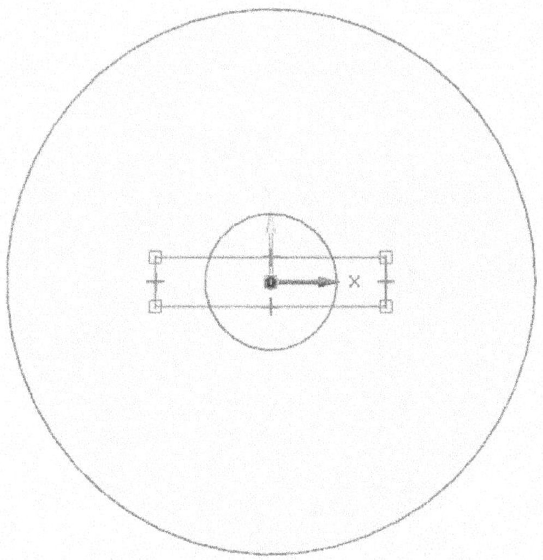

*Figure-111. Sketch after creating rectangle*

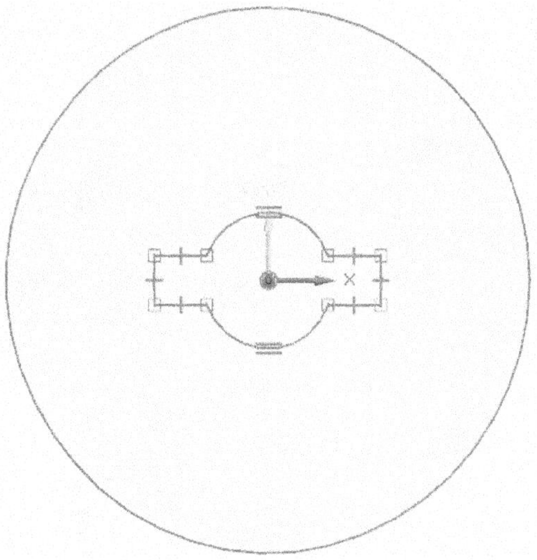

*Figure-112. Sketch after trimming*

- Click on the **Smart Dimension** tool from the **Dimension** panel in the **Home** tab of **Ribbon** and create the dimensions as shown in Figure-113.

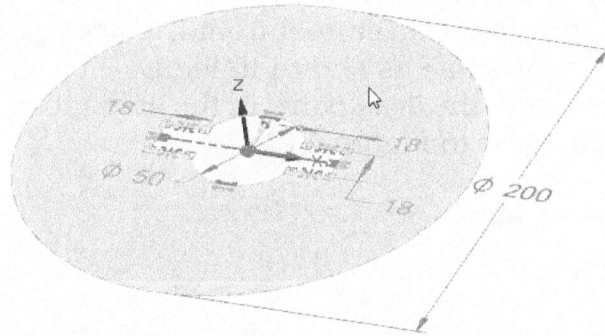

*Figure-113. Sketch of Model*

- Now, click on the **Extrude** tool from the **Solids** panel in the **Home** tab of **Ribbon**, and select the region between circle and inner section; refer to Figure-114. Right-click in the empty area. Preview of extrude feature will be displayed; refer to Figure-115.

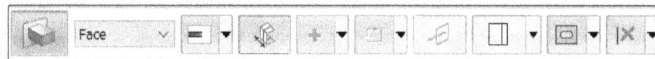

*Figure-114. Selecting region*

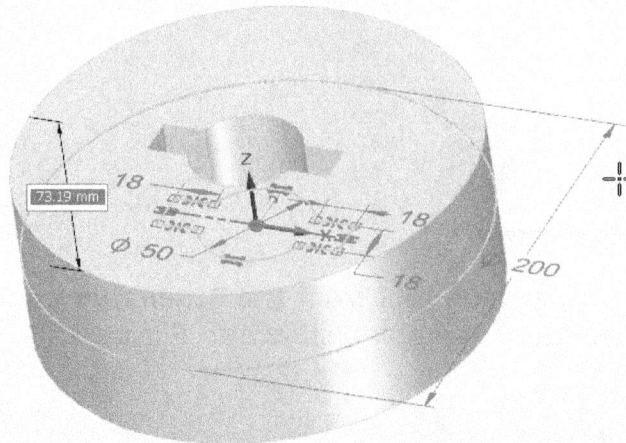

*Figure-115. Preview of extrude feature*

- Specify the height of extrude feature as **50** in dynamic input box and press **ENTER**. The **Extrude** feature will be created; refer to Figure-116.

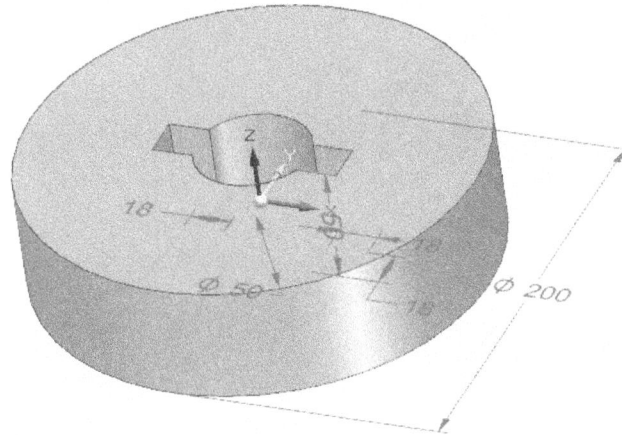

*Figure-116. Extrude feature created*

- Click on the **Thin Wall** tool from the **Solids** panel in the **Home** tab of **Ribbon**. You will be asked to select the faces to be removed and dynamic input box will be displayed to specify thickness of feature.
- Select the top face of feature and enter the thickness of feature as **10** in dynamic input box. The model will be created; refer to Figure-117.

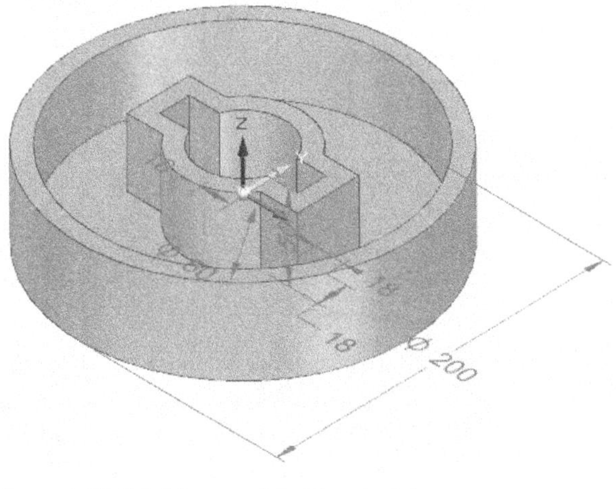

*Figure-117. Model created for Practical 1*

# PRACTICAL - 2

Create the model (isometric view) as shown in Figure-118. The dimensions of the model are given in Figure-119.

*Figure-118. Model for Practical 2*

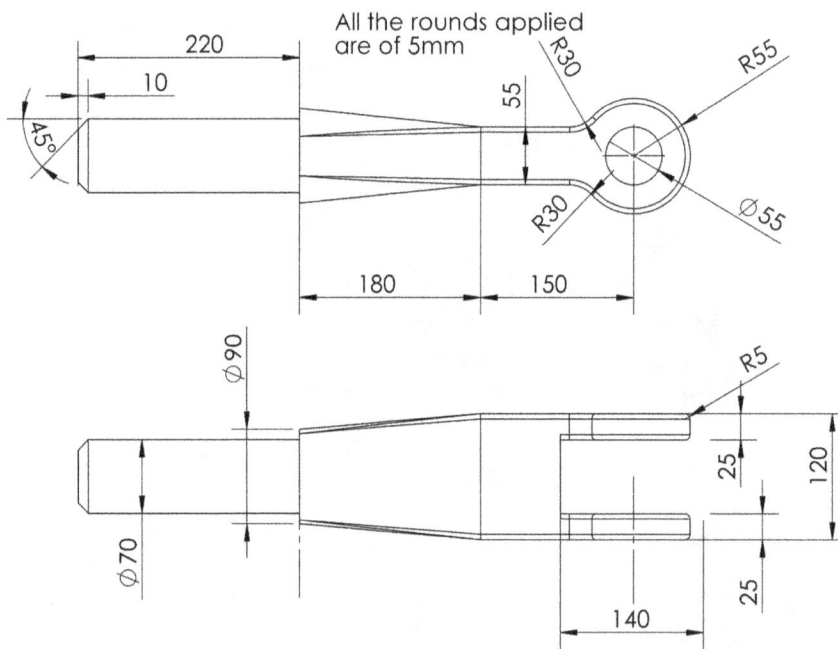

*Figure-119. Practical 2 drawing views*

## Creating first extrude feature

We will create this model in the ordered sketching environment. So, we will first switch to Ordered environment. The procedure to switch environment and create first extrude feature is given next.

- Start a part file in Solid Edge as discussed in previous practical.
- Right-click in the empty area of drawing window and select the **Transition to Ordered** option from the shortcut menu. The Ordered environment will be activated.
- Click on the **Sketch** tool from the **Sketch** panel in the **Home** tab of **Ribbon** and select the **Top** plane from the **Base Reference Planes** node in the **Path Finder**. Draw the sketch as shown in the Figure-120.

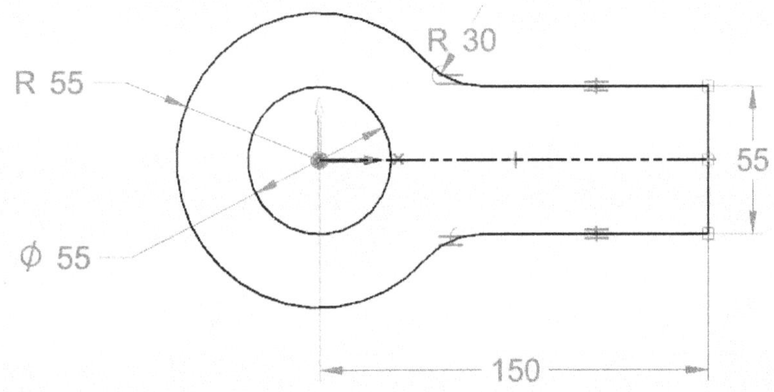

*Figure-120. Sketch created at top plane*

- Close the sketch environment by clicking on the **Close Sketch** tool from the **Close** panel in the **Ribbon**.
- Click on the **Extrude** tool from **Solids** panel in the **Ribbon**. The **Extrude Command Bar** will be displayed.
- Select the **Select from Sketch** option from the **Create-From Options** drop-down of the **Command Bar** and select the complete sketch section created earlier. After selecting sketch, press **ENTER**. You will be asked to specify the depth of extrude feature (you might need to move cursor to display depth options).
- Specify the depth value as **120** in the **Distance** edit box of the **Command Bar** and press **ENTER**. You will be asked to specify the side on which feature will be created with respect to sketch plane.
- Select the **Symmetric Extent** button from the **Command Bar** to keep sketch plane at center and symmetrically create the feature. The preview of the feature will be displayed. Click on the **Finish** button to create the feature; refer to Figure-121 and press **ESC** to exit the tool.

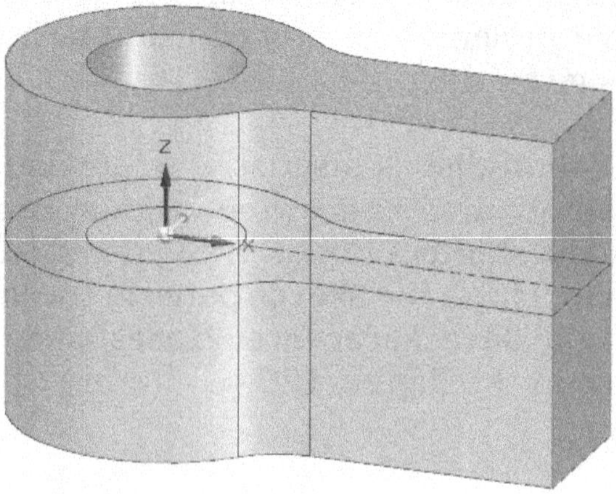

*Figure-121. Extrude feature created*

## Creating loft feature

- Click on the **Parallel** tool from the **More Planes** drop-down in the **Planes** panel of the **Ribbon** and select the side face of model as shown in Figure-122.
- Specify the offset distance as **180** and create the plane on the right side of selected face as shown in Figure-122. Press **ESC** to exit the tool.

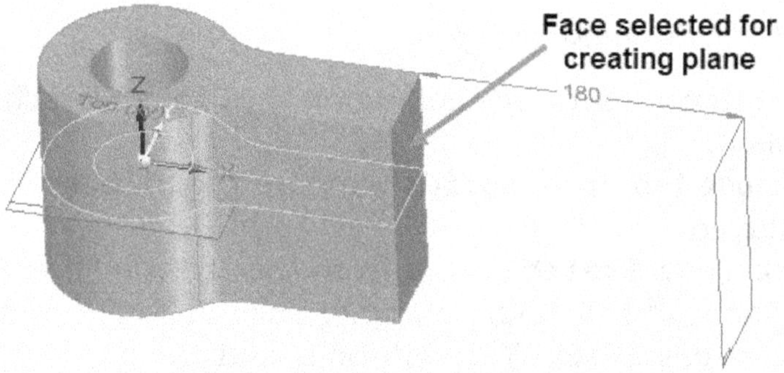

Face selected for creating plane

180

*Figure-122. Plane to be created*

- Select this plane and create the sketch as shown in Figure-123.

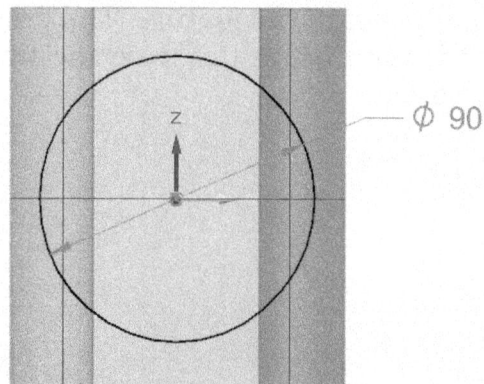

⌀ 90

*Figure-123. Sketch to be created*

- Click on the **Close Sketch** to exit from sketching and finish the sketch.
- Click on the **Loft** tool from the **Add** drop-down in the **Solids** panel of the **Ribbon**. The **Lofted Protrusion Command Bar** will be displayed.
- Select the **Chain** option from the **Select** drop-down in **Command Bar** and select the sketch created on new plane. Now, select the **Face** option from the **Select** drop-down and select the face as shown in Figure-124.

*Figure-124. Selection for loft feature*

- Click on the **Preview** button from the **Command Bar**. The preview of loft feature will be displayed. Click on the **Finish** button to create the feature; refer to Figure-125. Press **ESC** to exit the tool.

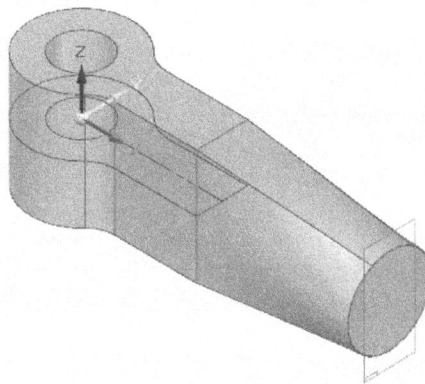

*Figure-125. Loft feature created*

## Creating Extrude Feature

- Click on the **Extrude** tool from the **Solids** panel in the **Ribbon** and select the **Coincident Plane** option from the **Create-From Options** drop-down of the **Command Bar**.
- Select the flat face of the loft feature recently created and draw a circle of **70** diameter at origin as shown in Figure-126.

*Figure-126. Circle created for extrude feature*

- Click on the **Close Sketch** tool from the **Close** panel in the **Ribbon** to exit sketch environment. Preview of extrude feature will be displayed and you will be asked to specify height of feature. Enter the height as **220** in the **Distance** edit box of **Command Bar** and click outside of feature to create feature. Preview of the feature will be displayed; refer to Figure-127.
- Click on the **Finish** button from the **Command Bar** to create the feature. Press **ESC** to exit the tool.

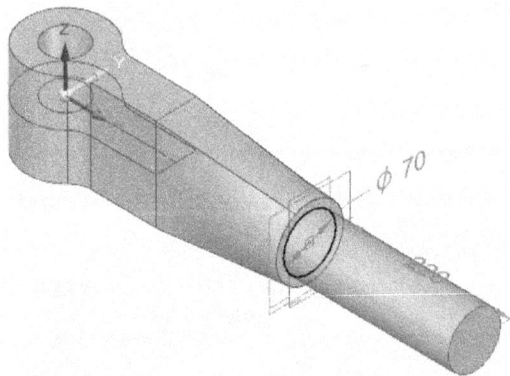

*Figure-127. Preview of model after extrusion*

## Creating Extrude Cut feature

- Click on the **Cut** tool from the **Solids** panel in the **Home** tab of **Ribbon** and select the left flat face of part. The sketching environment will be displayed.
- Draw a rectangle of **140** x **70** from the center of coordinate system; refer to Figure-128.

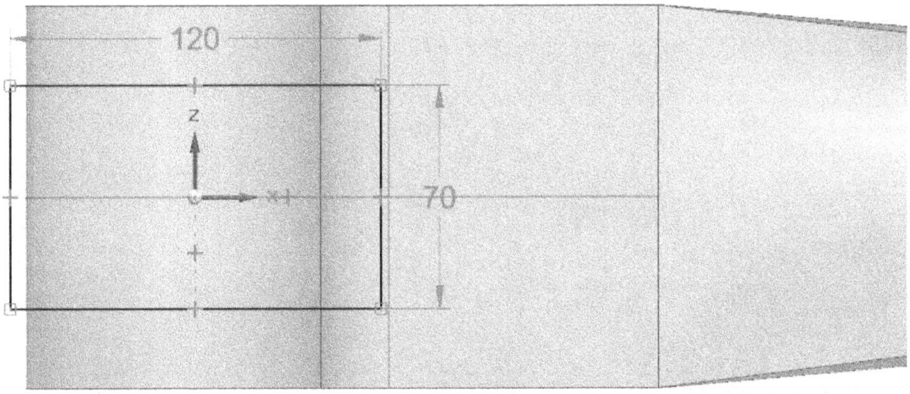

*Figure-128. Extrude cut sketch*

- Click on the **Close Sketch** tool from the **Close** panel in **Ribbon** and select **Symmetric Extent** button from the **Cut Command Bar** displayed. Preview of cut feature will be displayed as you move cursor left/right in drawing area.
- Specify the distance value as **180** in the **Distance** edit box of the **Command Bar** and press **ENTER**. Preview of cut feature will be displayed.
- Click on the **Finish** button to create the feature; refer to Figure-129.

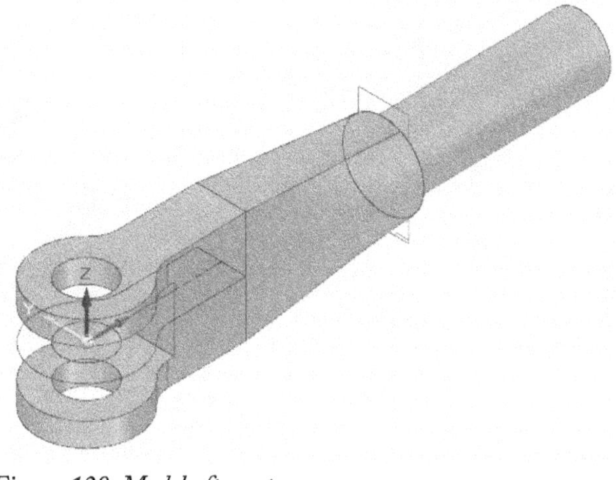

*Figure-129. Model after cut*

## Creating Round and chamfers

- Click on the **Round** tool and specify the radius as **5** in the **Radius** edit box of **Command Bar**.
- Select all the edges on which you want to create the round; refer to Figure-130.
- Click on the **OK** button from the **Command Bar** to preview the feature. Click on the **Preview** button and then click on the **Finish** button from **Command Bar** to create the feature. Press **ESC** to exit the tool.

*Figure-130. Preview of round feature*

• Similarly, click on the **Chamfer** tool from the **Round** drop-down in the **Solids** panel of the **Ribbon** and apply chamfer of 10 x 45 degree at the end face of model. Hide all the sketches and planes to display a neat model; refer to Figure-131.

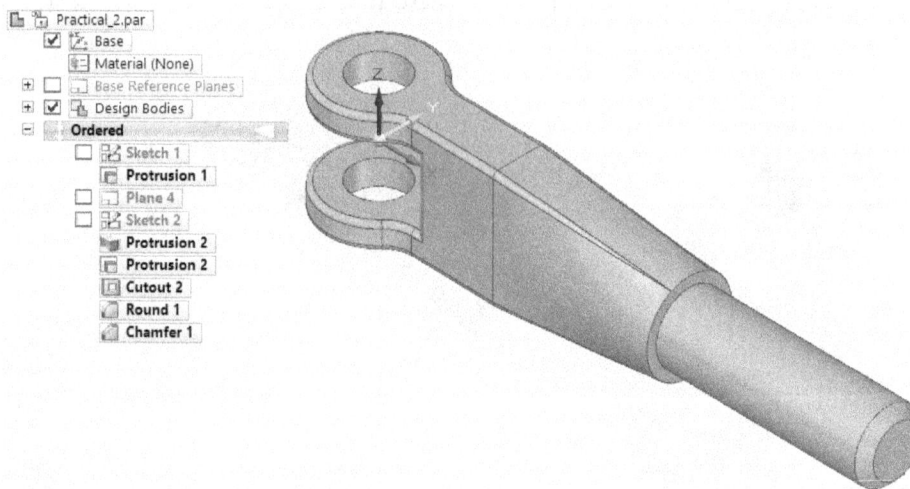

*Figure-131. Final Model for Practical 2*

• Save the file at desired location.

# Chapter 4

# Advanced Solid Modeling

## Topics Covered

The major topics covered in this chapter are:

- *Advanced Solid Modeling tools*
- *Creating Pattern*
- *Modifying tools*
- *Practical 1*
- *Practical 2*
- *Practice*

# INTRODUCTION

In previous chapter, you have learned about the basic solid modeling tools which are regularly needed in 3D modeling using any CAD software. In this chapter, you will learn about some advanced 3D features like Sweep, loft, helix and so on. The tools to create these features are available in drop-downs of **Solids**, **Pattern**, and **Modify** panels of **Ribbon**. These tools are discussed next.

# ADD DROP-DOWN

There are various tools in **Add** drop-down to create 3D features like Sweep, loft, fillet weld, and so on; refer to Figure-1. These tools are discussed next.

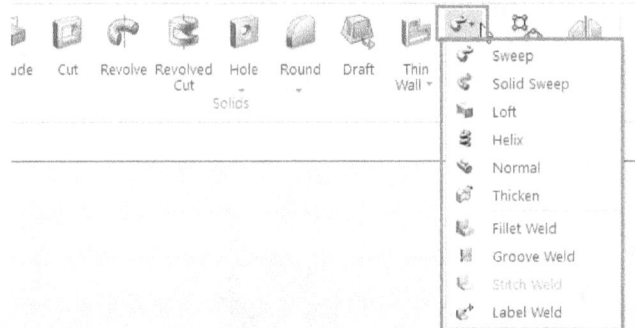

*Figure-1. Add drop-down*

# CREATING SWEEP FEATURE

The **Sweep** tool is used to create a solid volume by moving a sketch along the selected path. In other words, if you move a sketch along a path then the volume that is covered by moving sketch boundary is called sweep feature. Note that to use this tool, you must have a sketch section and a path. The steps to create sweep feature are given next.

- Click on the **Sweep** tool from the **Add** drop-down in the **Solids** panel of **Home** tab in the **Ribbon**. The **Sweep Options** dialog box will be displayed; refer to Figure-2. You can create mainly two type of sweep features; Sweep using single path and single cross-section or Sweep using multiple paths and multiple cross-sections. The procedures to create both type of sweep sections are discussed next.

## Sweep with Single Path and Single Cross-section

- Select the **Single path and cross section** radio button from the dialog box if you want to create sweep feature with single path and cross section.
- Select the **No merge** radio button from the **Face Merging** area of the dialog box if you do not want to merge faces while creating sweep feature. Select the **Full Merge** radio button to merge as many faces as possible while creating feature. Select the **Along path** radio button to merge faces along the path curve selected for sweep feature.
- Select the **Normal** radio button from the **Section Alignment** area of dialog box to keep section plane perpendicular to path curve. Select the **Parallel** radio button if you want the sections to be in parallel plane along the path while creating the feature.

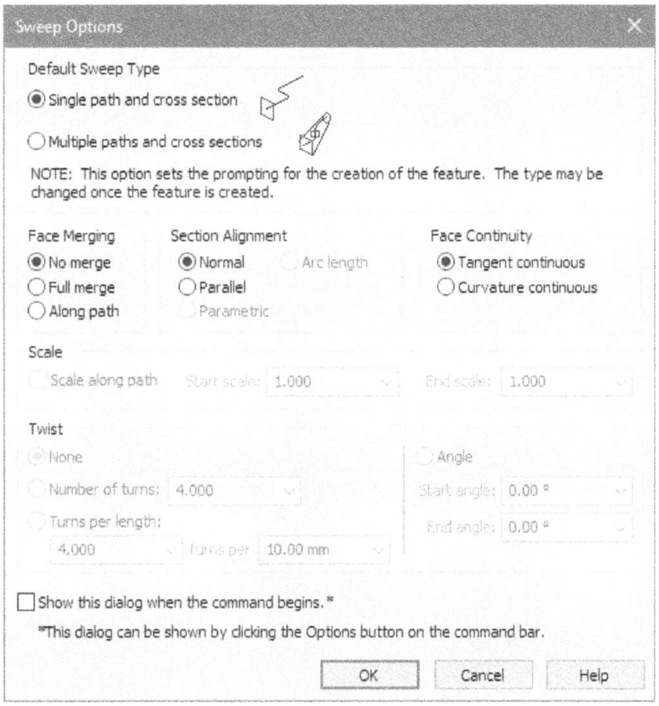

*Figure-2. Sweep Options Dialog Box*

- Select the **Tangent continuous** radio button or **Curvature continuous** radio button from the **Face Continuity** area of the dialog box to define how continuity will be followed for different sections of sweep feature.
- Click on the **OK** button from the dialog box to apply parameters.
- Select path from sketch along which you want to sweep the sketched profile. The path sketch can be an open or closed curve. After selecting path, click on the **OK** button from the **Command Bar**. The options in **Command Bar** will be modified and you will be asked to define cross section for sweep feature.
- Select the **Plane Normal To Curve** option from the drop-down and click on the curve created for path of sweep feature. A plane will get attached to cursor.
- Click at desired location on curve to place the plane. The sketching environment will displayed and you will be asked to draw section for sweep.
- Create a closed loop sketch for cross-section; refer to Figure-3.

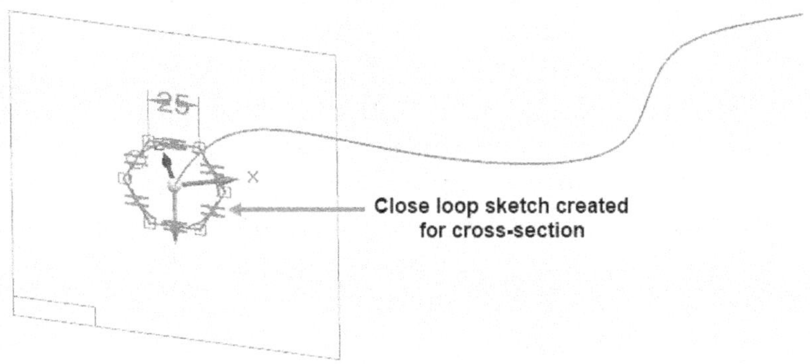

*Figure-3. Sketch for sweep cross-section*

- After creating sketch, close the sketching environment by clicking on the **Close Sketch** tool from the **Ribbon**. You will be asked to specify starting point of the cross-section.

- Click at desired point of cross-section to define start point and then right-click in the empty area. The sweep feature will be created; refer to Figure-4.

Note that if the sweep feature is not created and an error message is displayed then you need to first check the options specified in the **Sweep Options** dialog box and then make sure cross-section profile is a closed loop sketch created using plane perpendicular to path curve.

- Click on the **Finish** button from the **Command Bar** to create the feature and press **ESC** to exit the tool.

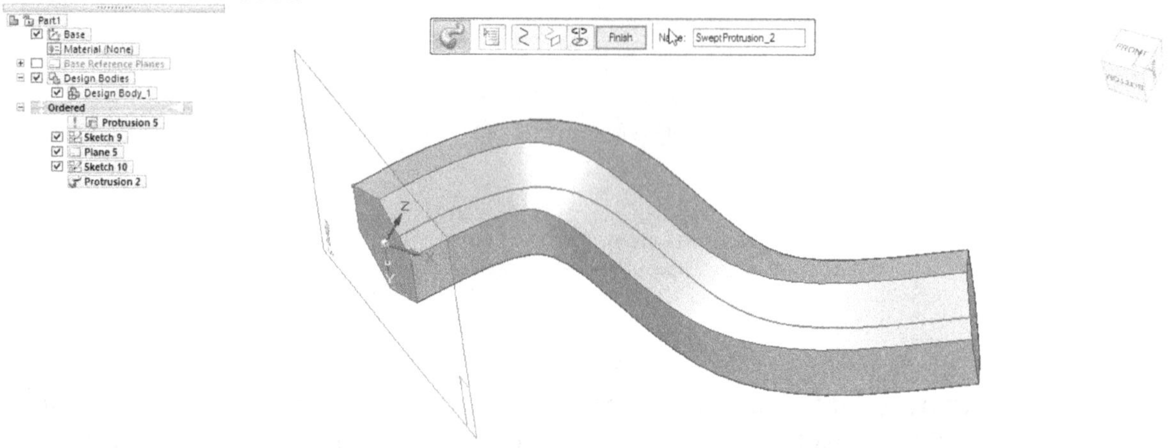

*Figure-4. Sweep protrusion created*

## Multi Section Sweep

The **Multi paths and cross sections** radio button in **Sweep Options** dialog box is used to create sweep feature using two or more sketch paths and cross sections at a time. The procedure to create multi section sweep feature is discussed next.

- Click on the **Sweep** tool from the **Add** drop-down in the **Solids** panel of the **Ribbon**. The **Sweep Options** dialog box will be displayed as discussed earlier.
- Select the **Multiple paths and cross sections** radio button from the dialog box to create feature with multiple paths and cross sections.
- Select the **Parametric** radio button from the **Section Alignment** area of the dialog box to create sweep feature up to the end points of path curves. Note that to accomplish this, the orientation of cross section will change along the path. Select the **Arc length** radio button from the dialog box if you want to vary the orientation of cross sections according to points on path curves. Nota that these points will have proportional arc length distance along the path curves.
- The other options in the dialog box are same as discussed earlier. Click on the **OK** button from the dialog box. The **Command Bar** will be displayed and you will be asked to select sketch curves for paths.
- Select the first path curve, right-click in the empty area and then select the next path curve. Repeat the step until all the path curves are selected.
- After selecting path curves, click on the **Next** button from the **Command Bar**. You will be asked to select the cross section sketches.
- One by one select the cross sections from their starting points. If you do not want the feature to be twisted then these starting points of sections should be in alignment; refer to Figure-5.

- After selecting sketch entities, click on the **Preview** button. The feature will be displayed; refer to Figure-6.

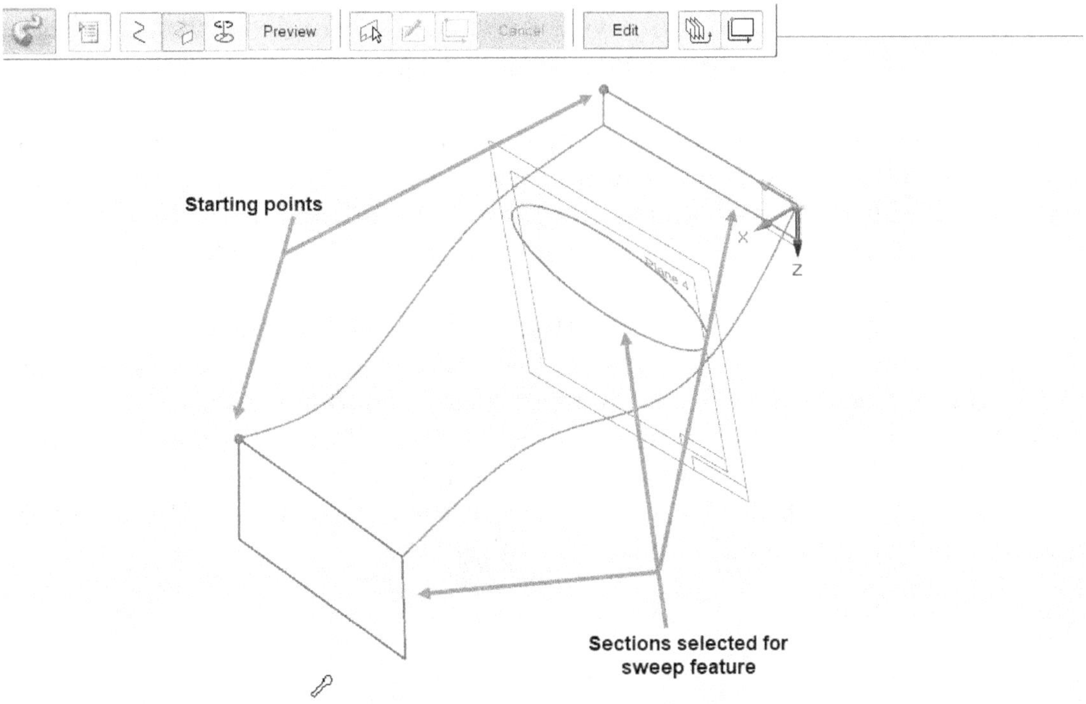

Figure-5. Multiple sections for sweep

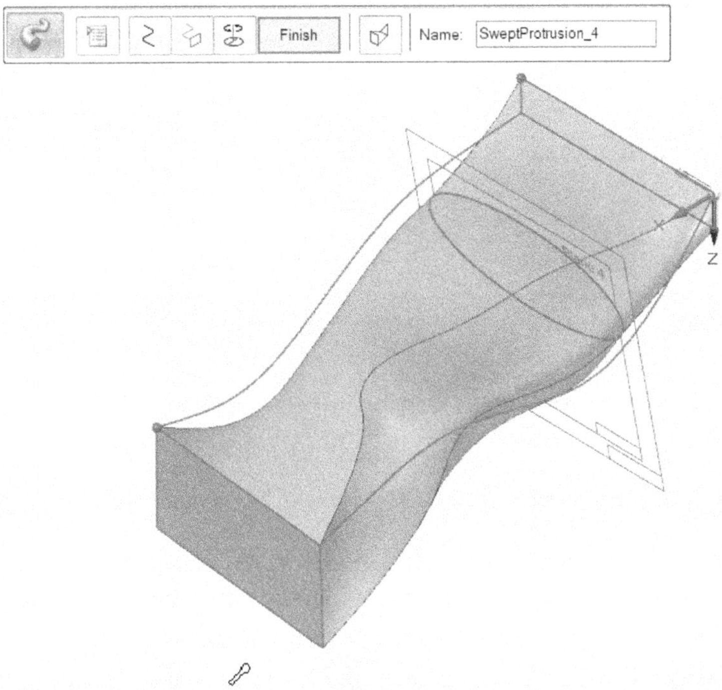

Figure-6. Multiple Cross Sections Sweep Protrusion Created

- Click on the **Finish** button from the **Command Bar** to create the feature. Press **ESC** to exit the tool.

## Twist Sweep

You can twist the section while sweeping along the path to create drill bit type of shape. Note that to use this option, you need to use single path and single cross section option. The procedure to create twist sweep feature is given next.

- Select the **Sweep** tool from the **Add** drop-down in the **Solids** panel of **Ribbon**. The **Sweep Options** dialog box will be displayed.
- Select the **Single path and cross section** radio button from the dialog box and click on the **OK** button from the dialog box. You will be asked to select sketch curve for path.
- Select the curve earlier created for path and right-click in the empty area. You will be asked to select sketch for cross section. Select or create sketch as discussed earlier. You will be asked to specify start point of cross section.
- Click at the desired point of cross section sketch and then right-click in the empty area. The preview of feature will be displayed.
- Now, click on the **Sweep Options** button from the **Command Bar**. The **Sweep Options** dialog box will be displayed as shown in Figure-7.

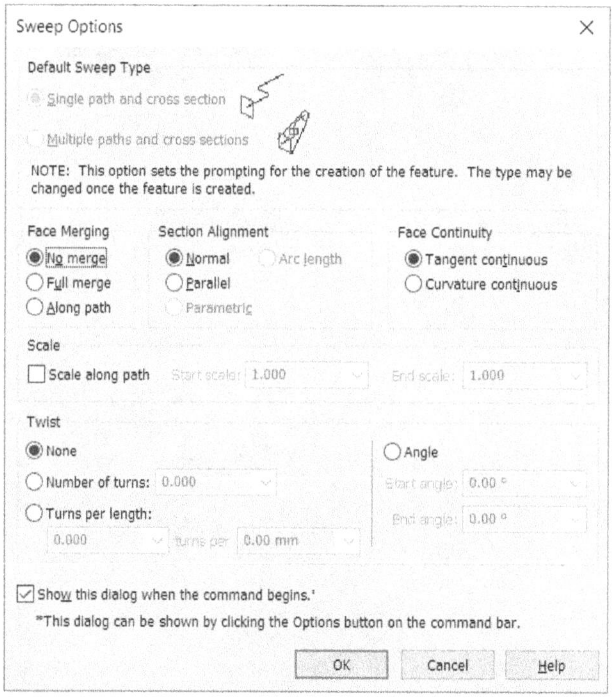

*Figure-7. Sweep Options dialog box with Twist options*

- Select the **Scale along path** check box if you want to increase/decrease the size of cross section while moving from one end of path curve to another. After selecting check box specify the desired scale factor values in the Start scale and End scale edit boxes.
- Select the **Number of turns** radio button from the **Twist** area if you want to specify how many turns will be applied on cross section along the path to create twisted sweep feature. Select the **Turns per length** radio button if you want to specify number of turns of cross section in specified length along the path curve. Select the **Angle** radio button if you want to specify start angle and end angle value of cross section. For example, if you want to give 10 turns along the full length of path curve then specify 0 in **Start angle** and 3600 in **End angle** edit boxes. After selecting the radio button, specify respective values in edit boxes of **Twist** area.

- After setting desired parameters, click on the **OK** button from the dialog box. Preview of twisted sweep feature will be displayed; refer to Figure-8.
- Click on the **Finish** button from **Command Bar** and press **ESC** to exit the tool.

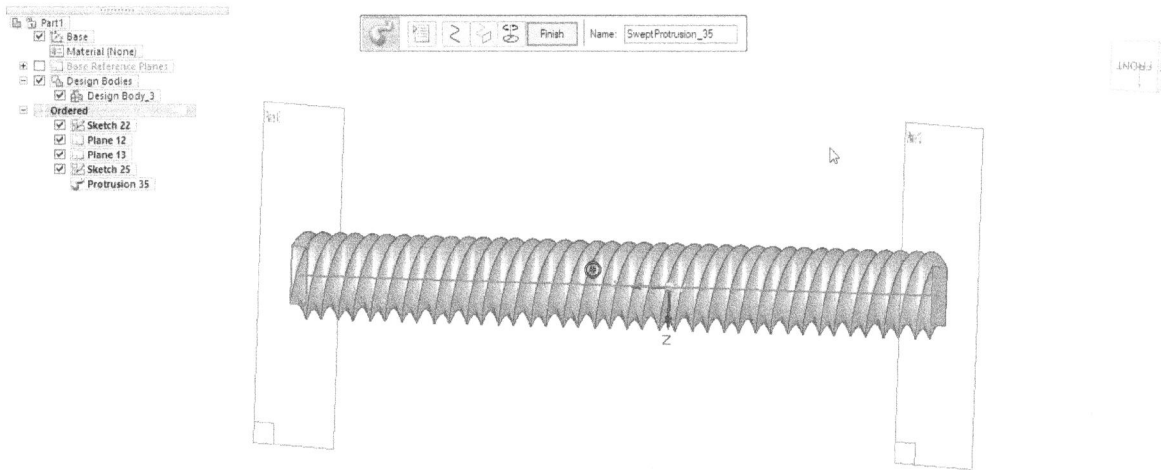

*Figure-8. Twist Sweep Created*

## SOLID SWEEP

This tool is same as **Sweep** discussed earlier but for cross section we will use solid body in place of sketch for creating the feature. The procedure to create solid sweep feature is given next.

- Click on the **Solid Sweep** tool from the **Add** drop-down in the **Solids** panel of the **Home** tab of the **Ribbon**. You will be asked to select sketch. After that **Solid Sweep Command Bar** will be displayed; refer to Figure-9.

*Figure-9. Solid Sweep Command Bar*

- Select the sketch curve for path and right-click in the empty area. You will be asked to select the body to be used for sweep.
- Select the desired solid body and press **ENTER**. You will be asked to select an edge, cylindrical face, or axis to be used as axis for revolving the solid body.
- Select the desired axis/edge and press **ENTER**. You will be asked to define the axis for direction of sweep.
- Select the desired entity; refer to Figure-10. After selecting the entities, press **ENTER**. The preview of feature will be displayed; refer to Figure-11.

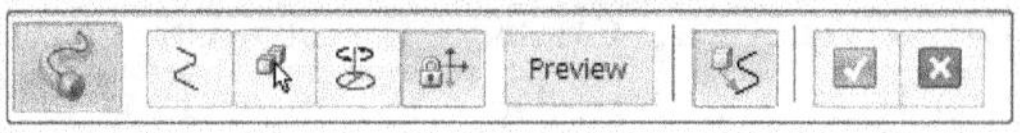

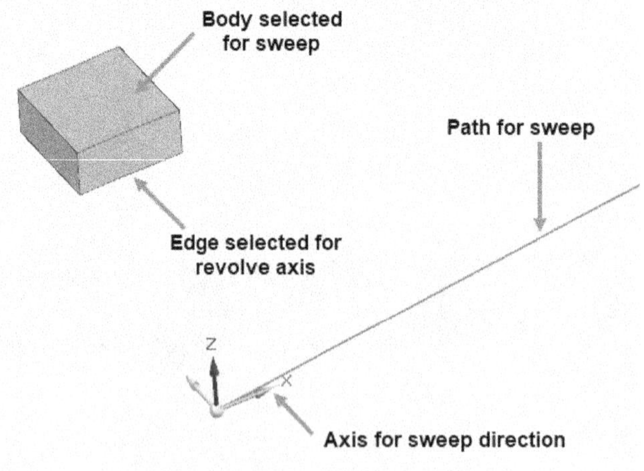

Figure-10. Entities selected for solid sweep

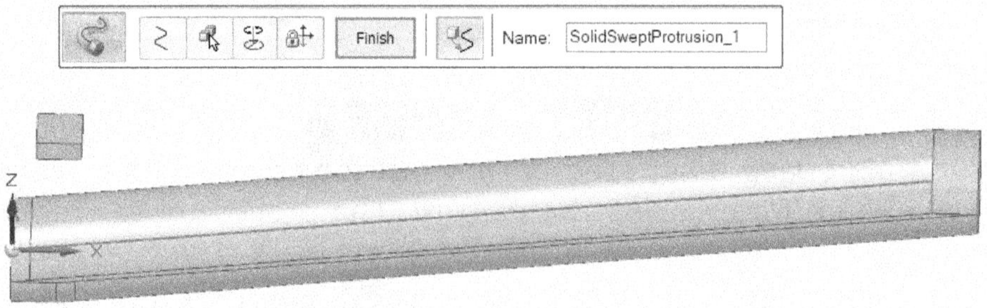

Figure-11. Preview of solid sweep feature

*   Click on the **Finish** button from **Command Bar** and press **ESC** to exit the tool.

## LOFT

The **Loft** tool is used to create solid by using a series of section. These sections define the resulting shape of feature. The procedure to create loft feature is discussed next.

*   Click on the **Loft** tool from the **Add** drop-down in the **Solids** panel of **Home** tab in the **Ribbon**. The **Lofted Protrusion Command Bar** will be displayed; refer to Figure-12 and you will be asked to select the sketch entities.

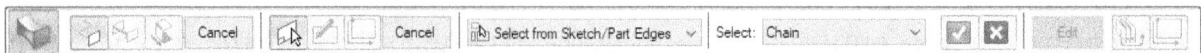

Figure-12. Lofted Protrusion Command Bar

*   One by one select the section sketches in sequence from their start point and right-click in empty area. The preview of feature will be displayed; refer to Figure-13.

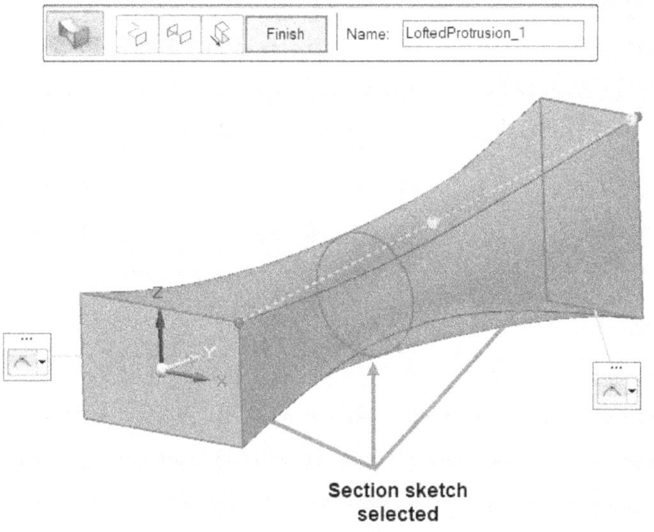

*Figure-13. Preview of loft feature*

- If you want to use a guide curve then click on the **Guide Curve Step** button from the **Command Bar** and select the desired curve. Right-click in the empty area to apply the settings.
- Click on the **Finish** button from the **Command Bar** and press **ESC** to exit the tool.

## HELIX

The **Helix** tool is used to create helical feature or spring like structures. The procedure to use this tool is discuss next.

- Click on the **Helix** tool from the **Add** drop-down in the **Solids** panel of **Home** tab in the **Ribbon**. The **Helix Command Bar** will be displayed; refer to Figure-14 and You will be asked to select existing sketch or create sketch after selecting plane.

*Figure-14. Helix Command Bar*

- For creating helix feature there must be an axis of revolution and a closed loop cross section in the sketch; refer to Figure-15. Select the desired plane and create the sketch.

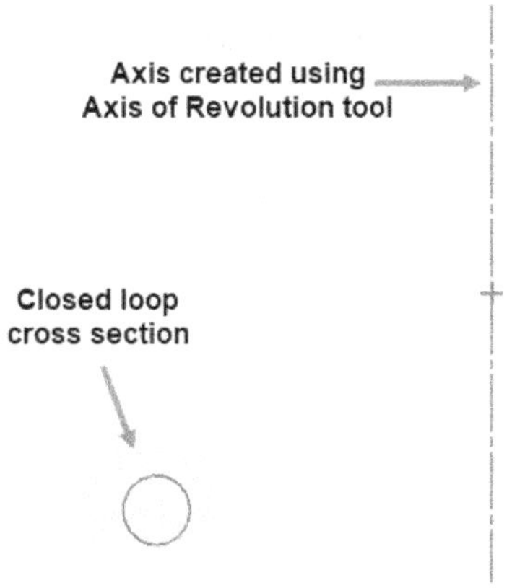

*Figure-15. Sketch for helix feature*

- After creating sketch, click on the **Close Sketch** tool from the **Close** panel in the **Ribbon**. You will be asked to specify start point of the feature.
- Select the desired point on the axis to specify start point. You will be asked to specify helix parameters and the **Command Bar** will be displayed as shown in Figure-16.

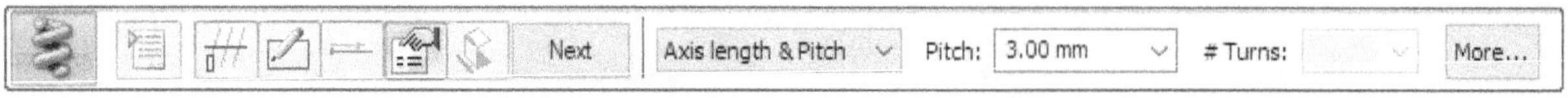

*Figure-16. Helix Customization Command Bar*

- Specify the desired parameters and press **ENTER**. You will be asked to specify start point of helix feature.
- Click at the end point of axis. Preview of the feature will be displayed.
- If you want to specify the extent of feature then click on the **Extent Step** button from the **Command Bar** and set desired parameters.
- After setting the parameters, click on the **Finish** button from the **Command Bar** to create feature and press **ESC** to exit the tool.

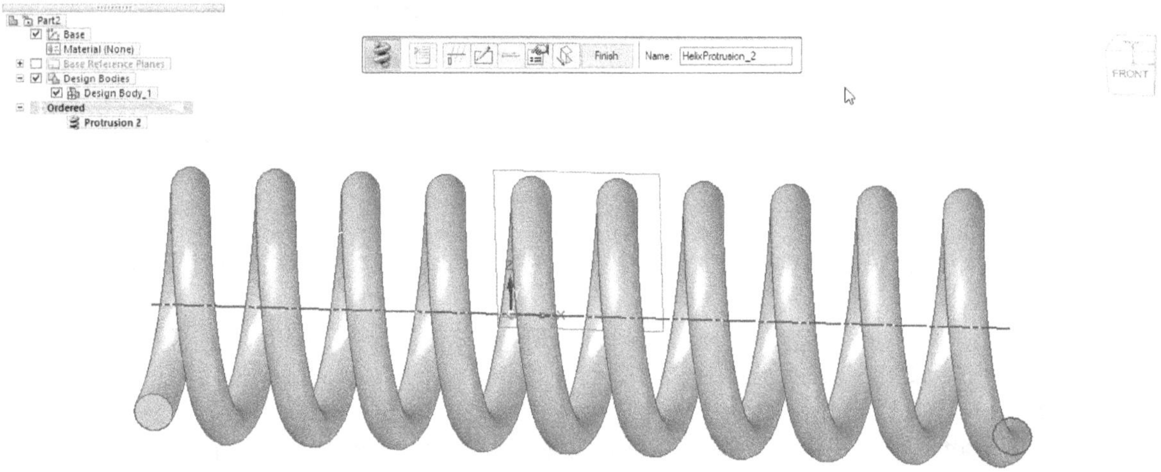

*Figure-17. Helix Created*

Before using the **Normal** tool, we need a sketch projected on the face of solid part. The tool to project sketch is a surfacing tool but we will discuss it here.

## PROJECTING SKETCH CURVES ON SURFACE/FACE

The **Project** tool in **Surfacing** tab is used to project curves of selected sketch onto a face/surface. The procedure to do so is given next.

- Click on the **Project** tool from the **Curves** panel in the **Surfacing** tab of the **Ribbon**. The **Project Command Bar** will be displayed; refer to Figure-18 and you will be asked to select the curves to be projected.

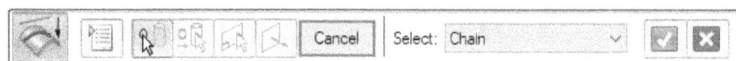

*Figure-18. Project Command Bar*

- Select the desired curves and right-click in the empty area. You will be asked to select the face on which curves are to be projected.
- Select the desired face and right-click in empty area. You will be asked to specify the direction in which the sketch curve will be projected.

- Click above or below the curve plane to define direction. The curves will be projected; refer to Figure-19.

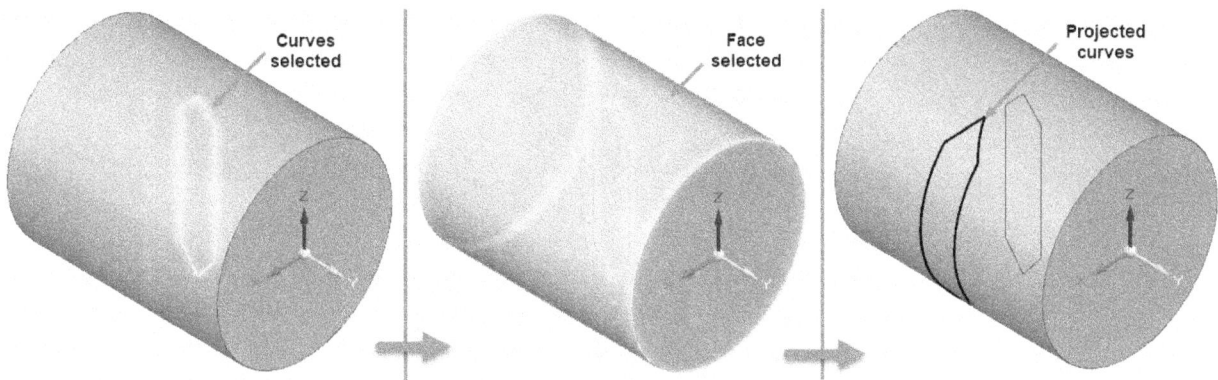

*Figure-19. Projecting curves on face*

- Click on the **Finish** button from the **Command Bar** to create the feature and press **ESC** to exit the tool.

# NORMAL

The **Normal** tool is used to create protrusion normal to a face using closed loop sketch section projected on the face. The procedure to use this tool is discuss next.

- Click on the **Normal** tool from the **Add** drop-down in the **Solids** panel of the **Home** tab in the **Ribbon**. **The Normal Protrusion Command Bar** will be displayed; refer to Figure-20 and you will be asked to select curves projected on a face.

*Figure-20. Normal Protrusion Command Bar*

- Select the desired curves and right-click in the empty area. The **Command Bar** will be displayed as shown in Figure-21 and you will be asked to specify which side of section will be used for protrusion.

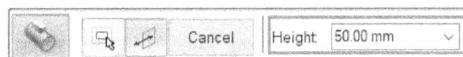

*Figure-21. Height edit box for protrusion height*

- Specify the desired value in the **Height** edit box of the **Command Bar** and then click inside or outside of the section to define the side to be used for protrusion. The protrusion will be created; refer to Figure-21.
- Click on the **Finish** button from the **Command Bar** and press **ESC** to exit the tool.

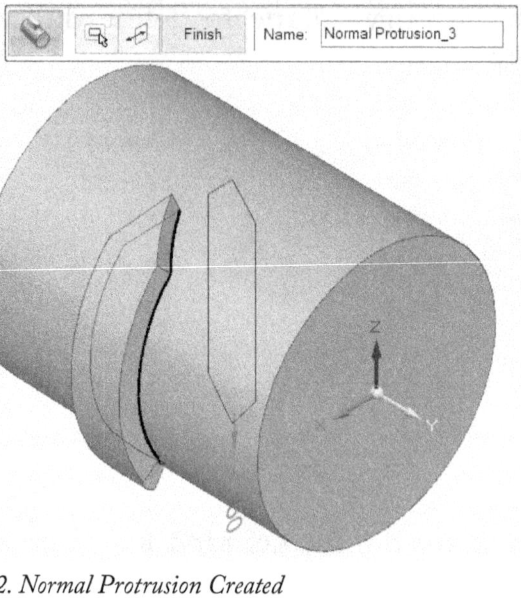

*Figure-22. Normal Protrusion Created*

## THICKEN

The **Thicken** tool is used to thicken selected surface or modify thickness of existing solid. Using this tool you can add material to model without modifying actual sketch. The steps to use this tool are discuss next.

* Click on the **Thicken** tool from the **Add** drop-down in the **Solids** panel of the **Home** tab in the **Ribbon**. The **Thicken Command Bar** will be displayed; refer to Figure-23. Note that this tool will be active only when there is a solid or surface body already created in the modeling area.

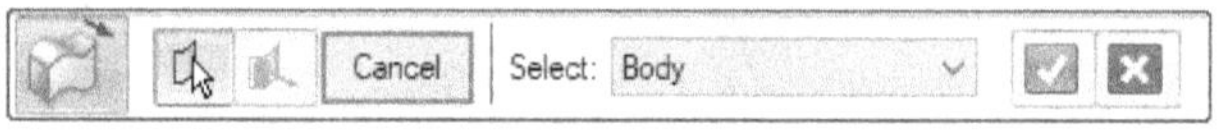

*Figure-23. Thicken Command Bar*

* Select the desired option from the **Select** drop-down to define scope of selection.
* Select the desired body/faces/surfaces to be thicken and right-click in the empty area.
* Specify distance value up to which you want to thicken selected objects in the **Distance** edit box of **Command Bar** and then click on desired side of surface/face/body to define direction of thickening. The preview of thicken feature will be displayed; refer to Figure-24.
* Click on the **Finish** button from the **Command Bar** and press **ESC** to exit the tool.

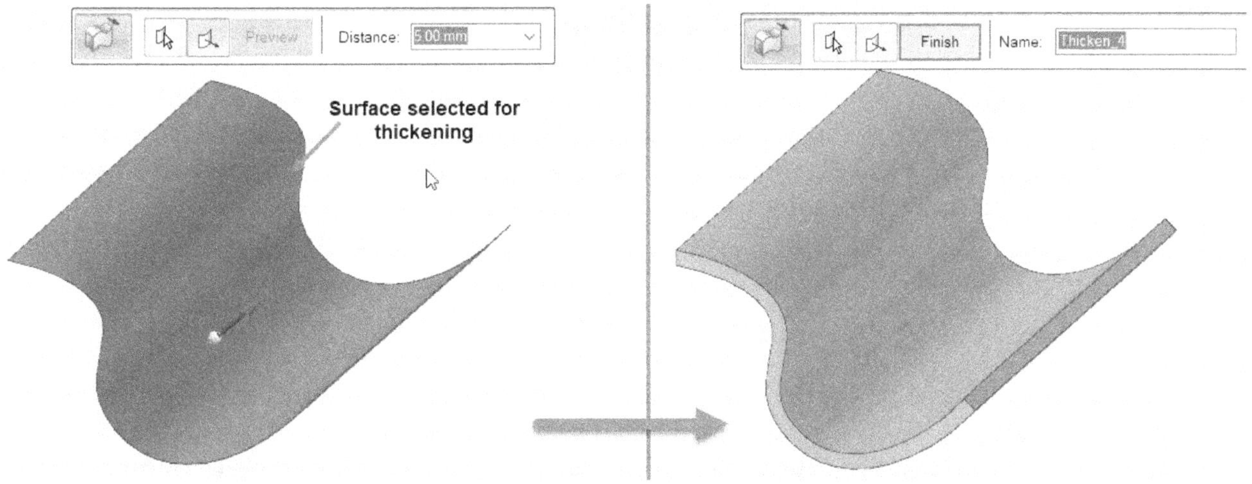

*Figure-24. Thicken Created*

# CREATING FILLET WELD

The **Fillet Weld** tool is used to create weld beam at intersection of two faces. The weld is applied during fabrication for permanent joint. The procedure to use this tool is given next.

• Click on the **Fillet Weld** tool from the **Add** drop-down in the **Solids** panel of **Home** tab in the **Ribbon**. The **Fillet Weld Options** dialog box will be displayed; refer to Figure-25.

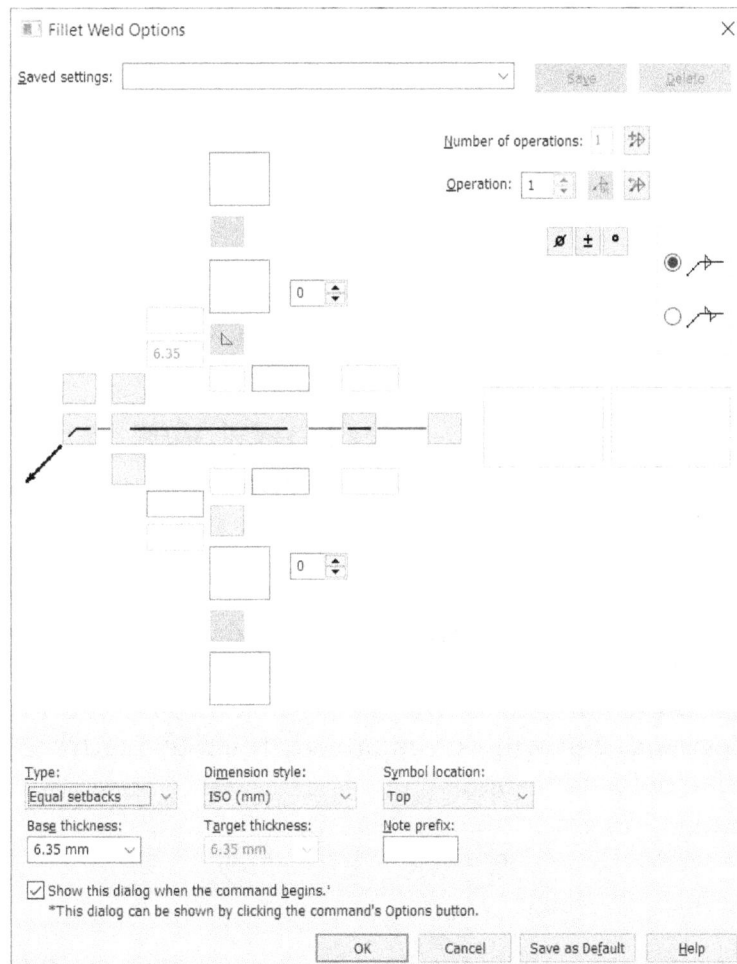

*Figure-25. Fillet Weld Options dialog box*

- Specify the desired parameters in the dialog box to define weld parameters. Note that tips for various options are displayed when you hover cursor on the options in the dialog box.
- After setting desired parameters in the dialog box, click on the **OK** button from the dialog box. The **Fillet Weld Command Bar** will be displayed; refer to Figure-26 and you will be asked to select faces for base set.

*Figure-26. Fillet Weld Command Bar*

- Select the desired face(s) and right-click in empty area. You will be asked to select faces of target set.
- Select the desired faces and right-click in empty area. Click on the **Preview** button from the **Command Bar**. The preview of weld will be displayed; refer to Figure-27.

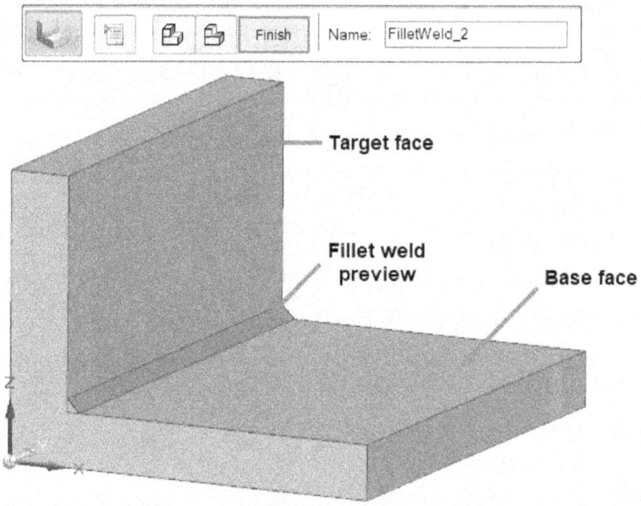

*Figure-27. Fillet Weld Preview*

- Click on the **Finish** button from the **Command Bar** to create the feature and press **ESC** to exit the tool.

## CREATING GROOVE WELD

The **Groove Weld** tool is used to create a weld bead joining two parallel or inclined intersecting faces. The procedure to use this tool is given next.

- Click on the **Groove Weld** tool from the **Add** drop-down in the **Solids** panel of **Home** tab in the **Ribbon**. The **Groove Weld Options** dialog box will be displayed similar to **Fillet Weld Options** dialog box discussed earlier.
- Specify the desired parameters in the dialog box and click on the **OK** button. The **Groove Weld Command Bar** will be displayed; refer to Figure-28 and you will be asked to select faces for base set.

*Figure-28. Groove Weld Command Bar*

- Select the desired face(s) and right-click in the empty area. You will be asked to select faces for target set.

- Select the desired face(s) and right-click in the empty area. You will be asked to select the top boundary of weld bead.

- Select the top edge of target/base face and right-click. You will be asked to define the bottom boundary of weld bead.

- Select the desired edge; refer to Figure-29 and press **ENTER**. Click on the **Preview** button from the **Command Bar**. Preview of weld bead will be displayed; refer to Figure-30.

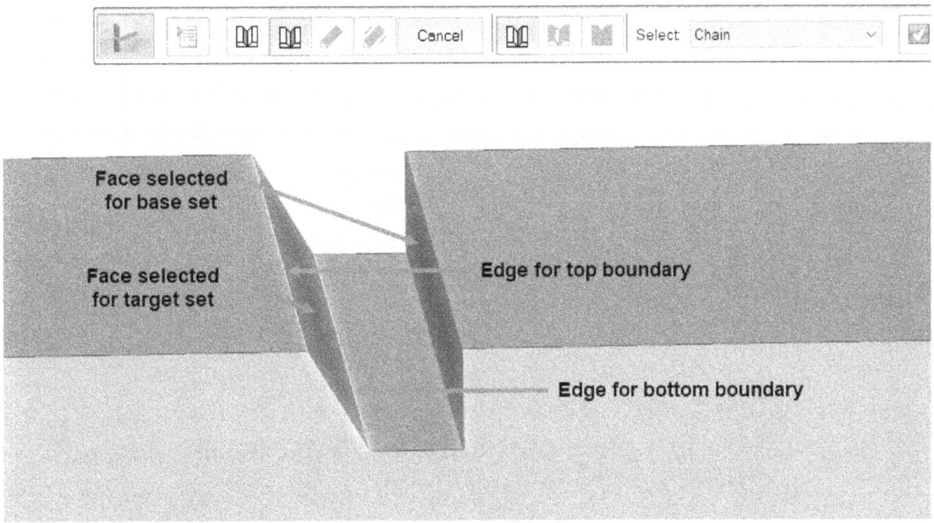

*Figure-29. Selection for Groove weld*

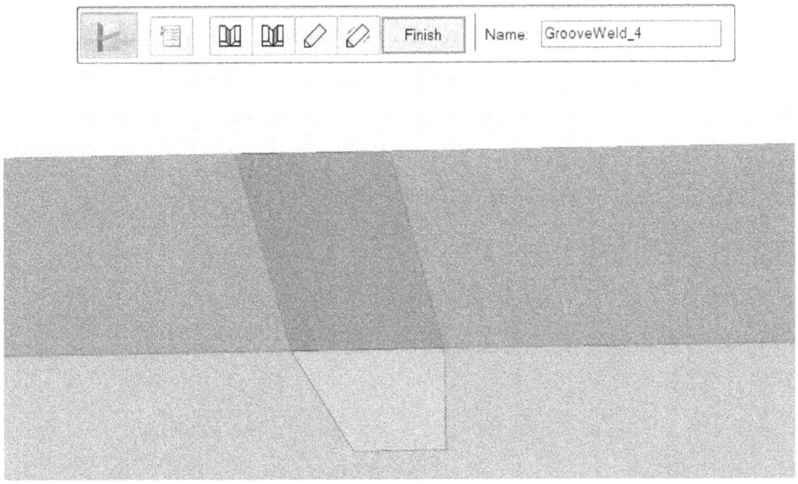

*Figure-30. Preview of groove weld bead*

- Click on the **Finish** button from the **Command Bar** to create the feature and press **ESC** to exit the tool.

## CREATING STITCH WELD

The **Stitch Weld** tool is used to add cutouts to a weld bead. The procedure to use this tool is given next.

- Click on the **Stitch Weld** tool from the **Add** drop-down in the **Solids** panel of **Home** tab in the **Ribbon**. The **Stitch Weld Options** dialog box will be displayed; refer to Figure-31.

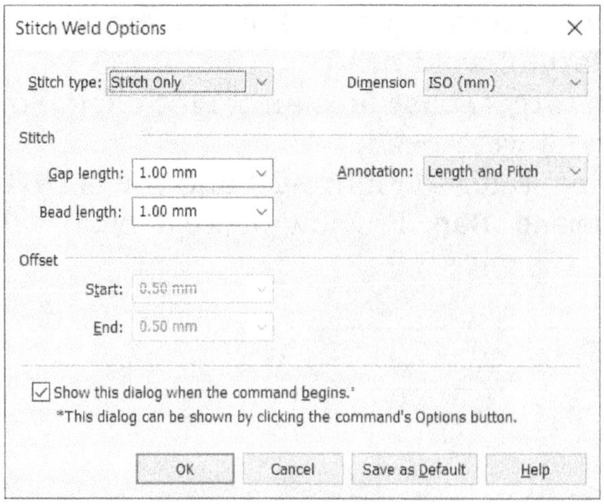

*Figure-31. Stitch Weld Options dialog box*

- Select the desired option from the **Stitch type** drop-down to define whether you want to create stitches in weld bead, offset the weld bead along with stitch, or want to offset the weld bead.
- Select the desired option from the **Dimension** drop-down to define the dimensioning style.
- Set the desired values for gap between two consecutive bead lengths and length of bead in respective edit boxes of the **Stitch** area in the dialog box. Select the desired option from the **Annotation** drop-down to define how stitch weld annotation will be created.
- If you have selected option for offset then specify the desired values of start and end offset in respective edit box. After setting parameters, click on the **OK** button from the dialog box. The **Stitch Weld Command Bar** will be displayed; refer to Figure-32 and you will be asked to select the edge of weld bead.

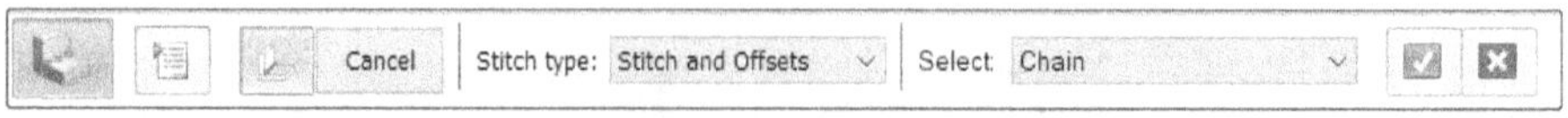

*Figure-32. Stitch Weld Command Bar*

- Select the desired edge and then right-click in empty area. The preview of stitch weld will be displayed; refer to Figure-33.

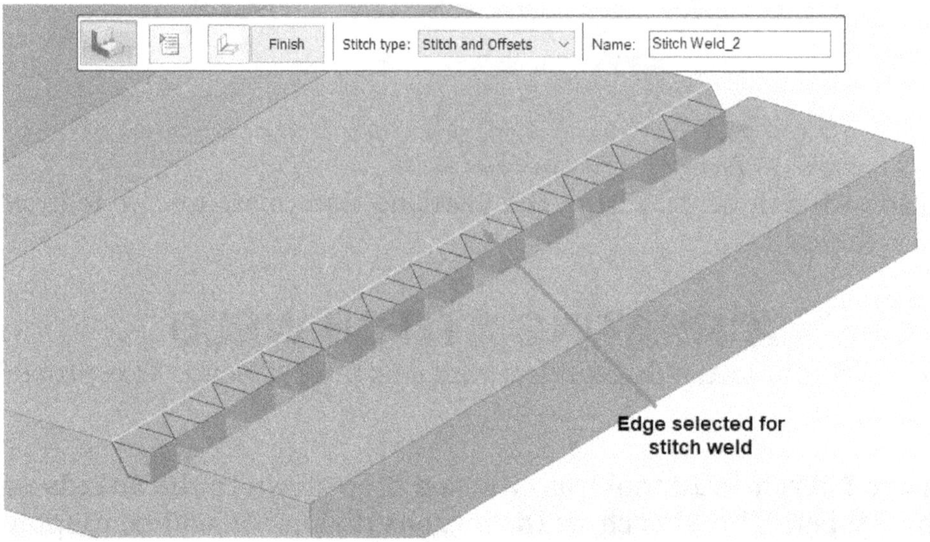

*Figure-33. Preview of stitch weld*

- Click on the **Finish** button from the **Command Bar** to create feature and press **ESC** to exit the tool.

# CREATING LABEL WELD

The **Label Weld** tool is used to create user defined weld label for selected edges. Using this tool, you can apply the weld annotation to any edge for fabrication. The procedure to use this tool is given next.

- Click on the **Label Weld** tool from the **Add** drop-down in the **Solids** panel of **Home** tab in the **Ribbon**. The **Label Weld Options** dialog box will be displayed similar to **Fillet Weld Options** dialog box as discussed earlier.
- Set the desired parameters in the dialog box and click on the **OK** button. You will be asked to select an edge or edge chain.
- Select the desired edges where you want to apply weld label and click on the **OK** button from the **Command Bar**. The preview of weld will be displayed with dark edges.
- Click on the **Finish** button to create the feature and press **ESC** to exit the tool.

# CUT FEATURES

There are five tools in available in the **Cut** drop-down to create different type of cut features; refer to Figure-94. These tools are discussed next.

*Figure-34. Cut drop-down*

## Sweep Cutout

The **Swept Cutout** tool is used to remove material from solid model by using swept feature. The **Swept Cutout** tool works in the same way as discussed for **Sweep** tool. The procedure to use this tool is given next.

- Click on the **Swept Cutout** tool from the **Cut** drop-down in the **Solids** panel of **Home** tab in the **Ribbon**. The **Sweep Options** dialog box will be displayed as discussed earlier.
- Specify the parameters as discussed for **Sweep** tool and click on the **OK** button from the dialog box. The **Swept Cutout Command Bar** will be displayed similar to **Swept Protrusion Command Bar**; refer to Figure-35.
- Select path and cross-section sketches which intersects with the solid body; refer to Figure-36.

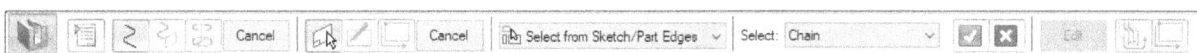

*Figure-35. Sweep Cutout Command Bar*

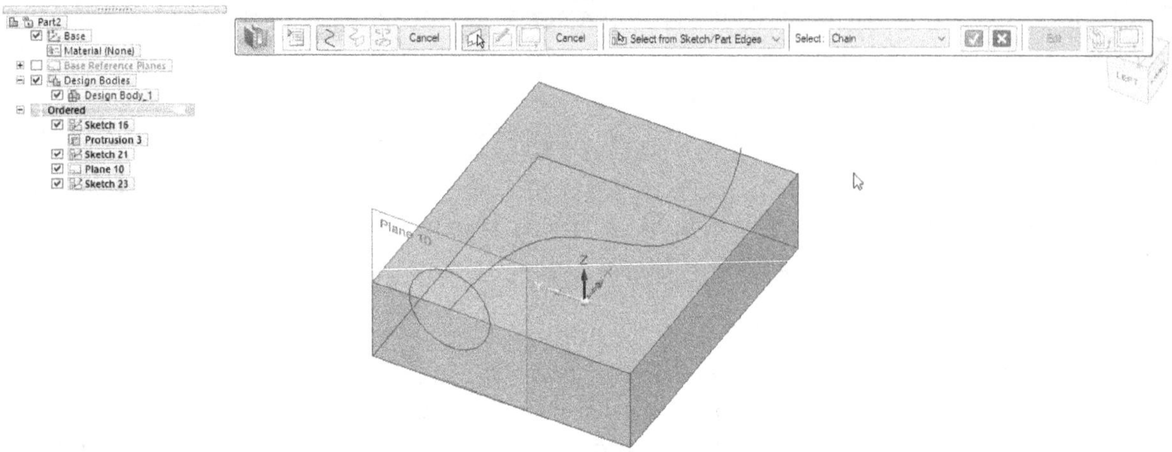

*Figure-36. Swept cutout sketch created*

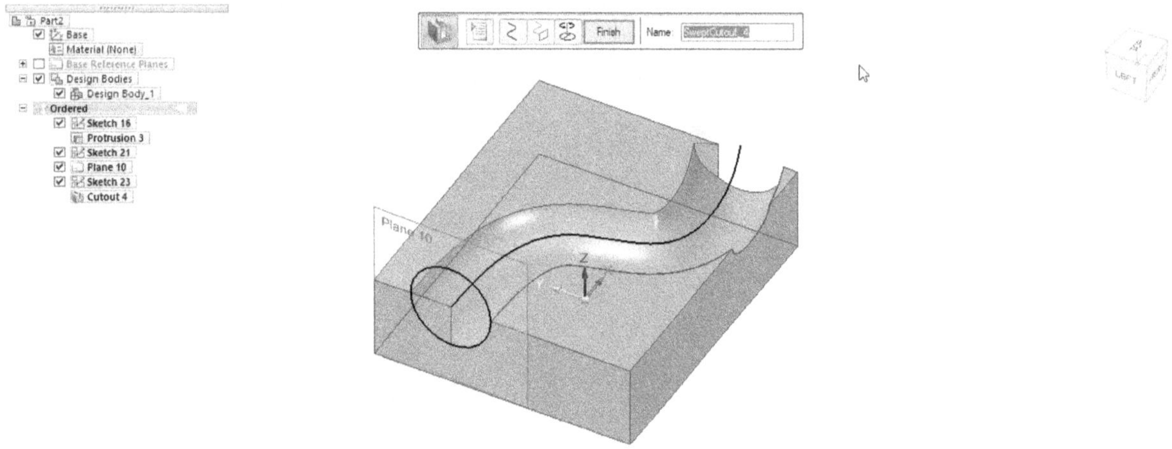

*Figure-37. Swept cutout feature created*

## Solid Sweep Cut

The **Solid Sweep Cutout** tool is used to remove material from a solid model by using another solid body and a path. Note that multiple bodies can be created by using the **Add Body** tool. The procedure to use this tool is discussed later in this chapter. After creating the bodies, select the body from which material will be removed from the **Path Finder** and right-click on it. The right-click shortcut menu will be displayed; refer to Figure-38. Select the **Activate Body** option from the shortcut menu to activate the base body. Now, click on the **Solid Sweep Cutout** tool from the **Cut** drop-down in the **Solids** panel of **Home** tab in the **Ribbon**. The **Solid Sweep Cutout Command Bar** will be displayed; refer to Figure-39. Rest of the steps are same as we used for Solid Sweep. Figure-40 shows an example of solid sweep cutout.

Figure-38. Activate Body

Figure-39. Solid Sweep Cutout Command Bar

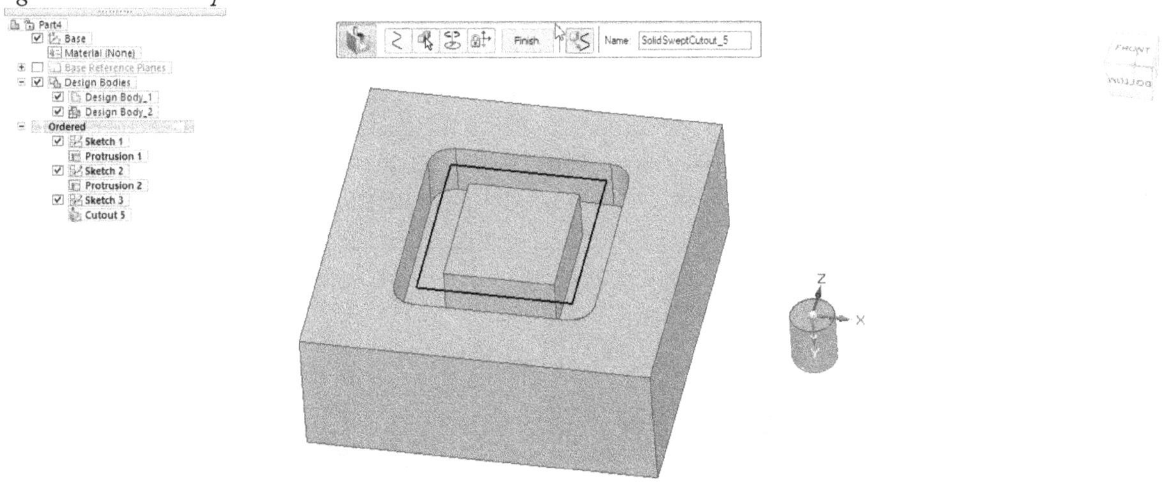

Figure-40. Solid Sweep Cutout Created

## Loft Cutout

The **Loft Cutout** tool is used to remove material by using multiple sections. The working steps are same as we discussed earlier for **Loft** tool. Refer to Figure-41, Figure-42, and Figure-43.

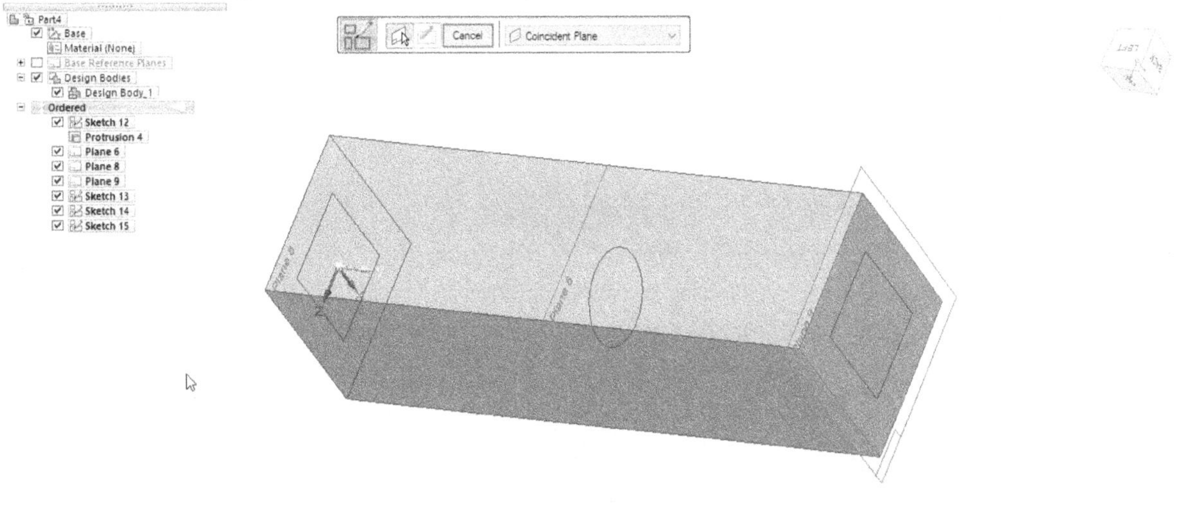

Figure-41. Loft Cutout Sketch sections

Figure-42. Loft Cutout Command Bar

Figure-43. Loft Cutout Created

## Helix Cut

The **Helix Cut** tool is used to create helix cut on solid model while removing material. This tool is available in the **Cut** drop-down. The procedure to create helix cut is similar to creating helix feature. Refer to Figure-44 and Figure-45.

Figure-44. Helix Cutout Command Bar

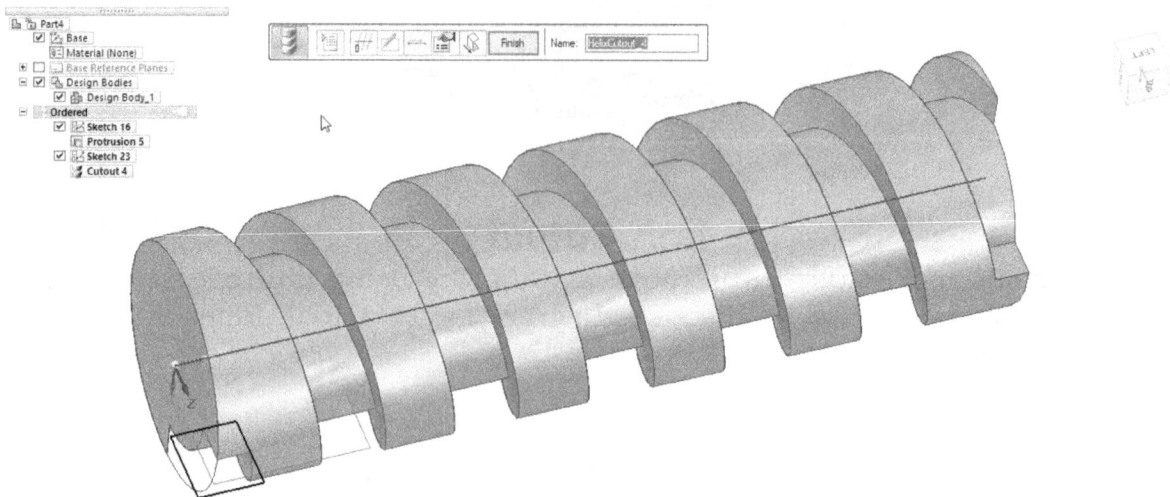

Figure-45. Helix Cutout Created

## Normal cut

The **Normal Cutout** tool is similar to **Normal** tool discussed earlier in **Add** drop-down. Using the **Normal Cutout** tool, you can remove material from the solid by using sketch projected on solid face; refer to Figure-46 and Figure-47.

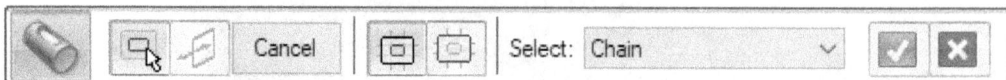

Figure-46. Normal Cutout Command Bar

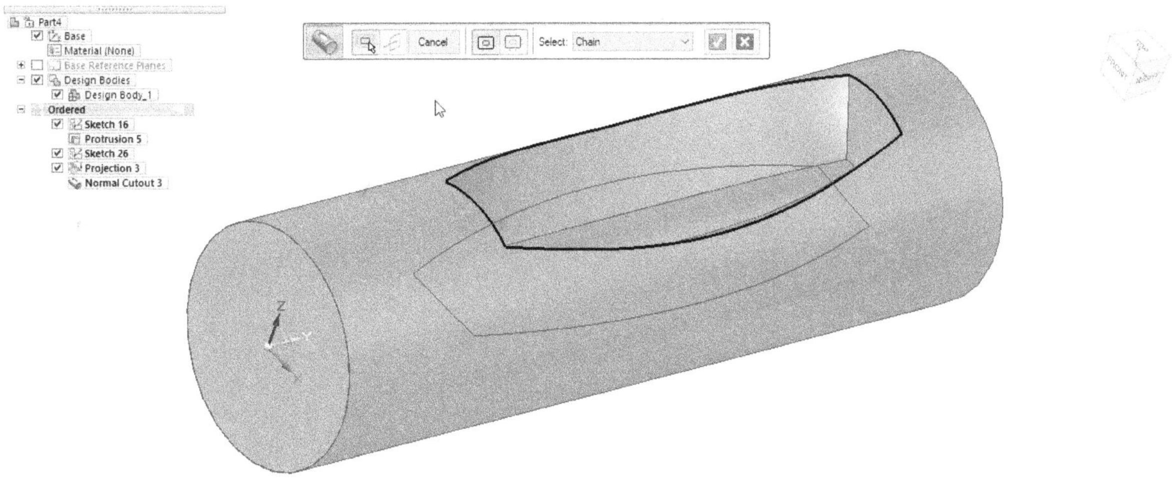

*Figure-47. Normal Cutout Created*

# ADD BODY

The **Add Body** tool is used to add another body in the model separate from existing bodies. The procedure to use this tool is discuss next.

- Click on **Add Body** tool from the **Solid** group. The **Add Body** dialog box will be displayed; refer to Figure-48.
- Select the **Add Part body** radio button to add a new part body. Select the **Add Sheet Metal body** radio button to create a new sheet metal body. You will learn about sheet metal model later in the book.

*Figure-48. Add Body Dialog Box*

- Select the **Add Part body** radio button and specify the desired names of existing body and new body in the respective edit boxes.
- Click on the **OK** button from the dialog box. New body will be activated.
- Create the desired solid using the solid modeling tools discussed earlier; refer to Figure-49.

*Figure-49. New Body Added*

Note that by default new body is activated after creation. So, if you now create any object then it will be created in new body. To activate any previous body, select it from the **Design Bodies** node in the **Path Finder**, right-click on it and then select the **Activate Body** option; refer to Figure-50.

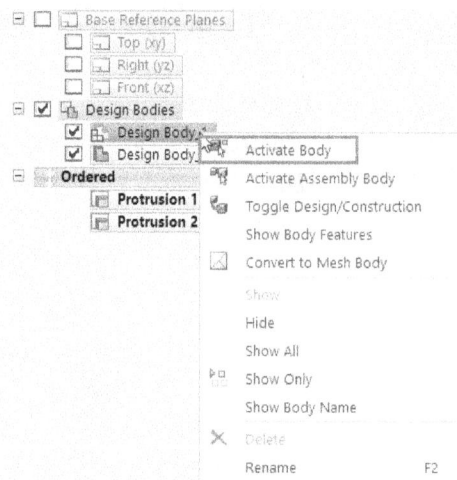

*Figure-50. Activate Body option*

## Enclosure

The **Enclosure** tool is used to enclose existing design body into box or cylinder. The procedure is discussed next.

*   Click on the **Enclosure** tool from the **Add Body** drop-down in the **Solids** panel of **Home** tab in the **Ribbon**. The **Enclosure Command Bar** will be displayed; refer to Figure-51.

*Figure-51. Enclosure Command bar*

*   Select the desired enclosure shape from the **Enclosure Shape** drop-down in the **Command Bar**.
*   Select the desired option from the **Selection Type** drop-down in the **Command Bar** and select the desired object for which enclosure is to be created.
*   After selecting object, right-click in the empty area. You will be asked to select a plane/face for defining normal direction of enclosure.

- Select the desired face/plane to give direction of enclosure and then specify the value in **Offset** edit box of **Command Bar** to increase side of enclosure; refer to Figure-52.
- Click on the **Finish** button from the **Command Bar** to create the feature and press **ESC** to exit the tool.

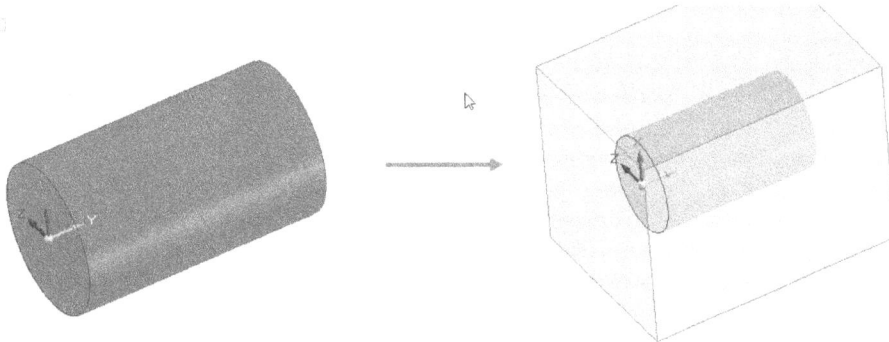

*Figure-52. Enclosure Created*

# Union

The **Union** tool is used to combine multiple bodies into single body. The procedure to use this tool is given next.

- Click on the **Union** tool from the **Add Body** drop-down of **Solids** panel in the **Ribbon**. The **Union Command Bar** will be displayed; refer to Figure-53.

*Figure-53. Union Command Bar*

- Select first body and then select the second body.
- Click on the **OK** button from the **Command Bar**. The two bodies will be converted into single entity which can be checked in **Path Finder**.

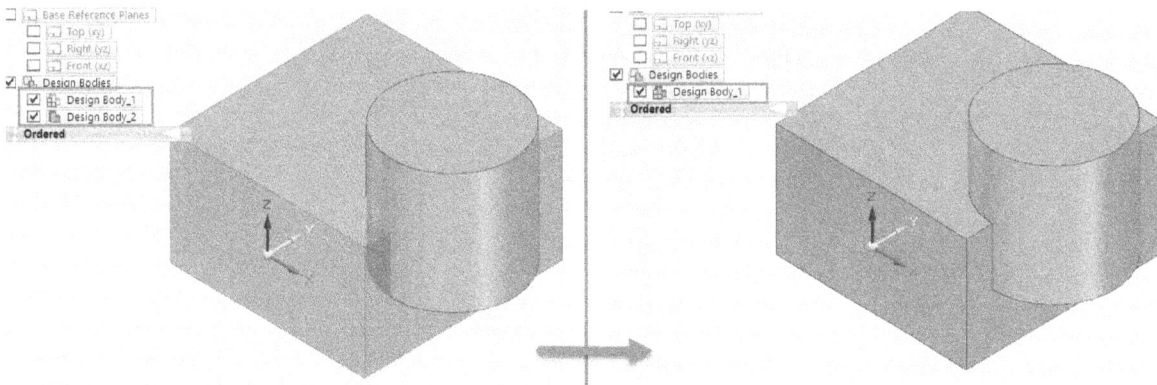

*Figure-54. Union Created*

# Subtract

The **Subtract** tool is used to remove material from main body using a tool body. Note that there must be at least two intersecting bodies for this tool to work. The procedure to use this tool is discuss next.

- Click on the **Subtract** tool from the **Add Body** drop-down in the **Solids** panel of **Home** tab in the **Ribbon**. The **Subtract Command Bar** will be displayed; refer to Figure-55.

*Figure-55. Subtract Command Bar*

- Click on the **Output Options** button from the **Command Bar** to modify output of operation. The **Subtract** dialog box will be displayed; refer to Figure-56.

*Figure-56. Subtract dialog box*

- Set the desired parameters in the dialog box and click on the **OK** button. You will be asked to select the target body.
- Select entity from which you want to subtract another body and right-click in the empty area. You will be asked to select the tool body.
- Select the second object to be used as tool for subtraction. Preview of the subtraction feature will be displayed; refer to Figure-57.

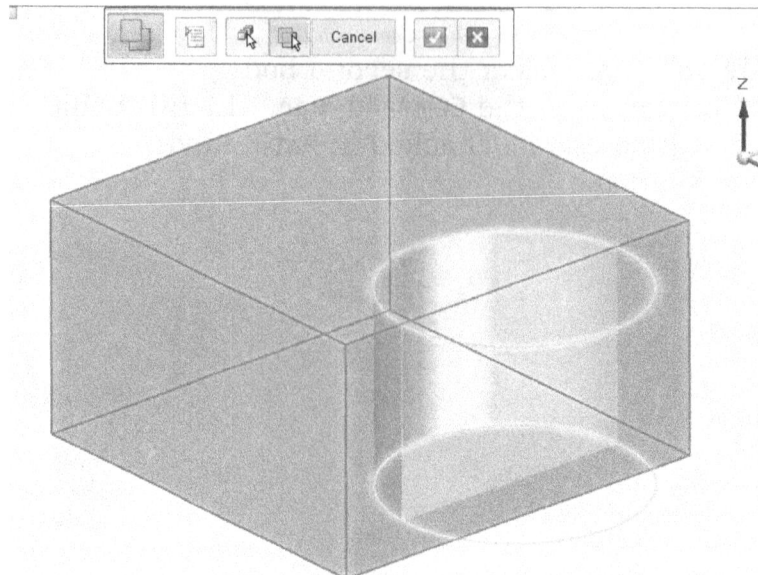

*Figure-57. Subtraction feature created*

- Click on the **Finish** button to create the feature and press **ESC** to exit the tool.

## Intersect

The **Intersect** tool is used to create common volume by using intersection of two solid bodies. The procedure of using this tool is discuss next.

- Click on the **Intersect** tool from **Add Body** drop-down. The **Intersect Command Bar** will be displayed; refer to Figure-58 and you will be asked to select the target body.

*Figure-58. Intersect Command Bar*

- Select the first entity which intersects with another body. You will be asked to select the tool body.
- Select the second entity. Preview of the feature will be displayed; refer to Figure-59.
- Click on the **OK** button from the **Command Bar** to create the feature and press **ESC** to exit the tool.

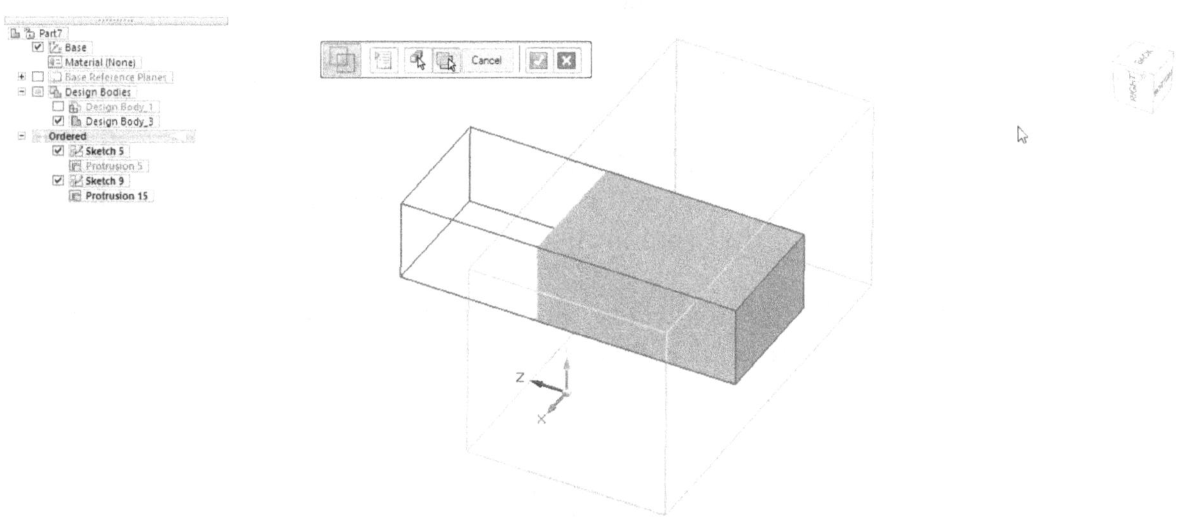

*Figure-59. Intersect Created*

## Split

The **Split** tool is used to break a solid body into two individual bodies. The procedure to use this tool is discussed next.

- Click on the **Split** tool from the **Add Body** drop-down in the **Solids** panel of **Home** tab in the **Ribbon**. The **Split Command Bar** will be displayed; refer to Figure-60.

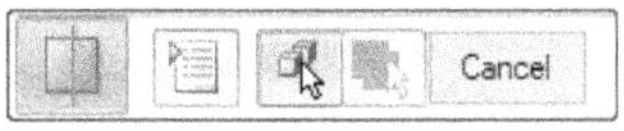

*Figure-60. Split Command Bar*

- Select entity which you want to split. You will be asked to select cutting tool (plane, face, body).
- Select the desired object. Preview of split feature will be displayed; refer to Figure-61.

*Figure-61. Split Of Solid Body Created*

- Click on the **OK** button from the **Command Bar**. The preview of feature will be displayed.
- Click on the **Finish** button from **Command Bar** and press **ESC** to exit the tool.

## Scale Body

The **Scale Body** tool is used to enlarge or diminish body by using specified scale factor. The procedure of using this tool is discussed next.

- Click on the **Scale Body** tool from **Add Body** drop-down in the **Solids** panel of **Home** tab in the **Ribbon**. The **Scale Body Command Bar** will be displayed; refer to Figure-62.

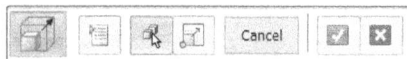

*Figure-62. Scale Body command bar*

- Select the desired body to be scaled up/down and right-click in the empty area. The **Command Bar** will be modified and you will be asked to specify the scale value in the **Scale Factor** edit box of **Command Bar**.
- Click on the **Preview** button to check preview of scaling; refer to Figure-63. Click on the **Finish** button to create the feature and press **ESC** key to exit the tool.

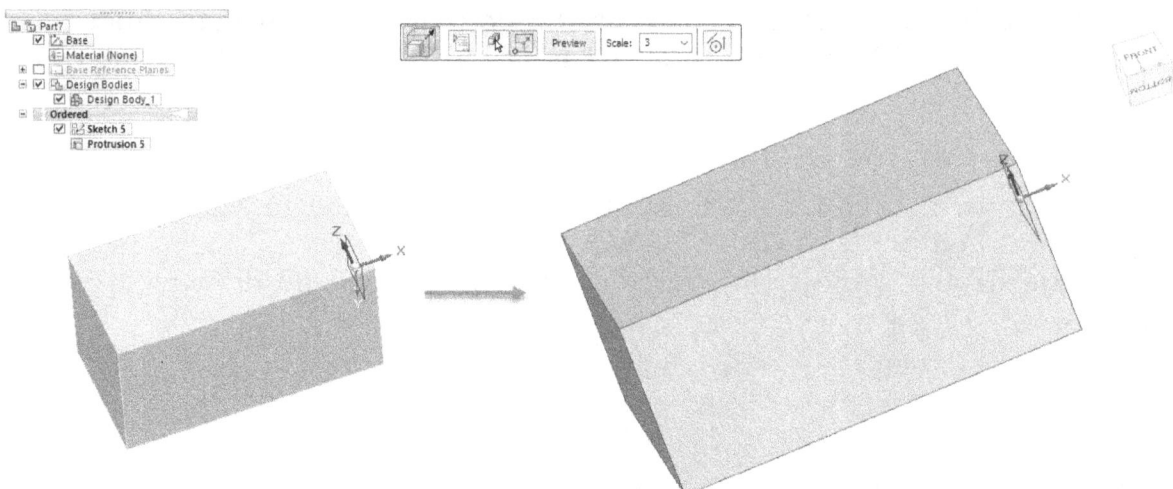

*Figure-63. Scale Body Created*

# PATTERN

We have already discussed **Pattern** tools in sketching but here we are going to discuss about creation of solid pattern. The tools to create pattern are available in **Pattern** drop-down of **Pattern** panel in the **Ribbon**; refer to Figure-64. Various tools in this drop-down are discussed next.

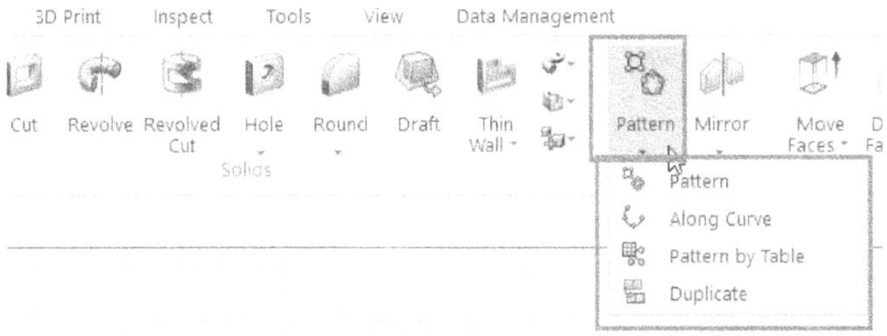

*Figure-64. Pattern drop down*

# CREATING PATTERN

The **Pattern** tool in **Pattern** drop-down is used to create multiple instances of selected body in rectangular as well as circular style. The procedure to use this tool is given next.

## Rectangular Pattern

In rectangular pattern style, the object instances are created in two perpendicular linear directions on a plane. This tool is similar to the sketch rectangular pattern. The procedure to create rectangular pattern is given next.

- Click on the **Pattern** tool from the **Pattern** drop-down in **Pattern** panel of the **Ribbon**. The **Pattern Command Bar** will be displayed; refer to Figure-65.

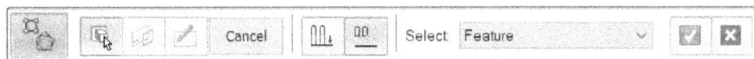

*Figure-65. Pattern Command Bar*

- Select all the features that you want to pattern and right-click in the empty area. You will be asked to select a planar face or plane for defining direction references of pattern.
- Select the desired planar face or plane. You are redirected to sketching environment.
- To create a rectangular plane, select the **Rectangular Pattern** button from **Feature** panel in the **Home** tab of **Ribbon**. The **Rectangular Pattern Command Bar** will be displayed; refer to Figure-66.

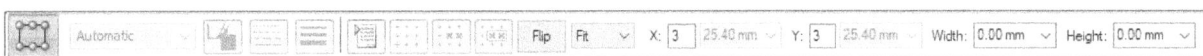

*Figure-66. Rectangular Pattern Command Bar*

- Click on the **Stagger Options** button from the **Command Bar** to define how instances will be placed in the rectangular boundary. The **Stagger Options** dialog box will be displayed; refer to Figure-67. Select the **Row** option from the drop-down to offset instances alternatively in rows or select the **Column** option to offset instances alternatively in columns. The preview of staggering will be displayed in the dialog box.

- Specify the desired value of stagger distance in the dialog box using the desired radio button. If you do not want pattern instances to go outside the boundary then clear the Include last column check box. Click on the **OK** button from dialog box to apply the changes.

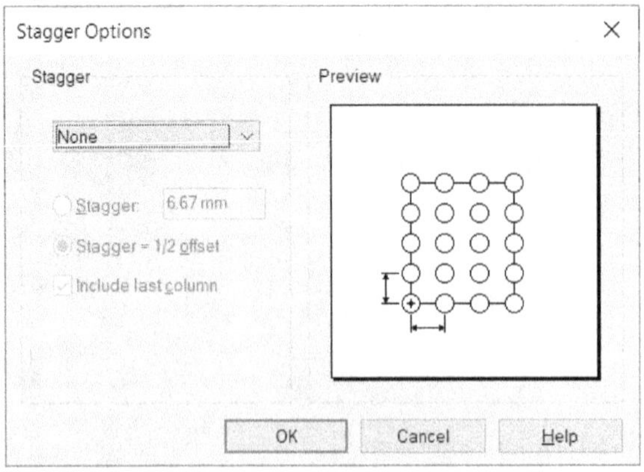

*Figure-67. Stagger Options dialog box*

- Specify the desired parameters in the **Command Bar** to define occurrences in X and Y directions. You can also specify the width, height and inclination angle for boundary rectangle in the **Command Bar**.
- Draw/place the rectangle to define area in which pattern will be created; refer to Figure-68 and click on the **Close Sketch** button from the **Close** panel in the **Ribbon**. Preview of pattern will be displayed.
- Click on the **Finish** button to create the pattern and press **ESC** key to exit the tool.

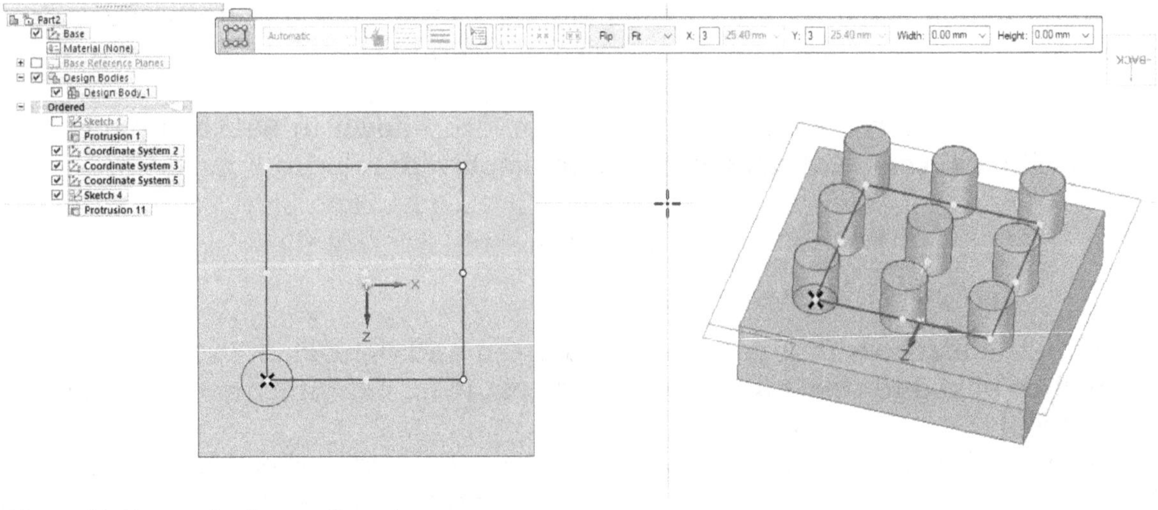

*Figure-68. Rectangular Pattern Created*

## Circular Pattern

You can use the **Pattern** tool to create circular style pattern in the same way as discussed for rectangular pattern. The procedure to create circular pattern is given next.

- Click on the **Pattern** tool from the **Pattern** drop-down of **Pattern** panel in the **Home** tab of **Ribbon**. The **Pattern Command Bar** will be displayed as discussed earlier.
- Select all the features that you want to pattern and click **OK** button from the **Command Bar**. You will be asked to select a sketching plane/face.
- Select the desired face/plane. You are redirected to sketching environment.

- Select the **Circular Pattern** button from the **Features** panel in the **Home** tab of **Ribbon**. The **Circular Pattern Command Bar** will be displayed; refer to Figure-69.

*Figure-69. Circular Pattern Command Bar*

- Set the desired options in the **Command Bar** as discussed earlier like radius of circle, count of instances, pattern type and so on.
- Draw a circle and specify the direction in which next instances will be created. Preview of pattern will be displayed in the form of sketch dots.
- Click on the **Close Sketch** tool from the **Close** panel in the **Home** tab of **Ribbon**. The preview of pattern will be displayed; refer to Figure-70.

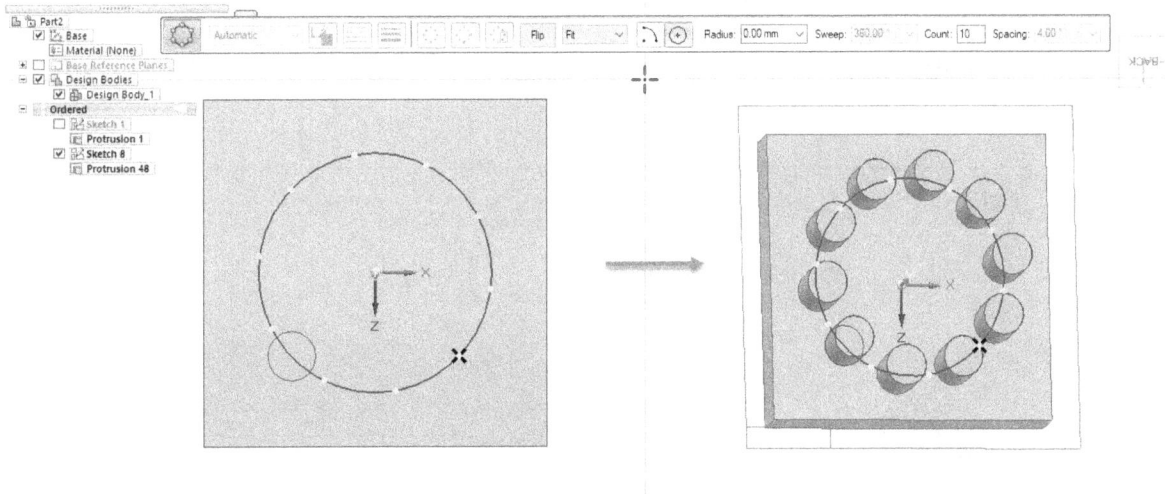

*Figure-70. Circular Pattern Created*

- Click on the **Finish** button from the **Command Bar** to create the feature and press **ESC** to exit the tool.

Note that in Synchronous environment, the **Rectangular** and **Circular** tools are separately available in the **Pattern** drop-down of **Pattern** panel to create the two types of patterns.

## Along Curve Pattern

The **Along Curve** tool is used to create multiple instances of a solid features along the selected path. To create along curve pattern, there should be a curve and solid body available in the modeling area. The procedure to create along curve pattern is given next.

- Click on the **Along Curve** tool from the **Pattern** drop-down in the **Pattern** panel of the **Home** tab in **Ribbon**. The **Along Curve Command Bar** will be displayed; refer to Figure-71.

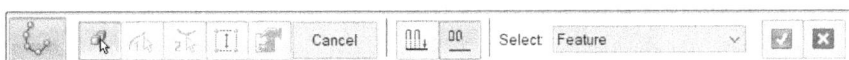

*Figure-71. Along Curve Command Bar*

- Select all the features that you want to pattern and right-click in empty area. You will be asked to select an edge chain or curve.

- Select curve path along which pattern is to be created and right-click in empty area. You will be asked to specify anchor point (start point) for pattern.
- Select the desired point on the path curve. If you want to offset the anchor point then specify the respective value in the **Offset** edit box of **Command Bar**.
- Click on the **OK** button from the **Command Bar**. Preview of pattern feature will be displayed with modified **Command Bar**; refer to Figure-72.
- Specify the desired values of number of instances in **Count** edit box and space between two consecutive instances of pattern in **Spacing** edit box of **Command Bar**.
- Note that you can also use a second curve path if you want to create pattern along two curves by selecting the **Path Curve Step** (2) button from the **Command Bar**.
- After setting desired parameters, click on **Next** button from **Command Bar**.
- If you want to suppress any instance in pattern then click on the **Suppress Occurrence** button from **Command Bar** and select the green dots for instances that you want to suppress from the preview.
- Click on the **Preview** button to check the preview; refer to Figure-73 and click on the **Finish** button to create the feature.

*Figure-72. Preview of curve pattern*

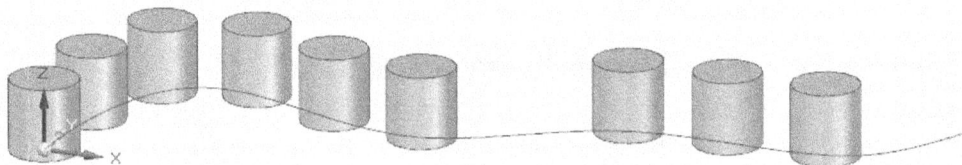

*Figure-73. Along Curve Pattern Created*

## Pattern by Table

The **Pattern by Table** tool is used to create multiple instances of a features as per the coordinates specified in the table. Note that you will need a Microsoft Excel Sheet file for specifying coordinates of pattern instances. The procedure to create pattern by table is given next.

- Click on the **Pattern by Table** tool from the **Pattern** drop-down in the **Pattern** panel of **Ribbon**. The **Pattern by Table Command Bar** will be displayed; refer to Figure-74.

*Figure-74. Pattern By Table Command Bar*

- Select the features which you want to pattern and right-click. You will be asked to select a coordinate system for reference.
- Select the desired coordinate system from the modeling area. The **Instance Table** dialog box will be displayed; refer to Figure-75. Click on the **Browse** button from the top area in the dialog box. The **Select Excel File** dialog box will be displayed; refer to Figure-76.

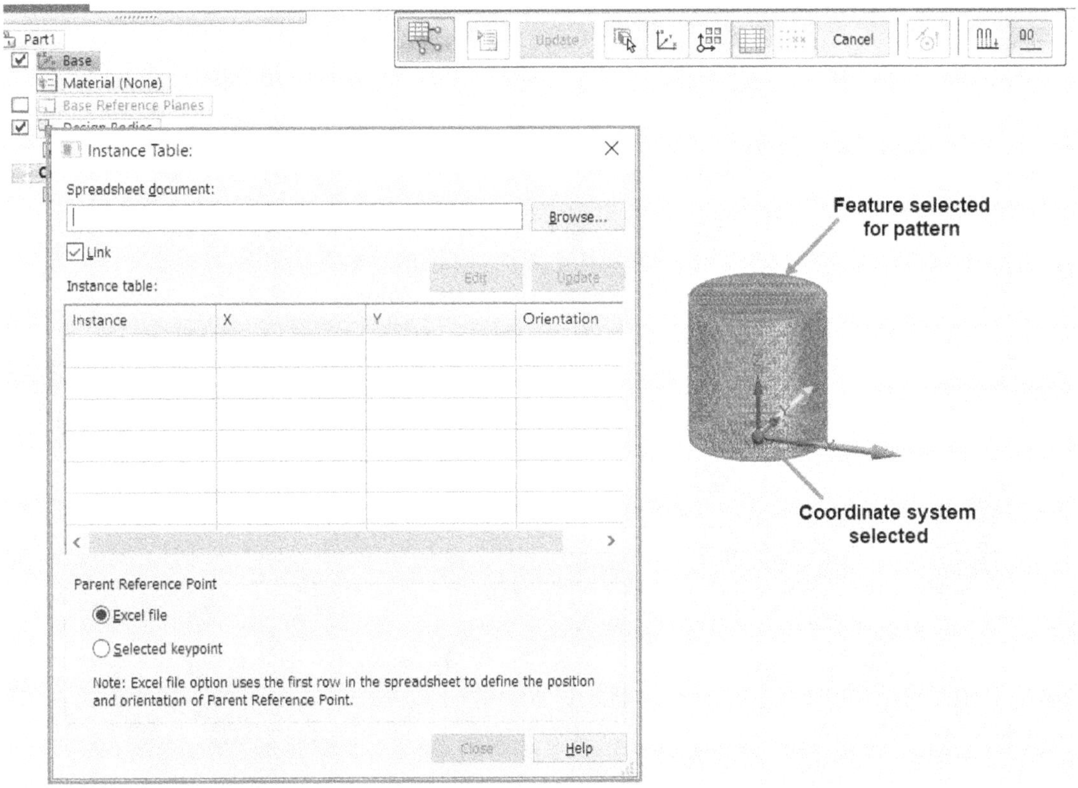

*Figure-75. Table dialog box*

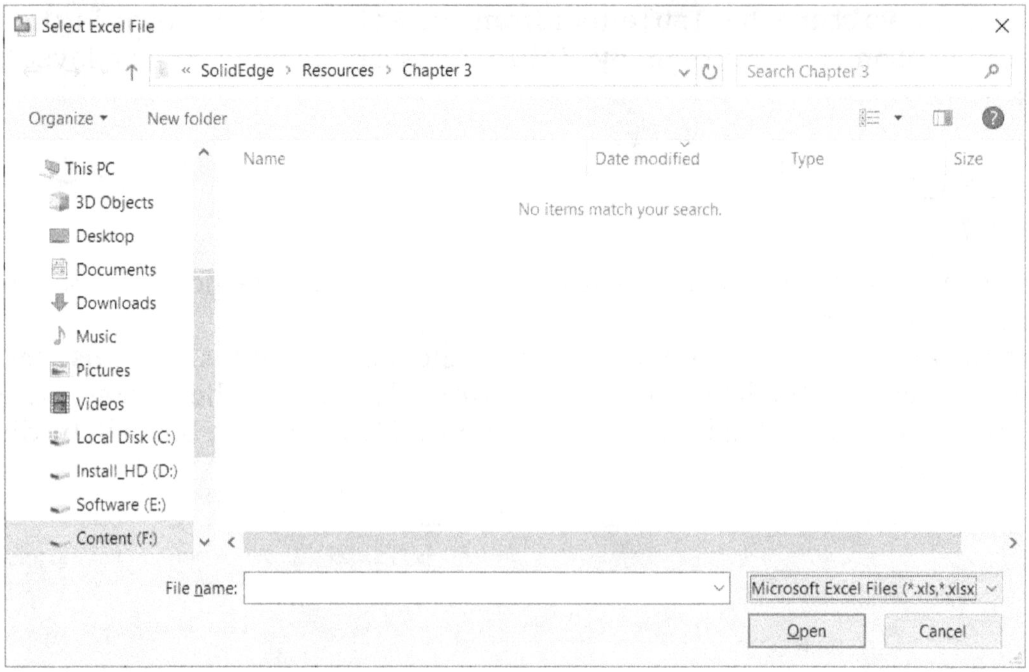

*Figure-76. Select Excel File dialog box*

- Select desired file in which you have specified coordinates for pattern instances. Preview of pattern will be displayed; refer to Figure-77.

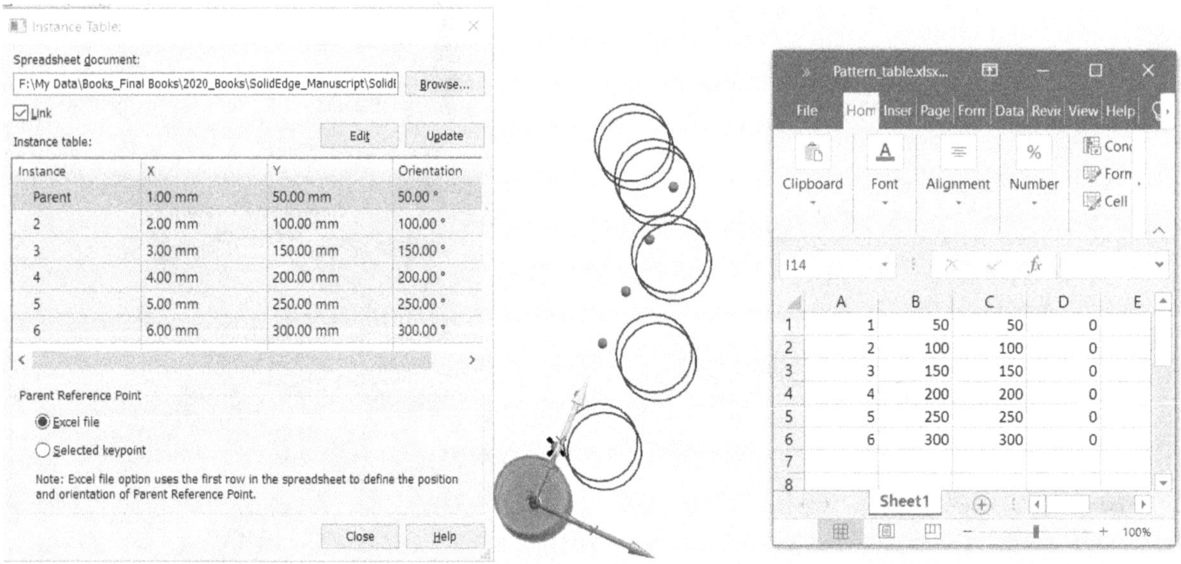

*Figure-77. Preview of Pattern by table*

- If you want to edit the table then click on the **Edit** button from the dialog box. The sheet will open in Microsoft Excel if installed.
- Click on the **Close** button from the dialog box to exit and click on the **Finish** button from the **Command Bar** to create the pattern feature.

## Duplicate Pattern

The **Duplicate** tool is used to create duplicate pattern from one body to another. There must be a reference coordinate on which duplicate pattern is to be created. The procedure of duplicate pattern is discussed next.

- Click on the **Duplicate** tool from the **Pattern** drop-down in the **Pattern** panel of **Home** tab in the **Ribbon**. The **Duplicate Command Bar** will be displayed; refer to Figure-78.

*Figure-78. Duplicate Pattern Command Bar*

- Select all pattern bodies you want to duplicate and press **ENTER**. You will be asked to coordinate system to be used as reference for copying the bodies.
- Select coordinate system of bodies as parent references and then select coordinate system with respect to which you want to place the duplicate bodies. The preview of duplicate pattern will be displayed in wire frame style. Click on the **Preview** button to check solid preview of feature; refer to Figure-79.

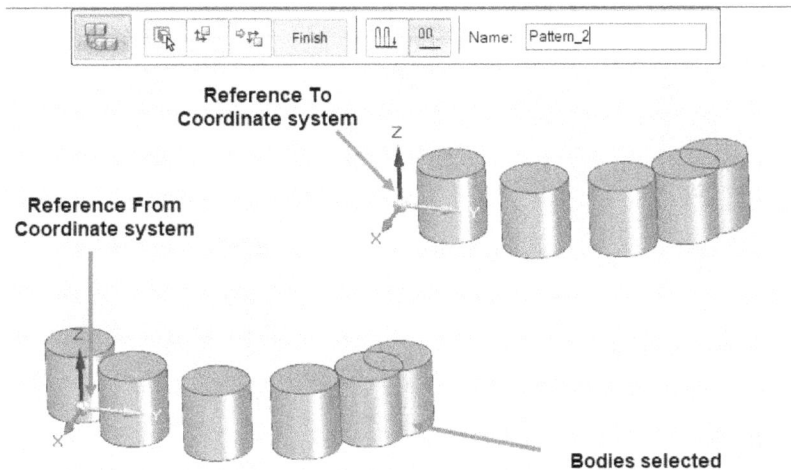

*Figure-79. Preview of Duplicate pattern*

- Click on the **Finish** button from **Command Bar** to create the feature.

# MIRROR

Earlier, we have discussed about working of **Mirror** tool for sketch elements. Now we will discuss the procedure to mirror solid part or features. The tools to perform mirror operation are available in the **Mirror** drop-down of **Pattern** panel in the **Home** tab of **Ribbon**; refer to Figure-80.

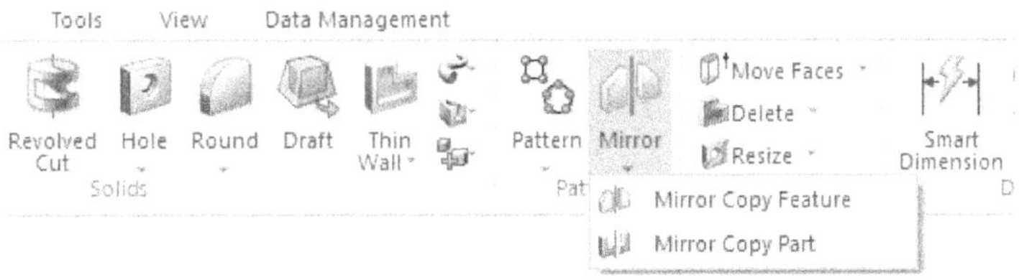

*Figure-80. Mirror drop-down*

## Mirror Copy Feature

The **Mirror Copy Feature** tool is used to create mirror copy of selected features of a solid body. The procedure to create mirror copy feature is given next.

- Click on the **Mirror Copy Feature** tool from the **Mirror** drop-down in the **Pattern** panel of **Ribbon**. The **Mirror Copy Feature Command Bar** will be displayed; refer to Figure-81.

*Figure-81. Mirror Copy Feature command bar*

- Select the features which you want to mirror copy and press **ENTER**. You will be asked to select a planar face/plane to be used as mirror.
- Select the desired plane about which you want to mirror selected feature. The preview of mirror feature will be displayed; refer to Figure-82.
- Click on the **Finish** button to create the feature.

*Figure-82. Preview of mirror copy feature*

## Mirror Copy Part

The **Mirror Copy Part** tool is used to mirror part or solid body. Note that all the features of selected part/body will be mirror copied. The procedure to use this tool is similar to **Mirror Copy Feature** tool.

## MOVING FACES

The **Move Faces** tool is used to move face at a finite extend. The tool is available in **Modify** panel of **Home** tab in **Ribbon**. The procedure to use this tool is discuss next.

- Click on **Move Faces** tool from **Move Faces** drop-down; refer to Figure-83. The **Move Face Command Bar** will be displayed; refer to Figure-84.

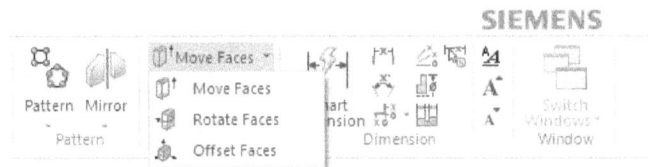

*Figure-83. Move Faces drop-down*

*Figure-84. Move Faces Command Bar*

- Select the desired option from the **Select** drop-down to filter selection of objects and then select the face(s) of model to be moved. After selecting faces, press **ENTER**. You will asked to select a direction reference along which the select faces will move.
- Select the desired edge/axis to define direction. You will be asked to specify start point for moving faces.
- Select desired reference point from the model and move it along direction earlier selected.
- Click at desired location to create the feature. Preview of feature will be displayed; refer to Figure-85.
- Click on the **Finish** button from the **Command Bar** to create feature and press ESC to exit the tool.

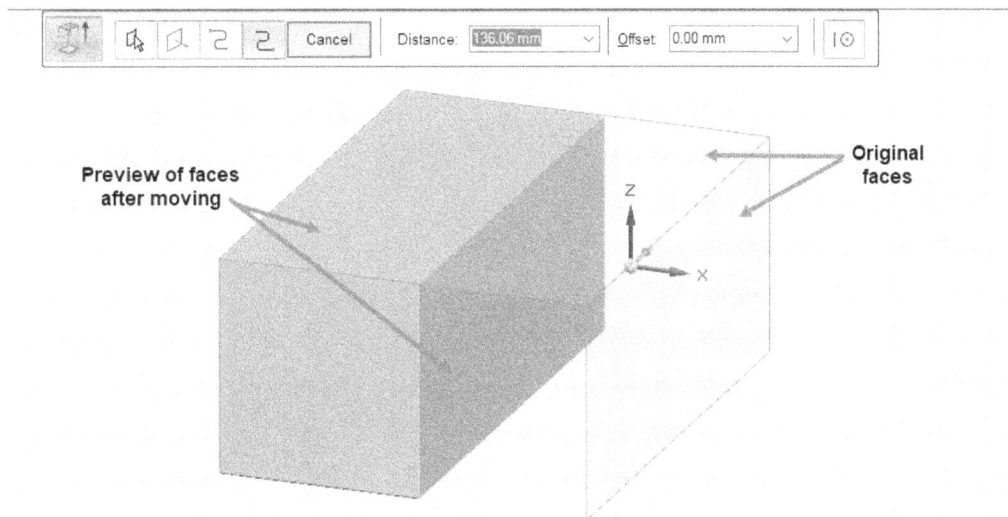

*Figure-85. Move Faces Created*

## ROTATE FACES

The **Rotate Face** tool is used to rotate face, feature and body along revolution of axis. The procedure given next.

- Click on the **Rotate Faces** tool from the **Move Faces** drop-down in the **Modify** panel of the **Ribbon**. The **Rotate Faces Command Bar** will be displayed; refer to Figure-86.

*Figure-86. Rotate Faces Command Bar*

- Choose selection type from the **Select** drop-down in the **Command Bar** and select the desired objects. After selecting objects, press **ENTER**. You will be asked to select an axis about which selected objects will be rotated.
- Select the desired edge/axis. The **Command Bar** will be modified and you will be asked to specify rotation angle.
- Specify the desired value in the **Angle** edit box of **Command Bar**; refer to Figure-87 and Figure-88.
- Click on the **Finish** button from the **Command Bar** to create the feature and press **ESC** to exit the tool.

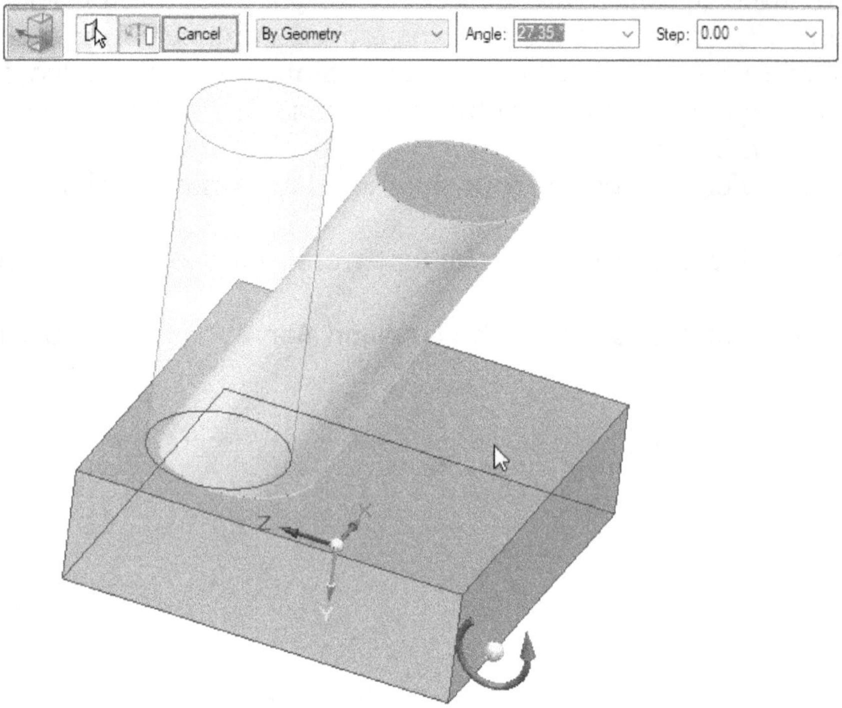

*Figure-87. Rotate Faces By Geometry Created*

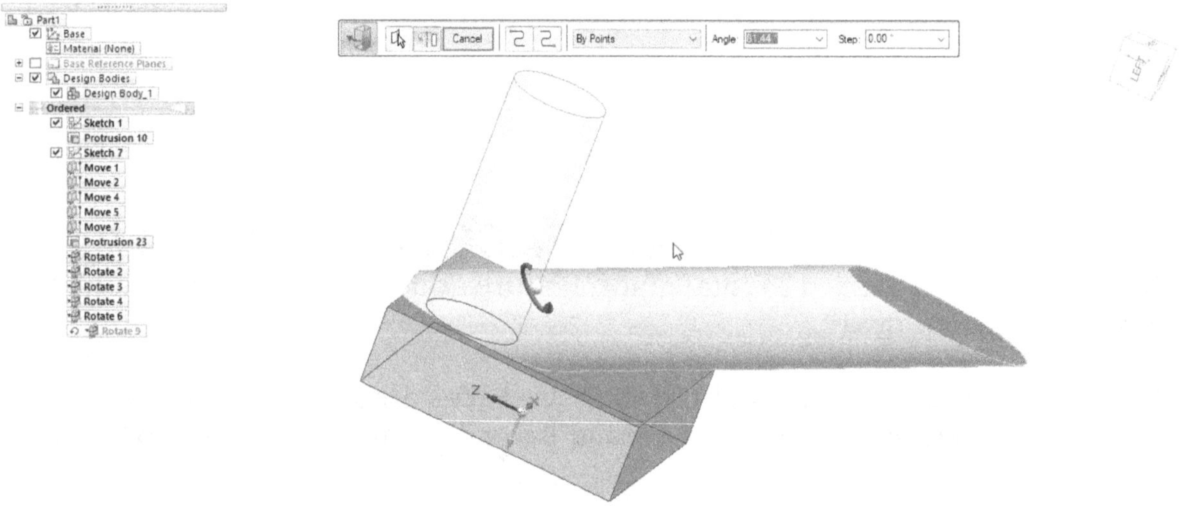

*Figure-88. Rotate Faces By Point Created*

## OFFSET FACES

The **Offset Faces** tool is used to offset faces to a finite extent. The procedure to use this tool is given next.

- Click on the **Offset Faces** tool from the **Move Faces** drop-down. The **Offset Faces Command Bar** will be displayed; refer to Figure-89.
- Select the desired option from the **Select** drop-down and select objects to be offset Press **ENTER** or right-click, you will be asked to specify the offset distance.

*Figure-89. Offset Faces Command Bar*

- Specify offset distance value in **Command Bar** and click to specify direction of offset. Preview of offset will be displayed; refer to Figure-90.
- Click on the **Finish** button to create the feature and press **ESC** to exit the tool.

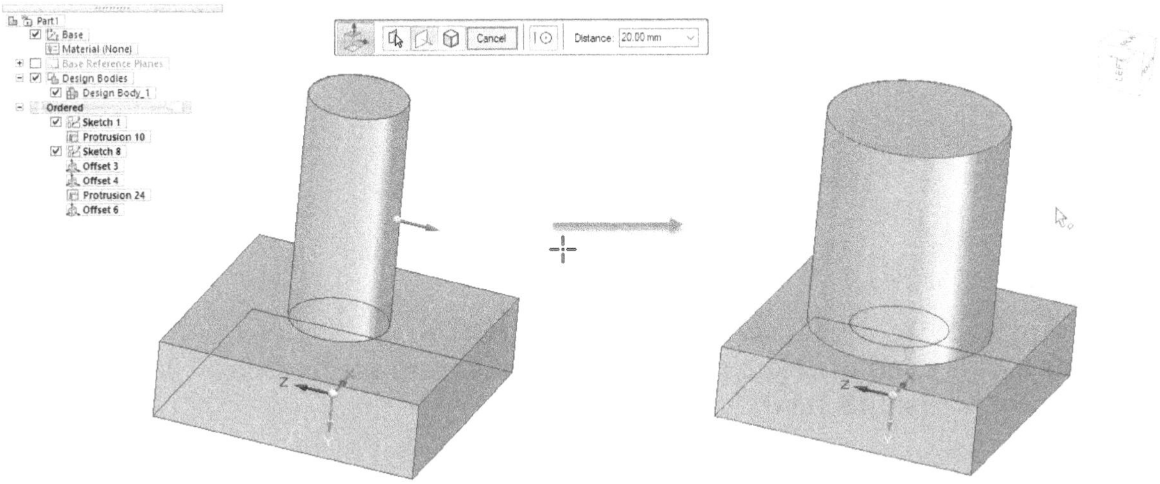

*Figure-90. Offset Faces Created*

# DELETE

The **Delete** tools are used to delete some unwanted elements of solid model. These tools are available in **Modify** panel; refer to Figure-91. Various tools in this drop-down are discussed next.

*Figure-91. Delete drop-down*

## Deleting Faces

The **Faces** tool is used to delete faces from solid body. The procedure to use this tool is discuss next.

- Click on the **Faces** tool from the **Delete Faces** drop-down in the **Modify** panel of **Home** tab in **Ribbon** in **Ordered** environment. The **Faces Command Bar** will be displayed; refer to Figure-92.

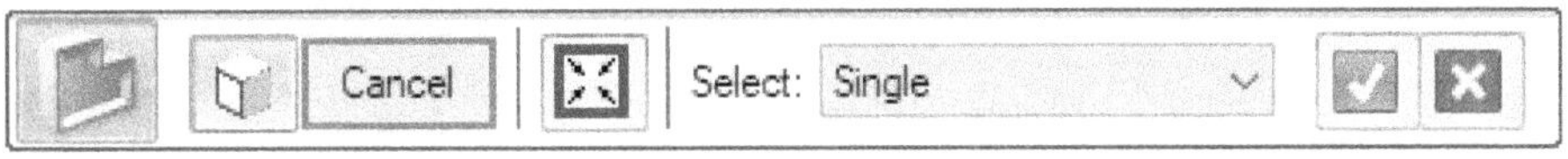

*Figure-92. Delete Faces Command Bar*

- Select desired option from the **Select** drop-down and select the entities from model to be deleted.
- After selecting faces, click on the **OK** button from the **Command Bar**. Preview of delete feature will be displayed; refer to Figure-93.
- Click on the **Finish** button from **Command Bar**. The selected faces will be deleted.

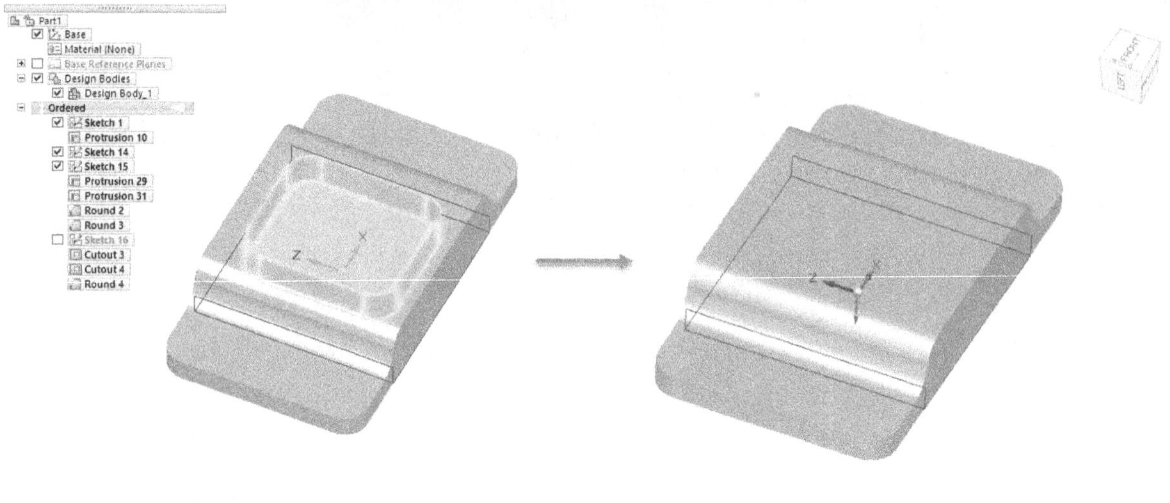

*Figure-93. Preview of faces deleted*

## Deleting Regions

The **Regions** tool is used to delete particular regions from solid surface. The procedure to use this tool is discuss next.

* Click on the **Regions** tool from the **Delete Faces** drop-down. The **Region Command Bar** will be displayed; refer to Figure-94.

*Figure-94. Regions Command Bar*

* Select desired option from the **Select** drop-down in the **Command Bar** and select edges bounding the faces to delete.
* After selecting edges, press **ENTER**. You will be asked to select faces to be deleted.
* Select the faces of the part to be delete and click on the **Preview** button. Preview of the feature will be displayed; refer to Figure-95.
* Click on the **Finish** button to create the feature.

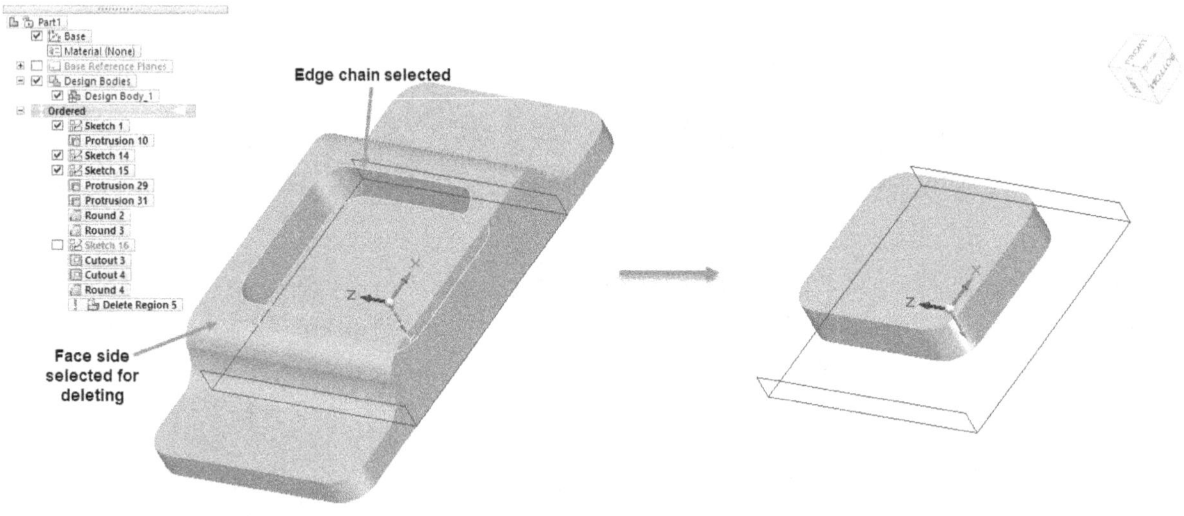

*Figure-95. Delete Regions Created*

## Deleting Holes

The **Holes** tool is used to delete holes from solid model. After activating this tool from the **Delete Faces** drop-down in **Modify** panel of **Ribbon**, select holes from solid model. Note that you can set the selection filter in **Command Bar** as discussed earlier;

refer to Figure-96. Click on the **OK** button from the **Command Bar**. Preview of deletion will be displayed; refer to Figure-97. Click on the **Finish** button to create the feature.

*Figure-96. Delete Holes Command Bar*

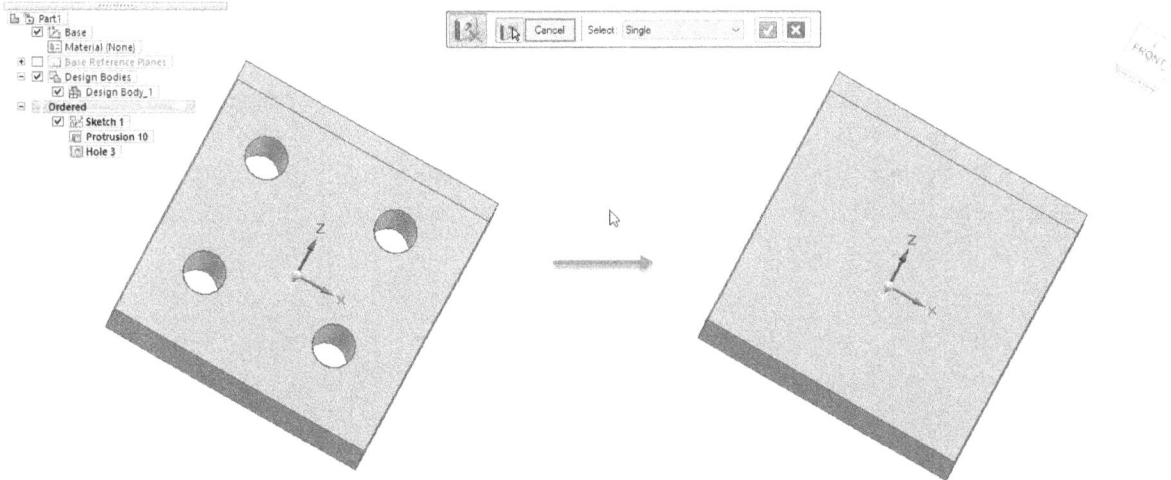

*Figure-97. Delete Holes Created*

## Deleting Rounds

The **Rounds** tool is used to delete rounds/fillets applied on solid model. The procedure is similar to deleting holes. The **Command Bar** and preview of deletion are shown in Figure-98 and Figure-99.

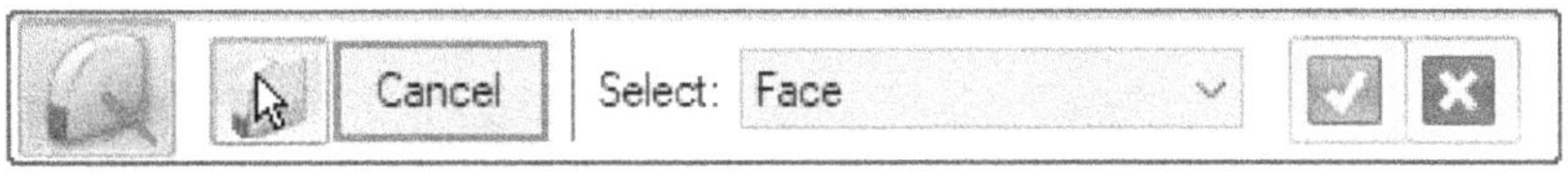

*Figure-98. Delete Rounds Command Bar*

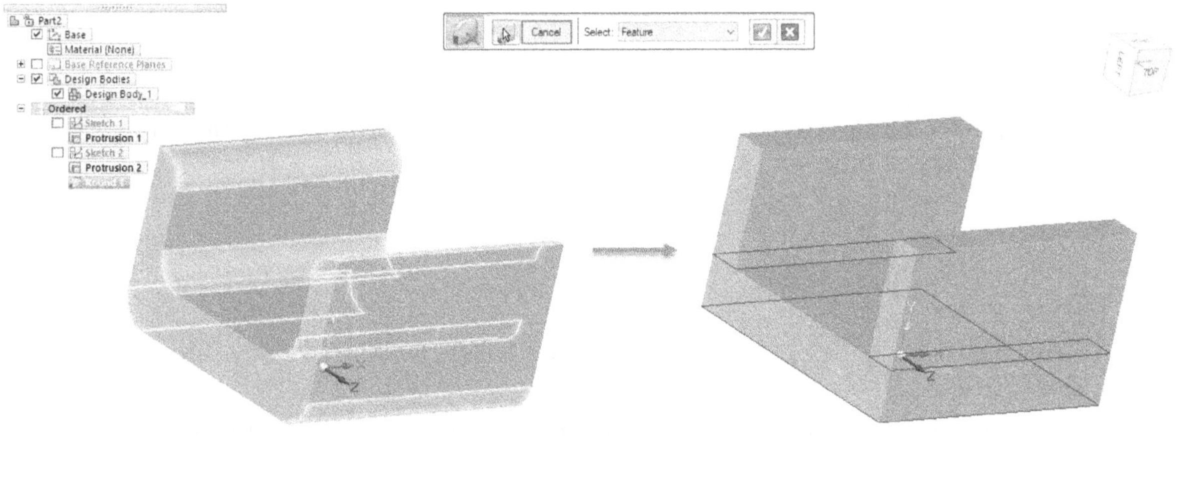

*Figure-99. Delete Rounds Created*

## Re-sizing Holes

The **Resize Holes** tool is used to change size of selected holes. The procedure to use this tool is given next.

- Click on the **Resize Holes** tool from the **Modify** panel in the **Home** tab of **Ribbon**. The **Resize Holes Command Bar** will be displayed; refer to Figure-100.

*Figure-100. Resize Holes Command Bar*

- Select the hole(s) from solid model and press **ENTER**. You will be asked to specify hole size in the **Command Bar**.
- Select the desired option from the **Standard** drop-down in **Command Bar** and enter the new size in edit box.
- Click on the **Preview** button to check preview of new holes; refer to Figure-101.
- Click on the **Finish** button to create the feature and press **ESC** to exit the tool.

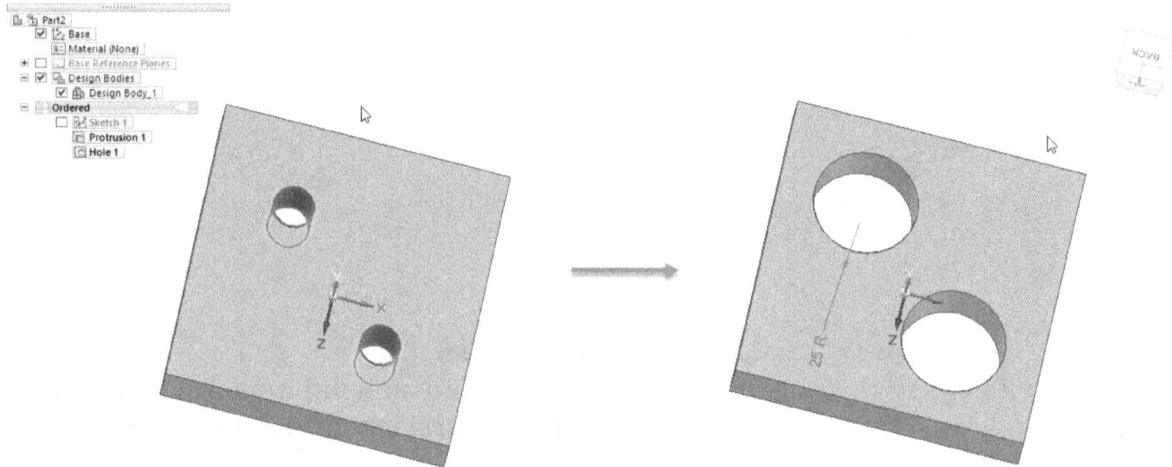

*Figure-101. Resize Holes Created*

## Resize Round

The **Resize Round** tool is used to change size of round/fillet applied on the model. The procedure to use this tool is discuss next.

- Click on the **Resize Round** tool from the **Modify** panel in the **Home** tab of **Ribbon**. The **Resize Round Command Bar** will be displayed; refer to Figure-102.

*Figure-102. Resize Round Command Bar*

- Select the round(s) to be modified and press **ENTER**. You will be asked to specify the size of round.
- Specify new round size value in the **New Radius** edit box of **Command Bar** and press **ENTER**. The preview of modified size will be displayed; refer to Figure-103.
- Click on the **Finish** button to create the feature.

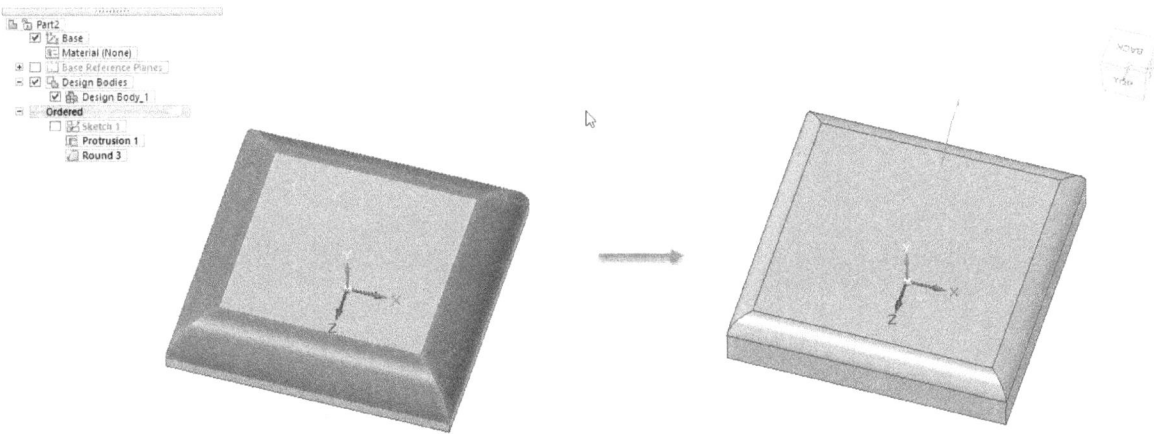

*Figure-103. Resize Rounds Created*

# PRACTICAL - 1

Create the model (isometric view) as shown in Figure-104. The dimensions and view are given in Figure-105.

*Figure-104. Model for Practical 1*

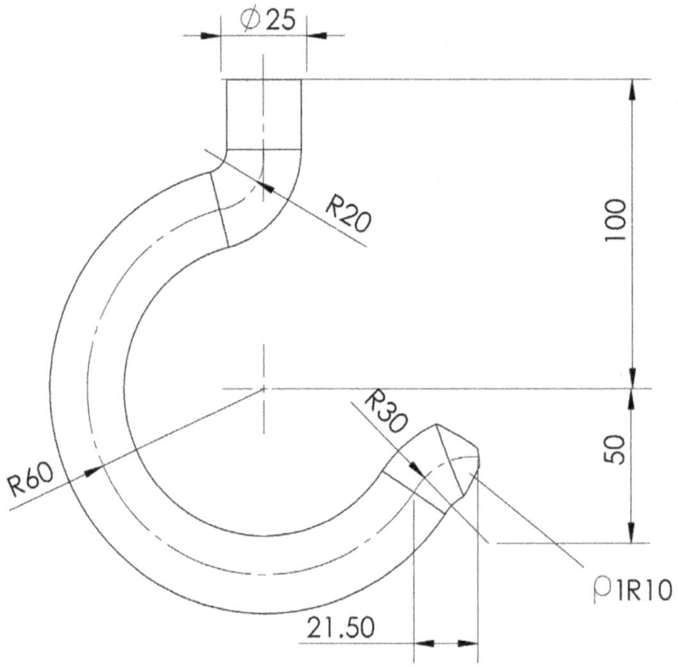

*Figure-105. Practical 1 drawing view*

## Creating Sketches for Hook

*   Switch to Ordered modeling style as discussed earlier.
*   Start a new sketch on **Front** plane.
*   Click on the **Circle by Center Point** tool and create the circle of diameter **120** taking coordinate system as center.
*   Draw a straight line starting from top quadrant point of circle and having length of approximately **40** (No need to be accurate).
*   Click on the **Arc by 3 Points** tool from **Arc** drop-down in the **Draw** panel of **Home** tab in the **Ribbon** and create a tangent arc at the bottom of the circle as shown in Figure-106.
*   Click on the **Fillet** tool  from **Draw** panel in **Ribbon** and apply fillet between the vertical line and the circle of radius **20**. Refer to Figure-106.

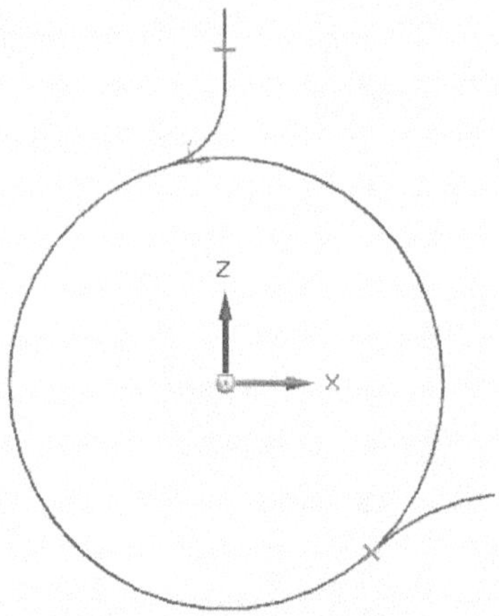

*Figure-106. Sketch after creating three points arc and fillet*

*   Click on the **Trim** tool from the **Draw** panel in the **Ribbon** and trim the portion between the straight line and the arc; refer to Figure-107.

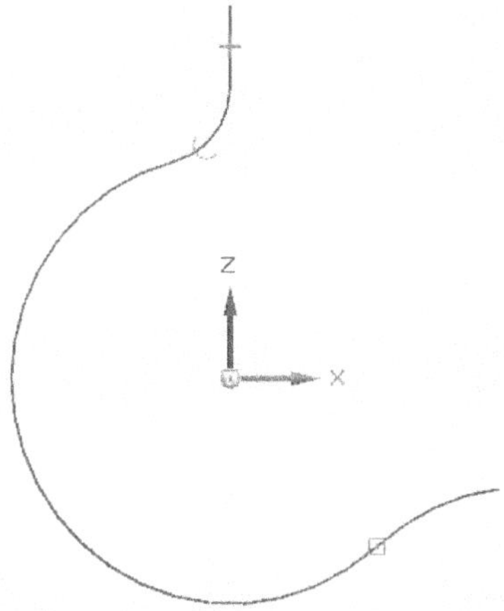

*Figure-107. Sketch after trimming circle*

- Click on the **Smart Dimension** tool from the **Dimension** panel in the **Home** tab of **Ribbon** and apply the dimensions as shown in Figure-108. Click on the **Close Sketch** tool from **Ribbon** and exit the sketching environment. We have created the sketch of the path. Now, we will create the section sketch.

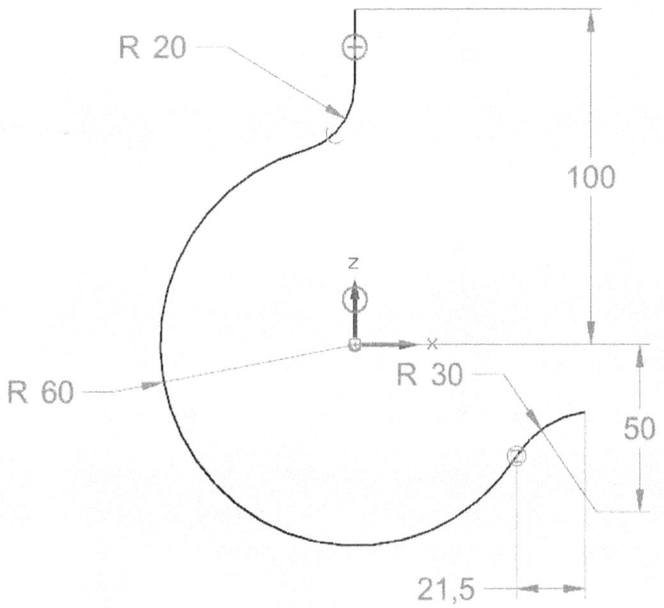

*Figure-108. Sketch after applying dimensions*

- Click on the **Normal to curve** tool from the **More Planes** drop-down in the **Planes** panel of the **Ribbon** and select the curve near straight vertical line. A plane will be displayed attached to cursor.
- Click near the start point of the curve to create the plane; refer to Figure-109.
- Click on the **Sketch** tool from the **Sketch** panel in the **Home** tab of **Ribbon** and select the newly created plane. The sketching environment will be displayed.

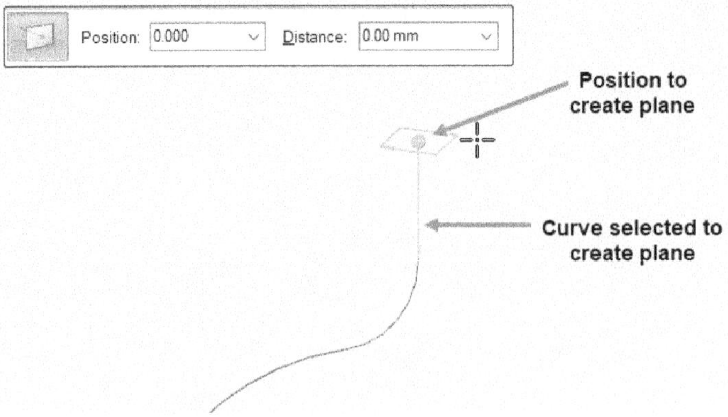

*Figure-109. Creating plane*

- Create a circle of diameter **25** using the coordinate system origin as center of circle. Close the sketch and finish sketch creation.

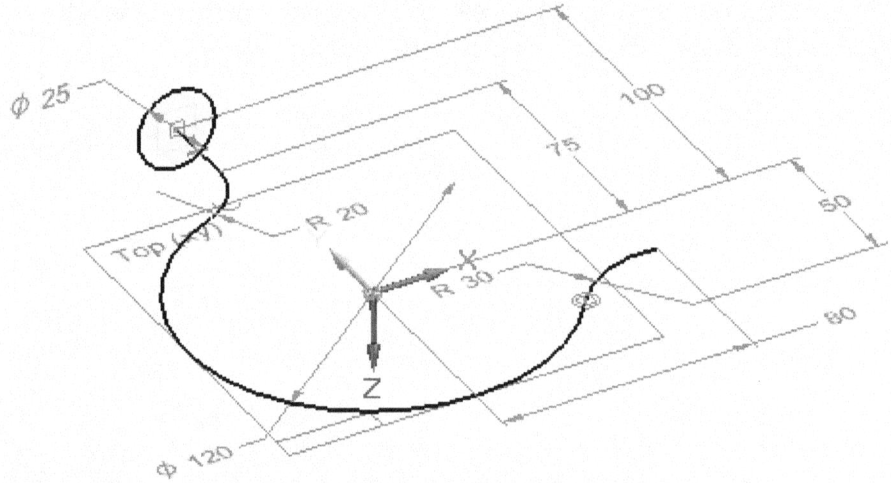

*Figure-110. Circle created on the normal to plane*

## Creating Sweep Feature

Now, we have all the sketches to create the sweep feature. The procedure to create the sweep features using these sketches is given next.

- Click on the **Sweep** tool from the **Add** drop-down in the **Solids** panel of the **Ribbon**. The **Sweep Options** dialog box will be displayed.
- Select the **Single path and cross section** radio button from the **Default Sweep Type** area of the dialog box. Select the **No merge**, **Normal**, and **Tangent continuous** radio buttons from the dialog box. Click on the **OK** button. You will be asked to select the curve for path.
- Select the open curve and right-click in the empty area. You will be asked to select sketch section for profile.
- Select the **Select from Sketch/Part Edges** option from the **Create-From Options** drop-down in the **Command Bar** and select the circle created earlier. Preview of the sweep feature will be displayed; refer to Figure-111.

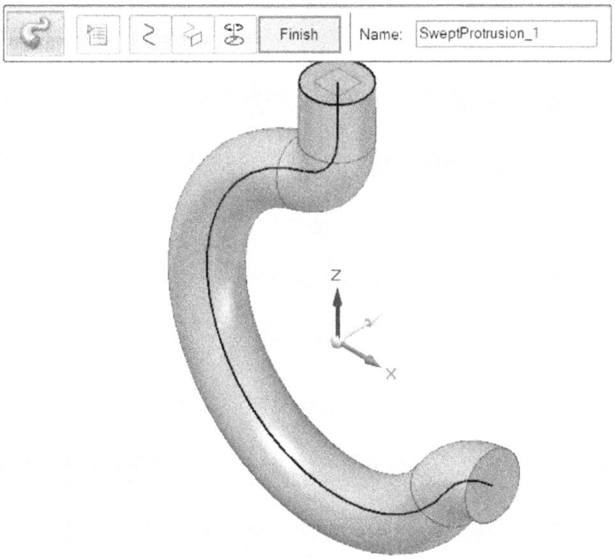

*Figure-111. Sweep feature created*

- Click on the **Finish** button from the **Command Bar** and press **ESC** to exit the tool.

## Applying Round at End

As we all know that hooks do not end with sharp edges. So, we need to apply round at the end. The steps to do so are given next.

- Click on the **Round** tool from the **Solid** panel in the **Home** tab of the **Ribbon**. The **Round Command Bar** will be displayed.
- Click on the **Round Options** button from the **Command Bar** and select the **Blend** radio button from the dialog box displayed. Click on the **OK** button from the dialog box to apply settings.
- Select the **Bevel** option from the **Shape** drop-down in the **Command Bar**. The options in **Command Bar** will be displayed as shown in Figure-112.

*Figure-112. Round Command Bar with Bevel options*

- Select the front face and side faces at the bottom end of hook. Preview of round will be displayed; refer to Figure-113.

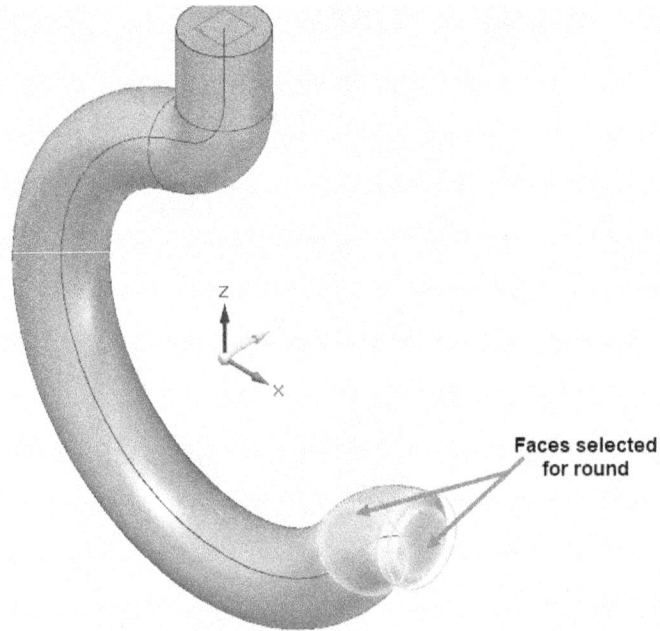

*Figure-113. Face selection for round*

- Specify the value as **10** in **Setback** edit box and **1** in the **Value** edit box of **Command Bar** and click on the **OK** button from the **Command Bar**. Click on the **Preview** button and then **Finish** button from the **Command Bar** to create the round feature. Press **ESC** to exit the tool.
- Hide the extra sketches and planes. The final model will be displayed; refer to Figure-114.

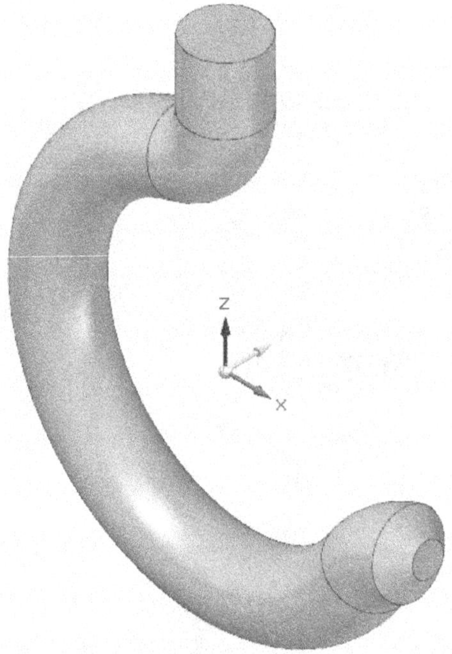

*Figure-114. Final Model*

# PRACTICE 1

Create the model as shown in Figure-115. The dimensions are given in Figure-116.

*Figure-115. Practice 1*

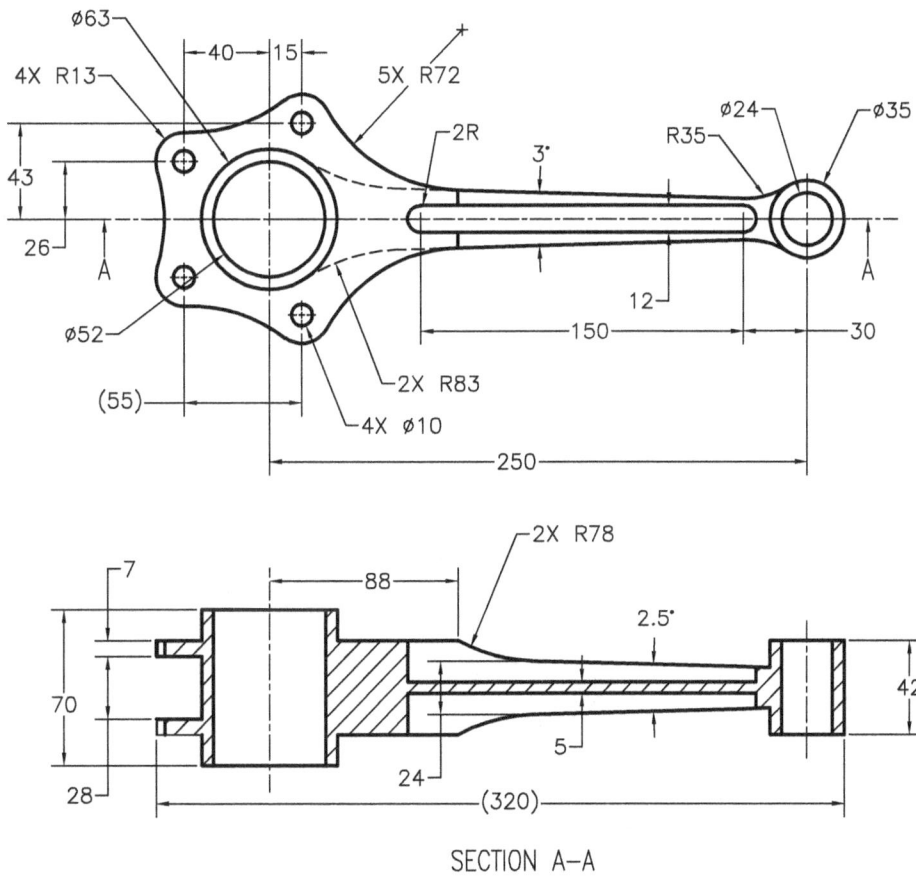

SECTION A–A

*Figure-116. Dimensions of the Practice 1 model*

# PRACTICE 2

Create the model using the drawings shown in Figure-117.

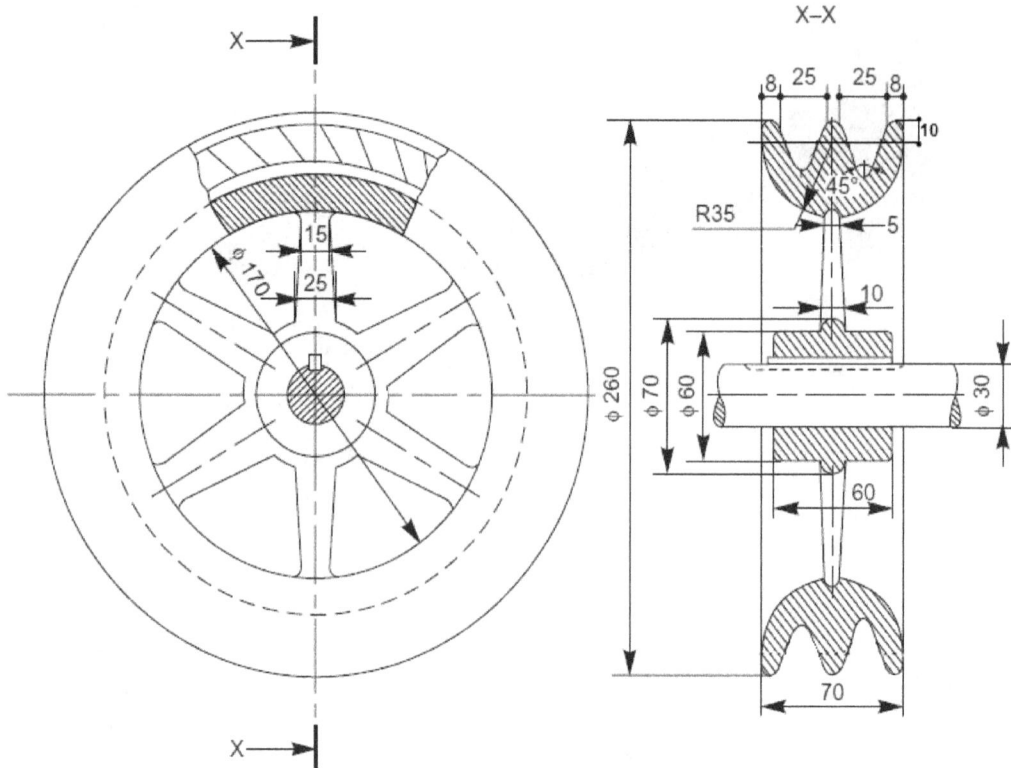

*Figure-117. Rope Pulley*

# PRACTICE 3

Create the model as shown in Figure-118. Dimensions are given in Figure-119. Assume the missing dimensions.

*Figure-118. Practice 3 model*

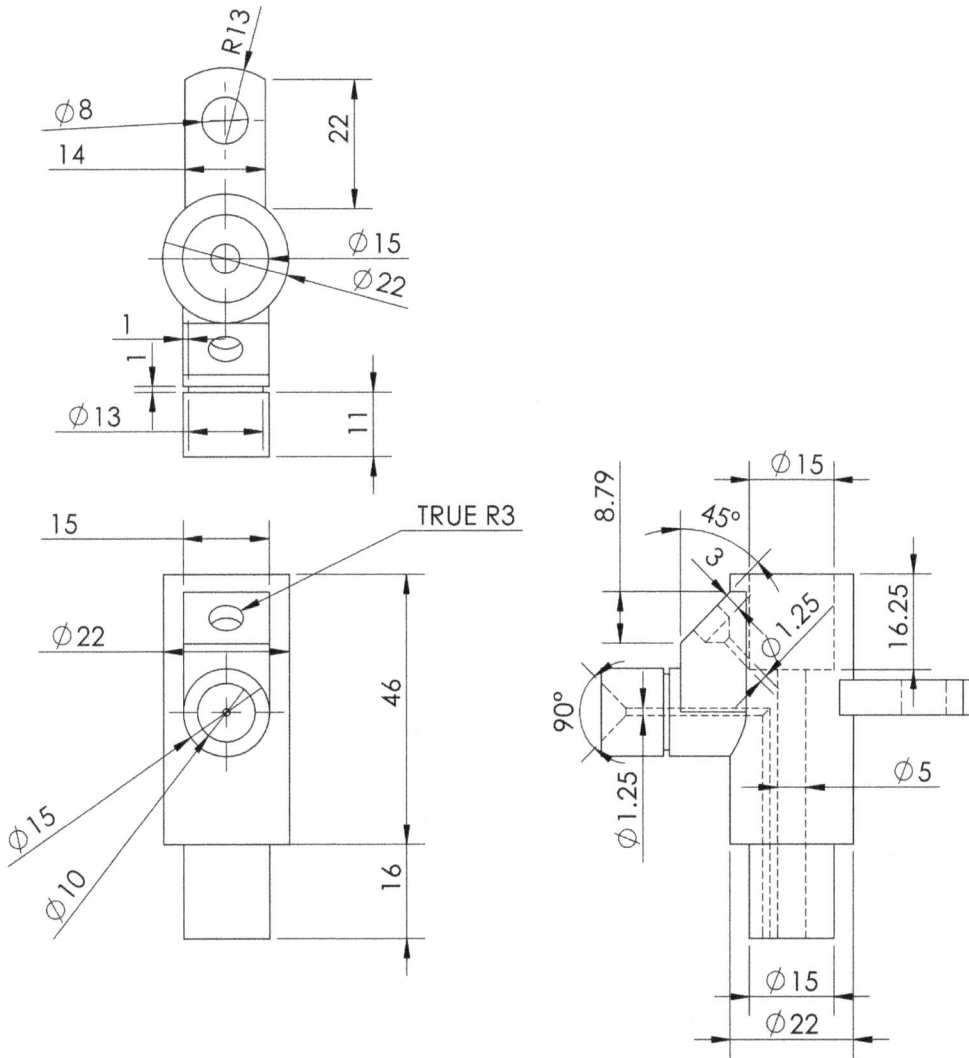

*Figure-119. Practice 3*

# PRACTICE 4

Create the model by using the dimensions given in Figure-120.

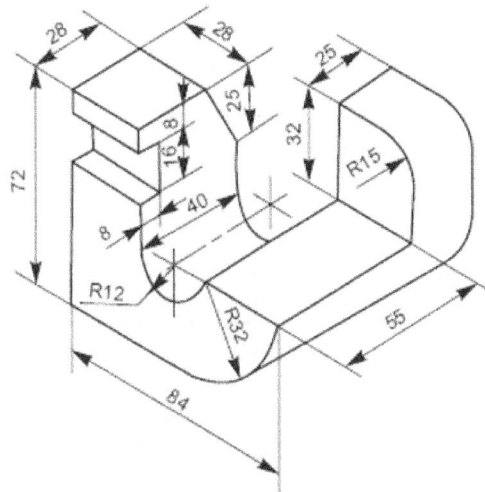

*Figure-120. Practice 4*

# PRACTICE 5
Create the model by using the dimensions given in Figure-121.

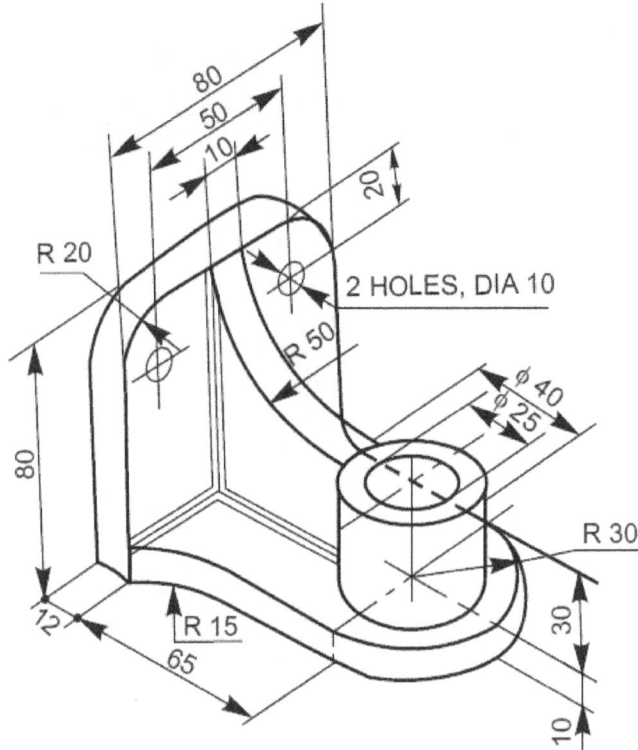

*Figure-121. Practice 5*

# PRACTICE 6
Create the model by using the dimensions given in Figure-122.

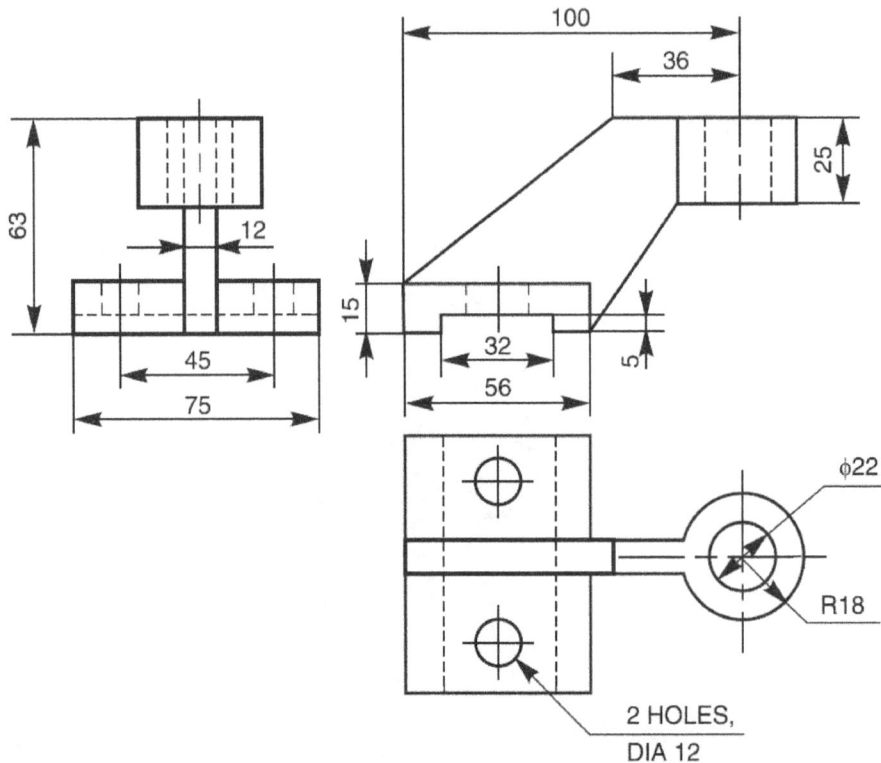

*Figure-122. Practice 6*

# PRACTICE 7

Create a ring nut with value of **D** as **5,6,8**, and **10** using equation and design table. Dimensions are given in Figure-123.

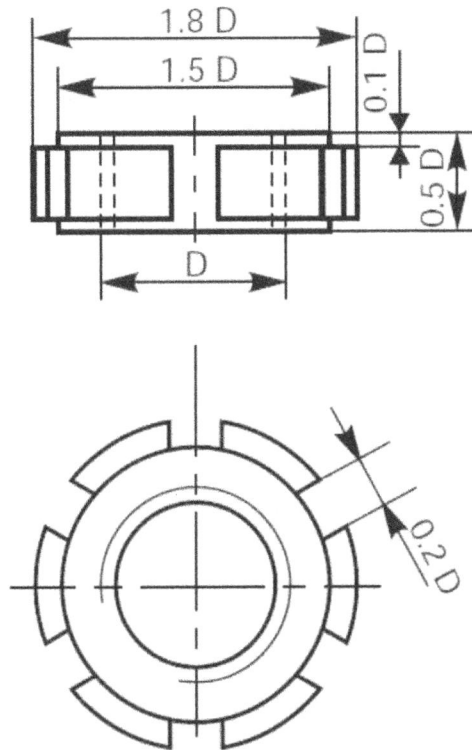

*Figure-123. Ring Nut*

# FOR STUDENT NOTES

# Chapter 5

# Selection Tools, 3D Printing, and Views

## Topics Covered

The major topics covered in this chapter are:

- *Selection Tools*
- *Introduction to 3D Printing*
- *3D Printing Tools*
- *Introduction to Views*
- *View Tools*

# INTRODUCTION

There are various features of objects in the model like face, vertex, body, part and so on. Sometimes, default selection does not allow you to select the desired feature of model. The tools in the **Select** group are used to define criteria for selecting objects in the model; refer to Figure-1. The **Select** group is available in multiple tabs of **Ribbon** like **Home** tab, **Features** tab, **PMI** tab, **Simulation** tab, and so on. Various tools in this group are discussed next.

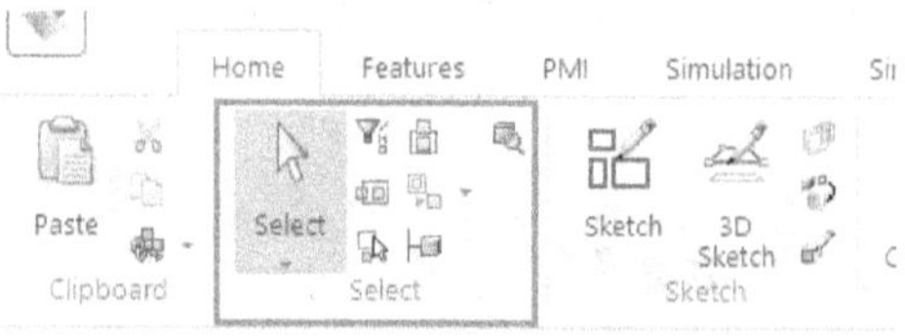

*Figure-1. Select group*

# SELECT DROP-DOWN

The tools in the **Select** drop-down are used to specify which feature of model takes priority while selecting objects. Note that some of the options in this drop-down will be available in the Assembly environment. The options in this drop-down are given next.

- Select the **Face Priority** option from the **Select** drop-down if you want to select faces from the model.
- Select the **Part Priority** option from the **Select** drop-down in the **Select** group to select parts from the assembly model.
- Select the **Normal Select Mode** option from the **Select** drop-down to use standard selection mode. In this selection mode, you can select multiple objects while holding the **SHIFT** key and remove objects from selection while holding the **CTRL** key during selection.
- After selecting an object from the model, select the **Add/Remove Mode** option from the **Select** drop-down if you want to add or remove objects from the selection.
- Similarly, you can use the **Add Mode** and **Remove Mode** options from the **Select** drop-down to add and remove objects from selection.
- Select the **Selection Manager Mode** option from the drop-down to use selection manager for selecting object. After selecting this option, select the desired face of synchronous model. The **Selection Manager** menu will be displayed; refer to Figure-2. Various options of this menu are discussed next.

  - Click on the **Deselect** option to deselect latest selected face.
  - Click on the **Clear Selection** option from the menu to remove all the faces from selection.
  - Click on the **3D Box Select** option from the menu to draw a 3D box. All the objects falling inside 3D box will be selected.
  - Hover the cursor on **Connected** option and select the desired option from the cascading menu. Select the **Connected** option to select all the faces connected to latest selected face. Select the **Interior Faces** option from the cascading menu to select all the connected interior faces. Select the **Exterior Faces** option from the cascading menu to select all the connected exterior faces.
  - Select the **Related Items** option to select the objects which are linked to selected object.

- Select the **Sets** option from the menu to select all the features of object whose face is selected. The **Quick Pick** dialog box will be displayed; refer to Figure-3. Select the desired feature to be selected from the dialog box.

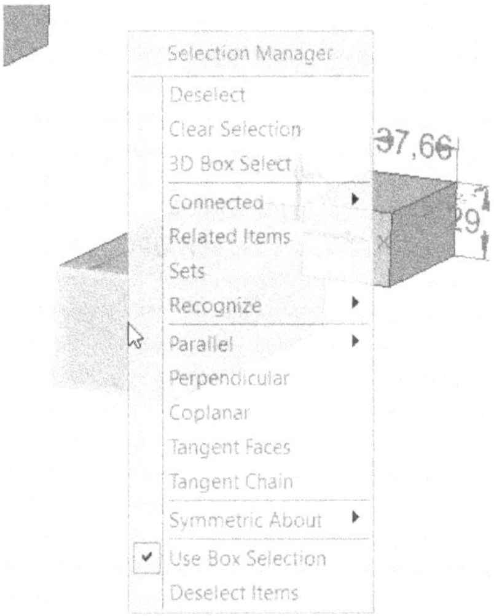

*Figure-2. Selection Manager menu*

*Figure-3. Quick Pick*
*dialog box*

- Select the desired option from the **Recognize** cascading menu in the **Selection Manager** menu to select respective features identified in selection.
- Select the desired option from the **Parallel** cascading menu to select faces parallel to selected face; refer to Figure-4. On selecting an option, you will be asked to draw a 3D box for selected parallel faces. Draw the desired box and all the faces parallel to selected face inside the 3D box will be selected.

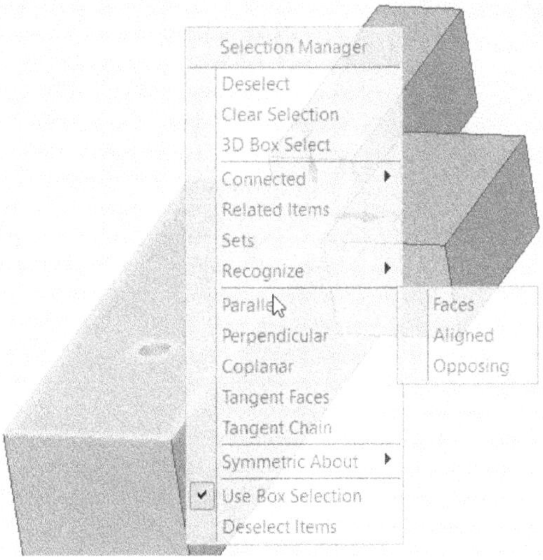

*Figure-4. Parallel cascading menu*

- Similarly, you can use the **Perpendicular**, **Coplanar**, **Tangent Faces**, and **Tangent Chain** options from the **Selection Manager** menu.
- Select the desired option from the **Symmetric About** cascading menu and then select the desired face. All the faces symmetric about selected plane will be also get selected.

- Clear the **Use Box Selection** option to use single object selection.
- Select the **Deselect Items** option from the menu to deselect the items.
- Press **ESC** to exit the **Selection Manager** menu.

## SELECTION FILTER

The **Selection Filter** tool is used to apply filter on selection like you can select only dimensions, annotations, model geometry and so on. On clicking the **Selection Filter** tool from the **Select** group of the **Home** tab in the **Ribbon**. The **Selection Filter** menu will be displayed; refer to Figure-5. Select the desired check boxes from the menu to define what type of objects can be selected.

*Figure-5. Selection Filter menu*

## OVERLAPPING SELECTION BOX

Select the **Overlapping** toggle button from the **Select** group of **Home** tab in the **Ribbon** to use a selection box which can be used to select objects which intersect with selection box and fall inside it; refer to Figure-6.

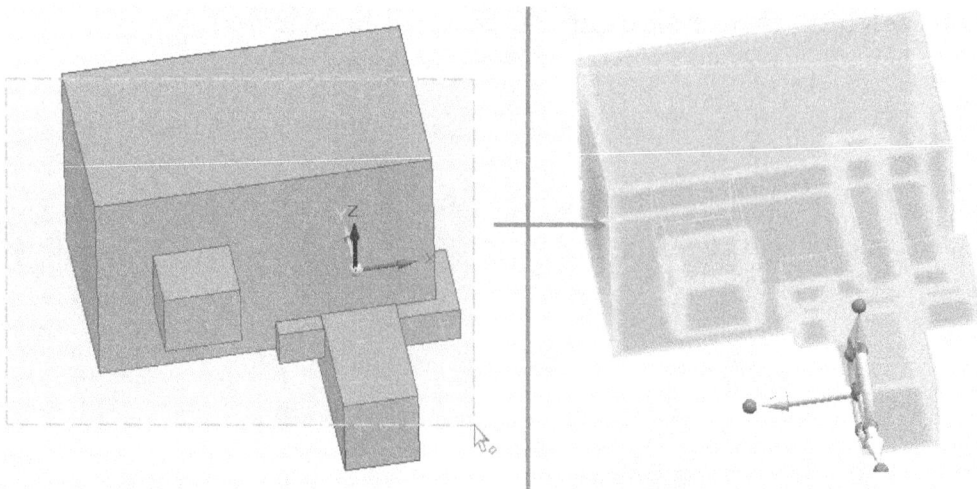

*Figure-6. Overlapping selection box*

Press **ESC** to exit the selection.

## 3D PRINTING

3D Printing also known as Additive Manufacturing is not a new concept as it was developed in 1981 but since then 3D Printing technology is continuously evolving. In early stages, the 3D printers were able to create only prototypes of objects using the polymers. But now a days, 3D printers are able to produce final products using metals, plastics and biological materials. 3D printers are being used for making artificial organs, architectural art pieces, complex design objects etc. Although, 3D Printing technique was created for manufacturing industry but now, it has found more applications in medical field.

In SolidEdge, there is a very simple and robust mechanism for 3D printing. The procedure of 3D Printing itself is not difficult but it is important to prepare your part well for 3D printing. We will first discuss the part preparation for 3D Printing and then we will use SolidEdge tools for performing 3D print.

# PART PREPARATION FOR 3D PRINTING

Part preparation is very important step for 3D Printing. If your part is not stable in semi molten state then it is less suitable for 3D printing. Stability of model is directly dependent on the material you are using for 3D Printing. We will learn more about part preparation but before that it is important to understand different type of processes available in 3D Printing.

## 3D Printing Processes

Not all 3D printers use the same technology. There are several ways to print and all those available are additive, differing mainly in the way layers are build to create the final object.

Some methods use melting or softening material to produce the layers. Selective laser sintering (SLS) and fused deposition modeling (FDM) are the most common technologies using this way of 3D printing. Another method is when we talk about curing a photo-reactive resin with a UV laser or another similar power source one layer at a time. The most common technology using this method is called stereolithography (SLA).

In 2010, the American Society for Testing and Materials (ASTM) group "ASTM F42 – Additive Manufacturing", developed a set of standards that classify the Additive Manufacturing processes into 7 categories according to Standard Terminology for Additive Manufacturing Technologies. These seven processes are:

1. Vat Photopolymerisation
2. Material Jetting
3. Binder Jetting
4. Material Extrusion
5. Powder Bed Fusion
6. Sheet Lamination
7. Directed Energy Deposition

Brief introduction to these processes is given next.

## Vat Photopolymerisation

A 3D printer based on the Vat Photopolymerisation method has a container filled with photopolymer resin which is then hardened with a UV light source; refer to Figure-7.

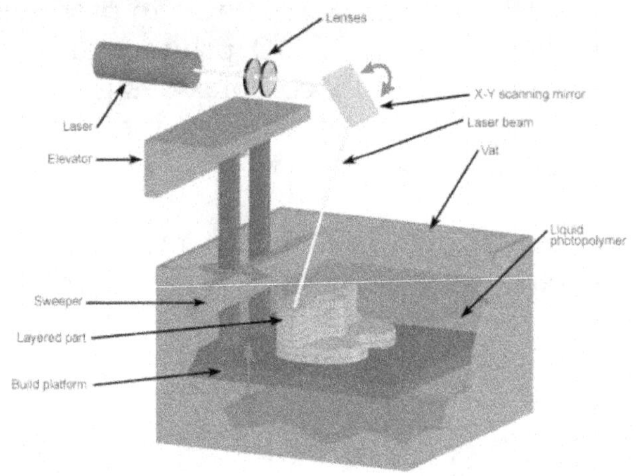

*Figure-7. 3D Printing via vat-photopolymerisation*

The most commonly used technology in this processes is Stereolithography (SLA). This technology employs a vat of liquid ultraviolet curable photopolymer resin and an ultraviolet laser to build the object's layers one at a time. For each layer, the laser beam traces a cross-section of the part pattern on the surface of the liquid resin. Exposure to the ultraviolet laser light cures and solidifies the pattern traced on the resin and joins it to the layer below.

After the pattern has been traced, the SLA's elevator platform descends by a distance equal to the thickness of a single layer, typically 0.05 mm to 0.15 mm (0.002″ to 0.006″). Then, a resin-filled blade sweeps across the cross section of the part, re-coating it with fresh material. On this new liquid surface, the subsequent layer pattern is traced, joining the previous layer. The complete three dimensional object is formed by this project. Stereolithography requires the use of supporting structures which serve to attach the part to the elevator platform and to hold the object because it floats in the basin filled with liquid resin. These are removed manually after the object is finished.

## Material Jetting

In this process, material is applied in droplets through a small diameter nozzle, similar to the way a common inkjet paper printer works, but it is applied layer-by-layer to a build platform making a 3D object and then hardened by UV light; refer to Figure-8.

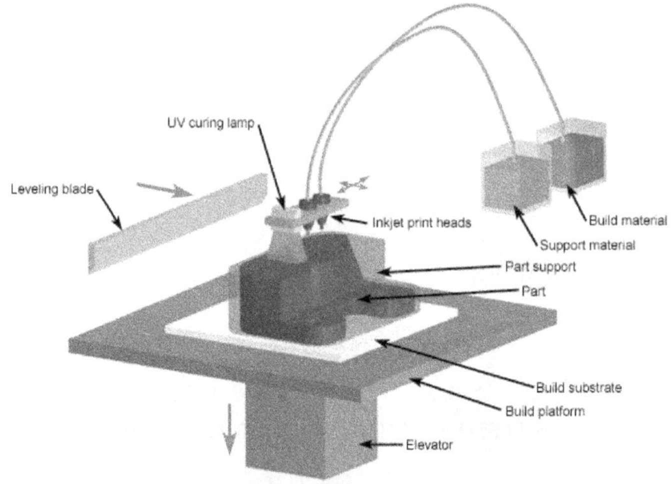

*Figure-8. 3D Printing via Material-Jetting*

## Binder Jetting

With binder jetting two materials are used: powder base material and a liquid binder. In the build chamber, powder is spread in equal layers and binder is applied through jet nozzles that "glue" the powder particles in the shape of a programmed 3D object; refer to Figure-9. The finished object is "glued together" by binder remains in the container with the powder base material. After the print is finished, the remaining powder is cleaned off and used for 3D printing the next object.

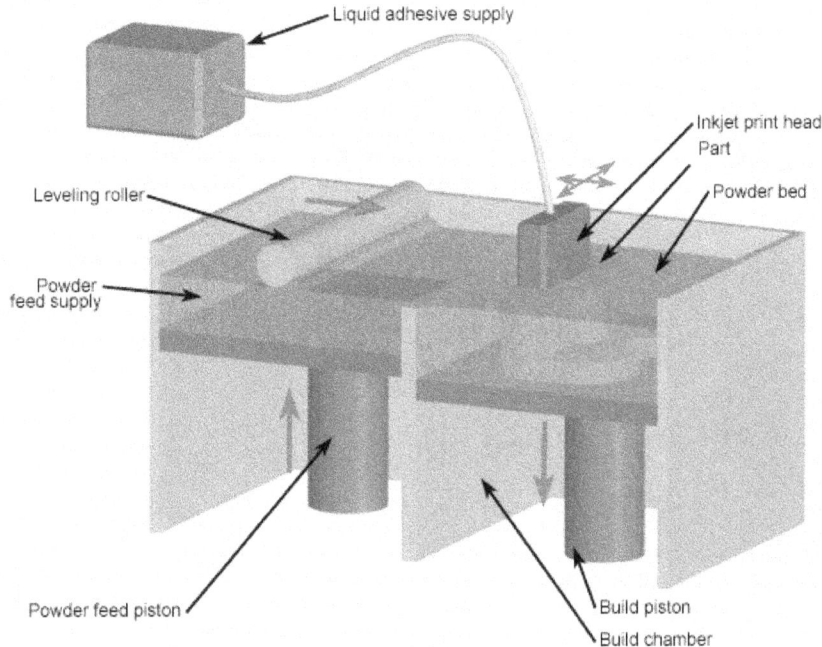

*Figure-9. 3D Printing via binder-jetting*

## Material Extrusion

The most commonly used technology in this process is Fused deposition modeling (FDM). The FDM technology works using a plastic filament or metal wire which is unwound from a coil and supplying material to an extrusion nozzle which can turn the flow on and off. The nozzle is heated to melt the material and can be moved in both horizontal and vertical directions by a numerically controlled mechanism, directly controlled by a computer-aided manufacturing (CAM) software package; refer to Figure-10. The object is produced by extruding melted material to form layers as the material hardens immediately after extrusion from the nozzle. This technology is most widely used with two plastic filament material types: ABS (Acrylonitrile Butadiene Styrene) and PLA (Polylactic acid) but many other materials are available ranging in properties from wood filed, conductive, flexible etc.

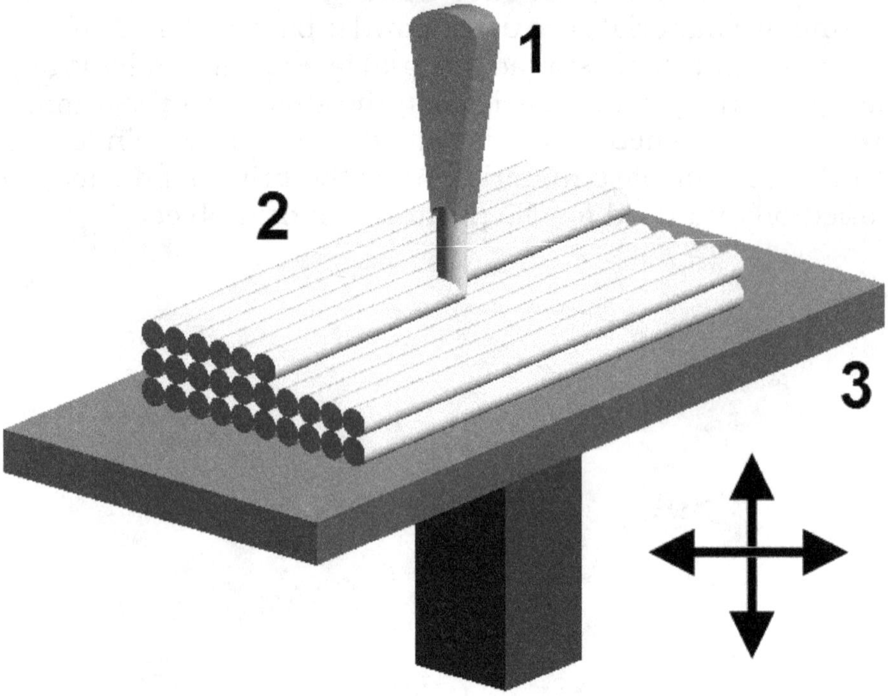

*Figure-10. 3D Printing via Fused Deposition Modeling*

In the above figure:
1 – nozzle ejecting molten material (plastic),
2 – deposited material (modelled part),
3 – controlled movable table.

## Powder Bed Fusion

The most commonly used technology in this processes is Selective laser sintering (SLS). This technology uses a high power laser to fuse small particles of plastic, metal, ceramic or glass powders into a mass that has the desired three dimensional shape. The laser selectively fuses the powdered material by scanning the cross-sections (or layers) generated by the 3D modeling program on the surface of a powder bed; refer to Figure-11. After each cross-section is scanned, the powder bed is lowered by one layer thickness. Then a new layer of material is applied on top and the process is repeated until the object is completed.

All untouched powder remains as it is and becomes a support structure for the object. Therefore there is no need for any support structure which is an advantage over SLS and SLA. All unused powder can be used for the next print.

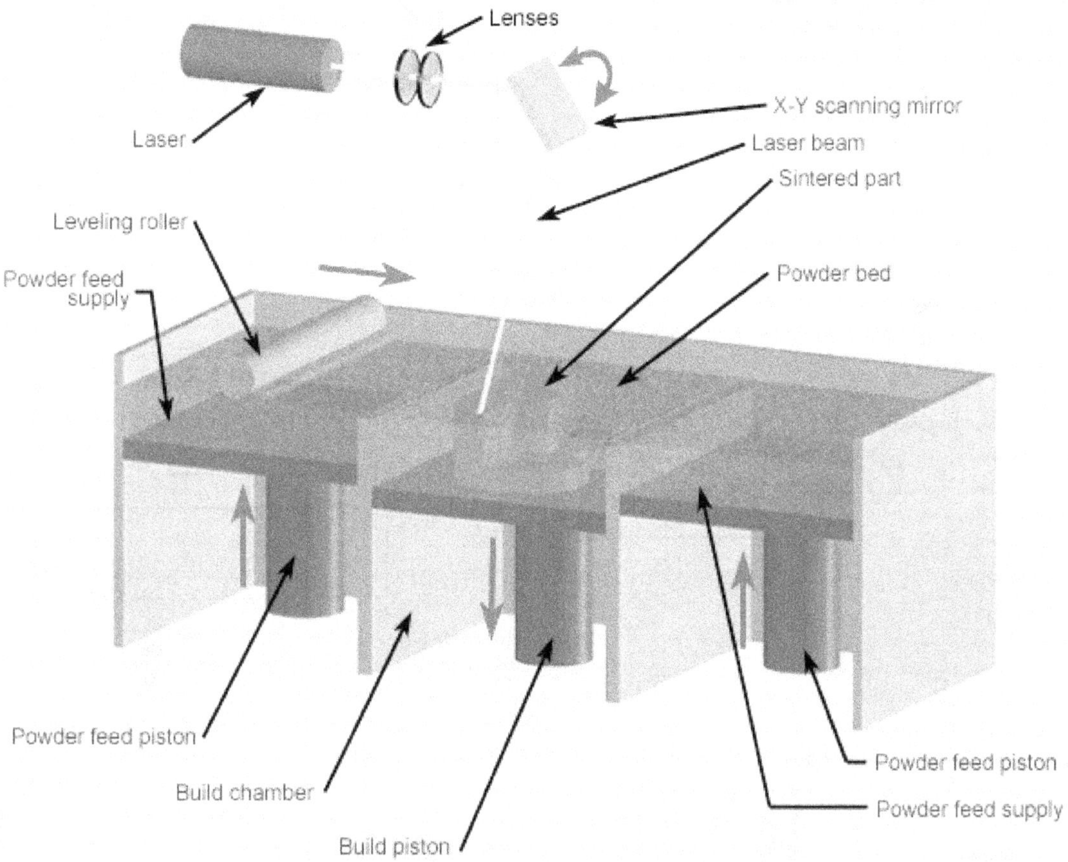

*Figure-11. 3D Printing via Selective Laser Sintering*

## Sheet Lamination

Sheet lamination involves material in sheets which is bound together with external force. Sheets can be metal, paper or a form of polymer. Metal sheets are welded together by ultrasonic welding in layers and then CNC milled into a proper shape; refer to Figure-12. Paper sheets can be used also, but they are glued by adhesive glue and cut in shape by precise blades.

## Directed Energy Deposition

This process is mostly used in the high-tech metal industry and in rapid manufacturing applications. The 3D printing apparatus is usually attached to a multi-axis robotic arm and consists of a nozzle that deposits metal powder or wire on a surface and an energy source (laser, electron beam or plasma arc) that melts it, forming a solid object; refer to Figure-13.

Ultrasonic
Oscillation

Rotating
Cylindrical
Sonotrode

Base Plate

Anvil

Metal Foils

Clamping
Force from
Sonotrode

Ultrasonic Oscillation

Foil/Foil Interface and
Solid State Bond

Reaction Force
from Anvil and
Base Plate

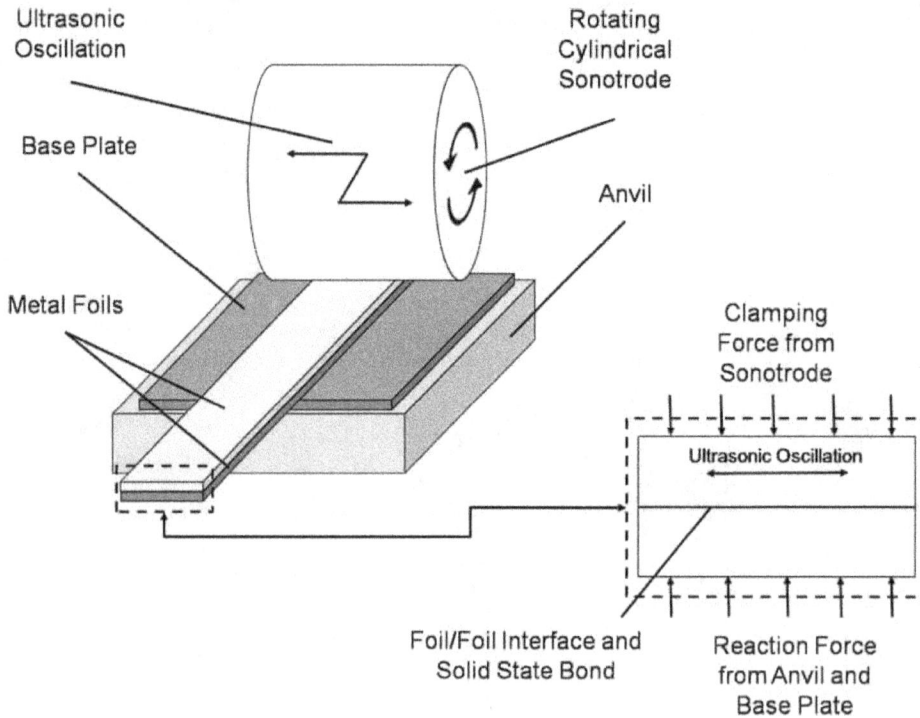

*Figure-12. 3D Printing via Ultrasonic Sheet Lamination.jpg*

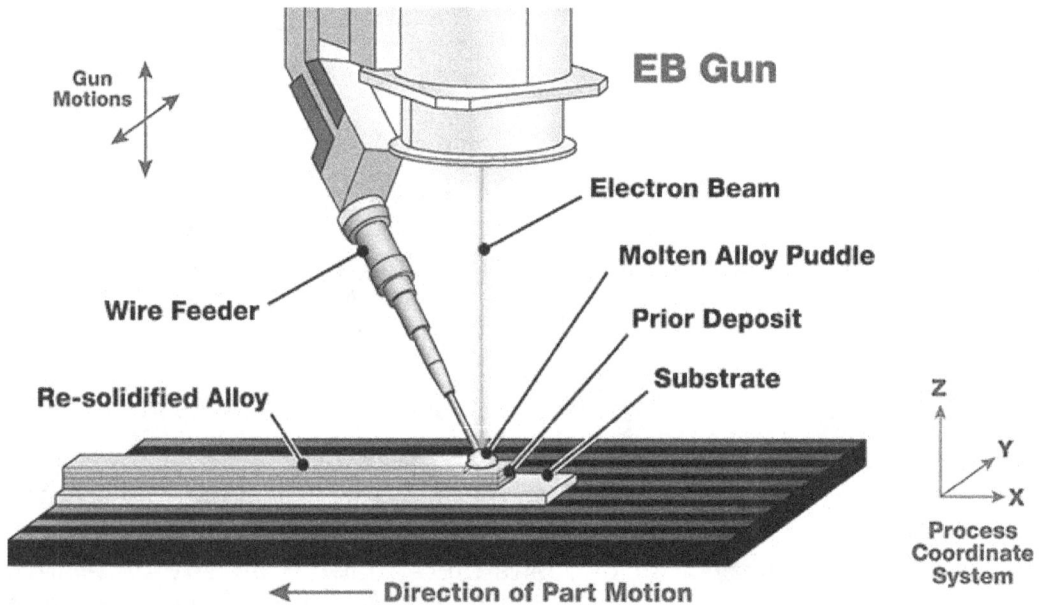

Gun
Motions

**EB Gun**

**Electron Beam**

**Molten Alloy Puddle**

**Wire Feeder**

**Prior Deposit**

**Re-solidified Alloy**

**Substrate**

Z

Y

X

**Process
Coordinate
System**

**◄──── Direction of Part Motion**

*Figure-13. 3D Printing via Direct Energy Deposition*

Now, you know about different 3D Printing techniques so it is clear from different processes that the main work of 3D printer is to solidify material at your will, through different techniques. Now, we will learn about the points to be taken care of while preparing model for 3D printing.

## Part Preparation for 3D Printing

Various important points to remember while preparing part for 3D printing are given next.

* You should avoid holes in the areas where model is not supported. Holes can cause material to flow out in various 3D printing processes.

- Use the Solid modeling tools. It does not mean that you cannot use surface modeling tools but you should avoid surfaces in your model. Thicken your surfaces after performing modeling operations.
- Make sure that you have not left any unwanted piece inside the model enclosure after performing boolean operations.
- Shell your model after creating it. You can 3D print solid model but if a hollow box can do your work then why to waste material on solid cube. Less material means cost efficient.
- If you want to write text on your model then check the specification of your printer for possible font range.
- The color and textures you apply on model in SolidEdge will not be exported for 3D printing so do not waste your time on them.
- 3D printing is closer to mesh modeling than solid modeling. So, check your mesh model by exporting solid model in .stl file.
- If your file has quite a bit of text and multiple emboss/engrave features, try exporting the file as a vector. Vector files are more appropriate for extremely complex files. Try .iges or .step.
- Orientation is particularly important when it comes to 3D printing in order to determine the interior and exterior of an object. Orient your part the same way as your want it in 3D Printing.
- Don't make a multi-body model for 3D printing. Your printer may die thinking what to do when two separate bodies overlap each other!!
- SolidEdge do not has capability to edit mesh model so you should prepare your model carefully to get desired mesh model.

## 3D PRINT PART PREPARATION IN SOLID EDGE

The tools to prepare part model for 3D printing are available in the **3D Print** tab of the **Ribbon**; refer to Figure-14. Various tools related to 3D printing in this tab are discussed next.

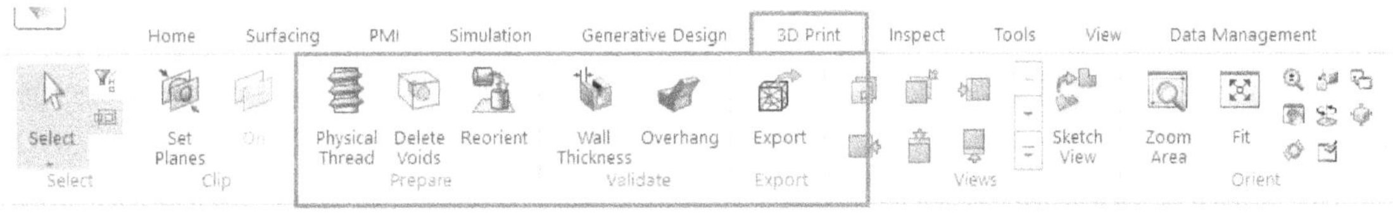

*Figure-14. 3D Print tab*

## Converting Cosmic Threads to Physical

The **Physical Thread** tool is used to convert cosmic threads into physical threads. The procedure to use this tool is given next.

- Click on the **Physical Thread** tool from the **Prepare** group of **3D Print** tab in the **Ribbon**. The **Physical Thread Command Bar** will be displayed; refer to Figure-15.

*Figure-15. Physical Thread Command Bar*

- Click on the **Select All** button from the **Command Bar** to select all the cosmic threads in the model or one by one click on the threads to be included.
- After selecting cosmic threads, click on the **OK** button from the **Command Bar**. The threads will be converted to physical threads and a message box will be displayed.

*   Click on the **OK** button from the message box and press **ESC** to exit the tool.

## Deleting Internal Voids

The **Delete Voids** tool is used to delete internal voids in the model. The procedure to use this tool is given next.

*   Click on the **Delete Voids** tool from the **Prepare** group of **3D Print** tab in the **Ribbon**. The **Delete Voids Command Bar** will be displayed; refer to Figure-16.

*Figure-16. Delete Voids Command Bar*

*   Click on the **Settings** button from the **Command Bar** to define colors of bodies with voids and without voids.
*   Click on the **Check** button from the **Command Bar** to check whether the part has internal voids or not. Like in Figure-17 we have created an intentional void.

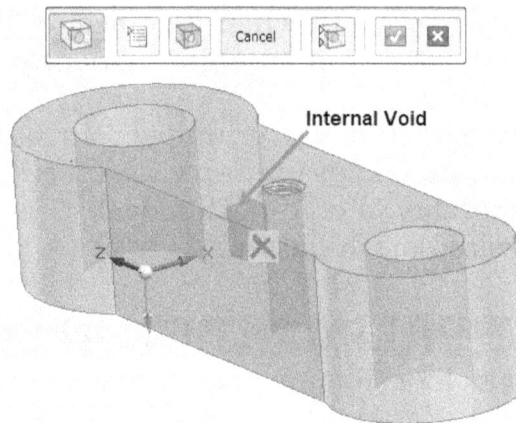

**Internal Void**

*Figure-17. Internal void*

*   Click on the **Delete** button in graphics area to delete desired void. After deleting the voids, press **ESC** to exit the tool.

## Reorienting Model for 3D Printing

The **Reorient** tool in **3D Print** tab is used to change the orientation of model as per the table of 3D Printer. The procedure to use this tool is given next.

*   Click on the **Reorient** tool from the **Prepare** group of the **3D Print** tab in the **Ribbon**. The **Reorient Command Bar** will be displayed with a preview box showing orientation of model in 3D Printing; refer to Figure-18.
*   Click on the **Settings** button from the **Command Bar** to define size of the 3D printing volume. The **Reorient Settings** dialog box will be displayed; refer to Figure-19. Specify the desired values of width, depth, and height of 3D printing volume as per your 3D printer specifications and select the resolution option from the **Display resolution** drop-down. Note that display resolution does not reflect the printing resolution. After setting desired parameters, click on the **OK** button.

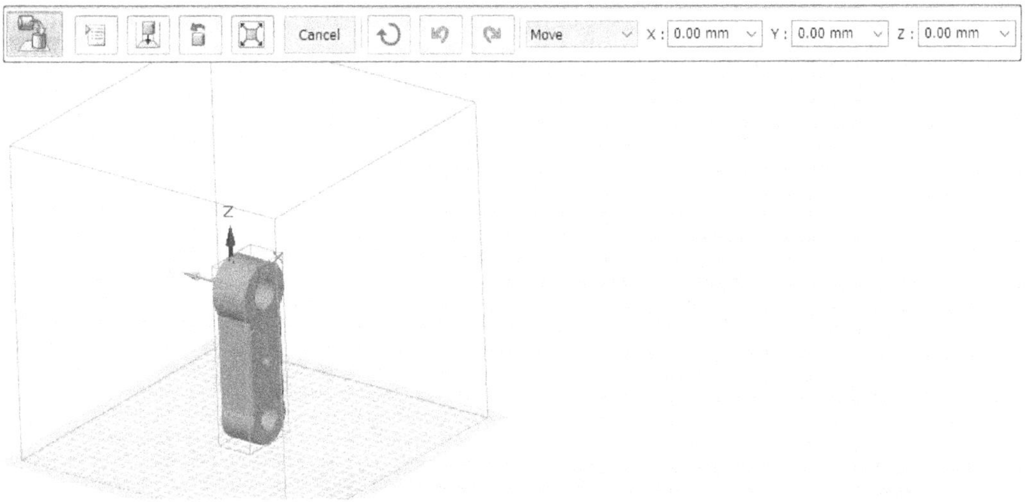

*Figure-18. Reorient Command Bar*

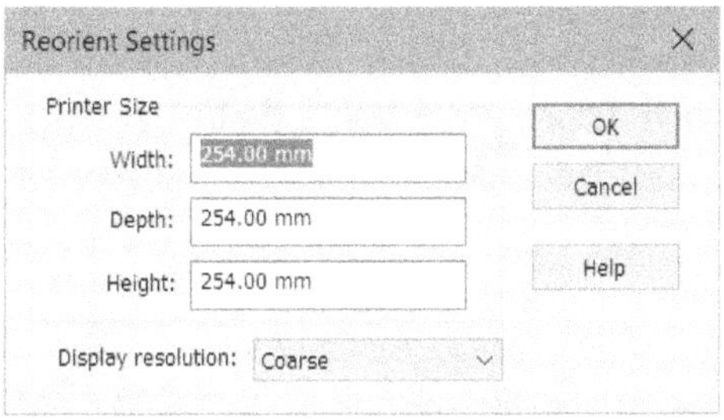

*Figure-19. Reorient Settings dialog box*

- Click on the **Settle** button from the **Command Bar** to place nearest flat face of model on the bed. Note that it might not be the most suitable orientation of model for 3D Printing.
- Click on the **Auto Orient** button from the **Command Bar** to orient model to next best fit for 3D printing using same base face.
- Click on the **Scale to Fit** button from the **Command Bar** to automatically increase/ decrease the size of model so that it fits in the printing volume.
- By default, the **Move** option is selected in the **Reorient** drop-down of the **Command Bar**. Specify the desired values in **X**, **Y**, and **Z** edit boxes to move the model in respective directions.
- Select the **Rotate** option from the **Reorient** drop-down of **Command Bar** to rotate the model. Specify the desired angle values in the **X**, **Y**, and **Z** edit boxes.
- After making desired changes, press **ESC** to exit the tool.

## Wall Thickness Validation

The **Wall Thickness** tool in **3D Print** tab is used to check if all the features in the model have valid thickness or not. The procedure to use this tool is given next.

- Click on the **Wall Thickness** tool from the **Validate** group of **3D Print** tab in the **Ribbon**. The **Wall Thickness Command Bar** will be displayed; refer to Figure-20.

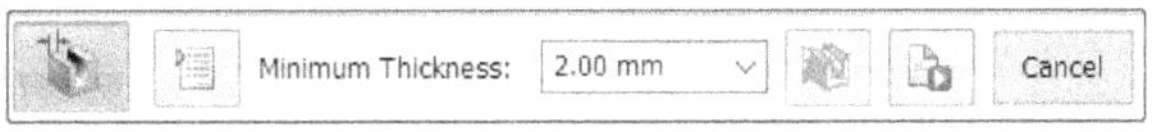

*Figure-20. Wall Thickness Command Bar*

- Click on the **Wall Thickness Settings** button from the **Command Bar** to specify colors for which are less thick and more thick than specified minimum thickness value.
- Specify the desired value of minimum thickness your 3D Printer can produce in the **Minimum Thickness** edit box of the **Command Bar**.
- Click on the **Calculate** button from the **Command Bar** to check whether all the features in the model have minimum thickness.
- Click on the **Toggle Display** button after running analysis to check features which have less thickness than required. Note that sometimes you may get false problematic areas so you need to carefully check the results.
- Press **ESC** to exit the tool.

# Checking Overhangs

The **Overhang** tool is used to check whether there are overhangs in the model at angle more than specified values. The procedure to use this tool is given next.

- Click on the **Overhang** tool from the **Validate** group of the **3D Print** tab in the **Ribbon**. The **Overhang Command Bar** will be displayed; refer to Figure-21.

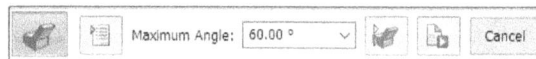

*Figure-21. Overhang Command Bar*

- Specify the desired value of overhang angle in the **Maximum Angle** edit box and click on the **Calculate** button. The problematic area of model will be displayed.
- Click on the **Toggle Display** button from the **Command Bar**. Only the problematic areas will be displayed.
- Press **ESC** to exit the tool.

# Exporting Model to STL File

The **Export** tool is used to generate an STL file using the 3D model prepared. Click on the **Export** tool from the **Export** group of **3D Print** tab in the **Ribbon**. The **3D Printing Export** dialog box will be displayed; refer to Figure-22. Specify the desired name of file and click on the **Save** button.

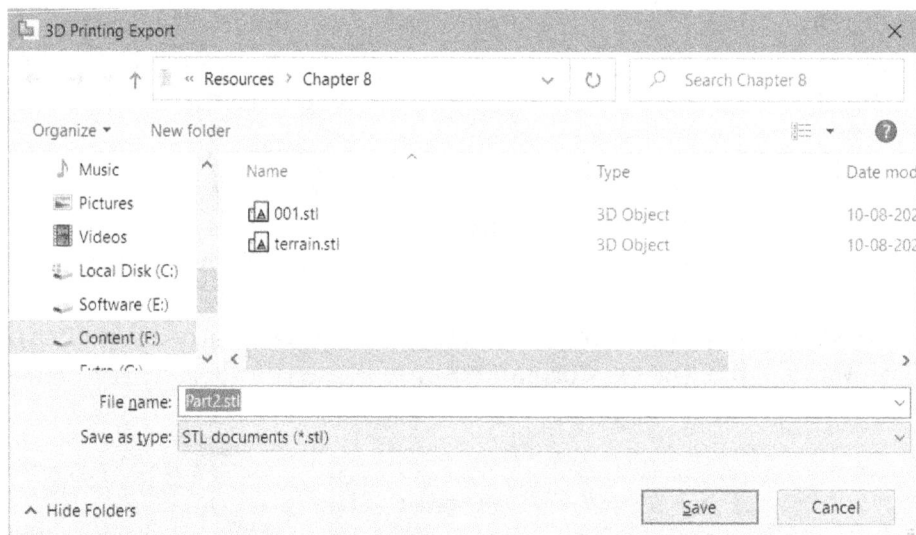

*Figure-22. 3D Printing Export dialog box*

The other options for 3D Printing has been discussed earlier in Chapter 1.

# VIEW TAB

The tools and options in the **View** tab are used to create and manage various views of the model. This tab is available in various environments like Part modeling, assembly, sheetmetal, and so on. The tools in this tab are discussed next.

# SHOWING PANES

The options in the **Panes** drop-down are used to display/hide various panes in the interface; refer to Figure-23. To display this drop-down, click on the **Panes** button from the **Show** group of the **View** tab in the **Ribbon**. Select the desired options from the drop-down to display respective interface elements.

*Figure-23. Panes drop-down*

# CONSTRUCTION DISPLAY OPTIONS

The **Construction Display** button in **Show** group is used to show/hide construction elements. The procedure to use this button is given next.

- Click on the **Construction Display** button from the **Show** group in the **View** tab of the **Ribbon**. The **Show All/Hide All** dialog box will be displayed; refer to Figure-24.
- Select the check boxes under **Show All** or **Hide All** columns for various construction objects to be displayed or hidden.
- After setting desired parameters, click on the **OK** button from the dialog box.

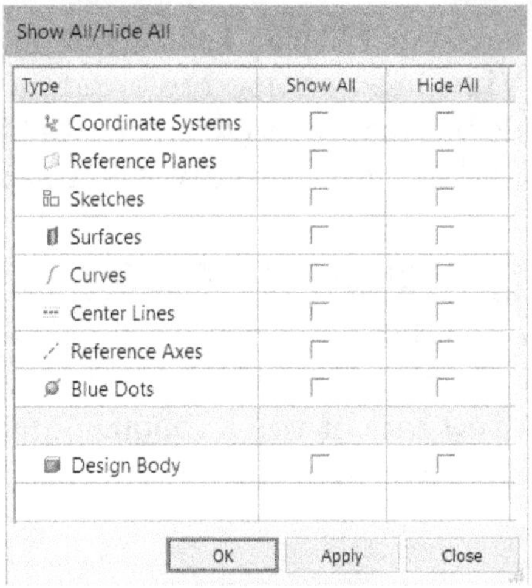

*Figure-24. Show All/Hide All dialog box*

# SETTING PLANES

The **Set Planes** tool is used to create reference planes for clipping the model. The procedure to use this tool is given next.

- Click on the **Set Planes** tool from the **Clip** group in the **Views** tab of the **Ribbon**. The **Set Planes Command Bar** will be displayed; refer to Figure-25.

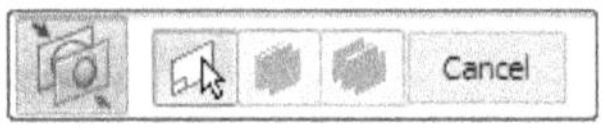

*Figure-25. Set Planes Command Bar*

- Select the desired plane or face to be used as reference for creating clipping planes. Note that the planes will be created parallel to selected face/plane.
- Click at the desired location to specify starting clipping plane. You will be asked to specify end clipping plane.
- Click at desired location to specify end clipping plane. The model between created clipping planes will be displayed when **Clipping On** button is active in the **Clip** group of **View** tab in the **Ribbon**; refer to Figure-26.
- Click on the **Finish** button from the **Command Bar** to create clipped model.

*Figure-26. Clipping*

# VIEWS

The buttons in the **Saved Views** box are used to set the orientation of parts and assemblies as per saved orientation styles; refer to Figure-27.

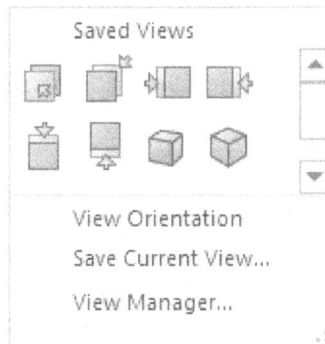

*Figure-27. Saved Views*

## Front View

The **Front View** button in the **Saved Views** box is used to change the view to front view. The hot key used for it is **CTRL+F**.

## Back View

The **Back View** button in the **Saved Views** box is used to change the view to back view. The hot key used for it is **CTRL+K**.

## Left View

The **Left View** button in the **Saved Views** box is used to change the view to left view. The hot key used for it is **CTRL+L**.

## Right View

The **Right View** button in the **Saved Views** box is used to change the view to right view. The hot key used for it is **CTRL+R**.

## Top View

The **Top View** button in the **Saved Views** box is used to change the view to top view. The hot key used for it is **CTRL+T**.

## Bottom View

The **Bottom View** button in the **Saved Views** box is used to change the view to bottom view. The hot key used for it is **CTRL+B**.

## Dimetric View

The **Dimetric View** button in the **Saved Views** box is used to change the view to dimetric view. In Dimetric view, two of the angles between the projection of axes are equal. The hot key used for it is **CTRL+J**.

## ISO View

The **ISO View** is used to change the view to ISO View. The ISO view is the method for visually representing three dimensional objects in two dimensions in technical and engineering drawing. It is an axonometric projection in which the three coordinate axes appear equally foreshortened and the angle between any two of them is 120 degree. The hot key used for it is **CTRL+I**.

## Trimetric View

The **Trimetric View** button in the **Saved Views** box is used to change the view to trimetric view. In Trimetric view, the projection of the angles between the axes are unequal. Thus, separate scales are needed to generate trimetric projection of an object. The hot key used for it is **CTRl+M**.

## Saving Current View

The **Save Current View** option in the **Saved Views** box is used to save current view of model in the **Saved Views** box. On clicking this button, the **New Named View** dialog box will be displayed; refer to Figure-28. Specify the desired name in the **New view** edit box and description of your view in the **Description** edit box of the dialog box. Click on the **OK** button from the dialog box to create the view. The new view will be displayed above **Save Current View** option from the **Saved Views** box.

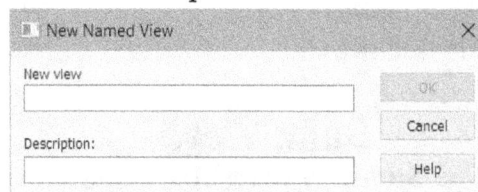

*Figure-28. New Named View dialog box*

## View Manager

The **View Manager** option in the **Saved Views** box is used to manager standard and custom views. Click on the **View Manager** option from the **Saved Views** box in the

**Views** group of **View** tab in the **Ribbon**. The **Named Views** dialog box will be displayed; refer to Figure-29. Select the desired view from the list and click on the **Apply** button to apply selected view. Select the desired view and click on the **Redefine to Current** button from the dialog box if you want to change the selected view based on how model is oriented in graphics area. Select the desired view and click on the **Delete** button. Click on the **Close** button from the dialog box to exit.

*Figure-29. Named Views dialog box*

# ORIENT

We have already discuss most of the tools in previous chapters. Here, we are going to discuss the view tools which have not been discussed earlier.

## Look at Face

The **Look at Face** tool is used to make the selected face parallel to the screen. To use this tool, click on the **Look at Face** tool from the **Orient** group of the **View** tab in the Ribbon and select the face that you want to make parallel to screen.

## Previous View

The **Previous View** tool is used to restore previous view. On selecting this tool, the part or assembly will be displayed in previous view.

## Spin About

The **Spin About** tool is used to rotate the model about a selected face. Click on this tool and select the face along which you want to rotate the model; refer to Figure-30.

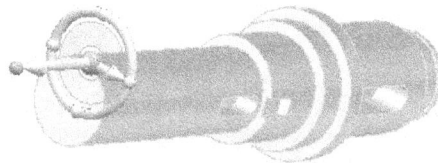

*Figure-30. Spin About*

## Wire Frame

The **Wire Frame** tool is used to apply wire frame display mode to the model. On selecting this tool the model changes to wire frame; refer to Figure-31.

*Figure-31. Wire Frame Model*

## Visible and Hidden Edges

The **Visible and Hidden Edges** tool is used to display visible and hidden edges of the model; refer to Figure-32.

*Figure-32. Visible and Hidden Edges*

## VISIBLE EDGES

The **Visible Edges** tool is used to display visible edges only in the model; refer to Figure-33.

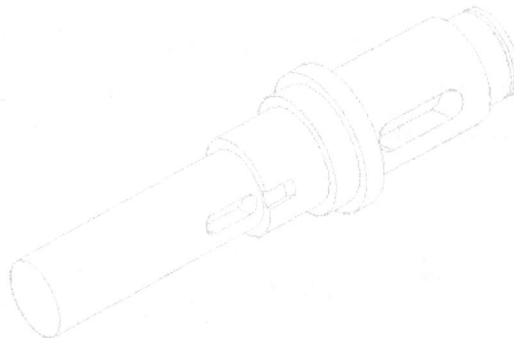

*Figure-33. Visible Edges*

## Shaded

The **Shaded** tool is used to display model in shaded display mode. On selecting this tool, the model display will change to shaded display; refer to Figure-34.

*Figure-34. Shaded*

## SHADED WITH VISIBLE EDGES

The **Shaded with Visible Edges** tool is used to display shaded model with dark visible edges; refer to Figure-35.

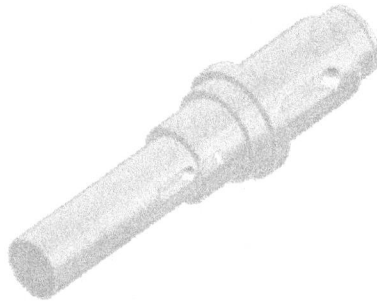

*Figure-35. Shaded with visible edges*

## FLOOR REFLECTION

The **Floor Reflection** tool is used to show reflection of model on the floor. Click on the **Floor Reflection** tool from the **Style** group of the **View** tab in the **Ribbon** to display floor reflection; refer to Figure-36.

*Figure-36. Floor Reflection*

## FLOOR SHADOW

The **Floor Shadow** tool is used to display shadow of model on the floor; refer to Figure-37.

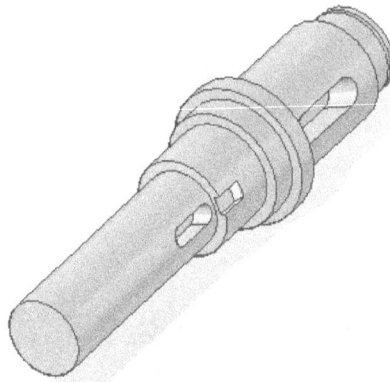

*Figure-37. Floor Shadow*

## HIGH-QUALITY RENDERING

The **High-Quality Rendering** button is used to improve the appearance of low face-count models and enables display of Bump Maps. This setting is an overrides to the active 3D view style.

## EDGE COLOR

Select the **Single Edge Color** button from the **Edge Color** drop-down of **Style** group in the **View** tab of the **Ribbon** to display all the edges in one color. You can set desired color for edges by using **Select Edge Color** tool in the **Edge Color** drop-down.

## SHARPEN

The **Sharpen** button in the **Style** drop-down is used to sharpen edges of the model. You can also select sharpening level by selecting desired option from the **Sharpness Set** cascading menu; refer to Figure-38.

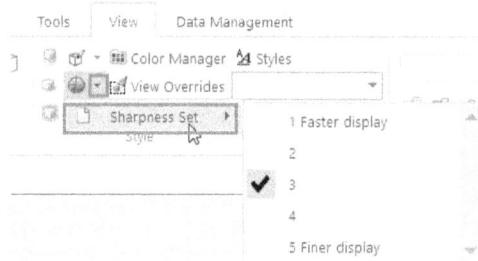

*Figure-38. Sharpness Set cascading menu*

## PERSPECTIVE VIEW

The **Perspective** button in the **Style** group of **View** tab in the **Ribbon** is used to display model in perspective view.

## COLOR MANAGER

The **Color Manager** tool is used to change color style for various objects in the model. The procedure to use this tool is given next.

- Click on the **Color Manager** tool from the **Style** group in the **View** tab of the **Ribbon**. The **Color Manager** dialog box will be displayed; refer to Figure-39.
- Select the **Use Solid Edge Options color settings** radio button to use the color scheme specified in **Solid Edge Options** dialog box.
- If you want to use individual part styles for various objects then select the **Use individual part styles** radio button. Select the desired color/material styles from the drop-downs for various objects in the **Base Styles** section of dialog box.
- After setting desired parameters, click on the **OK** button.

*Figure-39. Color Manager*

# VIEW OVERRIDES

The **View Overrides** tool is used to apply overrides to the active 3D view style. The procedure to use this tool is given next.

- Click on the **View Overrides** tool from the **Style** group in the **View** tab of **Ribbon**. The **View Overrides** dialog box will be displayed; refer to Figure-40.

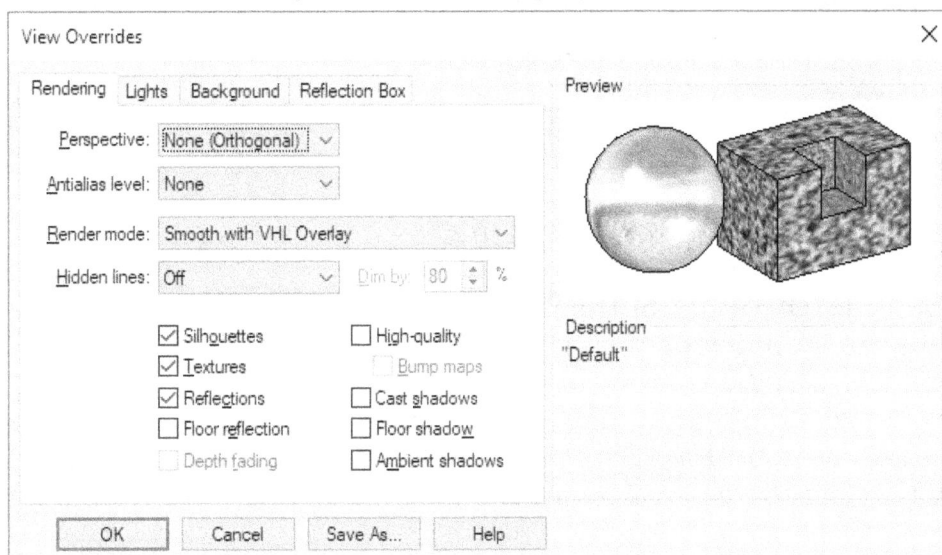

*Figure-40. View Overrides*

## Rendering Tab

- By default, the **Rendering** tab is selected in the dialog box. Set the desired options in this tab to define smoothness and style in the rendering.
- Select the desired option from the **Perspective** drop-down to set focus level of perspective view in the dialog box.
- Select the desired option from the **Antialias level** drop-down to specify edge quality of model.
- Select the desired option from the **Render mode** drop-down to specify rendering mode.
- Select the desired option from the **Hidden lines** drop-down to display hidden lines in the rendering of model.
- Select the desired check boxes from the dialog box to display various elements of rendering in the dialog box.

## Lights Tab

- Select the **Lights** tab in the dialog box to set the location, intensity, and color of light; refer to Figure-41.

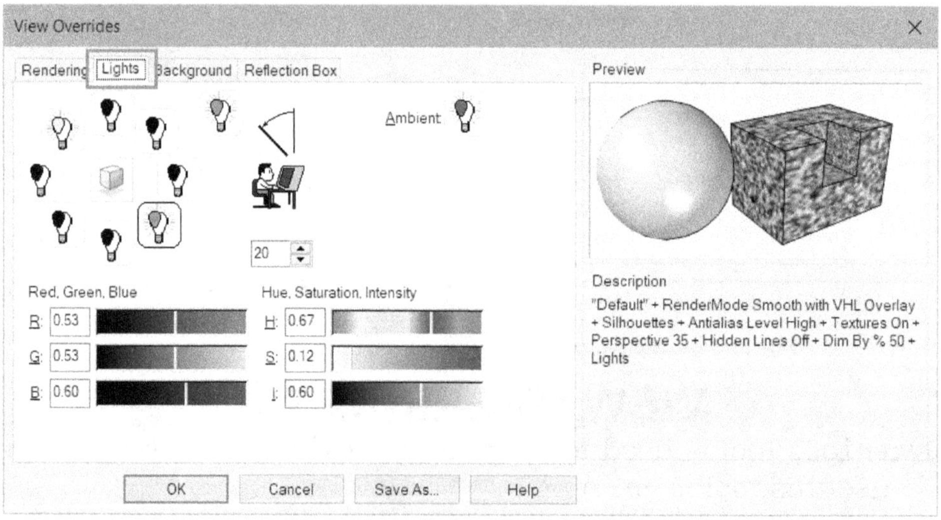

*Figure-41. Lights tab*

- Set the desired color and intensity using the sliders.
- Select the desired button for light from the dialog box to define direction and density of light in rendering.

## Background Tab

- Select the **Background** tab from the dialog box to apply color or image in the background of model.
- Select the **Solid** radio button and then select desired color from the drop-down next to the radio button.
- Select the **Gradient** radio button to use gradient background using two colors selected in the drop-downs. You can also select direction of gradient in the drop-down next to the radio button.
- Select the **Image** radio button to use desired image as background of the model and click on the **Browse** button. The **Open a File** dialog box will be displayed; refer to Figure-42. Select the desired image file and click on the **Open** button to select it.

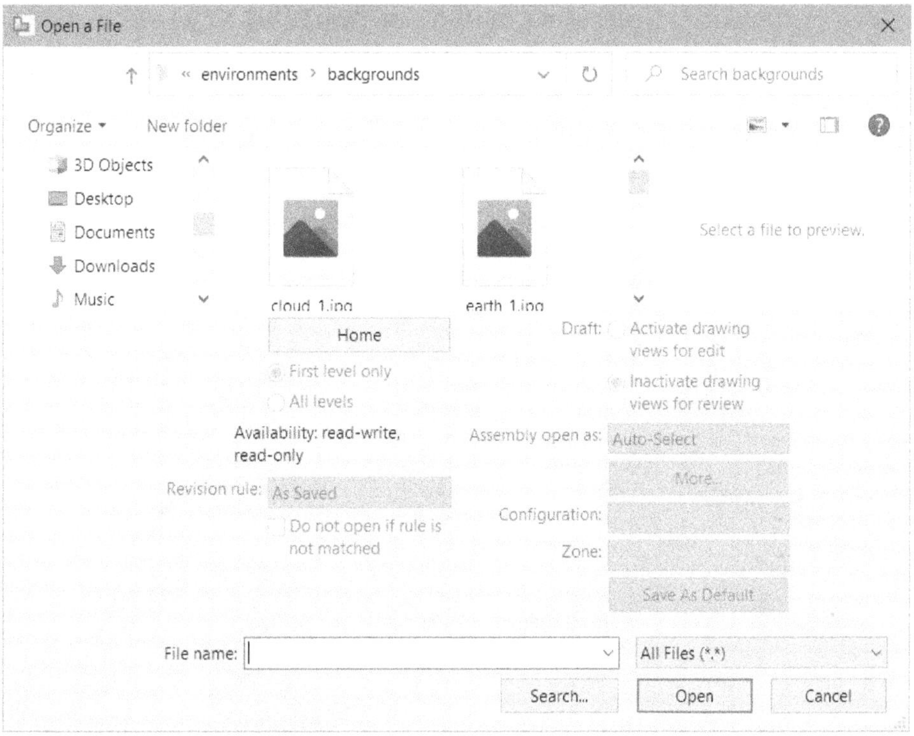

*Figure-42. Open a File dialog box*

- Select the **Mirror up/down** radio button to mirror the image background from up to down. Similarly, select the **Mirror left/right** radio button mirror image from left to right.

## Reflection Box Tab

- Select the **Reflection Box** tab from the dialog box to define location of reflection plane with respect to model.
- Select the desired button from the dialog box to define the plane.
- Select the **Use for background** check box to use reflection plane as background of model.
- After setting the desired parameters in the dialog box, click on the **OK** button.

## APPLYING APPEARANCES TO PART

The **Part Painter** tool is used to apply appearances to model like gold paint, brass paint, and so on. The procedure to use this tool is given next.

- Click on the **Part Painter** tool from the **Style** group in the **View** tab of the **Ribbon**. The **Part Painter Command Bar** will be displayed; refer to Figure-43.

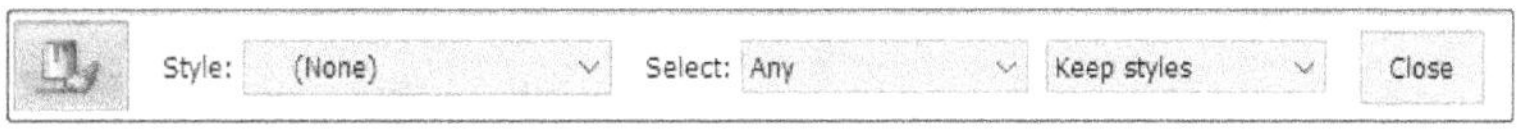

*Figure-43. Part Painter Command Bar*

- Select the desired color/appearance from the **Style** drop-down.
- Click in the **Select** drop-down to define the selection filter for selecting object. By default, **Any** is selected in the drop-down so you can select any of the object type from the model.

- Select the desired option from the **Currently Applied Styles** drop-down to define whether you want to replace the previous display style or you want to just apply the appearance and rest of the style will remain same.
- After setting desired options, click on the model to apply selected style.
- Click on the **Close** button from the **Command Bar** to exit the tool.

# Chapter 6

# Surfacing

## Topics Covered

The major topics covered in this chapter are:

- *Introduction to Surfacing*
- *Creating Curves for Surfacing*
- *Creating Surfaces*
- *Modifying Surfaces*
- *Surface Visualization Settings*
- *Displaying Section Curvature*

# SURFACING

The **Surfacing** is a separate world in the field of CAD. The complicated shapes which are difficult for solid modeling are most of the time easy for surfacing. Basic tools of surfacing are very similar to the solid creation tools discussed earlier like extrude, revolve, swept, and so on. But, there are some other tools that allow to modify the part shape freely in 3D. In Solid Edge, the tools to create surface models are available in the **Surfacing** tab of the **Ribbon** in the Part environment; refer to Figure-1. By default, most of the surfacing tools are not active. These tools will be active automatically when related objects are available in the drawing area.

*Figure-1. Surfacing Tab in Ribbon*

# KEY POINT CURVE

The **Keypoint Curve** tool is used to draw curve through selected key points or points which are created in sketch. The procedure to use this tool is given next.

*   Click on the **Keypoint Curve** tool from the **Keypoint** drop-down in the **Curves** panel of **Surfacing** tab in the **Ribbon**. The **Keypoint Curve Command Bar** will be displayed; refer to Figure-2.

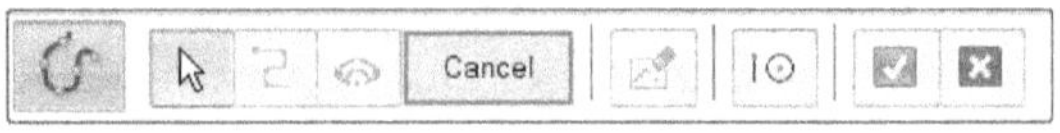

*Figure-2. Keypoint Curve Command Bar*

*   Click at desired locations to specify key points for curve and right-click in the empty area. Preview of the curve will be displayed; refer to Figure-3.

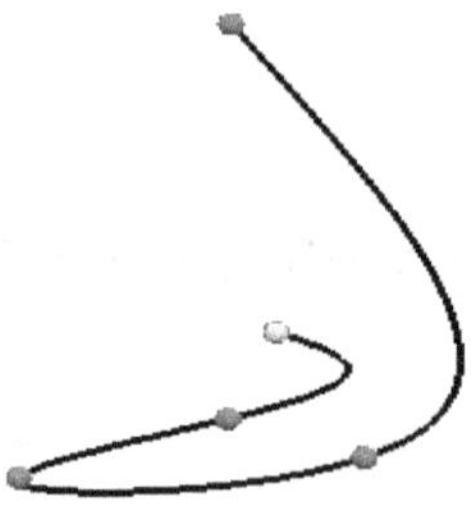

*Figure-3. Keypoints Curve*

*   If you want to modify the end condition of curve then click on the **End Conditions Step** button from the **Command Bar**. The options for end conditions will be displayed in the **Command Bar**; refer to Figure-4. Click on the **Closed** button from the **Command Bar** if you want to close start and end points of the curve.

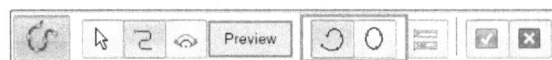

*Figure-4. Buttons for modifying end condtions*

*   If you want to modify the length of curve then click on the **Curve Length Step** button from the **Command Bar** and set the parameters as discussed earlier.

- Click on the **Preview** button from the **Command Bar** and then click on the **Finish** button to create the curve.

# CURVE BY TABLE

The **Curve by Table** tool is used to draw curve by using the coordinates specified in the table. The procedure is discuss next.

- Click on the **Curve by Table** tool from the **Keypoint** drop-down in the **Curves** panel of the **Surfacing** tab in the **Ribbon**. The **Insert Object** dialog box will be displayed as shown in Figure-5.
- Select the **Create from file** radio button if you want to use already existing Microsoft Excel file. If you want to create a new sheet then select the **Create new spreadsheet** radio button. After setting desired parameters, click on the **OK** button. The worksheet will open in Microsoft Excel. Make sure you have Microsoft Excel installed in your system.

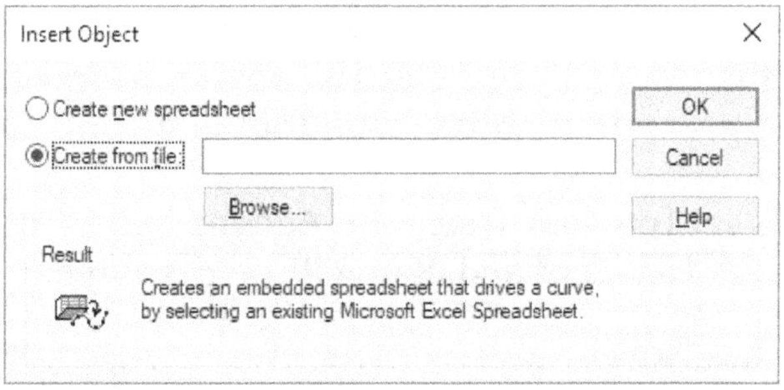

*Figure-5. Insert Object dialog box*

- Specify the desired values of X, Y, and Z coordinates of key points for curve; refer to Figure-6.

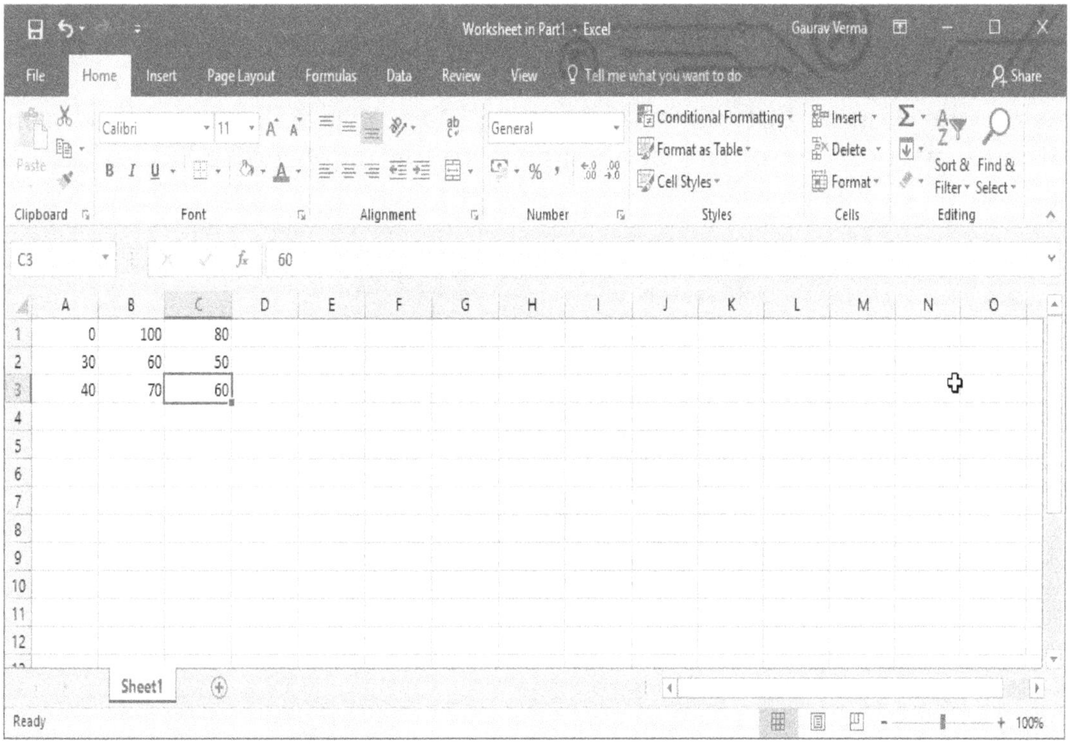

*Figure-6. Create New table in Excel file*

- After specifying parameters in the excel sheet, close the Microsoft Excel application. Preview of the curve will be displayed; refer to Figure-7.

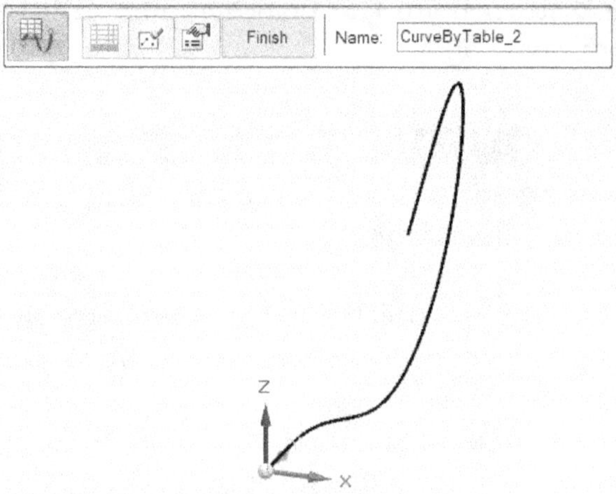

*Figure-7. Preview of curve by table*

*   Click on the **Finish** button to create the feature.

## INTERSECTION

The **Intersection** tool is used to create curve at the intersection of two surfaces or surface and plane. The procedure to use this tool is given next.

*   Click on the **Intersection** tool from the **Curves** panel in the **Surfacing** tab of **Ribbon**. The **Intersection Command Bar** will be displayed; refer to Figure-8.

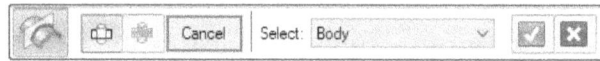

*Figure-8. Intersection Command bar*

*   Select the first set of surfaces or faces to be used for intersection curve while holding the **CTRL** key.
*   Select the second set of surfaces or face to be intersected. Preview of intersection curve will be displayed.
*   Click on the **Finish** button from **Command Bar**. You will notice that a curve is created at intersection of two bodies; refer to Figure-9. Press **ESC** to exit the tool.

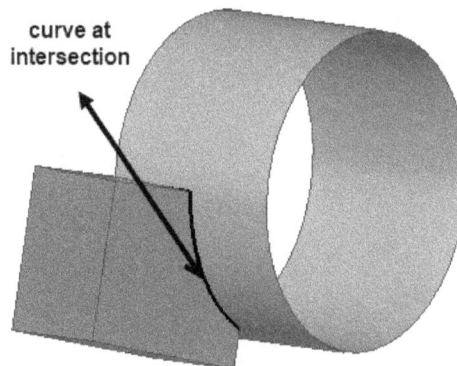

curve at intersection

*Figure-9. Intersection Curve*

# PROJECT

The **Project** tool is used to project one or more curve on to a surface or set of surfaces. You can also project the curve along a vector or along surface normals and a point onto a surface using this tool. The procedure to use this tool is discussed next.

- Click on the **Project** tool from the **Curves** group in the **Surfacing** tab of **Ribbon**. The **Project Command Bar** will be displayed; refer to Figure-10.

*Figure-10. Project Command Bar*

- **Project Curve Options** : The **Project Curve Options** button is used to set the projection to be either along a vector or normal to the selected surface. On clicking this button, the **Project Curve Options** dialog box will be displayed; refer to Figure-11. Select the desired radio button from the dialog box and click on the **OK** button.

*Figure-11. Project Curve Options*

- Select the curve to be projected.
- Select the surface or face onto which the curve would be projected.
- Specify the direction towards the surface and click on the **Finish** button from **Command Bar**; refer to Figure-12.

*Figure-12. Project*

# HELICAL CURVE

The **Helical Curve** is used to create helical or spiral curves. The procedure to use this tool is discussed next.

- Click on the **Helical Curve** tool from the **Curves** group in the **Surfacing** tab of the **Ribbon**. The **Helical Curve Command Bar** will be displayed; refer to Figure-13.

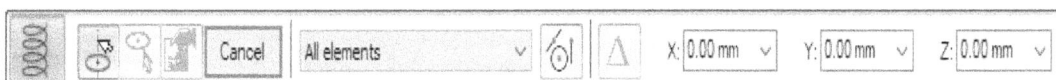

*Figure-13. Helical Curve Command Bar*

- Select the desired edge, circular face, cone face, or line to define axis of helical curve. The outer edge of helical curve will get attached to cursor and you will be asked to specify diameter of helical curve.
- Click at desired location to specify diameter of curve. The **Helical Curve Parameters** dialog box will be displayed; refer to Figure-14.
- Select the desired option from the **Type** drop-down of the dialog box to define type of helical curve. Select the **Constant Pitch** option from the drop-down if gap between two consecutive coils of helical curve are same; refer to Figure-15. Select the **Variable Pitch** option from the drop-down if you want to specify different pitch at start and end of the helical curve; refer to Figure-16. Select the **Compound** option from the drop-down if you want to specify varying pitch, number of turns, and diameter at different length points; refer to Figure-17. Select the **Spiral** option from the drop-down if you want to create spiral curve; refer to Figure-18.

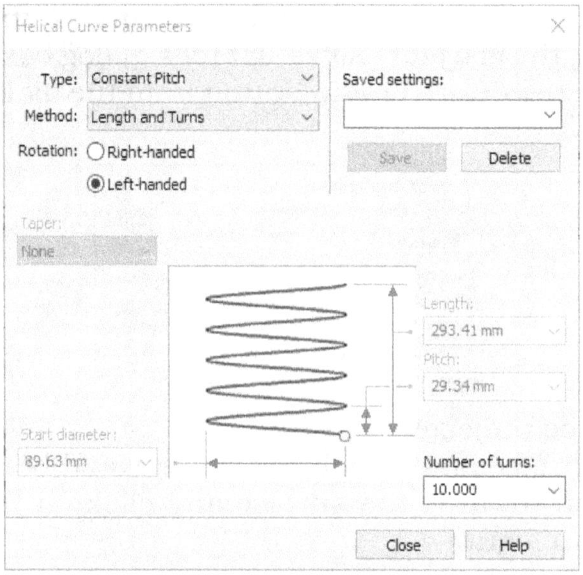

*Figure-14. Helical Curve Parameters*

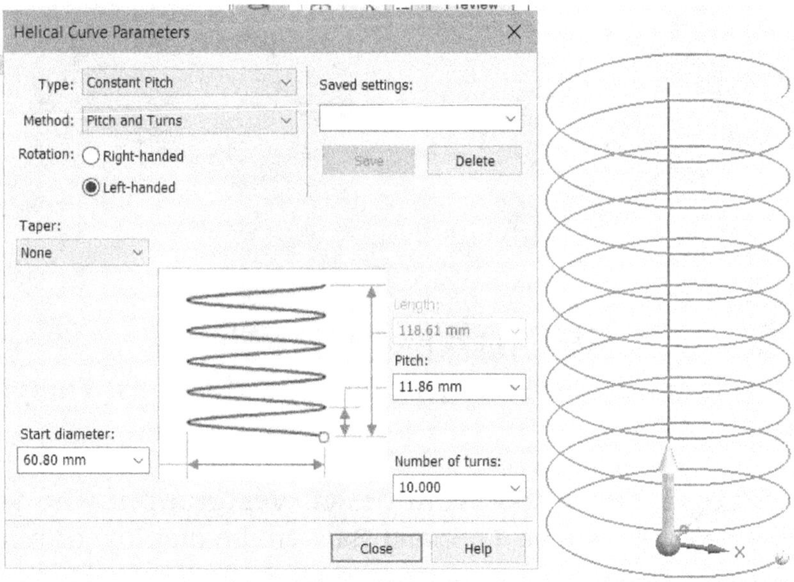

*Figure-15. Constant pitch helical curve*

*Figure-16. Variable pitch helical curve*

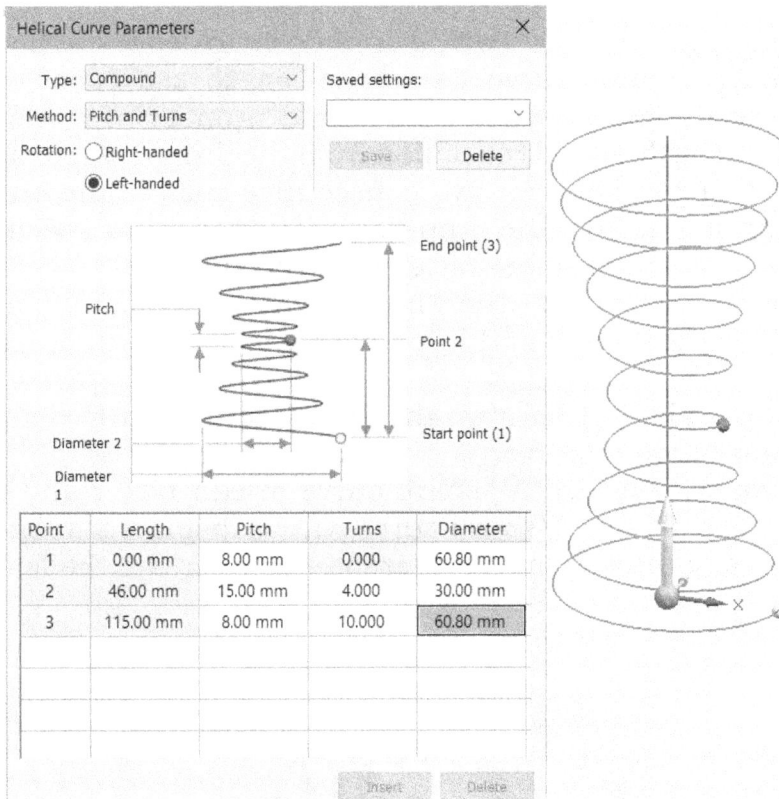

| Point | Length | Pitch | Turns | Diameter |
|-------|-----------|-----------|--------|----------|
| 1 | 0.00 mm | 8.00 mm | 0.000 | 60.80 mm |
| 2 | 46.00 mm | 15.00 mm | 4.000 | 30.00 mm |
| 3 | 115.00 mm | 8.00 mm | 10.000 | 60.80 mm |

*Figure-17. Compound helical curve*

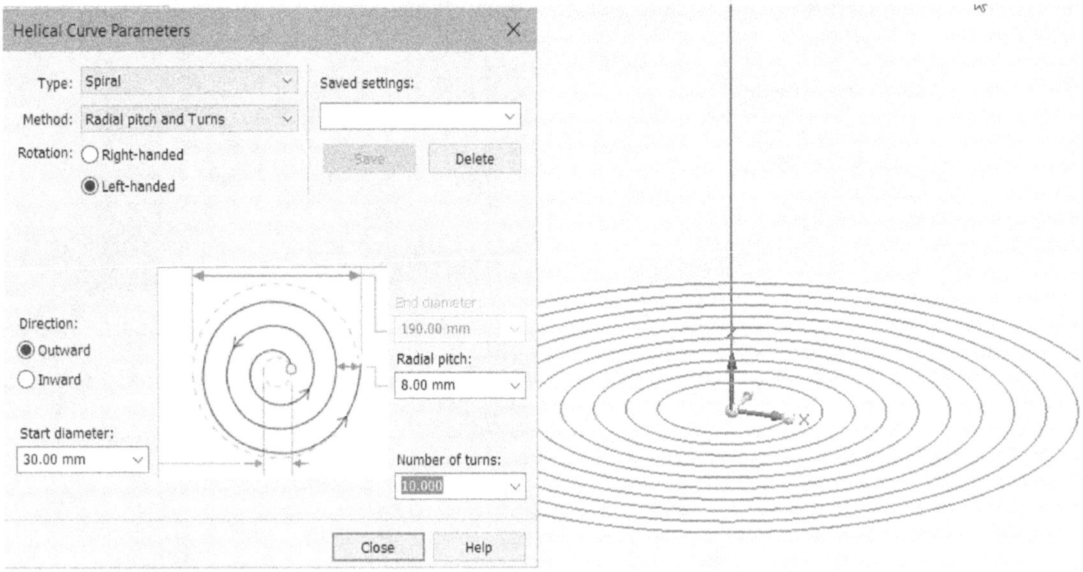

*Figure-18. Spiral curve*

- Select the desired option from the **Method** drop-down to define method for specifying parameters of helical/spiral curve. Select the **Length and Turns** option from the drop-down to specify total height of curve and number of turns in the helical curve. Select the **Length and Pitch** option from the drop-down to specify the total height of curve and gap between two consecutive coils of the curve. Select the **Pitch and Turns** option from the drop-down if you want to specify gap between two consecutive coils and number of turns in the curve.
- If you are creating spiral curve then in place of length parameter, you will be asked to specify End diameter and in place of pitch, you will be asked to specify radial pitch.
- Select the desired radio button from the **Rotation** section to define left handed or right-handed rotation in curve.
- Specify the parameters of helical/spiral curve in edit boxes and click on **Close** button. The preview of helical/spiral curve will be displayed; refer to Figure-19.
- Click on the **Finish** button from the **Command Bar** to create the feature.

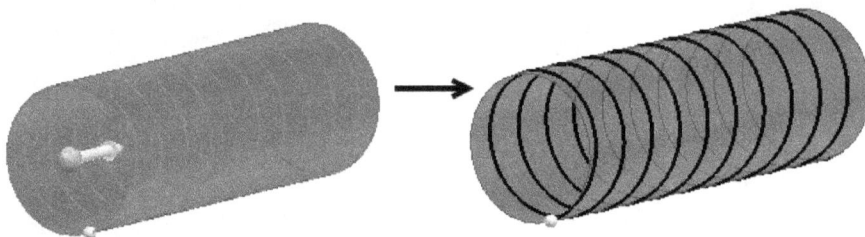

*Figure-19. Preview of Helical Curve*

# CROSS

The **Cross** tool is used to creates a curve at intersection of virtual extrusion of two curves. The procedure to use this tool is given next.

- Click on the **Cross** tool from the **Curves** group in the **Surfacing** tab of the **Ribbon**. The **Cross Command Bar** will be displayed; refer to Figure-20. Choose **Select from Sketch** option from the **Create-From Options** drop-down in the **Command Bar** if sketched curves are already created. You can also select an option to use plane for creating sketch from the drop-down as discussed earlier.

*Figure-20. Cross Command bar*

- Select first set of curves to be projected from the drawing area and then right-click. You will be asked to select second set of curves.
- Select the second set of curves to be projected and then right-click. The preview is shown in Figure-21.
- Click on the **Finish** button from the **Command Bar** to create the feature. Press **ESC** to exit the tool.

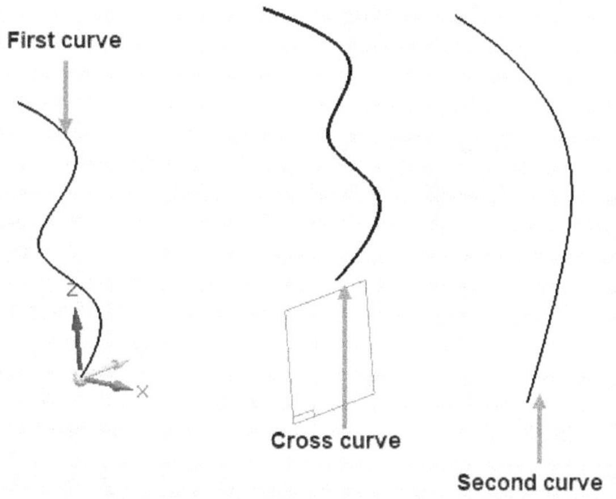

*Figure-21. Curve created by Using Cross tool*

## CONTOUR

The **Contour** tool is used to draws a curve directly on the selected surface. The procedure to use this tool is given next.

- Click on **Contour** tool from the **Curves** group in the **Surfacing** tab of the **Ribbon**. The **Contour Command Bar** will be displayed; refer to Figure-22.

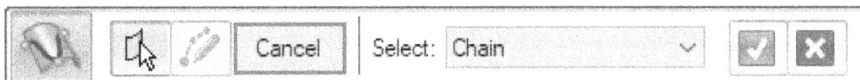

*Figure-22. Contour Command Bar*

- Select the surface on which you want to create curve and press **ENTER**. You will be asked to specify points of the curve.
- There are two type of curves that can be created on the surface, open curve and close curve. Select the **Open** button from the **Command Bar** to create open curves and select the **Close** button from the **Command Bar** to create closed curves.
- Click to specify the points on the surface. Preview of the curve will be displayed as you specify the points; refer to Figure-23.

- If you want to insert a point anywhere after creating the curve then click on the **Insert Point** button from the **Command Bar**.
- After specifying the points, click on the **OK** button from the **Command Bar** to create the curve.
- Specify the desired name of curve in the **Name** edit box of the **Command Bar** displayed and click on the **Finish** button. Press **ESC** to exit the tool.

*Figure-23. Contour Curve*

## ISOCLINE

The **Isocline** tool is used to create a curve on selected surface at specified angle from selected plane. The procedure to use this tool is discussed next.

- Click on the **Isocline** tool from the **Curves** group in the **Surfacing** tab of the **Ribbon**. The **Isocline Command Bar** will be displayed; refer to Figure-24.

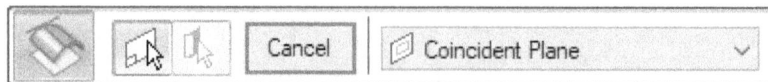

*Figure-24. Isocline Command Bar*

- Selects a reference plane or a planar face to be used for specifying vector angle. The plane should be perpendicular and intersecting with the surface.
- Select face or body on which isocline curve is to be created. The direction should be towards the face or bodies. The preview of isocline curves will be displayed; refer to Figure-25.
- Right-click to accept the selection and click on the **Finish** button to create curve.
- Press **ESC** to exit the tool.

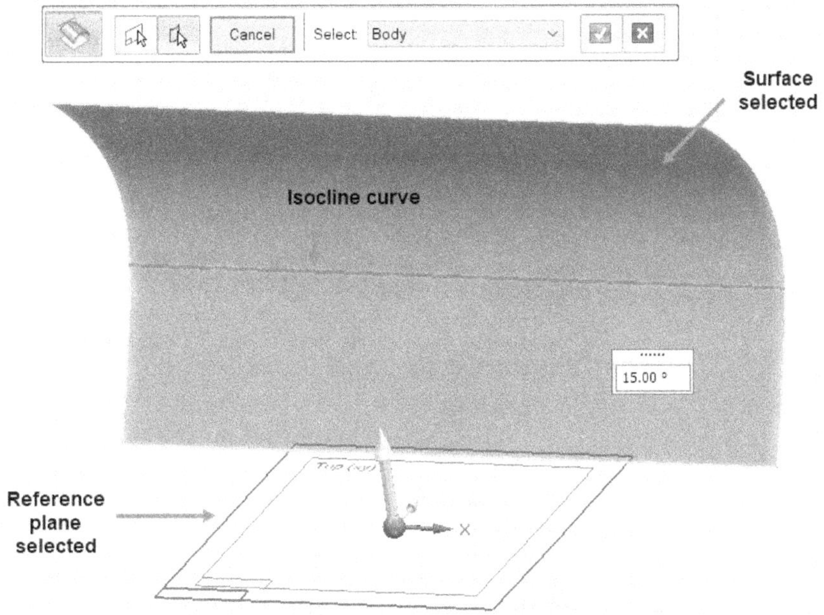

*Figure-25. Preview of Isocline*

# DERIVED

The **Derived** tool is used to create a curve using one or more selected curves/edges. The procedure to use this tool is given next.

- Click on the **Derived** tool from the **Curves** group in the **Surfacing** tab of the **Ribbon**. The **Derived Command Bar** will be displayed; refer to Figure-26.

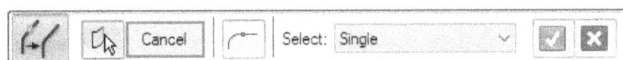

*Figure-26. Derived Command bar*

- Selects one or more curves or edges as the base for creating derived curves. Click **RMB** button to accept. Refer to Figure-27.

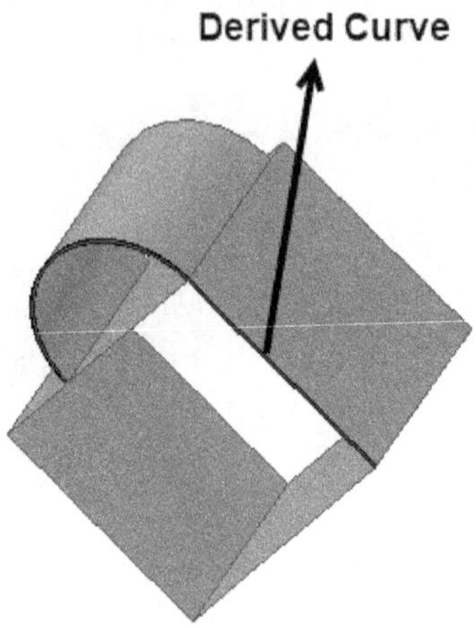

*Figure-27. Derived Curve*

*   **Single Curve** [icon] : If this option is turn **"ON"** from **Command Bar** then it creates a single curve from end-connected, tangent input elements.
*   Click on the **Finish** button from the **Command Bar** to create the curves. Press **ESC** to exit the tool.

# SPLIT [icon]

The **Split** tool is used to split a curve using intersecting geometry. The procedure is given next.

*   Click on the **Split** tool from the **Curve** group in the **Surfacing** tab of the **Ribbon**. The **Split Command Bar** will be displayed; refer to Figure-28.

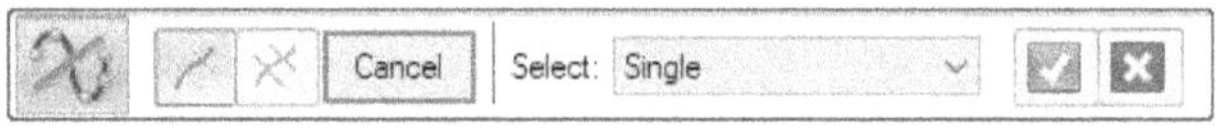

*Figure-28. Split Curve Command bar*

*   Select the curves to be split and right-click. You will be asked to select the intersecting geometry.
*   Select the desired option from the **Select** drop-down and then select respective intersecting geometry.
*   Click on the **OK** button from the **Command Bar**. The curve(s) will split at intersection points; refer to Figure-29.
*   Click on the **Finish** button to create feature and press **ESC** to exit the tool.

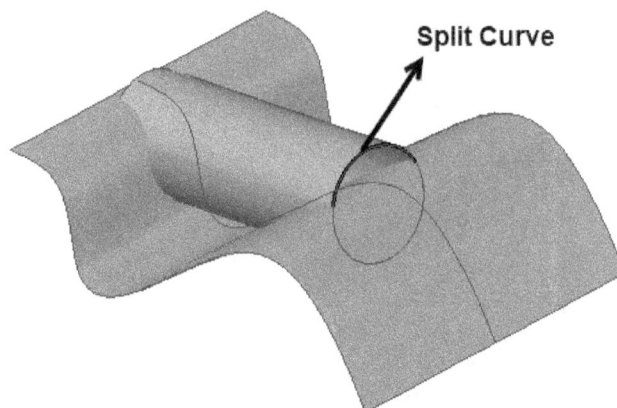

*Figure-29. Split Curve*

# INTERSECTION POINT

The **Intersection Point** tool is used creates point at the intersection of selected curves and edges. The procedure to use this tool is given next.

* Click on the **Intersection Point** tool from the **Curve** group in the **Surfacing** tab of the **Ribbon**. The **Intersection Point Command Bar** will be displayed; refer to Figure-30.

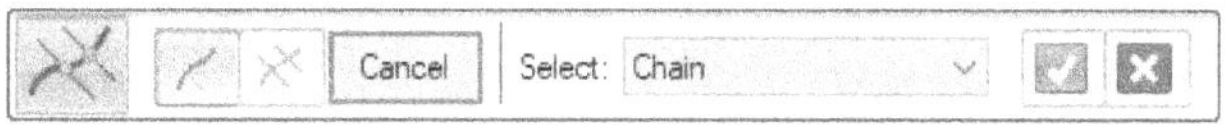

*Figure-30. Intersection point Command bar*

* Select a continuous primary curve to be used for generating intersection points and right-click in the drawing area. You will be asked to select an intersecting plane/axis/body.
* Select the desired intersecting object and right-click in the drawing area. The intersection points will be generated; refer to Figure-31.
* Click on the **Finish** button from the **Command Bar** to create the feature and press **ESC** to exit the tool.

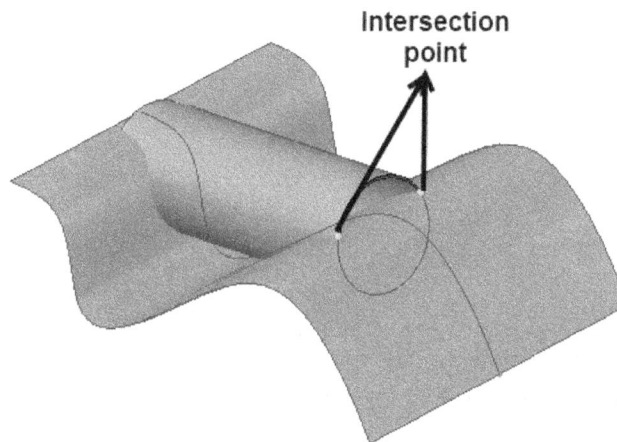

*Figure-31. Intersection Point*

# CREATING SURFACES

The tools to create surfaces are available in the **Surfaces** panel of the **Surfacing** tab in the **Ribbon**; refer to Figure-32. These tools are discussed next.

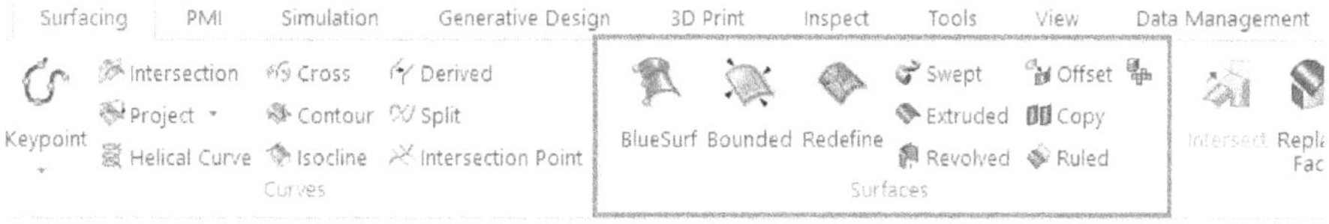

*Figure-32. Surfaces panel*

## Creating Surface using BlueSurf tool

The **BlueSurf** tool is used to surface by using selected edges or curves. You can also use a guide curve to control the shape of surface. Note that the guide curve should intersect with the cross-section curves and should not have shape turns. The procedure to use this tool for creating surface is discuss next.

- Click on the **BlueSurf** tool from the **Surfaces** group in the **Surfacing** tab of the **Ribbon**. The **BlueSurf Command Bar** will be displayed; refer to Figure-33.

Figure-33. BlueSurf Command Bar

- Select the first curve and then right-click in drawing area. You will be asked to select the second curve.
- Select the second curve and right-click in drawing area. Preview of surface will be displayed; refer to Figure-34.

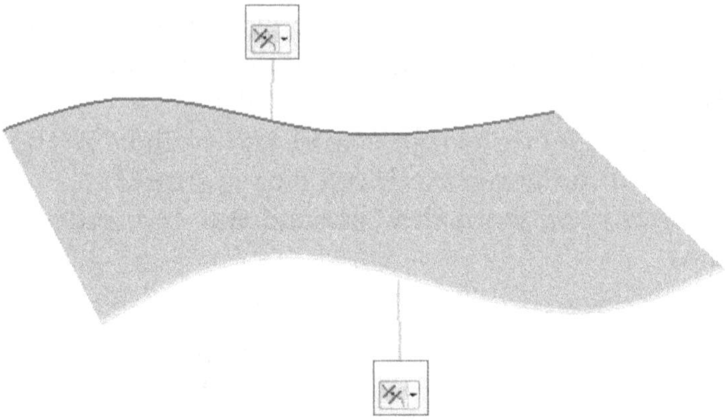

Figure-34. BlueSurf Preview

- **Tangency Control** : Click on the down arrow of the **Tangency Control** box. The options to control the shape of surface at the ends of the feature will be displayed; refer to Figure-35. Select the **Natural** option if you want to continue natural flow in surface based on selected curves. Select the **Normal to Section** option if you want the surface to be perpendicular to edge at start/end point. Select the **Parallel to Section** option to make the surface edge parallel to selected cross-section.

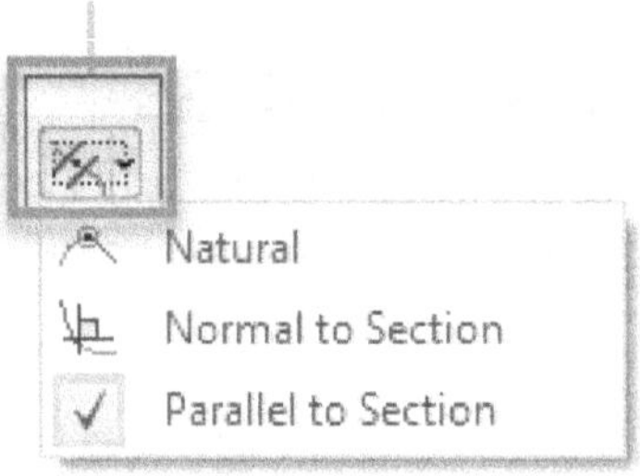

Figure-35. Tangency Control drop down

- Click on the **Guide Curve Step** button from the **Command Bar** and select the curve that you want to use as guide curve. Preview of the surface will be displayed.
- Right-click in the drawing area to accept the selection.
- Click on the **BlueSurf Options** button to modify standard and advanced parameters of the surface. The **BlueSurf Options** dialog box will be displayed; refer to Figure-36.

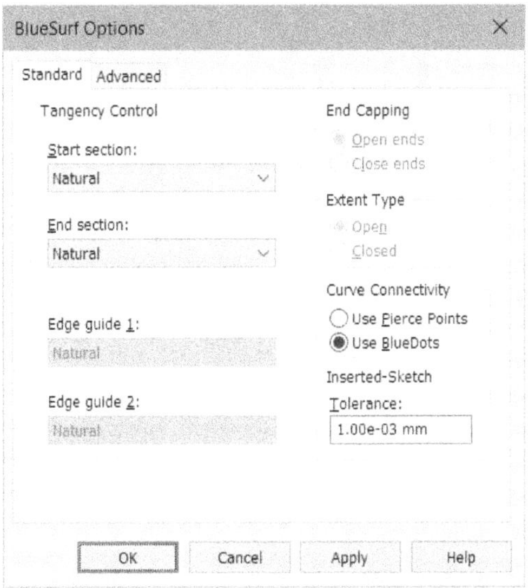

*Figure-36. BlueSurf Options dialog box*

- Set the tangency options as discussed earlier in the dialog box.
- Select the desired option from the **Curve Connectivity** area to define how curves of surface will be connected. Select the **Use Pierce Points** radio button to use connect relationship at intersections of cross-section and guide curves. When you move cross section curves or guide curves then connections will be automatically updated for new intersection points. Select the **Use BlueDots** radio button to modify the shape of surface by using the blue dots displayed on curves.
- Select the **Open ends** radio button from the **End Capping** area of the dialog box if you want to keep the end faces of surface open. Select the **Closed ends** radio button if you want to cap the end faces of the surface; refer to Figure-37.

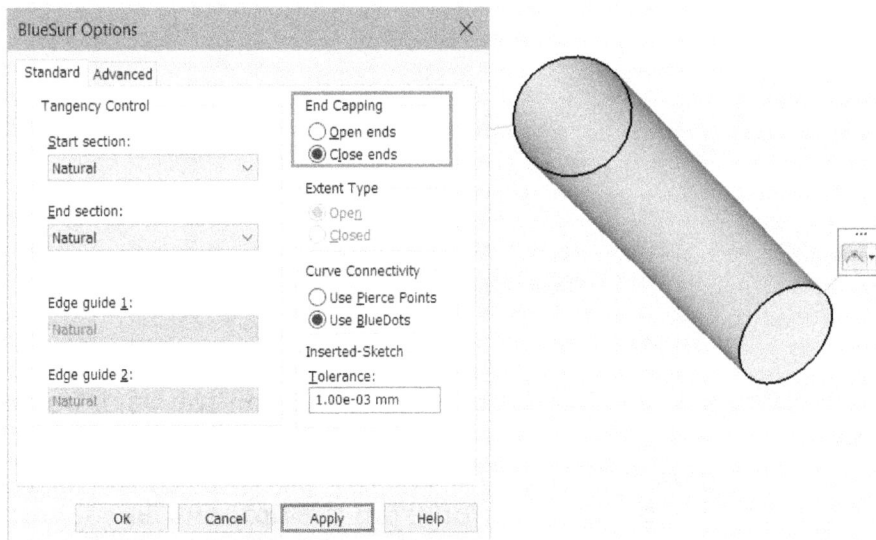

*Figure-37. Closed end capping*

- Select the desired option from the **Extent Type** area to define whether you want to create closed surface or open surface. If selected curvatures can form surface then select the **Closed** radio button from this area and click on the **Apply** button; refer to Figure-38.

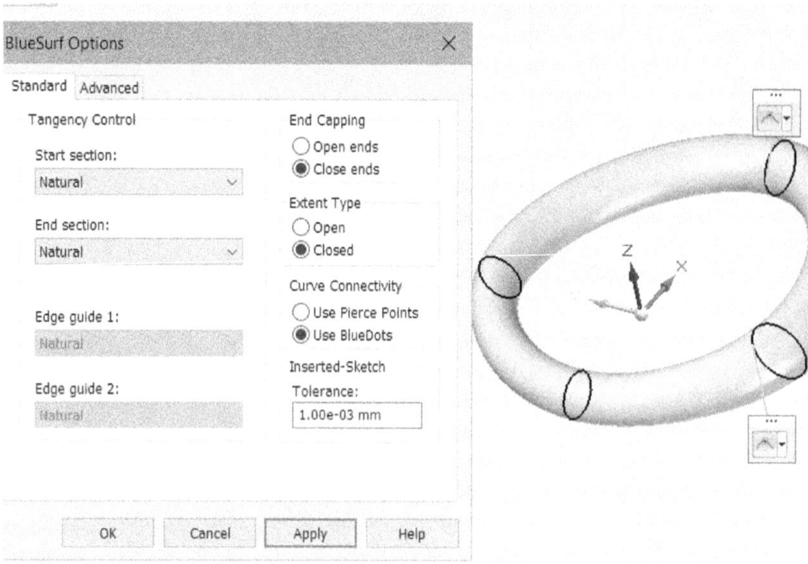

*Figure-38. Closed surface*

- Click on the **Advanced** tab in the dialog box if you want to reorder sections in the surface; refer to Figure-39.

*Figure-39. Advanced tab*

- To reorder a section, select it from the list and click on the **Up** or **Down** button from the dialog box.
- After setting desired parameters in the dialog box, click on the **OK** button.
- Click on the **Insert Sketch Step** button from the **Options Bar** if you want to add another sketch in the surface at intersection of a plane and curve.
- After setting desired parameters for surface, click on the **Next** button from the **Options Bar** and press **ENTER**. The surface will be created. Press **ESC** to exit the tool.

## Creating Surface using Bounded tool

The **Bounded** tool is used to create a surface within selected closed boundaries. The procedure to use this tool is discuss next.

• Click on the **Bounded** tool from the **Surfaces** group in the **Surfacing** tab of the **Ribbon**. The **Bounded Command Bar** will be displayed refer to Figure-40.

*Figure-40. Bounded command bar*

• Select the closed boundary to be used for creating surface. The preview of surface will be displayed; refer to Figure-41.

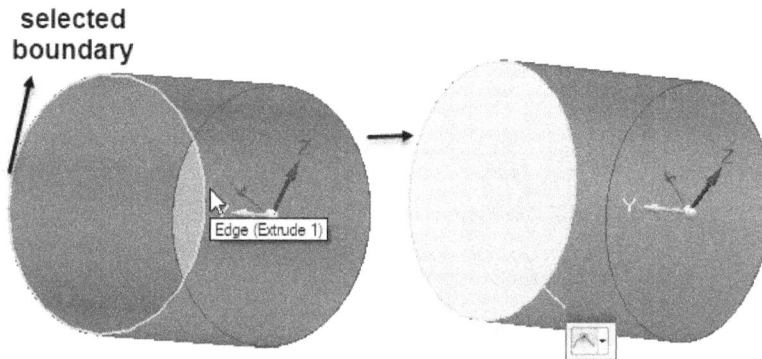

*Figure-41. Bounded surface created*

• Click on the **Bounded Surface Options** button from the **Command Bar** to define how created surface will be connected to curves and adjoining faces. The **Bounded Surface Options** dialog box will be displayed; refer to Figure-42.

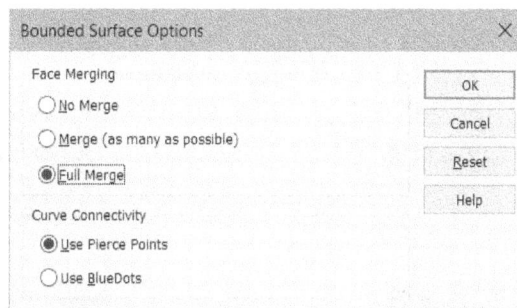

*Figure-42. Bounded Surface Options dialog box*

• Select the **No Merge** radio button from the **Face Merging** area of the dialog box to create new surface as new body. Select the **Merge (as many as possible)** radio button to merge new surface with connected surfaces. Select the **Full Merge** radio button to merge new surface with connected surfaces and form one single surface body.

• Select the desired radio button from the **Curve Connectivity** area of the dialog box to define which points will be used to control shape of surface. The function of these radio buttons has been discussed earlier. After setting desired parameters in the dialog box, click on the **OK** button.

• Click on the **Guide Curve Step** button from the **Command Bar** if you want to use a guide curve to define shape of bounded surface. The method to use a guide curve is same as discussed for **BlueSurf** tool.

• After setting desired parameters, click on the **Finish** button from the **Command Bar** to create surface. Press **ESC** to exit the tool.

## Creating Surface using Redefine tool

The **Redefine** tool is used to create a surface by using existing model faces and surfaces. The procedure is given next.

- Click on the **Redefine** tool from the **Surfaces** group in the **Surfacing** tab of the **Ribbon**. The **Redefine Command Bar** will be displayed; refer to Figure-43.

*Figure-43. Redefine Command Bar*

- **Options** 📋 : Click on the **Options** button from the **Command Bar** to display the redefine surface options. The **Redefine Surface** dialog box will be displayed; refer to Figure-44. Specify the parameters as discussed earlier. Select the **Replace faces on solid body** check box to replace the selected faces with a new surface. After setting desired parameters, click on the **OK** button from the dialog box.

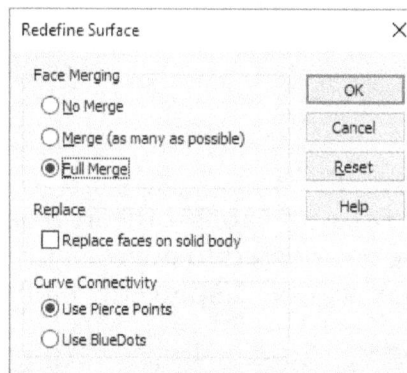

*Figure-44. Redefine Surface dialog box*

- Select the faces of the model which you want to use for creating surfaces and right-click in the drawing area. The surfaces will be created; refer to Figure-45.

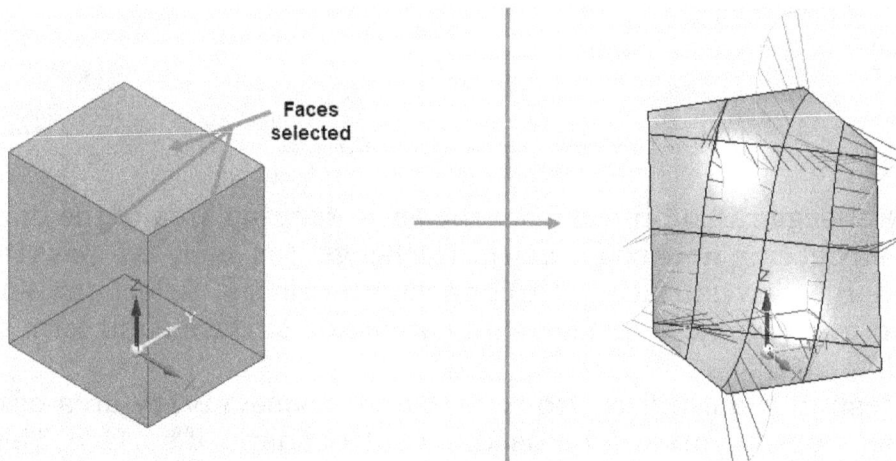

*Figure-45. Redefine surface created*

- Click on the **Finish** button from the **Command Bar** to create the surface and press **ESC** to exit the tool.

## Creating Surface using Swept tool

The **Swept** tool in surfacing is similar to the **Sweep** tool used in solid creation. The **Swept** tool in surfacing is used to sweep a section along the selected path to form a surface. The steps to use this tool are given next.

- Click on the **Swept** tool from **Surfaces** group in the **Surfacing** tab of **Ribbon**. The **Sweep Options** dialog box will be displayed; refer to Figure-46. Set the desired parameters as discussed in chapter related to 3D Modeling and click on the **OK** button. The **Swept Command Bar** will be displayed; refer to Figure-47.

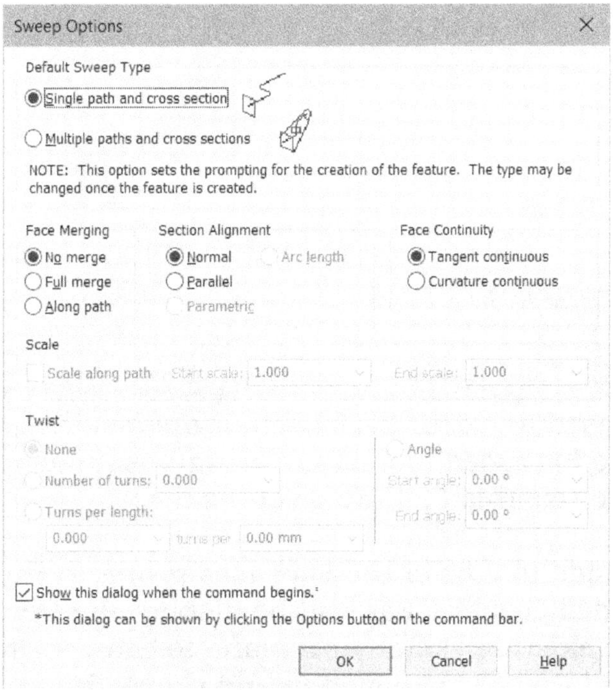

*Figure-46. Sweep Options dialog box*

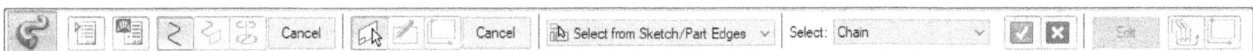

*Figure-47. Swept command bar*

- Select desired curve to be used as path for sweep. You will be asked to select a sketch/curve for section of sweep.
- Select the open or close sketch for the section. Preview of the swept surface will be displayed; refer to Figure-48.
- Click on the **Finish** button to create the feature.

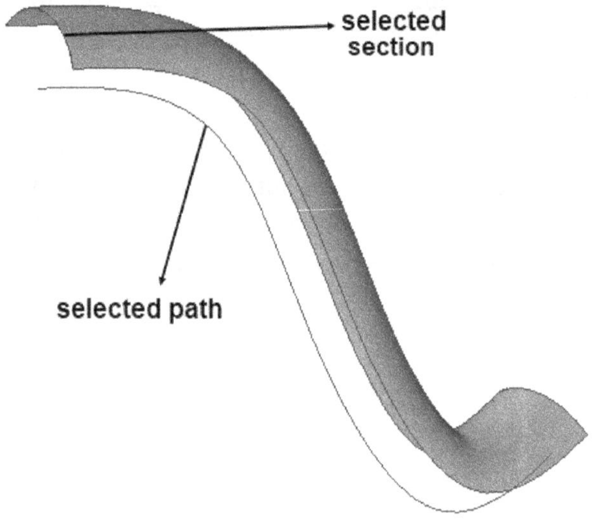

*Figure-48. Swept Surface*

## Creating Surface using Extruded tool

The **Extruded** tool is used to extrude a closed or open sketch to the specified height to from a surface. The **Extruded** tool of surfacing is similar to **Extrude** tool in solid modeling. The steps to use this tool are given next.

• Click on the **Extruded** tool from the **Surfaces** group in the **Surfacing** tab of the **Ribbon**. The **Extruded Command Bar** will be displayed; refer to Figure-49. You are asked to select the existing sketch or a plane to draw sketch.
• Select the desired closed or open sketch and right-click in the drawing area. The preview of the surface will be displayed with extent of extrude feature attached to the cursor; refer to Figure-50.

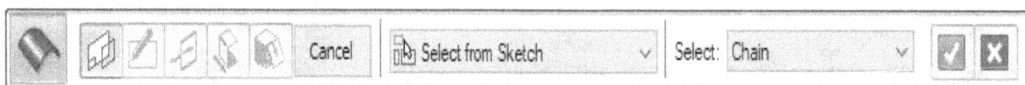

*Figure-49. Extruded Command bar*

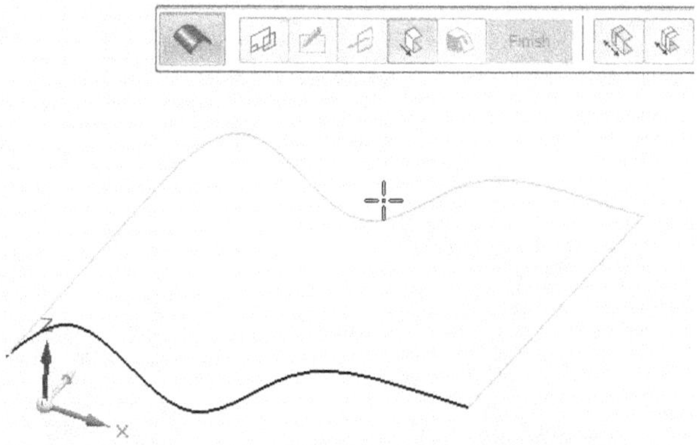

*Figure-50. Creating extruded surface*

- Specify the desired value of extent in the **Distance** edit box of the **Command Bar** to define size of feature.
- You will be asked to specify the side on which you want to create feature with respect to selected section. Click on the desired side. Preview of feature will be displayed; refer to Figure-51.

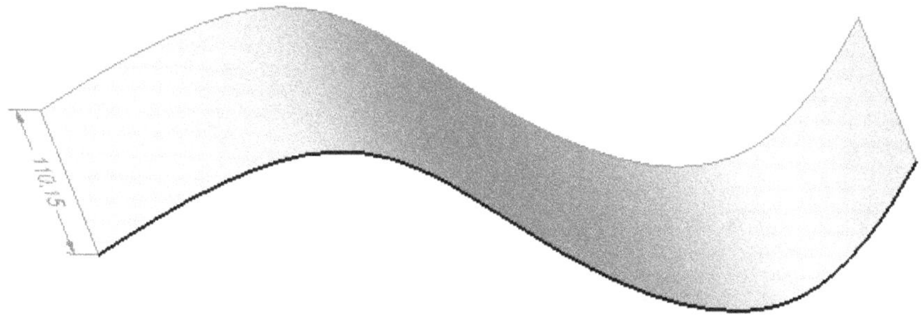

*Figure-51. Extruded Surface*

- Click on the **Finish** button from the **Command Bar** to create the feature. Press **ESC** to exit the tool.

## Revolved

The **Revolved** tool is used to revolve a close or open sketch to the specified angle with respect to selected reference to create a surface. The **Revolved** tool in surfacing is similar to **Revolve** tool in solid creation. The steps to use this tool are given next.

- Click on the **Revolved** tool from the **Surfaces** group in the **Surfacing** tab of the **Ribbon**. The **Revolved Command Bar** will be displayed; refer to Figure-52 and you will be asked to select an existing sketch or a plane to draw sketch.

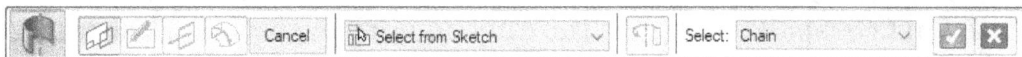

*Figure-52. Revolved Command Bar*

- Select the desired open or closed sketch that you want to revolve and right-click in the drawing area. You will be asked to select revolution axis.
- Select the desired line or edge to be used as revolution axis. You will be asked to specify extent of feature.
- Specify the angle of revolution in the **Angle** edit box of the **Command Bar** and click in the drawing area. Preview of the surface will be displayed; refer to Figure-53.
- Click on the **Finish** button from the **Command Bar** to create the feature and press **ESC** to exit the tool.

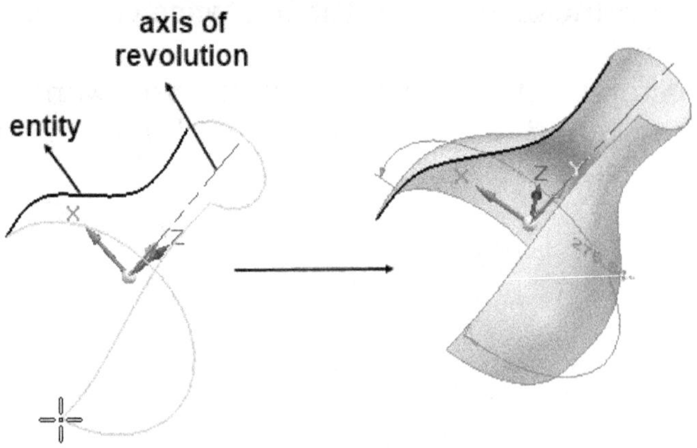

*Figure-53. Revolved Surface*

## Offset

The **Offset** tool is used to create a surface at an offset distance from selected face/surface. The steps to create offset surface are given next.

- Click on the **Offset** tool from the **Surfaces** group in the **Surfacing** tab of the **Ribbon**. The **Offset Command Bar** will be displayed; refer to Figure-54.

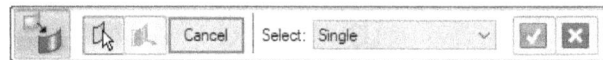

*Figure-54. Offset command bar*

- Select the face/surface which you want to use for creating the offset surface and right-click in the drawing area. You will be asked to specify direction of offset and distance value.
- Specify the desired value of distance in the **Distance** edit box of **Command Bar**.
- Select the **Show Boundaries** button from the **Command Bar** if you want to include internal cuts of surface while offsetting. By default, the **Remove Boundaries** button is selected in **Command Bar** and so, the cuts on surface are not created in the offset surface.
- After specifying the parameters, click on the desired side of base surface to create offset surface. Preview of the surface will be displayed; refer to Figure-55.
- Click on the **Finish** button to create the offset surface. Press **ESC** to exit the tool.

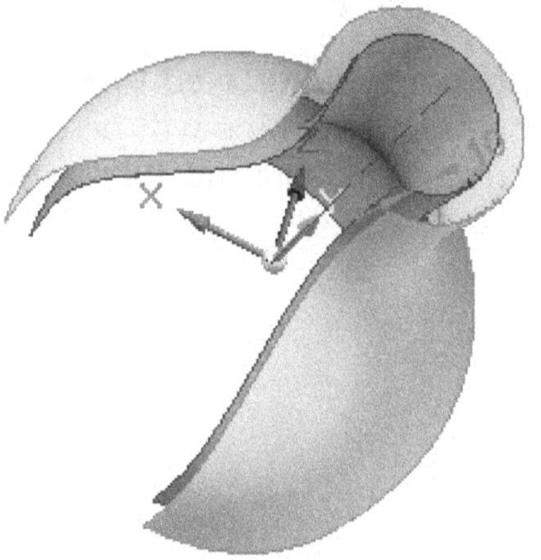

*Figure-55. Offset Surface*

## Copy

The **Copy** tool is used to create surface derived from one or more input faces. The select faces may or may not be adjacent to each other. You can specify whether any internal or external boundaries are removed on the new copy of the surface. The procedure to use this tool is given next.

- Click on **Copy** tool from the **Surfaces** group in the **Surfacing** tab of the **Ribbon**. The **Copy Command Bar** will be displayed; refer to Figure-56.

*Figure-56. Copy Command Bar*

- Select the faces/surfaces to be copied and right-click in the drawing area to accept selection. Preview of faces will be displayed; refer to Figure-57

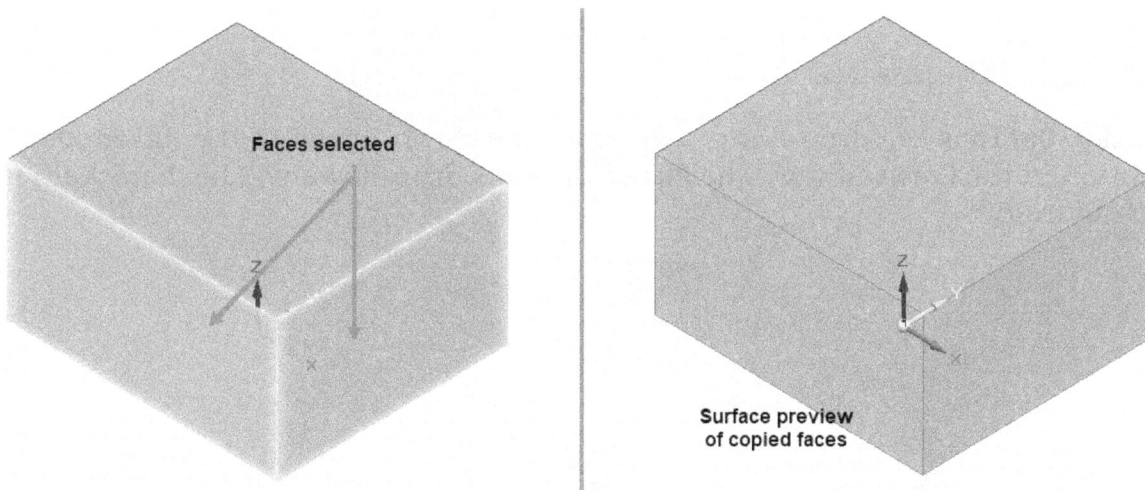

*Figure-57. Faces copied as surfaces*

## Ruled

The **Ruled** tool is used to create a surface that extends out in a specified direction from the selected edges. This type of surface becomes very important when creating parting surface for molding/casting. The steps to use this tool are given next.

- Click on the **Ruled** tool from the **Surfaces** group in the **Surfacing** tab of the **Ribbon**. The **Ruled Command Bar** will be displayed; refer to Figure-58.

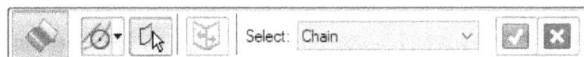

*Figure-58. Ruled command bar*

- Select the desired edge of surface to be used for creating ruled surface. The preview of surface will be displayed; refer to Figure-59.

**Selected edge**

Distance | 10.00 mm

*Figure-59. Ruled*

- **Ruled Options** 🔌 drop-down : Click on the down arrow of the **Ruled Options** button in the **Command Bar**. The **Ruled Options** drop-down will be displayed; refer to Figure-60.

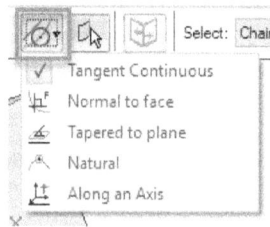

Select: Chain
- ✓ Tangent Continuous
- Normal to face
- Tapered to plane
- Natural
- Along an Axis

*Figure-60. Ruled drop-down*

- Select the **Tangent Continuous** option from the drop-down if you want to create an extension of surface tangent to base surface at selected edge.
- Select the **Normal to face** option from the drop-down if you want to create new surface perpendicular to base face at selected edge.
- Select the **Tapered to plane** option from the drop-down if you want to create surface at an angle to reference plane at selected edge. After selecting this option, select the desired reference plane for specifying angle. Preview of surface will be displayed; refer to Figure-61.
- Select the **Natural** option to create extension surface in natural flow of base surface.
- Select the **Along an Axis** option to create extension surface along selected axis. You will be asked to select an axis. Select the desired axis or edge to define direction of surface extension. Preview of the surface will be displayed; refer to Figure-62.

*Figure-61. Preview of surface at angle*

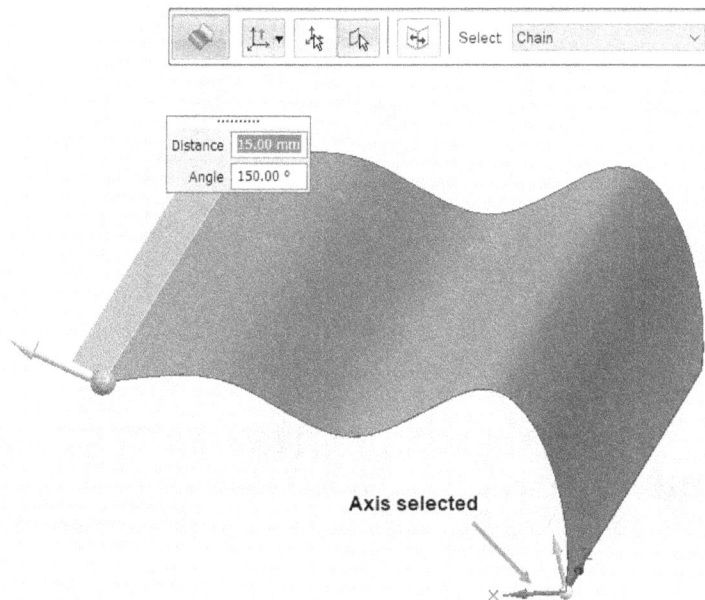

*Figure-62. Preview of surface at angle using axis*

- After setting desired parameters, click on the **OK** button from the **Command Bar** to create the surface.
- Press **ESC** to exit the tool.

## Blank Surface

The **Blank Surface** is used to creates a flattened surface from the selected model faces. The steps to use this tool are given next.

- Click on the **Blank Surface** tool from **Surfaces** group in the **Surfacing** tab of the **Ribbon**. The Material Table dialog box will be displayed if material has not been defined earlier.
- Select the desired material from the table and click on the **Apply to Model** button from the dialog box. The **Blank Surface Command Bar** will be displayed; refer to Figure-63.

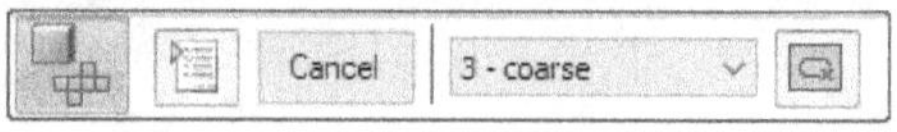

*Figure-63. Blank Surface command bar*

- Select the desired faces/surfaces from the model and specify the direction.
- Click on the **Preview** button from the **Command Bar**. The preview of surface will be displayed; refer to Figure-64.

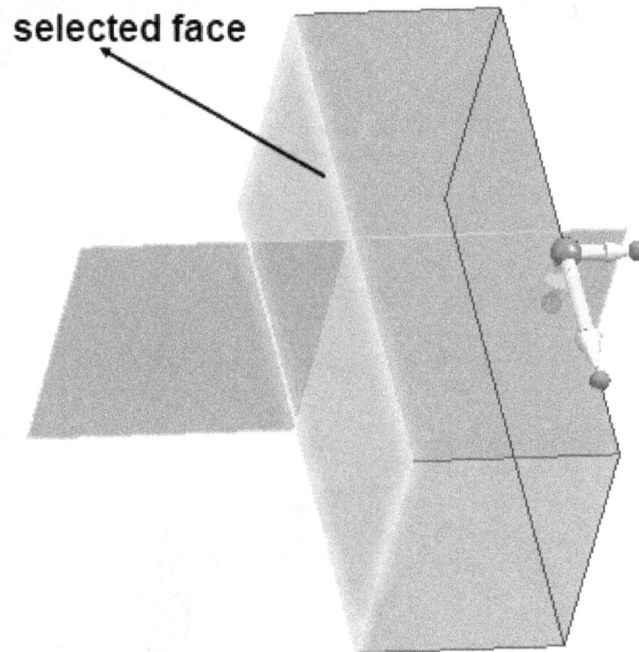

selected face

*Figure-64. Blank Surface*

# MODIFYING SURFACES

The tools in the **Modify Surfaces** group of **Surfacing** tab in the **Ribbon** are used to modify surfaces; refer to Figure-65. These tools are discussed next.

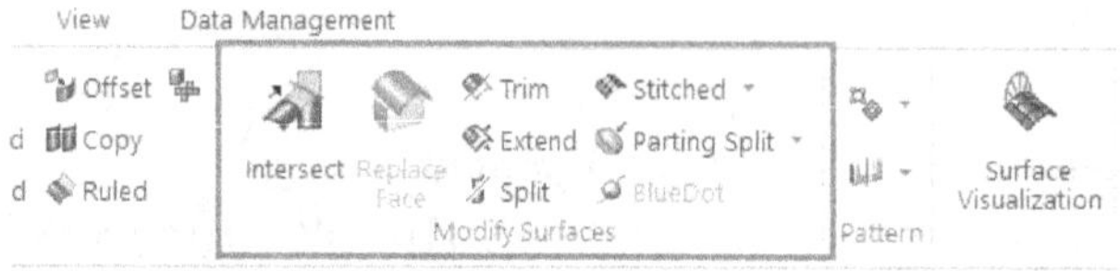

*Figure-65. Modify Surfaces group*

## Intersect

The **Intersect** tool is used to extend or trim selected surfaces to a common intersection. Using this tool, you can also create bodies using the areas fully enclosed by surfaces. The procedure to use this tool is discuss next.

- Click on the **Intersect** tool from the **Modify Surfaces** group in the **Surfacing** tab of **Ribbon**. The **Intersect Command Bar** will be displayed; refer to Figure-66.

*Figure-66. Intersect command bar surfacing*

- Select the two or more intersecting surfaces and then right-click in the drawing area. You will be asked to select the sections to be removed.
- Click on the desired sections of surfaces to be removed; refer to Figure-67. You can also extend the surfaces to intersect by using this tool; refer to Figure-68.

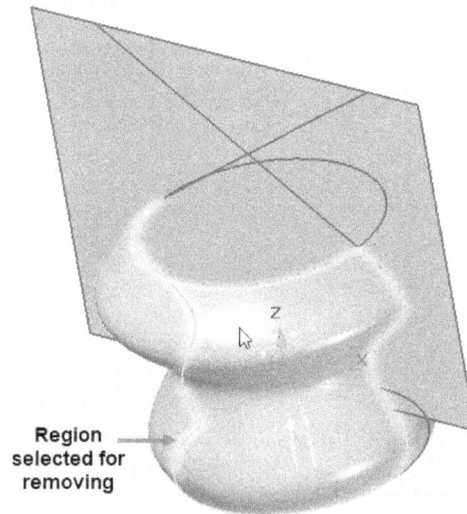

*Figure-67. Selecting region to trim*

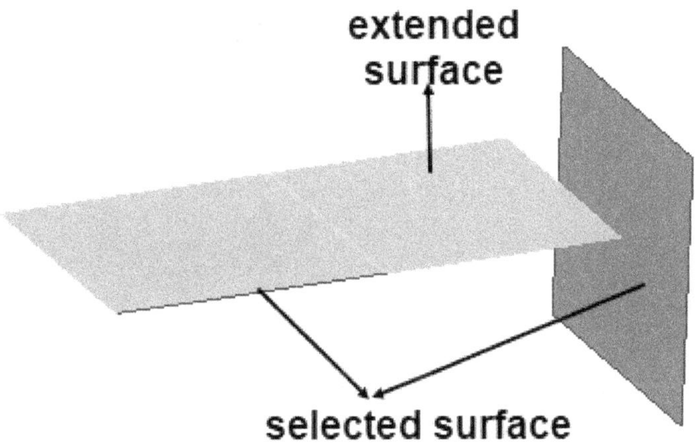

*Figure-68. Intersect Extend surface*

- If selected surfaces form a closed region then you can use this tool to convert selected region into solid. After selecting all the surfaces, select the **Create design bodies** option from the **Intersect Options** drop-down in the **Command Bar**; refer to Figure-69 and right-click in the drawing area. The **Volume Regions** dialog box will be displayed with preview of solid region; refer to Figure-70.

*Figure-69. Create design bodies option*

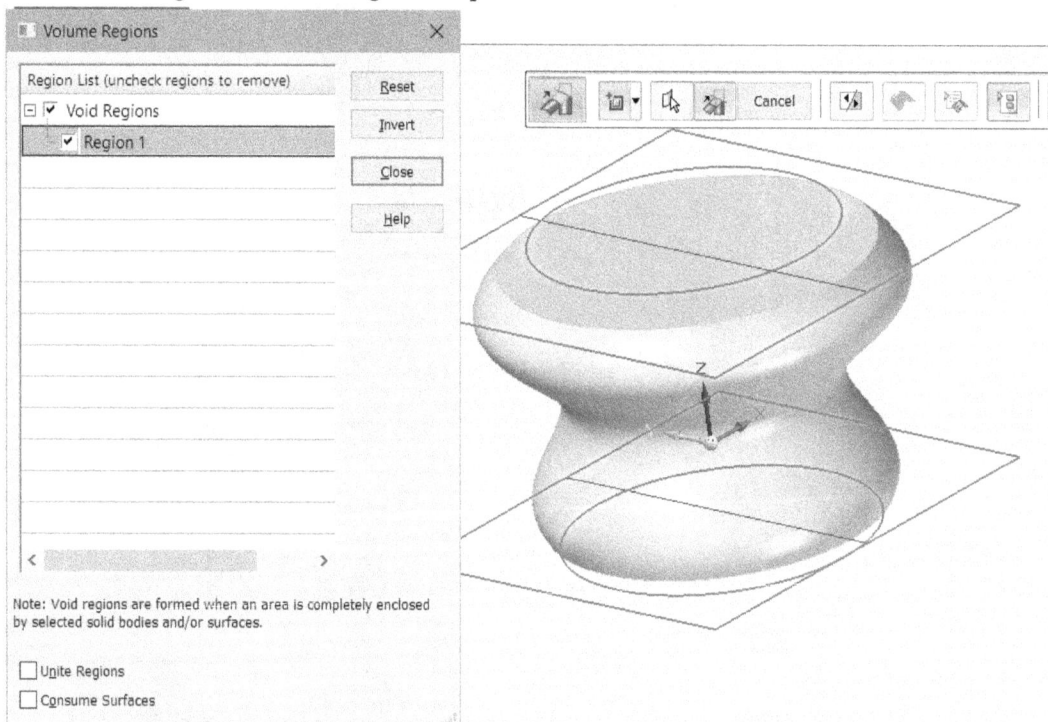

*Figure-70. Volume Regions dialog box*

- Select the **Unite Regions** check box to unite all the selected surfaces as single body.
- Select the **Consume Surfaces** check box if you want to remove selected surfaces after creating design body.
- Click on the **Close** button from the **Volume Regions** dialog box to exit and click on the **OK** button from the **Command Bar** to accept selection. The design body will be created.
- Click on the **Finish** button from the **Command Bar** and press **ESC** to exit the tool.
- Similarly, you can use the **Auto-trim** option from the **Intersect Options** drop-down of the **Command Bar** to automatically trim surfaces to form closed region; refer to Figure-71. Note that using this option will not form solid.

*Figure-71. Auto trim option*

## Replace Face

The **Replace Face** tool is used to replace selected faces of a solid with a surface. When replacing more than one face, the faces being replaced cannot touch each other. The procedure is discuss next.

- Click on the **Replace Face** too from the **Modify Surfaces** group in the **Surfacing** tab of the **Ribbon**. The **Replace Face Command Bar** will be displayed; refer to Figure-72.

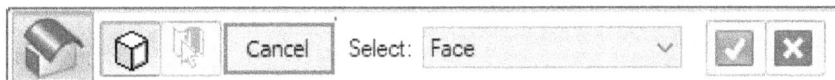

*Figure-72. Replace Face command bar*

- Select the part face to be replaced and right-click in the drawing area.
- Select the surface which should replace the part face and click on **Finish** button. The replacement of face will be applied; refer to Figure-73. Press **ESC** to exit the tool.

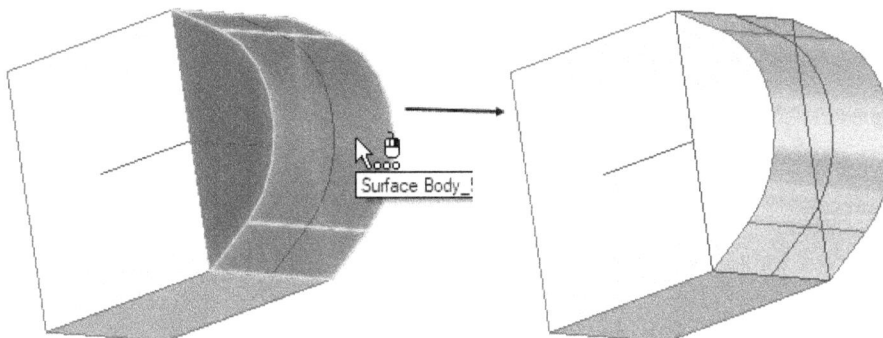

*Figure-73. Replace Face*

## Trim

The **Trim** tool is used to trims a surface with a selected curve, surface, or plane. The procedure is discuss next.

- Click on the **Trim** tool from the **Modify Surfaces** group in the **Surfacing** tab of **Ribbon**. The **Trim Command Bar** will be displayed; refer to Figure-74.

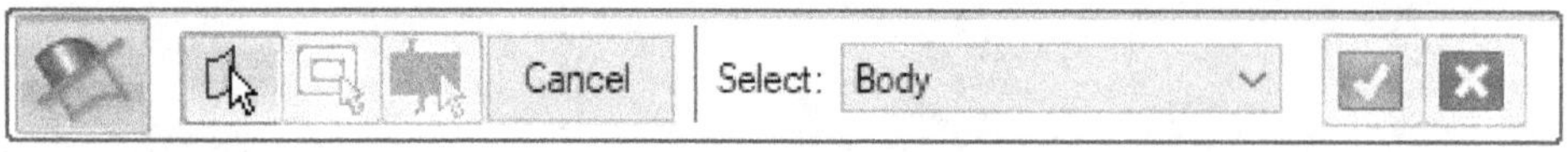

*Figure-74. Trim Command Bar*

- Selects the surfaces to be trimmed and right-click in the drawing area.
- Select curves, planes, surfaces, and sketch elements to be used as trim tool, and right-click in the drawing area. You will be asked to select the regions to be removed.
- Select the regions formed by trim tool that will be trimmed and right-click in the drawing area. The preview of trimming will be displayed; refer to Figure-75.
- Click on the **Finish** button to trim surfaces and press **ESC** to exit the tool.

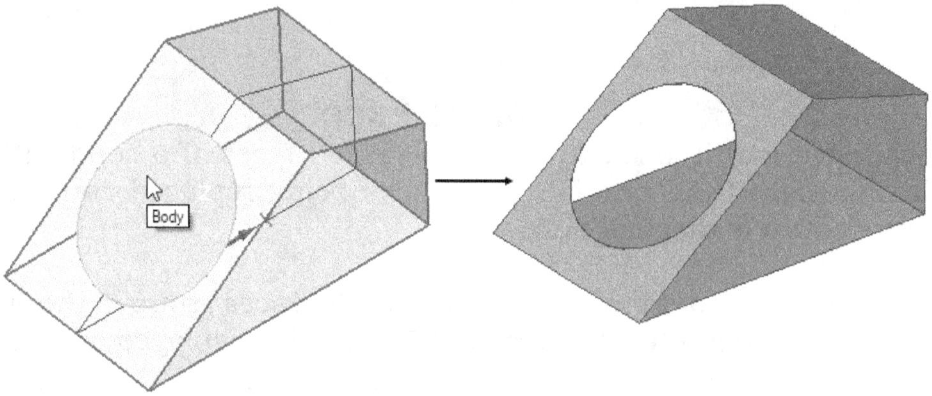

*Figure-75. Preview of Trim*

## Extend

The **Extend** tool is used to extends a surface along selected edges. The procedure is given next.

- Click on the **Extend** tool from the **Modify Surfaces** group in the **Surfacing** tab of the **Ribbon**. The **Extend Command Bar** will be displayed; refer to Figure-76.

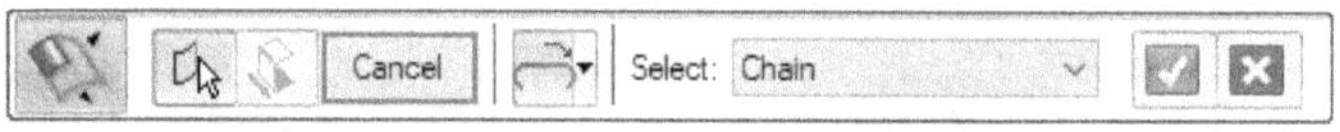

*Figure-76. Extend command bar*

- Select edges of surface to be extend and right-click in the drawing area. Selected end of surface will get attached to cursor and you will be asked to specify the extent of surface.
- Click at desired location or enter the value in the **Distance** edit box of the **Command Bar**. After specifying the value in edit box, click in the drawing area. The preview of extend surface will be displayed; refer to Figure-77.

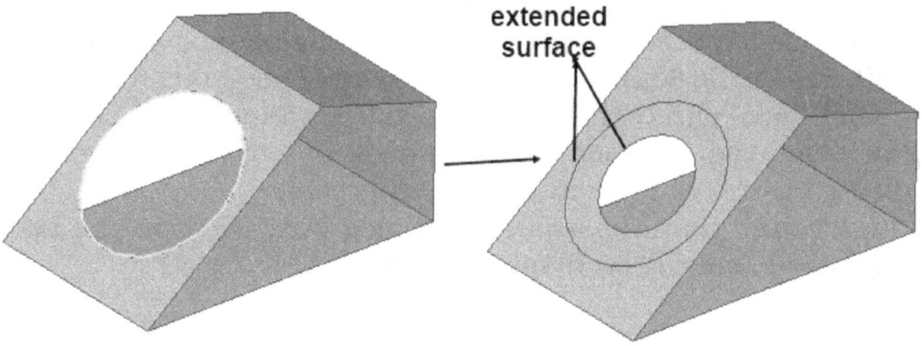

*Figure-77. Extend preview*

- Click on the **Finish** button from the **Command Bar** to create feature and press **ESC** to exit the tool.

## Split

The **Split** tool is used to split faces/surfaces using sketch, curve, surface, or reference planes. The procedure to use this tool is discuss next.

- Click on the **Split** tool from **Modify Surface** group in the **Surfacing** tab of the **Ribbon**. The **Split Command Bar** will be displayed; refer to Figure-78.

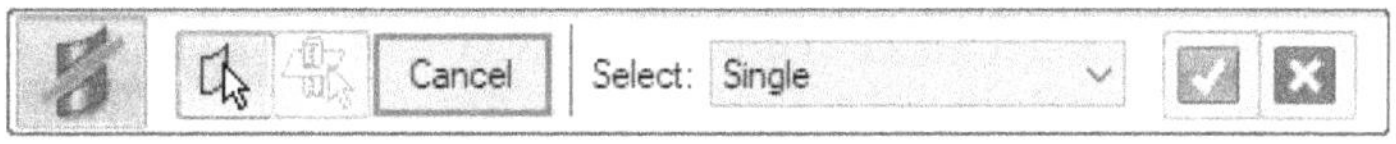

*Figure-78. Split Command Bar (Surfacing)*

- Select the surface to be split and right-click in the drawing area. You will be asked to select geometry being used as splitting tool.
- Select the geometry to be used as splitting tool for the surface/face and right-click in the drawing area. Preview of split faces/surfaces will be displayed; refer to Figure-79.

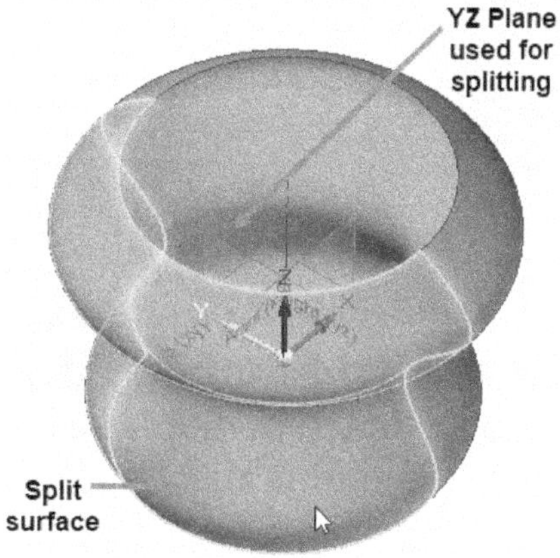

*Figure-79. Preview of split surface*

- Click on the **Finish** button from the **Command Bar** to perform splitting and press **ESC** to exit the tool.

## Stitched

The **Stitched** tool is used to creates a construction surface by stitching multiple sheet surfaces. The procedure to use this tool is discuss next.

- Click on the **Stitched** tool from the **Modify Surfaces** group in the **Surfacing** tab of the **Ribbon**. The **Stitched Surface Options** dialog box will be displayed; refer to Figure-80. Select the **Heal stitched surfaces** check box to stitch and heal the surfaces which have gap within specified tolerance range. Select the **Show this dialog box when the command begins** check box to display **Stitched Surface Options** dialog box every time you invoke this tool.

- After setting desired parameters in the dialog box, click on the **OK** button. The **Stitched Command Bar** will be displayed; refer to Figure-81.

*Figure-80. Stitched Surface Options*

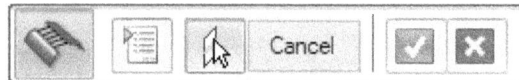

*Figure-81. Stitched command bar*

- Select the surfaces to be stitched together and click on the **Preview** button from the **Command Bar**. The preview of stitch operation will be displayed. Note that if selected surfaces form a closed region then a message box will be displayed asking you whether you want to convert stitched surfaces into solid; refer to Figure-82. Click on the **Yes** button if you want to create a solid body and click on the **No** button if you want to create a surface body after stitching.

- Click on the **Finish** button from the **Command Bar** to create the feature and press **ESC** to exit the tool.

*Figure-82. Message box for stitched surfaces*

## Show Non-Stitched Edges

The **Show Non-Stitched Edges** tool is used to display all non-stitched edges of selected surfaces. The procedure to use this tool is discuss next.

- Click on the **Show Non-Stitched Edges** from **Stitched** drop-down. The **Show Non-Stitched Edges Command Bar** will be displayed; refer to Figure-83.

*Figure-83. Show Non-Stitched Edges*
*command bar*

- Select the model to observe the non-stitched edges; refer to Figure-84. Click on the **Reset** button from **Command Bar** to start fresh selection and click on the **Close** button from the **Command Bar** to exit the tool.

*Figure-84. Show non stitched edges preview*

## Parting Split

The **Parting Split** tool is used to create parting lines along the silhouette edges of the part. This tool is useful in creating mold die design for the part. The procedure to use this tool is discuss next.

- Click on the **Parting Split** tool from the **Modify Surfaces** group in the **Surfacing** tab of the **Ribbon**. The **Parting Split Command Bar** will be displayed; refer to Figure-85 and you will be asked to select a plane for creating split lines on part.

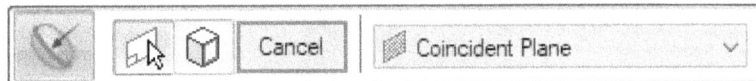

*Figure-85. Parting Split command bar*

- Select the desired plane type from the drop-down in the **Command Bar** and select/ define the plane to be used for creating split line. You will be asked to select the faces/surfaces to be split.
- Selects the face/surface on which split lines will be created and click on the **Preview** button. Preview of the lines will be displayed; refer to Figure-86.
- Click on the **Finish** button from the **Command Bar** to create the splitting lines and press **ESC** to exit the tool.

Parting split face

*Figure-86. Preview of Parting split*

## Parting Surface

The **Parting Surface** is used to create parting surface for mold tools. The procedure to use this tool is discuss next.

• Click on the **Parting Surface** tool from the **Parting Split** drop-down in the **Modify Surfaces** panel of the **Surfacing** tab in the **Ribbon**. The **Parting Surface Command Bar** will be displayed refer to Figure-87 and you will be asked to select a reference plane.

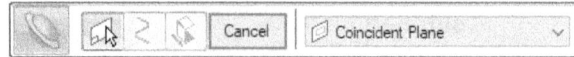

*Figure-87. Parting Surface Command Bar*

• Select the plane to be used as direction reference for split surface. Note that surface will be created along the selected plane/face. You will be asked to sketch, edge, or curve to be used for creating split surface.
• Select the desired curve. Generally, we use parting split curve for creating split surface. After selecting curve, right-click in the drawing area. You will be asked to specify the extent of surface.
• Specify the desired direction value in the **Distance** edit box of the **Command Bar** and click on the direction curve to generate face. The preview of parting surface will be displayed; refer to Figure-88.

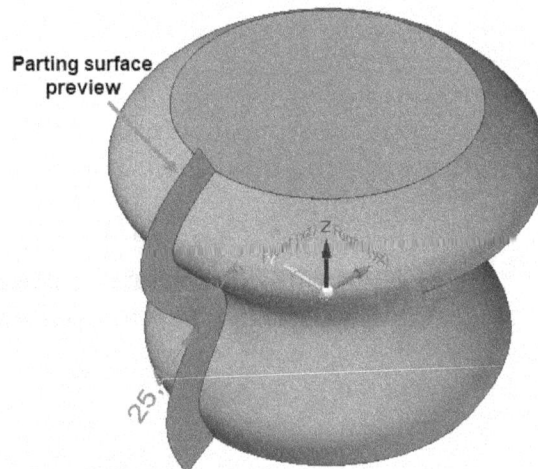

*Figure-88. Parting Surface preview*

• Click on the **Finish** button from the **Command Bar** to create surface and press **ESC** to exit the tool.

## Offset Edge

The **Offset Edge** tool is used to split a face by offsetting an existing edge. The procedure to use this tool is discuss next.

• Click on the **Offset Edge** tool from the **Parting Split** drop-down in the **Modify Surfaces** group of **Surfacing** tab in the **Ribbon**. The **Offset Edge Command Bar** will be displayed; refer to Figure-89.

*Figure-89. Offset Edge
Command Bar*

- Select the desired option from the drop-down to define what type of entity you want to select for offsetting.
- Select the face/edge/loop to offset edges. You will be asked to specify the distance up to which the edges will be offset in dynamic input box. The preview of offset will be displayed; refer to Figure-90.
- Right-click in the drawing area to create the edge offset and press **ESC** to exit the tool.

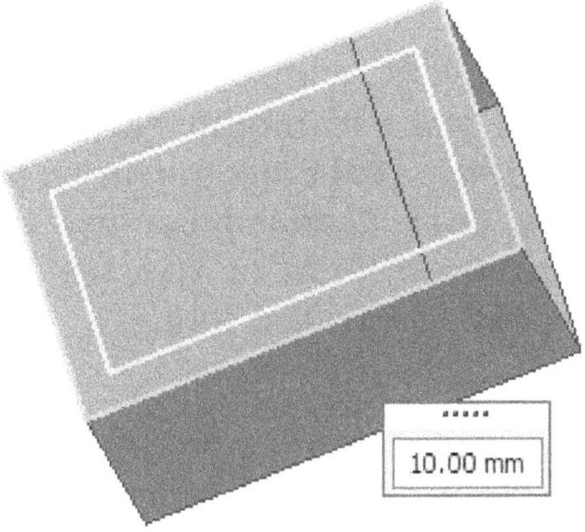

*Figure-90. Preview of Offset Edge*

## BlueDot

The **BlueDot** tool is used to create bi-directional connect point. You can create a BlueDot by identifying the two curves to connect. No more than two curves can be BlueDot connected. The first curve you select will move to intersect the second curve. The second curve is unchanged (it maintains its location and shape). There are several solutions possible when creating a BlueDot. The solution is dependent on how you select the curves. Each curve has four select zones (two endpoints, a midpoint, curve). The endpoints and midpoint display a red dot for selection. If you select a midpoint or endpoint, that point on the curve will move to the zone you select on the second curve. You should experiment with the BlueDot command to better understand the several solutions. The procedure is given next.

- Click on the **BlueDot** tool from the **Modify Surfaces** group in the **Surfacing** tab of the **Ribbon**. You will be asked to select the curves.
- One by one select the two curves to be joined by BlueDot. The curves will be joined by BlueDot; refer to Figure-91.

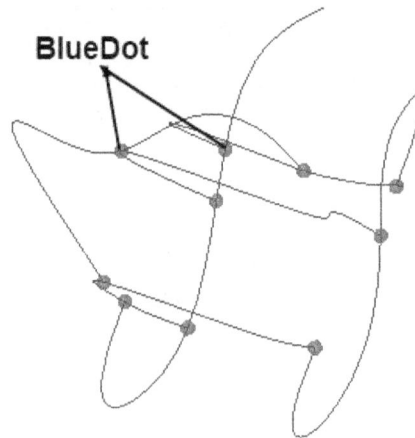

*Figure-91. Creating BlueDot by Using curve*

Press **ESC** to exit the tool. If you want to modify the position of joint then click on the blue dot and select the **Dynamic Edit** button from the mini toolbar displayed; refer to Figure-92. The **Dynamic Edit Command Bar** will be displayed where you can specify the respective new location of joint or you can dynamically drag the joint. After changing position, press **ESC** to exit the tool.

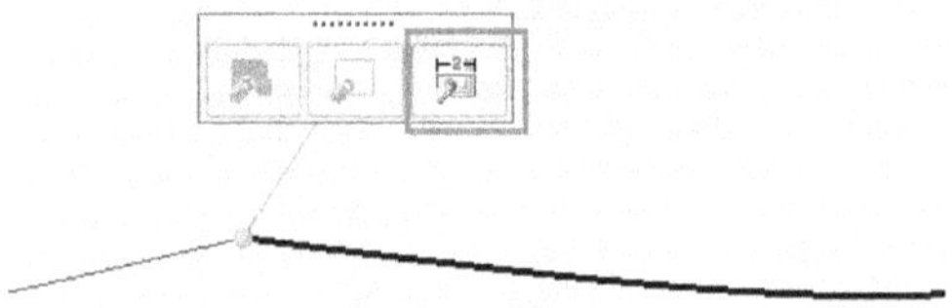

*Figure-92. Dynamic Edit button*

# SURFACE VISUALIZATION SETTINGS

The **Surface Visualization** tool is used to specify settings related to how the surface will be displayed in the drawing area. Using this tool, you can define parameters related to mesh density and curvature comb of surface. The procedure to use this tool is given next.

*   Click on the **Surface Visualization** tool from the **Inspect** group in the **Surfacing** tab of the **Ribbon**. The **Surface Visualization Settings** dialog box will be displayed; refer to Figure-93.
*   Select the **Show surface visualization** check box to display specified surface visualization on selected surfaces.
*   Select the **Direction 1** and **Direction 2** check boxes from the **Mesh density** area of the dialog box to display mesh density in X and Y directions. After selecting a check, you can define color of mesh lines in the drop-down below it.
*   Using the sliders, you can increase or decrease the density of mesh lines.
*   Similarly, you can set parameters for curvature comb in the **Curvature comb** area of the dialog box.
*   After setting desired parameters in the dialog box, select the surfaces to display surface visualization; refer to Figure-94.

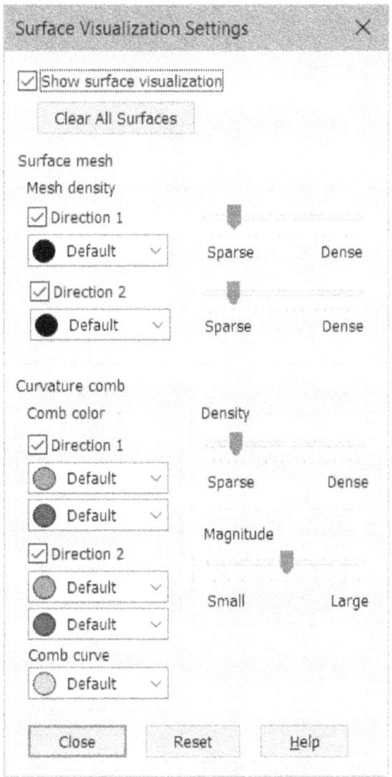

*Figure-93. Surface Visualization
Settings dialog box*

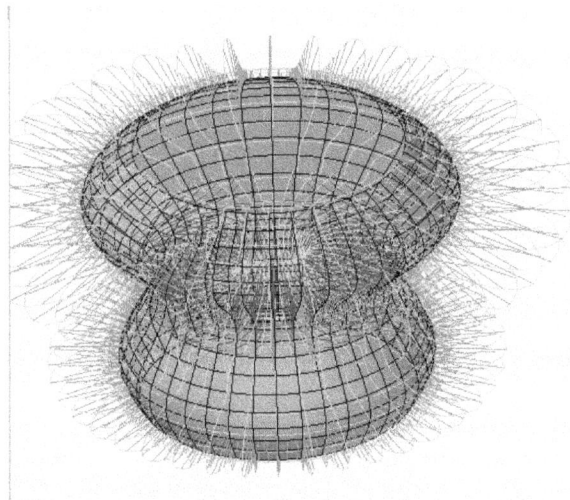

*Figure-94. Displaying surface visualization*

*   Click on the **Clear All Surfaces** button to clear all selected surfaces and start a new selection set.
*   Click on the **Close** button from the dialog box to exit tool.

## DISPLAYING SECTION CURVATURE

The Section curvature is used to display curvature of surface at intersection with selected reference plane. The procedure to display section curvature is discussed next.

*   Click on the **Section Curvature Settings** tool from the **Inspect** group of **Surfacing** tab in the **Ribbon**. The **Section Curvature Settings** dialog box will be displayed with **Section Curvature Settings Command Bar**; refer to Figure-95.

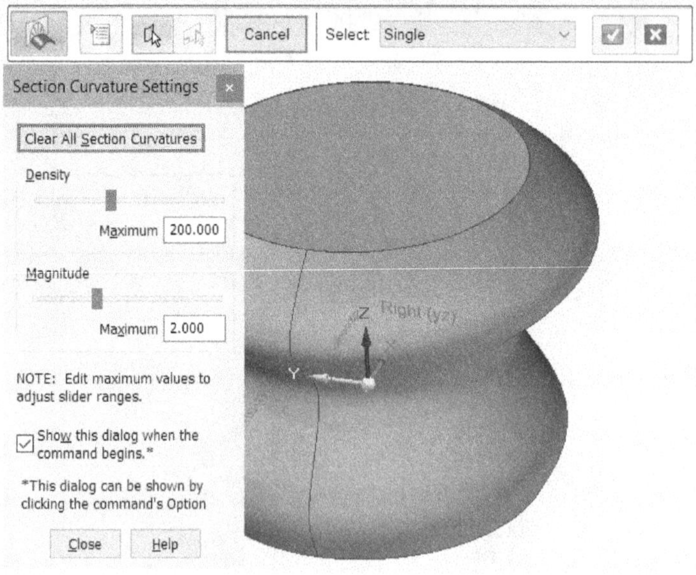

*Figure-95. Section Curvature Settings dialog box and Command Bar*

- Set the desired density and magnitude of curvature using the sliders in the dialog box. After setting parameters, click on the **Close** button from the dialog box.
- Select the desired surfaces/faces for which you want to display curvature and right-click in the drawing area. You will be asked to select a section plane that intersects with the surfaces/faces.
- Select the desired planar face or plane. Preview of curvature will be displayed and you will be asked to specify distance of curvature plane with respect to selected reference plane/face.
- Specify the desired value in the **Distance** edit box of the **Command Bar** or click at desired location. If you have specified value in the edit box then you need to click on the desired side of reference plane to place the curvature comb.
- Click on the **Finish** button from the **Command Bar** to create plot.
- Select or clear the **Show** check box from the **Section Curvature** area of the **Inspect** group in the **Surfacing** tab of the **Ribbon** to display or hide the section curvature respectively.

Similarly, you can use the **Reflective Plane** options in the **Inspect** group of **Surfacing** tab in **Ribbon** to display reflection of surfaces. The other tools of **Surfacing** tab have been discussed earlier.

# PRACTICAL 1

Create the model of helmet glass as shown in Figure-96. The dimensions of the model are given in Figure-97.

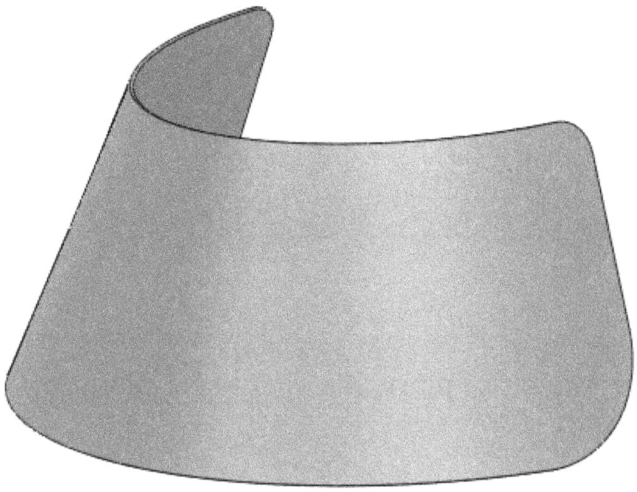

*Figure-96. Practical 1 model*

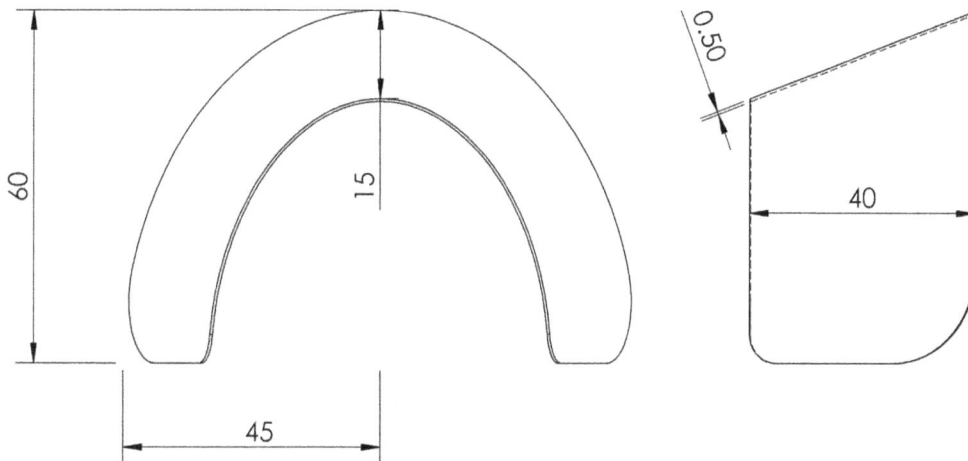

*Figure-97. Practical 1 drawing*

## Starting New Document

- Start Solid Edge if not started yet.
- Click on the **ISO Metric Part** tool from the **New** cascading menu of the **Application** menu. The Part modeling environment will be displayed in **Synchronous** mode.
- Select the **Ordered** radio button from the **Model** group in the **Tools** tab of the **Ribbon**. The ordered modeling mode will be activated.

## Creating Sketches

- Click on the **Sketch** tool from the **Sketch** group in the **Home** tab of **Ribbon**. You will be asked to select a sketching plane.
- Select the XY plane from the drawing area and create the sketch as shown in Figure-98.

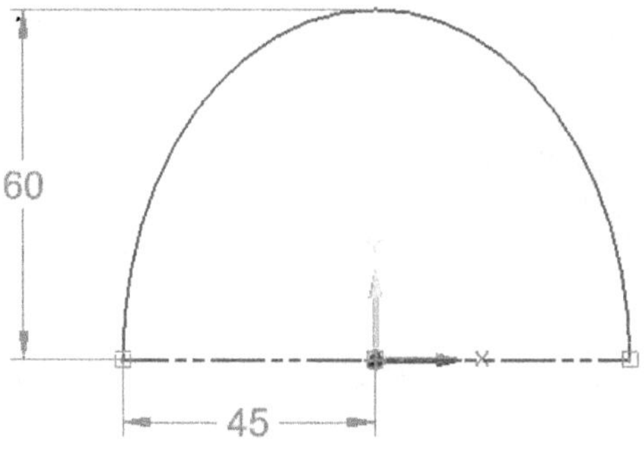

*Figure-98. Sketch 1 for practical*

- Similarly, create the sketch at plane parallel to XY plane by **40** distance as shown in Figure-99 and Figure-100.

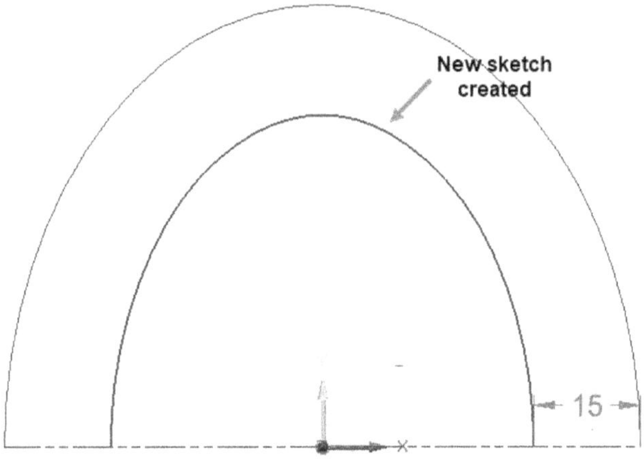

*Figure-99. Sketch 2 for practical*

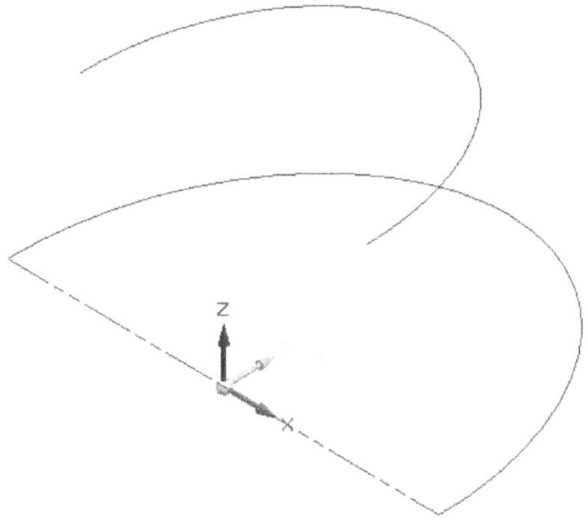

*Figure-100. Sketch for loft surface*

## Creating Surface using sketches

- Click on the **BlueSurf** tool from the **Surfaces** group in the **Surfacing** tab of the **Ribbon**. The **BlueSurf Command Bar** will be displayed.
- Select the bottom curve near the end point as shown in Figure-101 and right-click in the drawing area. You will be asked to select the next curve.

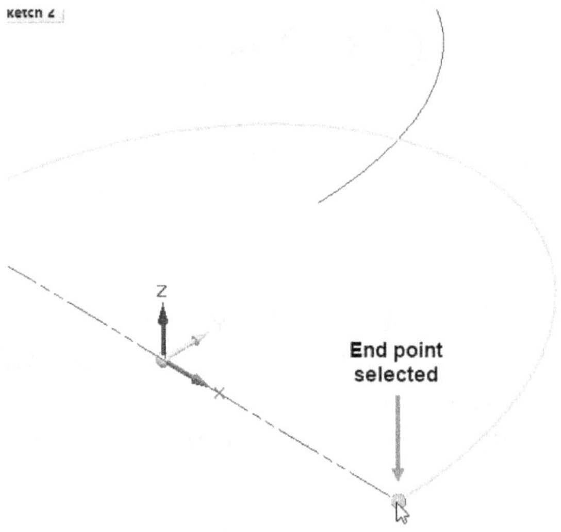

*Figure-101. End point of curve selected*

- Select the corresponding point on next curve; refer to Figure-102 and right-click in the drawing area. Preview of the surface will be displayed; refer to Figure-103.

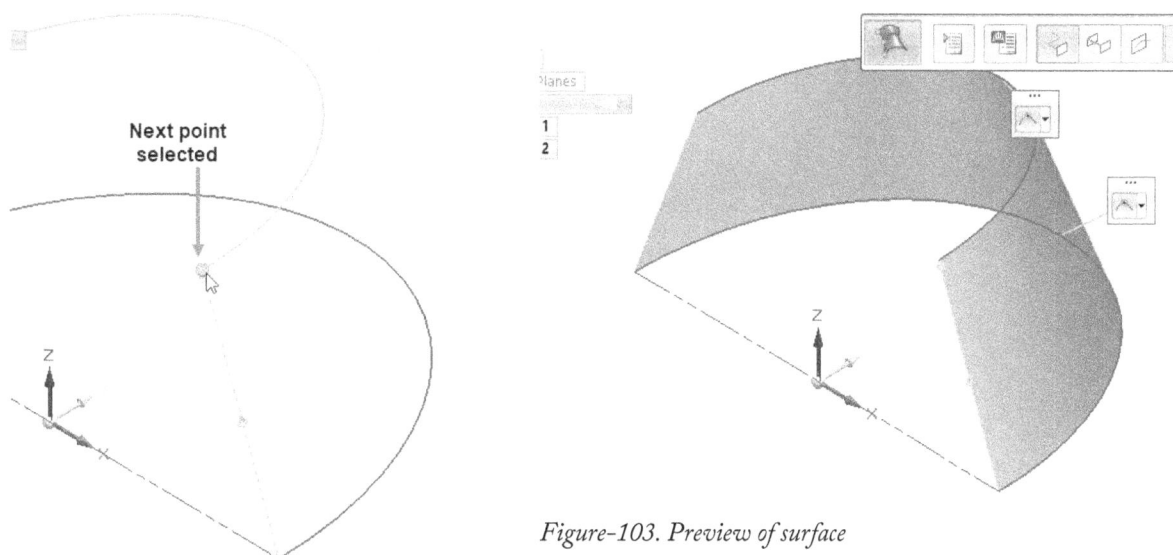

Figure-103. Preview of surface

Figure-102. Selecting next curve point

- Click on the **Next** button and then **Finish** button from the **Command Bar**. The surface will be created. Press **ESC** to exit the tool.

## Applying Thickness to Surface

- Click on the **Thicken** tool from the **Add** drop-down in the **Solids** group of **Home** tab in the **Ribbon**. The **Thicken Command Bar** will be displayed and you will be asked to select the surface body to be thickened.
- Select the surface body recently created and specify thickness value as **0.50** in the edit box of **Command Bar**; refer to Figure-104.

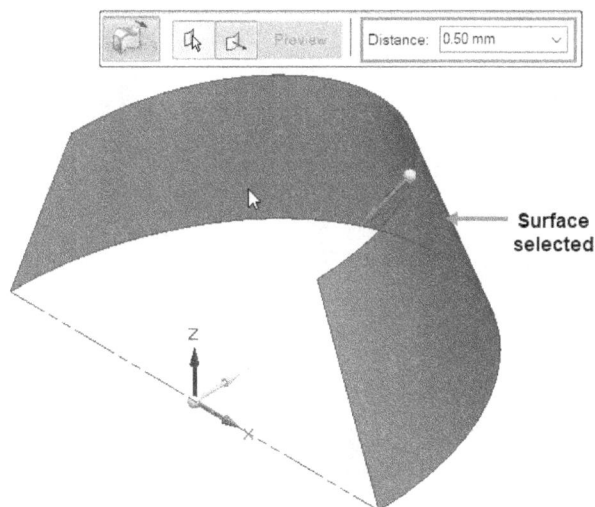

Figure-104. Specifying parameters for thicken

- Click towards the inside of surface to define direction of thickening. Preview of thickened feature will be displayed.
- Click on the **Finish** button from the **Command Bar** to create the feature and press **ESC** to exit the tool.

## Applying Round

- Click on the **Round** tool from the **Round** drop-down in the **Solids** group of **Home** tab in the **Ribbon**. The **Round Command Bar** will be displayed.
- Specify the radius value as **10** in the **Radius** edit box of **Command Bar** and select the corner edges of model. Preview of round will be displayed; refer to Figure-105.

*Figure-105. Preview of rounds*

- Right-click in the drawing area to accept selection.
- Click on the **Preview** button to check preview of round and then click on the **Finish** button to create round feature. Press **ESC** to exit the tool.

# PRACTICE 1

Create the surface model of tank as shown in Figure-106. The dimensions of the model are given in Figure-107.

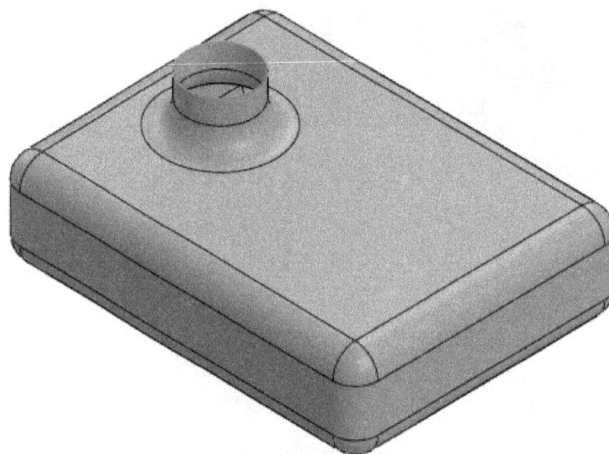

*Figure-106. Practice 1 model*

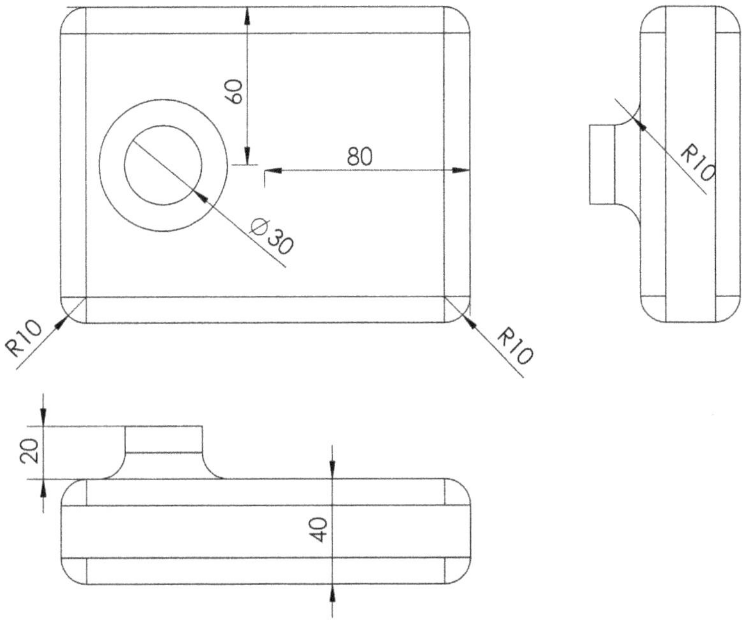

*Figure-107. Practice 1 drawing*

# PRACTICE 2

Create the surface model of car bumper as shown in Figure-108. The dimensions of the model are given in Figure-109. **Assume the missing dimensions. (Hint:** You will need vertex pulling in freestyle after making the base component.**)**

*Figure-108. Practice 2 model*

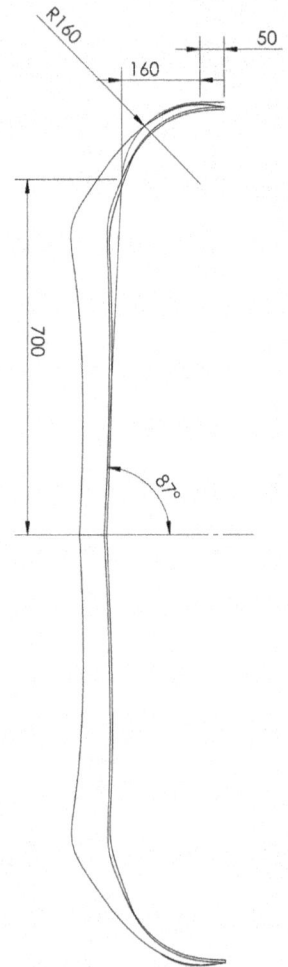

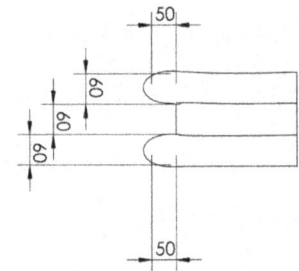

*Figure-109. Car Bumper*

# Chapter 7

# Product Manufacturing Information

## Topics Covered

The major topics covered in this chapter are:

- *Introduction to Product Manufacturing Information*
- *Setting Planes for PMI*
- *Applying PMI Dimensions*
- *Dimension Styles*
- *Applying Annotations*
- *Setting Annotation Font and Size*
- *Generating Model and Section Views*

# INTRODUCTION

PMI (Product Manufacturing Information) is non geometric data included with 3D model for critical digital inspection reporting. Rather than relying on 2D drawings or digital documents to convey engineering or manufacturing data, it can all be embedded within the original CAD file. In some software, it is also called Model Based Definition. PMI may include the following information:

- GD&T (Geometric dimensions & tolerances).
- Bill of materials (BOM).
- Surface finish.
- Weld symbols.
- Material specifications.
- Metadata & notes.
- History of engineering change orders.
- Legal/proprietary/export control notices.
- Other definitive digital data.

The tools to apply PMI are available in the **PMI** tab of the **Ribbon**; refer to Figure-1. Various tools in this tab are discussed next.

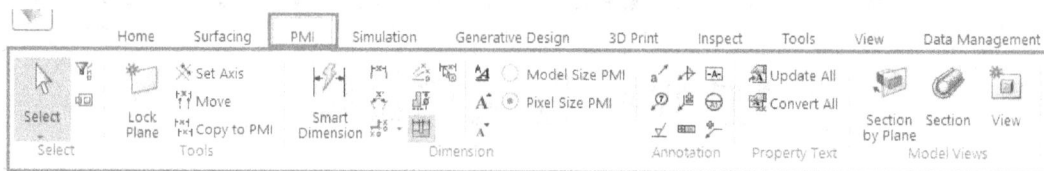

*Figure-1. PMI tab of Ribbon*

# LOCKING DIMENSION PLANE

When we annotate a 3D model using dimensions then there is a need of plane to be used for placing the annotations. We select the planes which are parallel to feature for annotating the model. The procedure to lock a plane for annotation is given next.

- Click on the **Lock Plane** tool from the **Tools** group in the **PMI** tab of the **Ribbon**. The **Lock Dimension Plane Command Bar** will be displayed; refer to Figure-2.

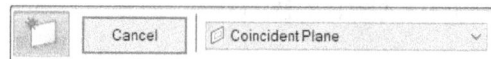

*Figure-2. Lock Dimension Plane Command Bar*

- Select the desired option from the drop-down in the **Command Bar** to select or create lock plane. Create a plane as discussed earlier or select a planar face to be used as annotation plane. The selected plane will be assigned as dimension plane.

# SETTING DIMENSION AXIS

The **Set Axis** tool is used to define axis for PMI dimensions at specified angle. This tool is useful when create inclined dimensions.

- Click on the **Set Axis** tool from the **Tools** group in the **PMI** tab of the **Ribbon**. You will be asked to select a line or key points to define the dimension axis.

- Select two key points on the model or select the desired line to define dimension axis. The axis will be created. You can check the axis when you next time click on the **Set Axis** tool; refer to Figure-3.

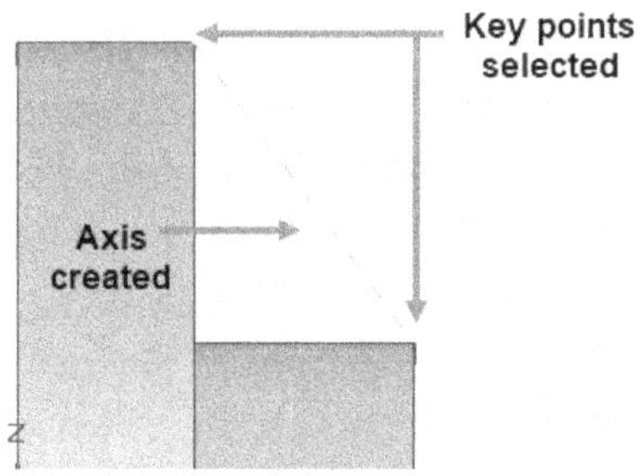

*Figure-3. Dimension axis created*

Other tools of **Tools** group will be discussed later in this chapter.

## CREATING DIMENSION

The tools to create dimensions for PMI are similar to the tools discussed earlier for dimensioning in sketches. Here, we will discuss the use of **Smart Dimension** tool. You can apply the other tools in the same way.

- Click on the **Smart Dimension** tool from the **Dimension** group in the **PMI** tab of the **Ribbon**. You will be asked to select the entities to be dimensioned and the **Smart Dimension Command Bar** will be displayed; refer to Figure-4.

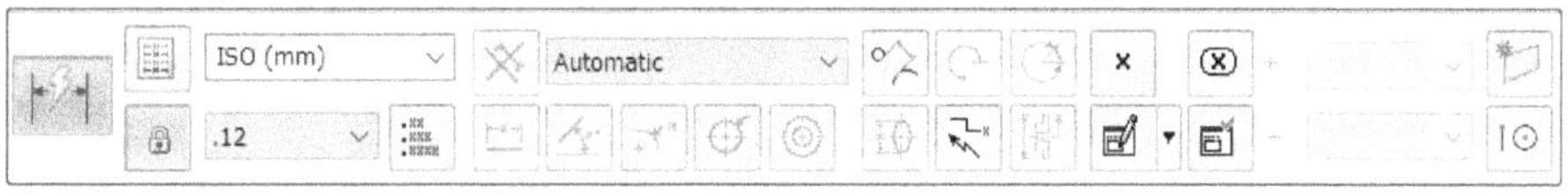

*Figure-4. Smart Dimension Command Bar*

- Select the desired option from the **Dimension Style** drop-down to define unit and style of dimensions. By default, **ISO(mm)** option is selected in our case.
- Select the desired option from the **Dimension Round-Off** drop-down to define accuracy level of dimension.
- If you want to define accuracy for primary and secondary dimensions separately then click on the **Advanced Round-Off** button from the **Command Bar**. The **Round-Off** dialog box will be displayed; refer to Figure-5. Set the desired precision levels in drop-downs of this dialog box and click on the **OK** button.

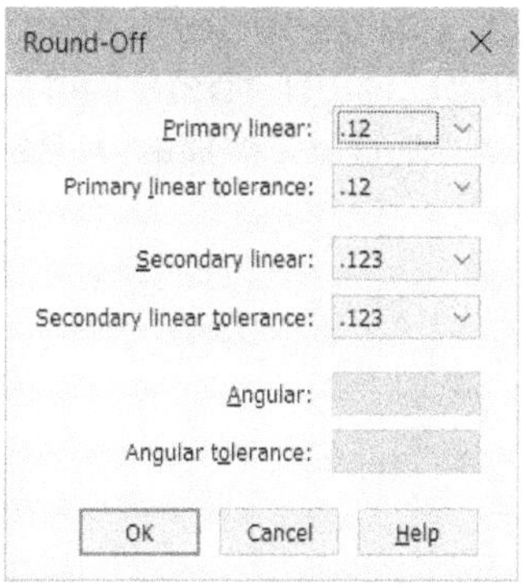

*Figure-5. Round-Off dialog box*

- Select the desired option from the **Orientation** drop-down to define how dimension will be oriented. Select the **Horizontal/Vertical** option from the drop-down if you want to create a dimension aligned horizontally or vertically. Select the **By 2 Points** option from the drop-down to create a dimension directly measuring selected two points; refer to Figure-6. Select the **Use Dimension Axis** option from the drop-down if you want to create dimension along dimension axis created earlier. You can create a dimension axis while the **Smart Dimension** tool is active by clicking on the **Dimension Axis** button in the **Smart Dimension Command Bar**. The procedure to create dimension axis has been discussed earlier. When you have selected the **Use Dimension Axis** option from the drop-down then any dimension you create will be aligned to the axis; refer to Figure-7. Select the **Automatic** option from the drop-down to automatically orient the dimensions based on selection.

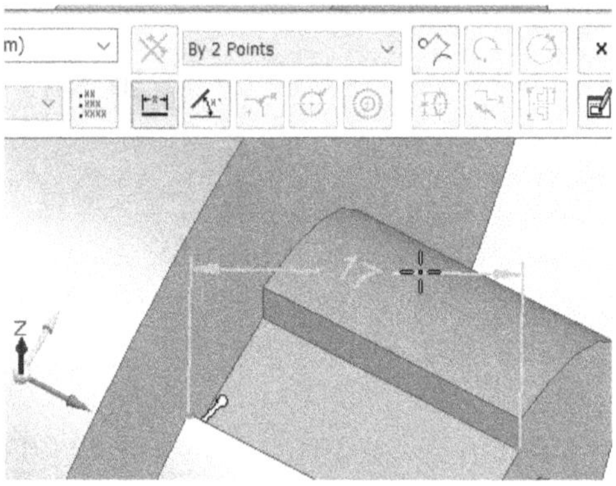

*Figure-6. Dimension by 2 points*

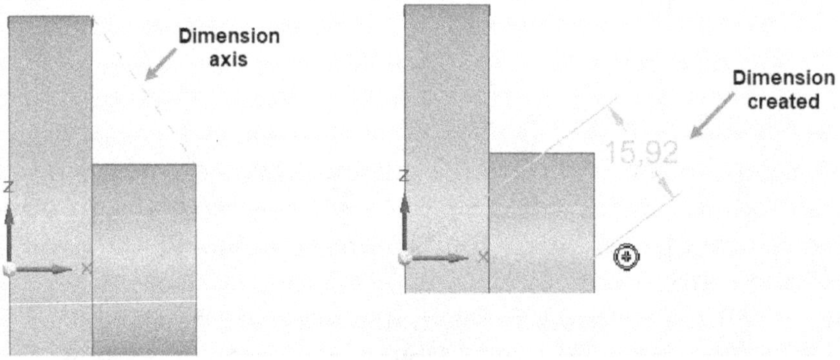

*Figure-7. Creating dimension using dimension axis*

- Select the **Tangent** toggle button to dimension objects using their tangent points. On activating this button, when you select a circle or arc then its tangent point will be used to dimensioning in place of default center point.
- If you are creating a radial dimension then **Radius**, **Diameter**, and **Concentric Diameter** buttons will be active in the **Command Bar**; refer to Figure-8. Select the

**Radius** button if you want to apply radius dimension to selected entity. Select the **Diameter** button to create diameter dimension.

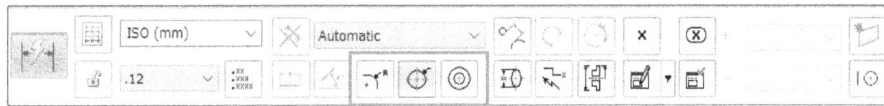

*Figure-8. Buttons for radial dimensioning*

- Select the **Diameter Normal** button from the **Smart Dimension Command Bar**. The diameter dimension will be created perpendicular to face of model; refer to Figure-9.

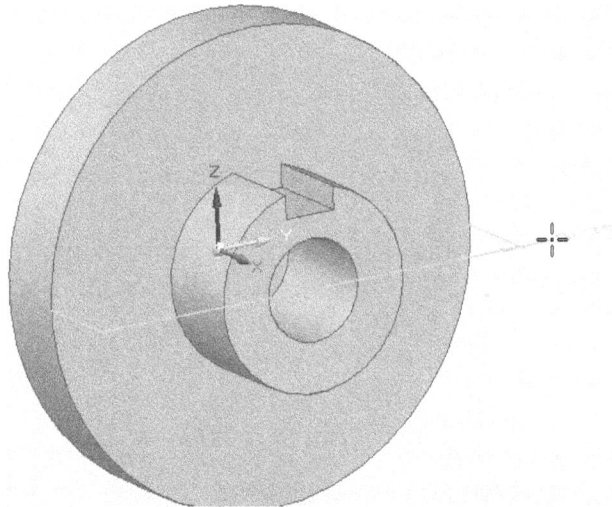

*Figure-9. Normal diameter dimension*

- Select the **Jog** button from the **Command Bar** to convert radial dimension into jogged line dimension; refer to Figure-10.

*Figure-10. Jogged line dimension*

- Select the **Diameter - Half/Full** button from the **Command Bar** to display diameter dimension in half or full length.

- If you are dimensioning an arc in the model then you can create the arc length or arc angle dimension by using the **Length** or **Angle** buttons from the **Command Bar**; refer to Figure-11.

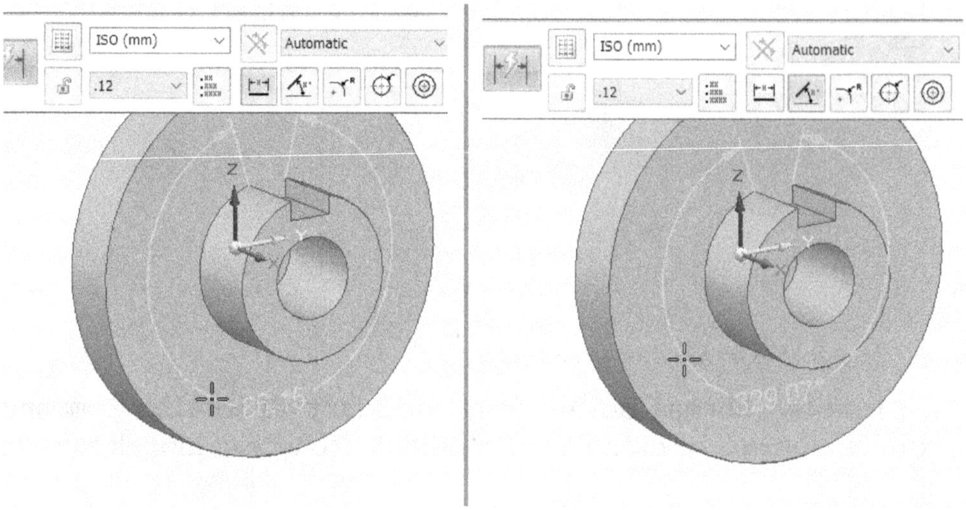

*Figure-11. Length and angle dimension of arc*

- If you are creating an angular coordinate dimension then you can use **Counterclockwise** button to measure dimensions counterclockwise. Similarly, you can use the **Major-Minor** button from **Command Bar** to switch between major and minor angle dimension.
- If you want to add symbols and prefixes to the dimension then click on the **Prefix** button from the **Command Bar**. The **Dimension Prefix** dialog box will be displayed; refer to Figure-12. Click in the **Prefix**, **Suffix**, **Superfix**, and **Subfix** edit boxes of the dialog box and specify the desired values. You can also add symbols in these edit boxes by selecting the respective buttons from the Symbol area of the dialog box. After specifying desired parameters, click on the **OK** button from the dialog box.

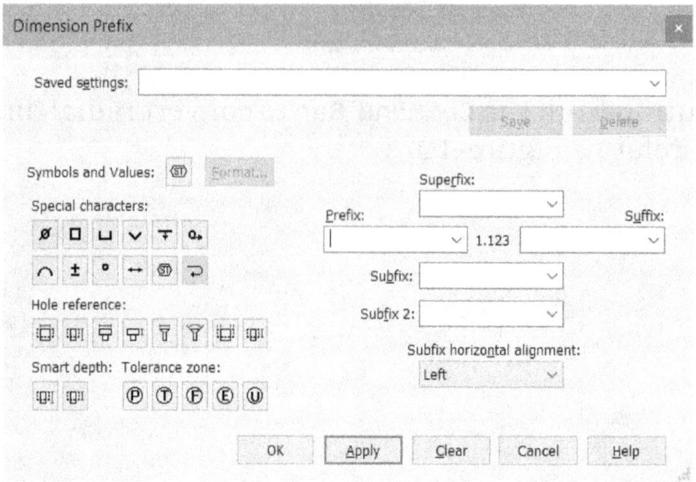

*Figure-12. Dimension Prefix dialog box*

- Click on the **Dimension Type** button from the **Command Bar** to define type of dimension. A list of options will be displayed as shown in Figure-13.

*Figure-13. Dimension Type options*

## Nominal Dimension Type

Select the **Nominal** option from the list if you want to specify only nominal value of dimension. This option is selected by default.

## Unit Tolerance Dimension Type

*   Select the **Unit Tolerance** option from the list if you want to specify upper and lower tolerances of dimension. The options to specify upper and lower limit of dimension will be displayed in the **Command Bar**; refer to Figure-14.

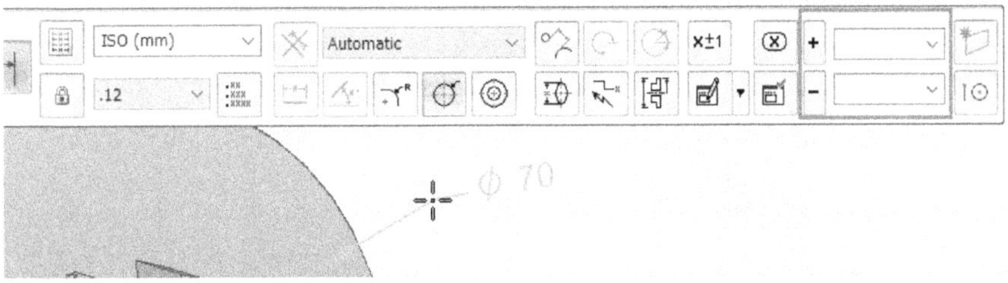

*Figure-14. Options to specify upper and lower limits of dimension*

*   If you want to change the sign of tolerance value then click on the button for sign.
*   Specify the desired values in tolerance edit boxes of the **Command Bar**.

## Alpha Tolerance Dimension Type

*   Select the **Alpha Tolerance** option from the **Dimension Type** list. The options to specify alphanumeric tolerance will be displayed in the **Command Bar**.
*   Specify the parameters as discussed for **Unit Tolerance** dimension type. Note that you can also specify alphabets along with numerical values in the tolerance edit boxes.

## Class Dimension Type

*   Select the **Class** option from the list of dimension type if you are applying dimensions to holes and shafts.
*   Select the desired option from the **Fit Type** drop-down to define how tolerance value will be displayed with dimension.
*   Select the desired tolerance class for shaft/hole from respective drop-down in the **Command Bar**; refer to Figure-15.

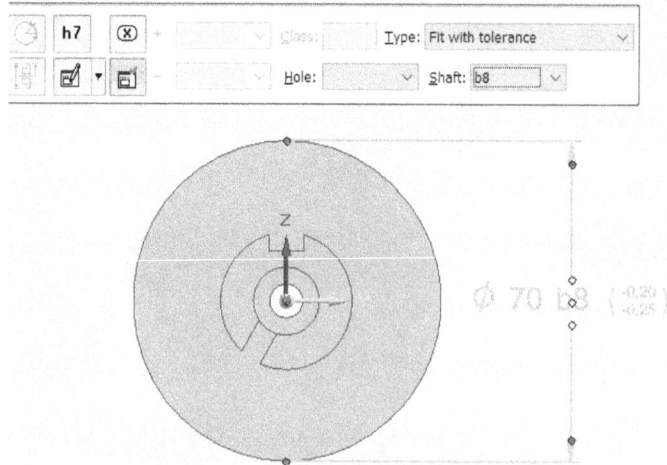

*Figure-15. Applying class tolerance*

## Limit Dimension Type

• Select the **Limit** option from the **Dimension Type** list if you want to display dimensions by their maximum and minimum values after adding/subtracting the tolerance values from nominal value.

• Specify the desired tolerance values in the **Upper Tolerance** and **Lower Tolerance** edit boxes of the **Command Bar**; refer to Figure-16.

*Figure-16. Limit tolerance type dimensions*

## Basic Dimension Type

• Select the **Basic** option from the **Dimension Type** list if you want to specify theoretical value of dimension. Note that theoretical value is impossible to achieve repeatedly in manufacturing so by specifying basic value, you want to say that all the other dimensions should be measured in reference to specified basic value dimension. The title block tolerance does not apply to basic dimensions. The tolerance associated with a basic dimension usually appears in a feature control frame or a note. Basic dimensions have a box around their values.

## Reference Dimension Type

• Select the **Reference** option from the **Dimension Type** list if you want to add a dimension which can be accessed already by other dimensions in the drawing; refer to Figure-17.

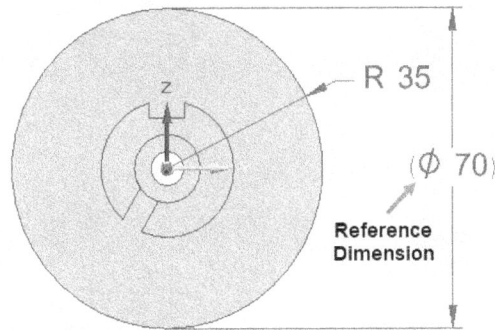

*Figure-17. Reference dimension*

## Feature Callout Dimension Type

* Select the **Feature Callout** option from the Dimension Type list if you want to dimension features like holes which have callouts associated with them.
* Select the round edge of hole or other feature which has callout associated with it. The callout dimension will be attached to cursor; refer to Figure-18.

*Figure-18. Callout dimension attached to cursor*

## Blank Dimension Type

* Select the **Blank** option from the Dimension Type list if you want to create a dimension without dimension value. Note that although the dimension value will not be displayed in blank dimension type but prefix and other parameters specified in the **Dimension Prefix** dialog box will be displayed.
* Select the entities to be dimensioned as discussed earlier.

* Select the **Inspection** button from the **Command Bar** if you want the selected dimension inspected after manufacturing.
* Click on the **Lock Dimension Plane** button from the **Command Bar** to select a plane/face used as base reference for placing dimensions. The **Lock Dimension Plane Command Bar** will be displayed.
* Select the desired plane/face for locking dimensions.
* Click on the **Keypoints** button from the **Command Bar** to select desired points while creating the dimension. The list of keypoints will be displayed; refer to Figure-19.

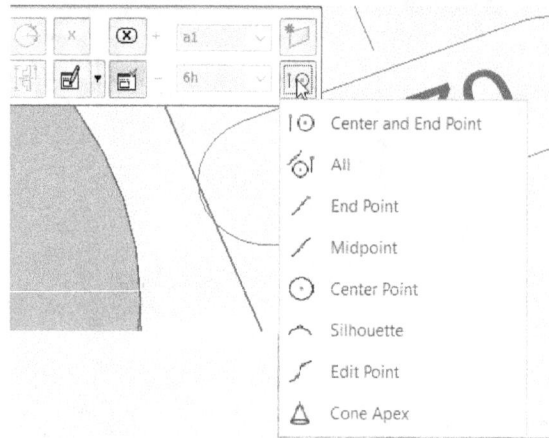

*Figure-19. Keypoints options*

- Select the desired key point from the list and you will be able to snap to that type of key points only in the model while creating dimensions.

## DIMENSION STYLES

The **Styles** tool in **Dimension** group of the **PMI** tab in **Ribbon** is used to create and manage dimension styles and parameters for various annotation features. The procedure to use this tool is given next.

- Click on the **Styles** tool from the **Dimension** group in the **PMI** tab of the **Ribbon**. The **Styles** dialog box will be displayed. The options of this dialog box have been discussed earlier in Chapter 2.

## MODEL SIZE AND PIXEL SIZE PMI

There are two methods to define size of annotations in PMI, using model text size or using Pixel size. Select the **Model Size PMI** radio button from the **Dimension** group in the **Ribbon** if you want to use size parameters specified in the **Styles** dialog box. Select the **Pixel Size PMI** radio button from the **Dimension** group of **PMI** tab in **Ribbon** if you want to increase or decrease the size of dimension dynamically. On selecting the **Pixel Size PMI** radio button, the **Increase PMI Font** and **Decrease PMI Font** buttons will become active in the **Dimension** group of **Ribbon**. Using these buttons, you can increase or decrease the size of annotations and dimensions. Note that size of all the dimensions will increase or decrease in same proportion.

## CREATING CALLOUT ANNOTATION

The **Callout** tool in **Annotation** group is used to create user defined text annotation with leader. The procedure to use this tool is given next.

- Click on the **Callout** tool from the **Annotation** group in the **PMI** tab of the **Ribbon**. The **Callout Properties** dialog box will be displayed; refer to Figure-20.

## General Tab

- Specify the desired text in the **Callout text** and **Callout text 2** edit boxes. You can also add symbols in the edit boxes by using the buttons in **Special characters** area of the dialog box.

- Click on the **Property Text** button from the **Reference** area to add properties used in current document in the edit boxes. On clicking this button, the **Select Property Text** dialog box will be displayed; refer to Figure-21.

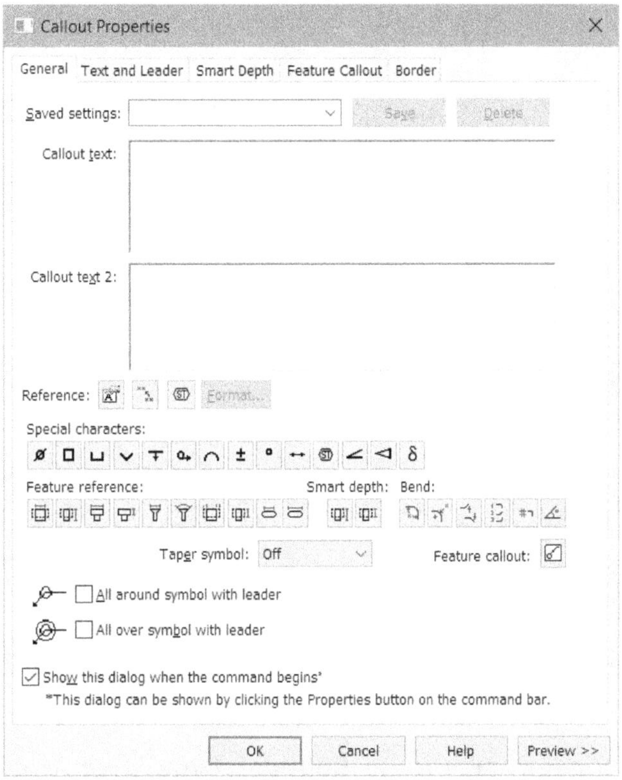

*Figure-20. Callout Properties dialog box*

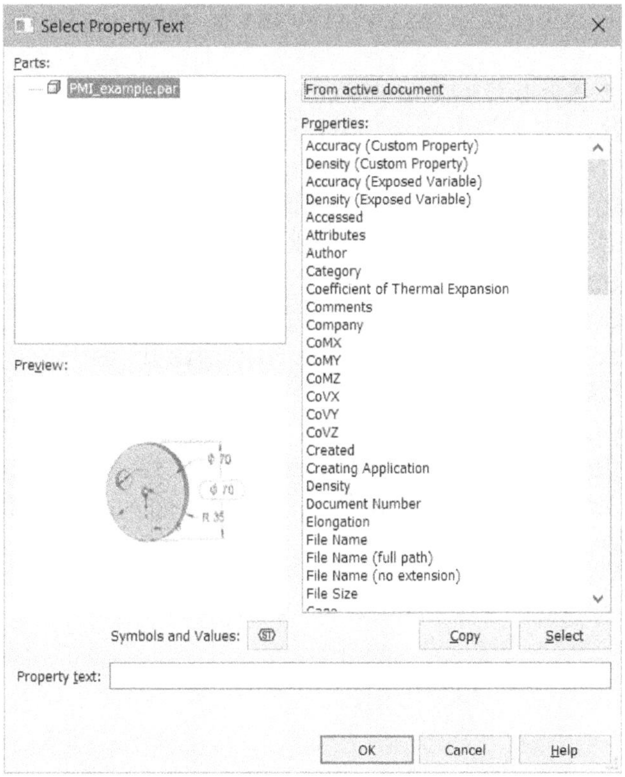

*Figure-21. Select Property Text dialog box*

- Double-click on the desired property from the **Properties** list of the dialog box. The property parameter will be added in the edit box. Click on the **OK** button from the dialog box.

- Click on the **Reference Text** button from the **Reference** area of the dialog box. The **Select Reference Text** dialog box will be displayed; refer to Figure-22.

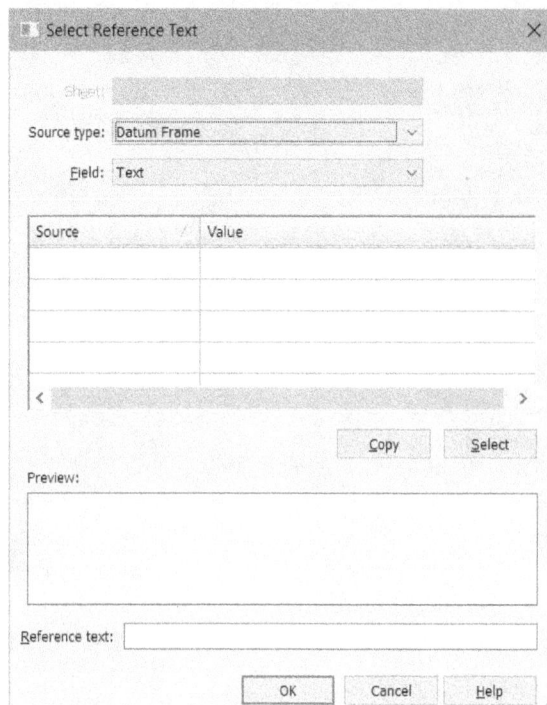

*Figure-22. Select Reference Text dialog box*

- Select the desired option from the **Source type** drop-down to select source of reference text.
- Double-click on the reference text you want to add in the callout and click on the **OK** button.
- Click on the **Select Symbols and Values** button from the **Reference** area to add GD&T symbols and other drawing parameters in the callout. The **Select Symbols and Values** dialog box will be displayed; refer to Figure-23. Expand the desired category of symbol or parameter in the dialog box and double-click on the symbol/ parameter to be added. Click on the **OK** button from the dialog box to add the symbol in callout text.
- You can select the standard feature references, smart depths of features, and bend parameters by selecting respective buttons from the **General** tab of **Callout Properties** dialog box.
- Select the desired option from the **Taper symbol** drop-down to add a taper symbol in the callout if selected feature is taper. Apart from the **Off** option which hides taper symbol, you can select the **Left** or **Right** option to place symbol at the left or right in the callout respectively.

*Figure-23. Select Symbols and Values dialog box*

- Select the **All around symbol with leader** check box to apply the callout annotation all around selected entity.
- Select the **All over symbol with leader** check box to add all over symbol with the leader.

## Text and Leader Tab

The options in the **Text and Leader** tab of the dialog box are used to specify parameters related to text size, leader lines, projection lines, color, terminator, and so on; refer to Figure-24. Set the parameters as desired in this tab. The options in this tab have been discussed earlier.

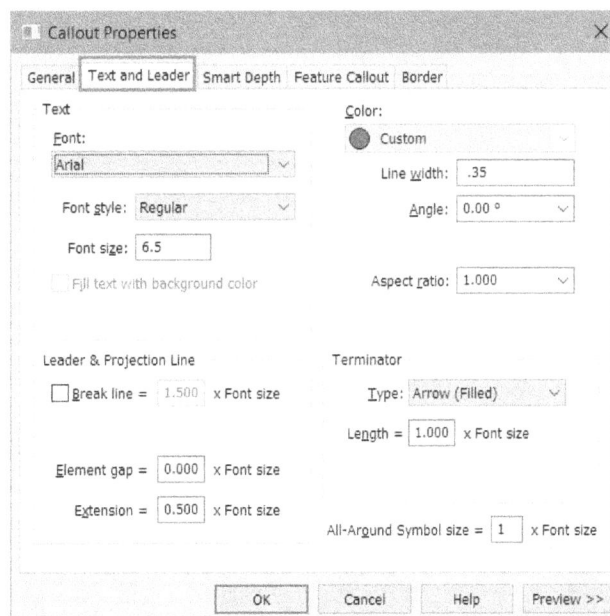

*Figure-24. Text and Leader tab of Callout Properties dialog box*

## Smart Depth Tab

The options in the **Smart Depth** tab are used to add smart depth parameters in the callout; refer to Figure-25. Click in the desired edit box and specify the desired parameters. You can use the buttons to add parameters.

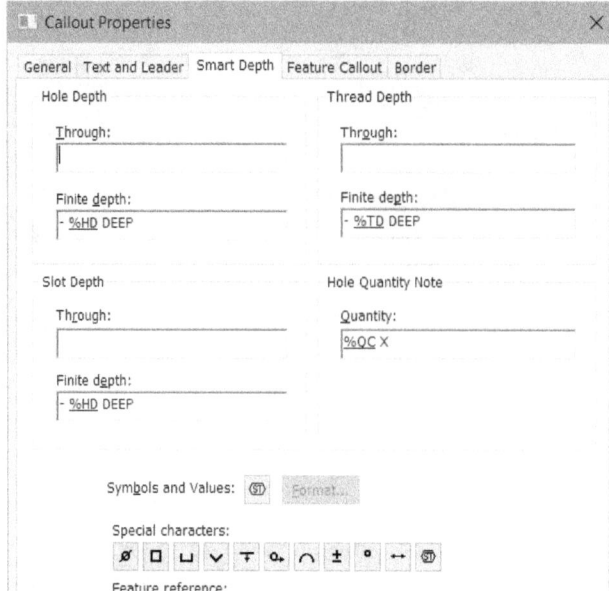

*Figure-25. Smart Depth tab*

## Feature Callout Tab

The options in the **Feature Callout** tab are used to define the parameters to be added in the standard hole and slot callouts; refer to Figure-26.

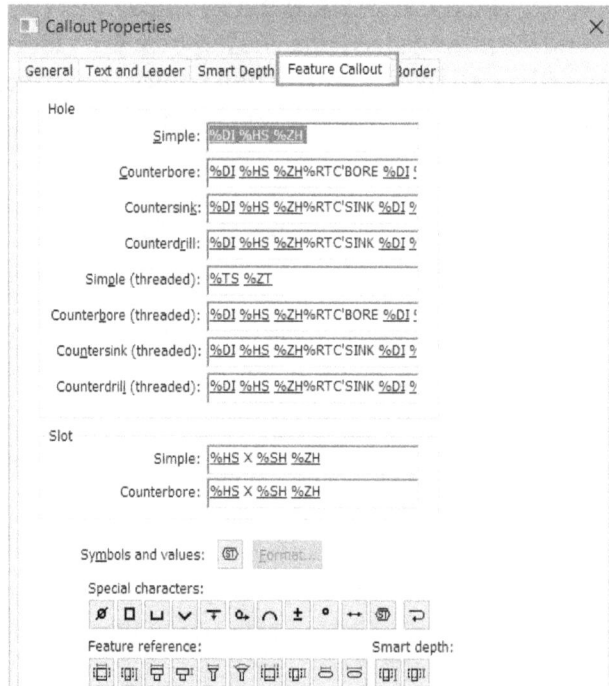

*Figure-26. Feature Callout tab*

## Border Tab

The options in the **Border** tab are used to thickness of callout outlines, gap between border and text, and text position in the callout; refer to Figure-27.

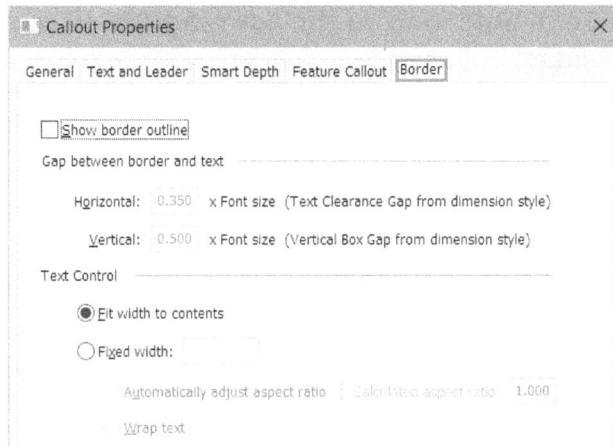

*Figure-27. Border tab*

- Select the **Show border outline** check box to display border line in callout; refer to Figure-28.

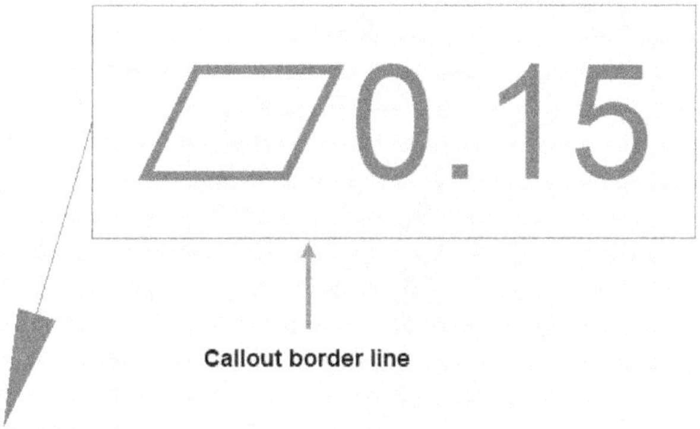

*Figure-28. Callout border line preview*

- Specify the desired parameters in the edit boxes of the **Gap between border and text** area to define gap between border of callout and text inside it.
- Select the desired radio button from the **Text Control** area of the dialog box to define the width of frame.
- After specifying the parameters, click on the **OK** button from the dialog box. The **Callout Command Bar** will be displayed; refer to Figure-29.

*Figure-29. Callout Command Bar*

- Specify the desired angle value in the **Angle** edit box to define orientation of callout with respect to leader landing.
- Select the **Break Line** button from the **Command Bar** to add a landing line with leader in annotation.
- Click on the **Horizontal Alignment** button from the **Command Bar** and select the desired option from the drop-down list displayed to define where text will be placed in callout box.
- Select the **Position** button to define position of text with respect to landing line in annotation. Select the desired option from the drop-down list displayed on clicking this button. Note that the **Position** button is active only when **Break Line** button is selected in the **Command Bar**.
- Select the **Underline Text** button to underline the text in callout.

- Select the **Parallel Text** button from the **Command Bar** to make callout box parallel to selected face and edge.
- Click on the **Invert Text** button from the **Command Bar** to invert the text of callout.
- Toggle on/off the **Show Border** button to display or hide the borders of callout.
- Click on the **Text Control** button from the **Command Bar** and select the desired option from the drop-down list to define how text will be arranged in the callout box.
- Click on the **Referenced Geometry** button from the **Command Bar** to select reference geometry callout.
- After setting desired parameters, click on the edge or face to which you want to add callout. The callout will get attached to the cursor; refer to Figure-30.

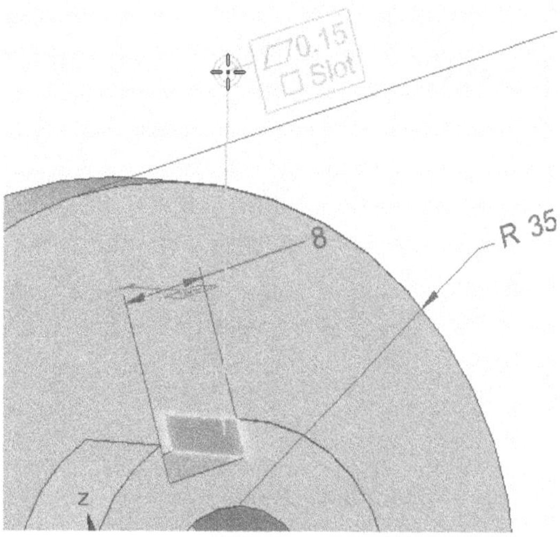

*Figure-30. Callout attached to cursor*

- Click at the desired location to place the callout. Press **ESC** to exit the tool.

## CREATING BALLOON ANNOTATION

The **Balloon** tool in the **Annotation** group is used to create balloon annotation on the model. The procedure to use this tool is given next.

- Click on the **Balloon** tool from the **Annotation** group in the **PMI** tab of the **Ribbon**. The **Balloon Command Bar** will be displayed; refer to Figure-31.

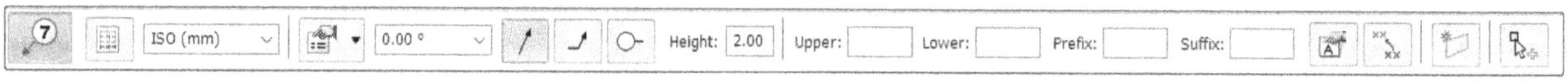

*Figure-31. Balloon Command Bar*

- Specify the desired alphanumeric values in the **Upper**, **Lower**, **Prefix**, and **Suffix** edit boxes to define balloon parameters.
- Click on the **Shape** button from the **Command Bar** and select the desired option from the list to define shape of balloon annotation.
- Set the other parameters as discussed earlier and click on the face to which you want to apply annotation. The balloon annotation will get attached to the cursor; refer to Figure-32.

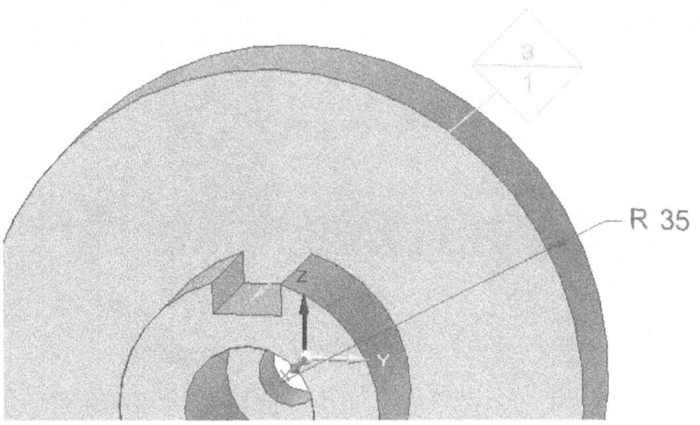

*Figure-32. Balloon annotation attached to cursor*

- Click at the desired location to place the balloon. Press **ESC** to exit the tool.

## CREATING SURFACE TEXTURE SYMBOL

The **Surface Texture Symbol** tool is used to apply surface texture symbol on selected face. The procedure to use this tool is given next.

- Click on the **Surface Texture Symbol** tool from the **Annotation** group of the **PMI** tab in the **Ribbon**. The **Surface Texture Symbol Properties** dialog box will be displayed; refer to Figure-33.

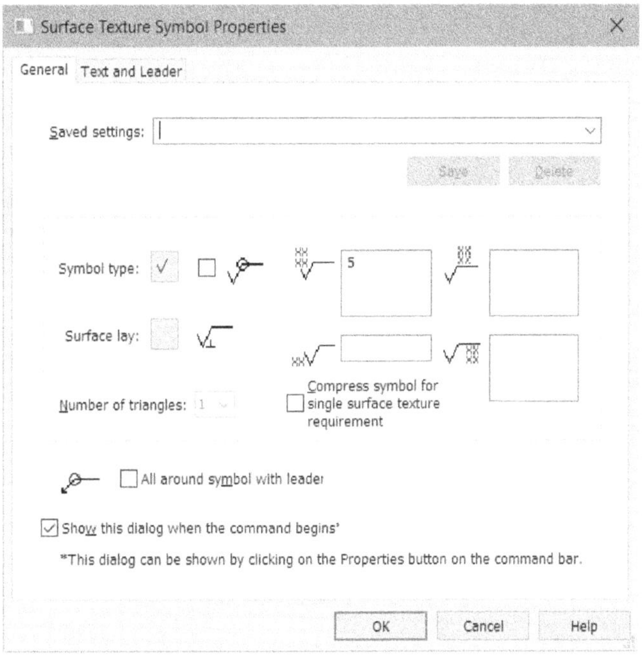

*Figure-33. Surface Texture Symbol Properties dialog box*

- Click on the button from the **Symbol type** area to select symbol for surface finish. The list of symbols will be displayed as shown in Figure-34.

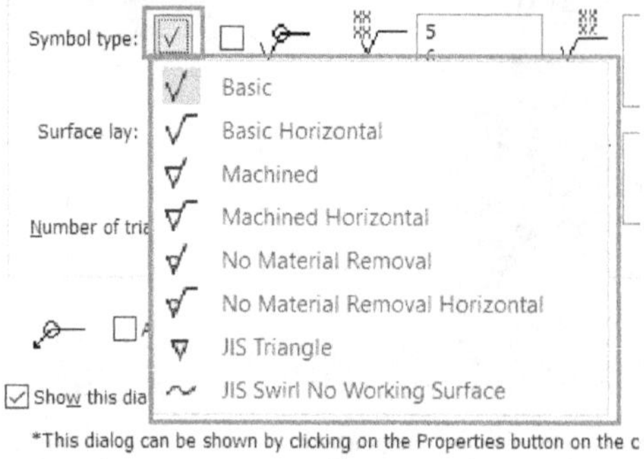

*Figure-34. Surface finish symbols*

- Select the desired symbol from the list and specify the parameters in edit boxes next to the button.
- Click on the **Surface lay** button from the dialog box and select the desired option to define direction of surface finish.
- Set the desired parameters of surface finish in the edit boxes, and click on the **OK** button from the dialog box. You will be asked to specify the location for symbol.
- Click at desired face, surface, edge, or other entity. The surface symbol will get attached to cursor.
- Click at desired location to place the symbol. Press **ESC** to exit the tool.

# CREATING WELD SYMBOL

The **Weld Symbol** tool is used to create weld symbol in PMI and drawing. The procedure to use this tool is given next.

- Click on the **Weld Symbol** tool from the **Annotation** group in the **PMI** tab of the **Ribbon**. The **Weld Symbol Properties** dialog box will be displayed; refer to Figure-35.

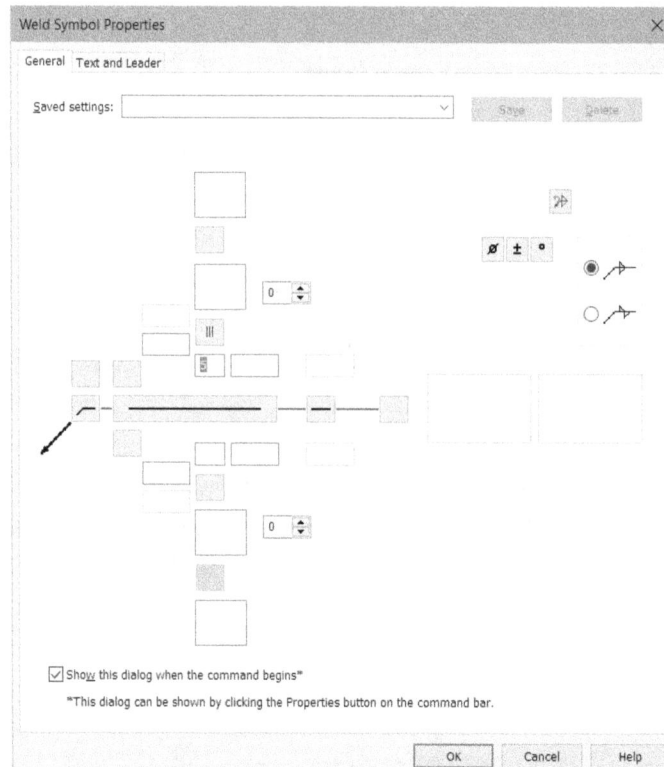

*Figure-35. Weld Symbol Properties dialog box*

- Specify the desired parameters in the edit boxes of dialog box for weld symbol and click on the **OK** button.
- The **Weld Symbol Command Bar** will be displayed; refer to Figure-36 and you will be asked to specify the location of weld symbol.

*Figure-36. Weld Symbol Command Bar*

- Click at desired location to specify weld bead location. The symbol will get attached to cursor.
- Click at desired location to place the symbol.

## CREATING EDGE CONDITION

The **Edge Condition** tool is used to annotation tolerance conditions for edges. The procedure to use this tool is given next.

- Click on the **Edge Condition** tool from the **Annotation** group in the **PMI** tab of **Ribbon**. The **Edge Condition Properties** dialog box will be displayed; refer to Figure-37.

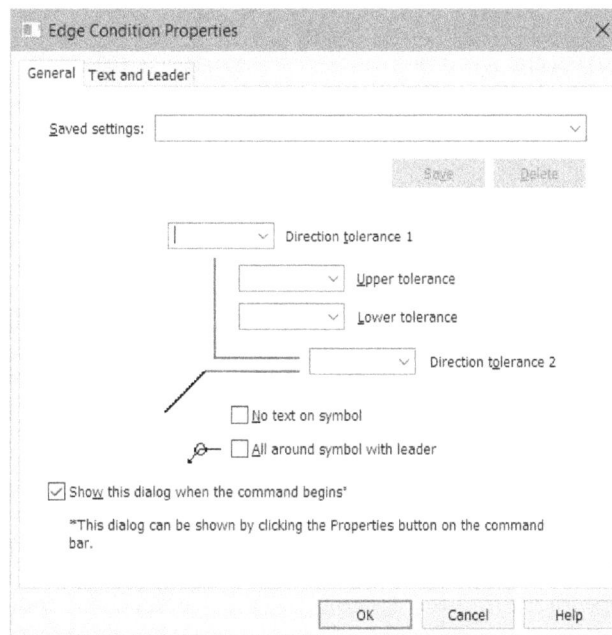

*Figure-37. Edge Condition Properties dialog box*

- Specify the desired tolerances values in the edit boxes and set the other parameters as discussed earlier.
- Click on the **OK** button from the dialog box. The **Edge Condition PropertyManager** will be displayed; refer to Figure-38.

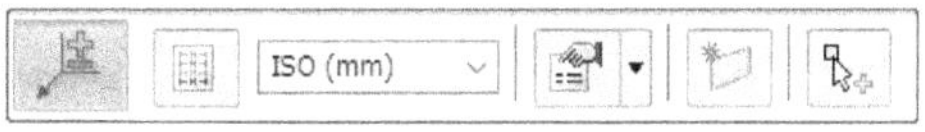

*Figure-38. Edge Condition Command Bar*

- You can set desired parameters in the **Command Bar** as discussed earlier and click at desired edge to which want to apply annotation. The symbol will get attached to cursor; refer to Figure-39.

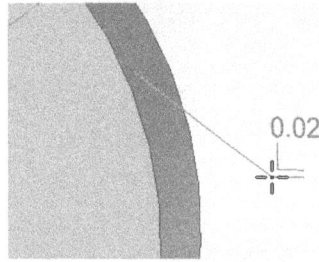

*Figure-39. Symbol attached to cursor*

• Click at desired location to place the symbol and press **ESC** to exit the tool.

## CREATING DATUM FRAME AND FEATURE CONTROL FRAME ANNOTATION

The **Feature Control Frame** tool is used to apply GD&T annotations to selected features. Geometrical tolerance is defined as the maximum permissible overall variation of form or position of a feature.

Geometrical tolerances are used,
(i) to specify the required accuracy in controlling the form of a feature.
(ii) to ensure correct functional positioning of the feature.

Figure-40 shows a geometric symbol with meaning of each box.

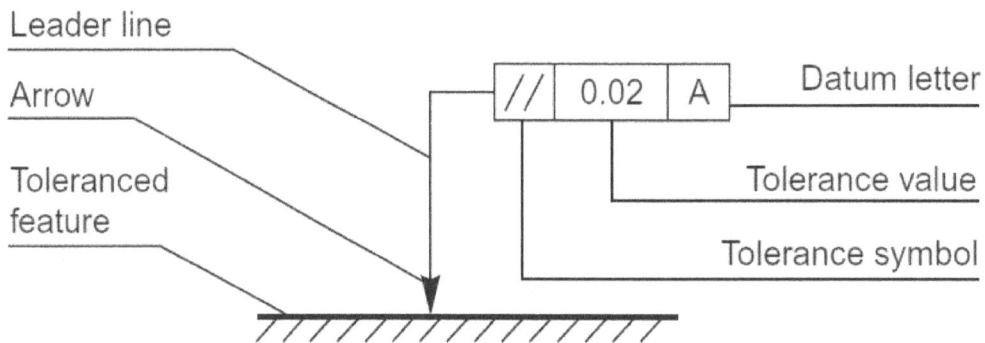

*Figure-40. Geometric Tolerance symbol*

Before we start using the tools in Solid Edge, it is important to understand the meaning of each Tolerance Symbol.

Meaning of various geometric tolerance symbols are given in Figure-41.

| Characteristics to be toleranced | | Symbols |
|---|---|---|
| Form of single features | Straightness | ——— |
| | Flatness | ▱ |
| | Circularity (roundness) | ◯ |
| | Cylindricity | ⌭ |
| | Profile of any line | ⌒ |
| | Profile of any surface | ⌓ |
| Orientation of related features | Parallelism | // |
| | Perpendicularity (squareness) | ⊥ |
| | Angularity | ∠ |
| Position of related features | Position | ⊕ |
| | Concentricity and coaxiality | ◎ |
| | Symmetry | ⩵ |
| | Run-out | ↗ |

*Figure-41. Meaning of geometric Tolerance symbol*

Figure-42 and Figure-43 shows the use of geometric tolerances in real-world.

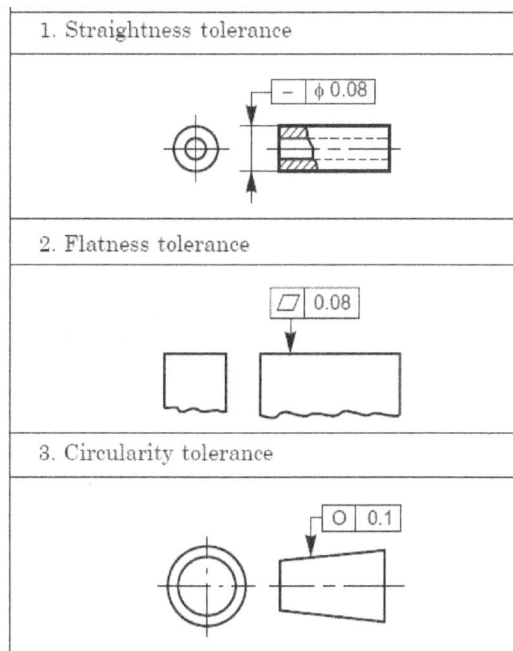

*Figure-42. Use of geometric tolerance 1*

| 4. Cylindricity tolerance | 8. Concentricity and coaxiality tolerance |
|---|---|
| | |
| 5. Parallelism tolerance | 9. Symmetry tolerance |
| | |
| 6. Perpendicularity tolerance | 10. Radial run-out |
| | |
| 7. Angularity tolerance | 11. Axial run-out |

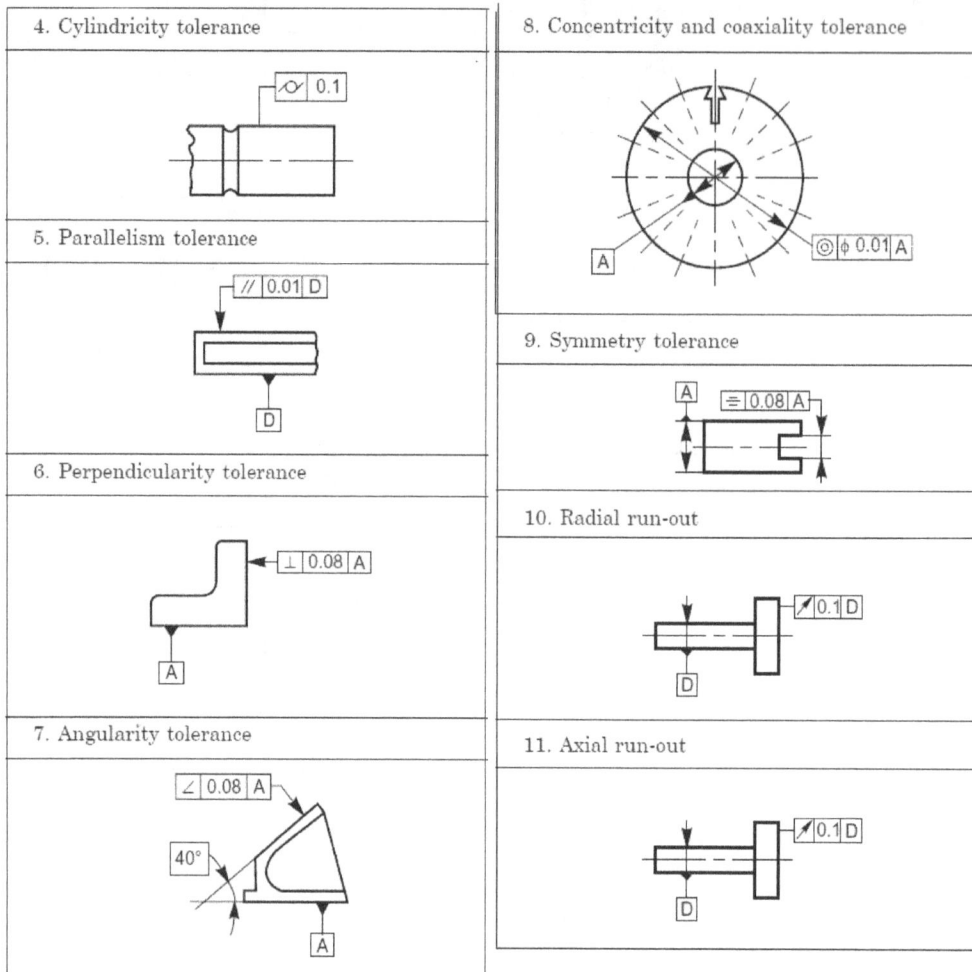

*Figure-43. Use of geometric tolerance 2*

Note that in applying most of the Geometrical tolerances, you need to define a datum plane. So, first we will create a datum Frame and then we will apply Feature Control Frame.

## Creating Datum Frame

The **Datum Frame** tool is used to define reference face/plane for applying geometric dimensioning and tolerance. The procedure to use this tool is given next.

- Click on the **Datum Frame** tool from the **Annotation** group in the **PMI** tab of the **Ribbon**. The **Datum Frame Command Bar** will be displayed; refer to Figure-44.

*Figure-44. Datum Frame Command Bar*

- Specify desired text for datum frame in the **Text** edit box of **Command Bar**.
- Click on the **Datum Frame Shape** button from the **Command Bar**. A list frame shapes will be displayed. Select the desired shape for boundary of datum frame.
- Select the desired face/surface/element to be marked as datum frame. The frame will get attached to cursor; refer to Figure-45.

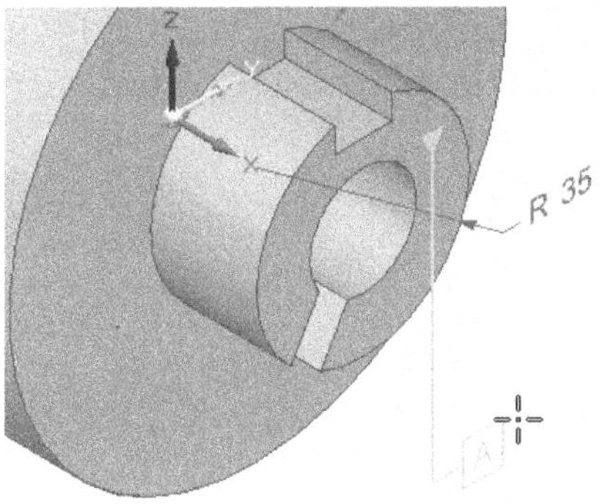

*Figure-45. Datum frame attached to cursor*

- Click at the desired location to place the symbol.
- Press **ESC** to exit the tool.

# Creating Feature Control Frame

The **Feature Control Frame** tool is used to create GD&T frame. The procedure to use this tool is given next.

- Click on the **Feature Control Frame** tool from the **Annotation** group in the **PMI** tab of the **Ribbon**. The **Feature Control Frame Properties** dialog box will be displayed; refer to Figure-46.

*Figure-46. Feature Control Frame Properties dialog box*

- Click in the first edit box of the **Content** area and specify GD&T symbol.
- Click on the **Divider** button from the dialog box to create division in frame box.
- Specify the value of tolerance in the edit box and create the divider.
- Click on the **Reference Text** button from the dialog box. The **Select Reference Text** dialog box will be displayed.

- Double-click on the desired datum reference value from the list and click on the **OK** button from the dialog box. The preview of symbol will be displayed; refer to Figure-47.

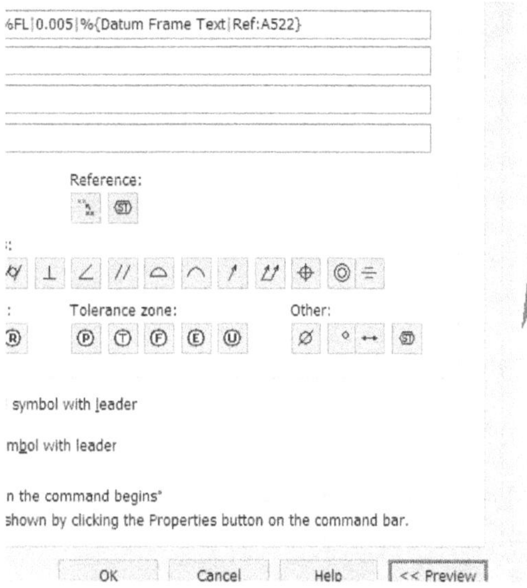

*Figure-47. Preview of Feature Control Frame*

- You can add material conditions, tolerance zones, and other symbols in the frame, in the same way as discussed earlier.
- Click on the **OK** button from the dialog box. The **Feature Control Frame Command Bar** will be displayed and you will be asked to select the entity to be annotated.
- Click at the desired face/surface/entity. The control frame will be attached to cursor; refer to Figure-48.

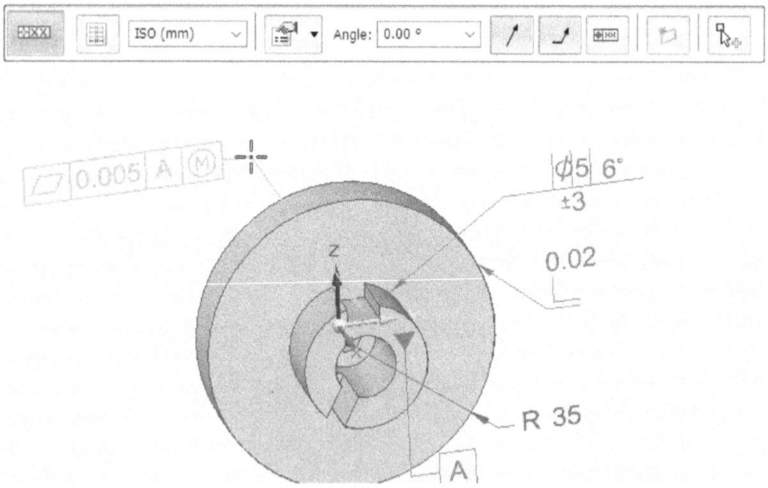

*Figure-48. Feature control frame attached to cursor*

- Click at desired location to place the symbol and press **ESC** to exit the tool.

## CREATING DATUM TARGET

The **Datum Target** tool is used to create datum target annotation for a datum point. The procedure to use this tool is given next.

- Click on the **Datum Target** tool from the **Annotation** group in the **PMI** tab of the **Ribbon**. The **Datum Target Command Bar** will be displayed; refer to Figure-49.

*Figure-49. Datum Target Command Bar*

- Specify the desired parameters in the edit boxes of the **Command Bar**.
- Click on the **Show Datum Area** button to show dimension with datum target symbol.
- Click at desired entity to apply the annotation. The datum target symbol will get attached to cursor.
- Click at the desired location to place the annotation. Press **ESC** to exit the tool.

## CREATING LEADER

The **Leader** tool is used to create leader without text. The leader is used to mark target for a text or annotation. The procedure to use this tool is given next.

- Click on the **Leader** tool from the **Annotation** group in the **PMI** tab of the **Ribbon**. The **Leader Command Bar** will be displayed; refer to Figure-50.

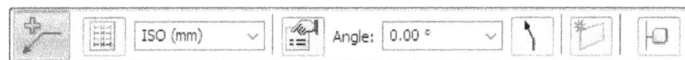

*Figure-50. Leader Command bar*

- Click at desired location to specify start point of leader. The end point will get attached to cursor.
- Click at desired location to specify end point of leader. Press **ESC** to exit the tool.

## CREATING SECTION BY PLANE

The **Section by Plane** tool is used to create 3D section of model using section plane. The procedure to use this tool is given next.

- Click on the **Section by Plane** tool from the **Model Views** group of the **PMI** tab in the **Ribbon**. The **Section by Plane Command Bar** will be displayed with preview of section; refer to Figure-51 and Figure-52.

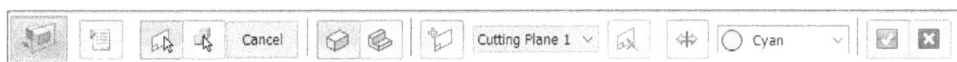

*Figure-51. Section by Plane Command Bar*

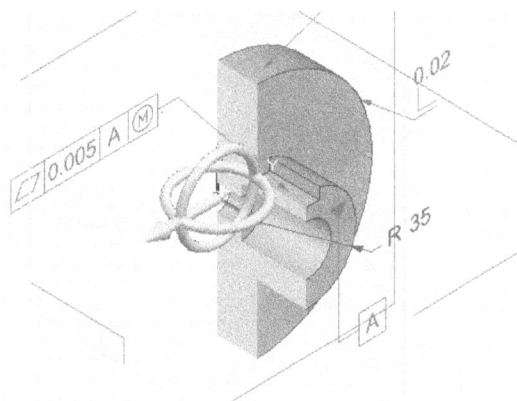

*Figure-52. Preview of section*

- Click on the **Section by Plane Options** button from the **Command Bar**. The **Section by Plane Options** dialog box will be displayed; refer to Figure-53.

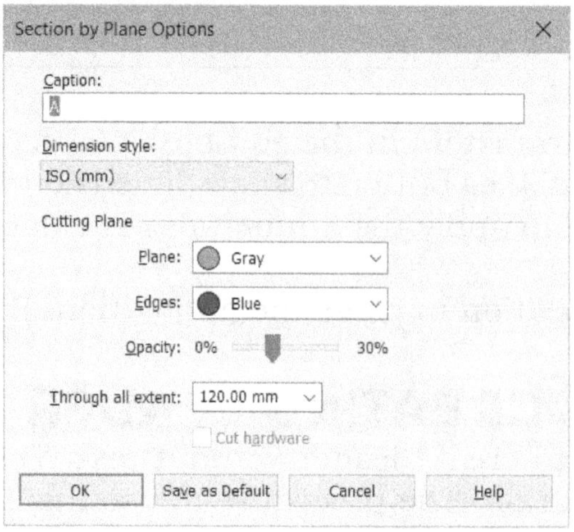

*Figure-53. Section by Plane Options dialog box*

- Specify the desired parameters in the dialog box to define name, plane color, edge color, and so on. Click on the **OK** button from the dialog box.
- Select the **Through All** or **Bounded** button from the **Command Bar** to define scope of section.
- Click on the **Add Cutting Plane** button from the **Command Bar** to add more standard cutting plane. If you want to delete a cutting plane then select the plane from the **Cutting Plane List** drop-down and click on the **Delete Cutting Plane** button from the **Command Bar**.
- You can flip the cut direction by using the **Flip Cut Direction** button from the **Command Bar**.
- After setting desired parameters, click on the **OK** button from the **Command Bar** and then click on the **Finish** button. The 3D section will be created; refer to Figure-54.

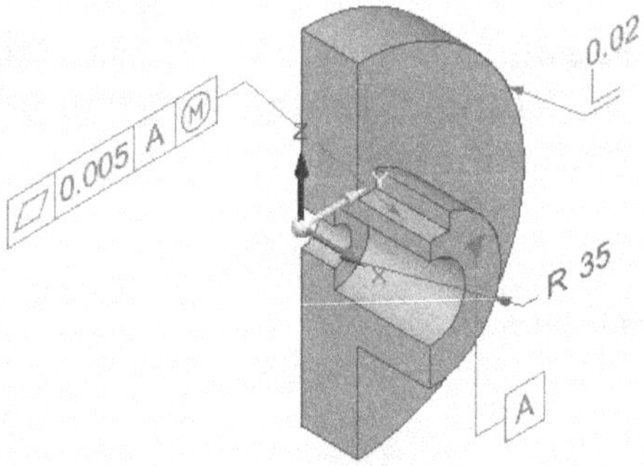

*Figure-54. 3D section created*

## CREATING USER DEFINED SECTION

The **Section** tool is used to create section using selected/created planes. The procedure to use this tool is given next.

- Click on the **Section** tool from the **Model Views** group in the **PMI** tab of the **Ribbon**. The **Section Command Bar** will be displayed; refer to Figure-55.

*Figure-55. Section Command bar*

- Select the desired plane that you want to use for creating section. You will be asked to create sketching boundary and the sketching environment will be activated.
- Create a closed sketch section; refer to Figure-56 and click on the **Close Sketch** button from the **Ribbon**. You will be asked to specify whether you want to section inside or outside of the sketch.

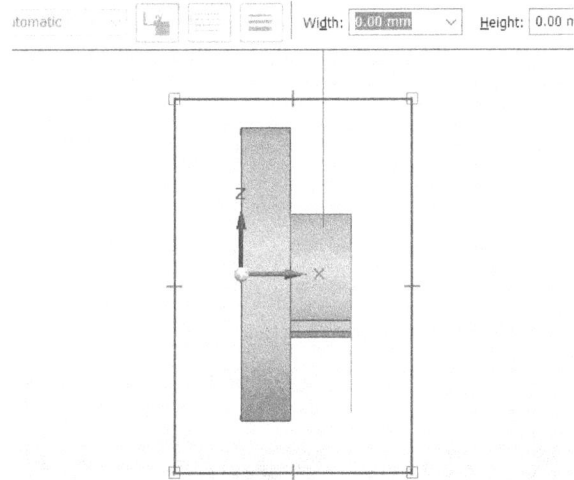

*Figure-56. Sketch section created*

- Click inside the sketch to create section. You will be asked to specify extent of section region.
- Enter the desired distance value in the **Distance** edit box or click at the desired location; refer to Figure-57.

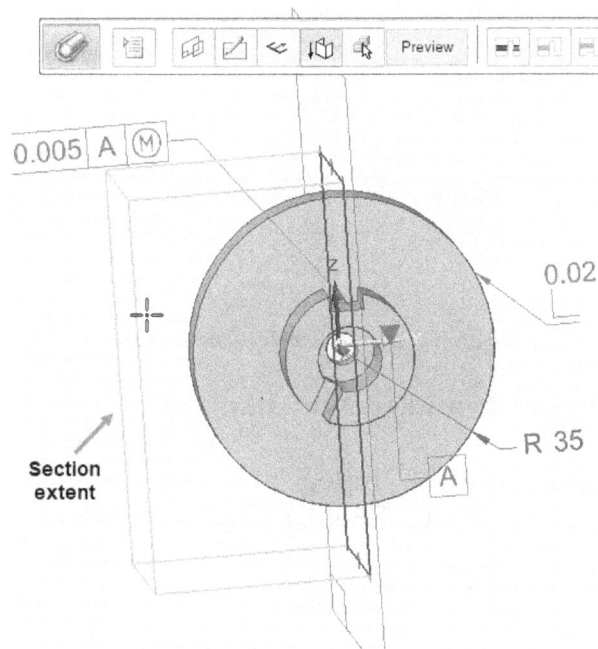

*Figure-57. Section extent*

- Click on the **OK** button from the **Command Bar**. The preview of section will be displayed; refer to Figure-58.

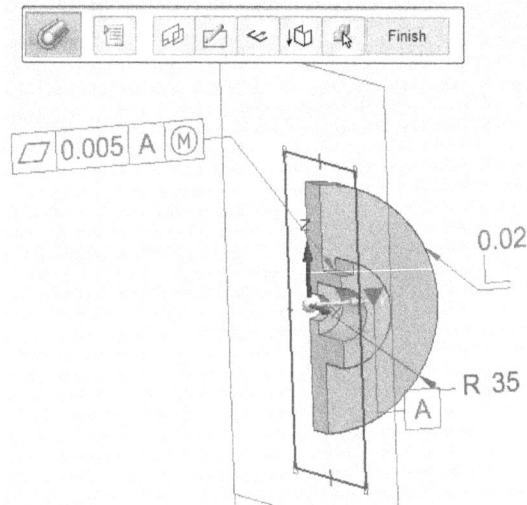

*Figure-58. Preview of section by plane*

*   Click on the **Finish** button from the **Command Bar** to create the section.

## CREATING MODEL VIEWS

The **View** tool is used to create model views using current model display. The procedure to use this tool is given next.

*   Click on the **View** tool from the **Model Views** group in the **PMI** tab of the **Ribbon**. The **Model View Options** dialog box will be displayed; refer to Figure-59.

*Figure-59. Model View Options dialog box*

*   Set the desired render mode and section display options in respective drop-downs of the dialog box.
*   Click on the **OK** button from the dialog box. The model view will be created and added in the **Model Views** node of **Feature Tree**; refer to Figure-60.

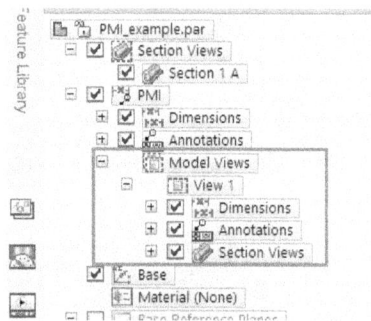

*Figure-60. Model view created*

# PRACTICAL 1

Create the model and apply the model based annotations as shown in Figure-61.

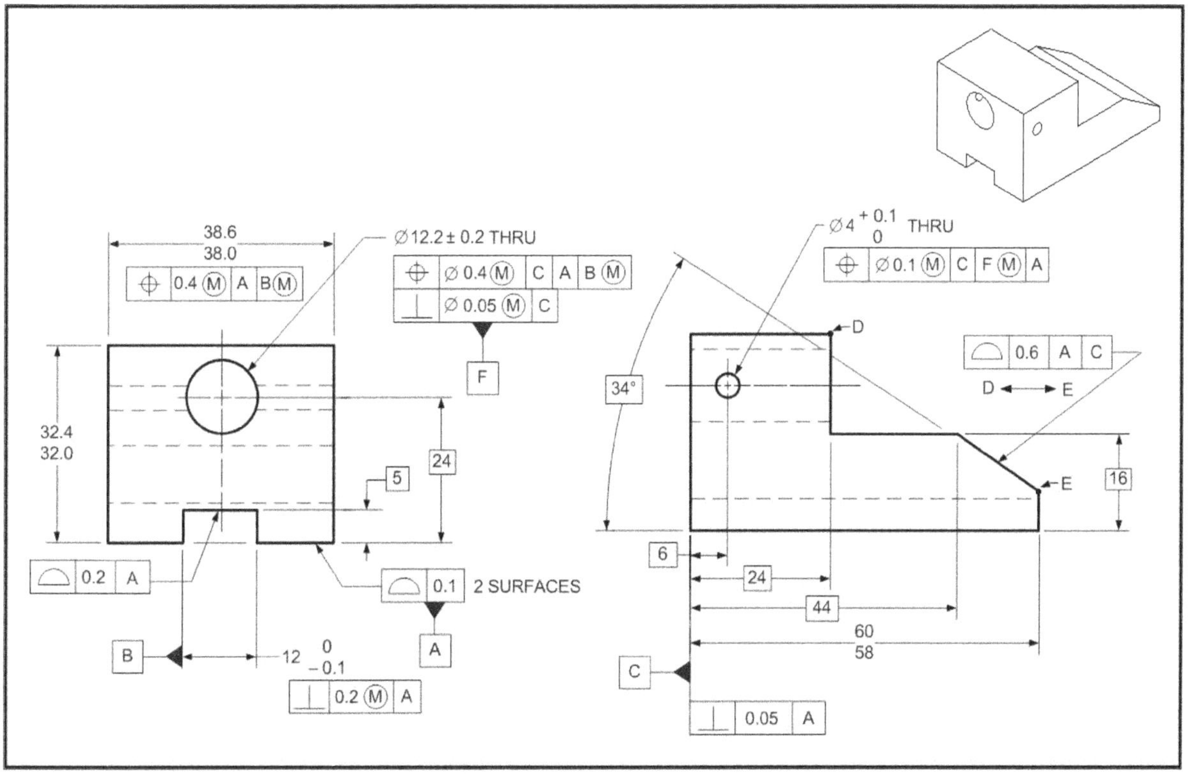

*Figure-61. Practical*

## Starting a Document

- Start Solid Edge by using Start menu or desktop icon, if not started yet.
- Click on the **ISO Metric Part** option from the **New** cascading menu of the **Application** menu. A new part file document will open.
- Select the **Ordered** radio button from the **Model** group in the **Tools** tab of the **Ribbon** to activate ordered mode.

## Creating Extrude Feature

- Click on the **Extrude** tool from the **Solids** group in the **Home** tab of **Ribbon**. The **Extrude Command Bar** will be displayed and you will be asked to select a sketching plane.
- Select the Front (XZ) plane and create the sketch as shown in Figure-62.
- Apply the dimension and then modify them as per the drawing to apply Basic dimension box; refer to Figure-63.

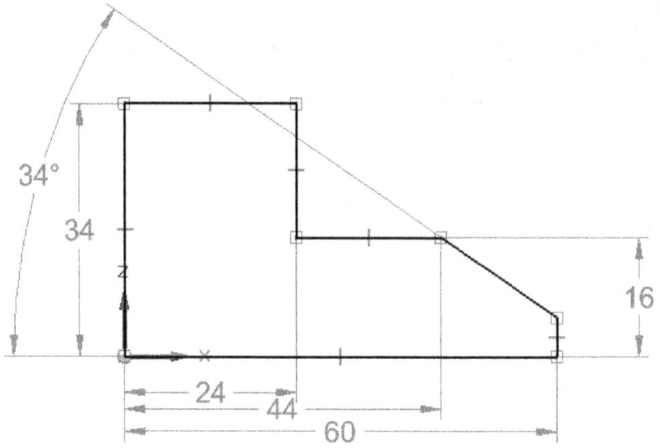

*Figure-62. Base sketch for Extrude*

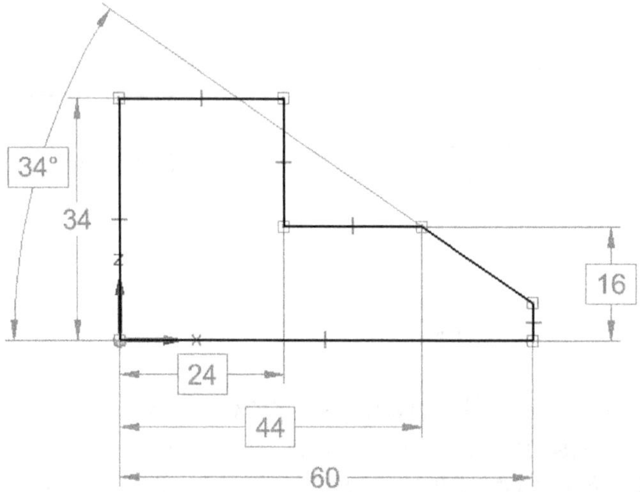

*Figure-63. Sketch after applying basic dimension symbols*

- After creating sketch, click on the **Close Sketch** tool from the **Close** group in the **Home** tab of **Ribbon**. You will be asked to specify extent of extrude feature.
- Specify the value **38** in the **Distance** edit box of **Command Bar** and press **ENTER**. You will be asked to specify the side on extrude feature will be created with respect to sketching plane.
- Click behind the sketching plane to specify side of feature and click on the **Finish** button. The feature will be created.

## Creating Extrude Cut Feature

- Click on the **Cut** tool from the **Solids** group in the **Home** tab of **Ribbon**. The **Cut Command Bar** will be displayed and you will be asked to select a sketching plane.
- Select the back face of model as shown in Figure-64 and orient the sketching plane parallel to screen.

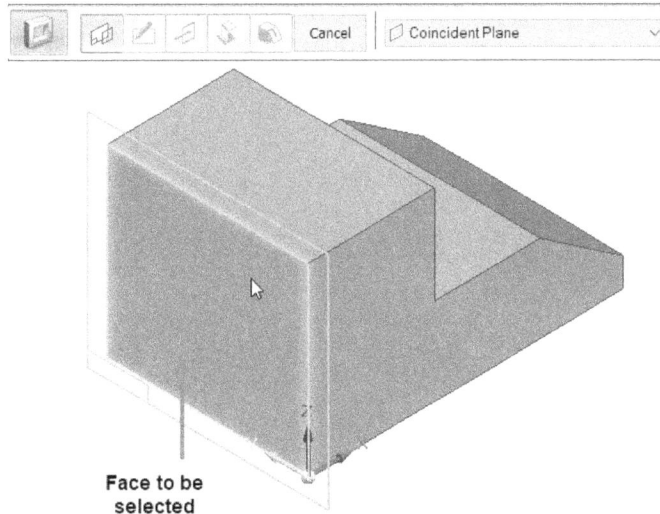

*Figure-64. Face selected for cut feature*

• Create the sketch for cut as shown in Figure-65 and apply the dimensions.

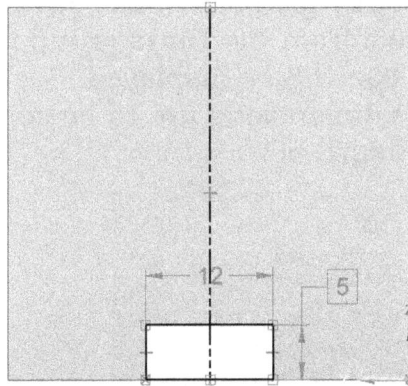

*Figure-65. Sketch for cut feature*

• Click on the **Close Sketch** button from the **Ribbon** to exit sketching environment. You will be asked to specify extent of cut feature.
• Change the orientation of model for clear view, select the **Through All** button from the **Command Bar**, and click on the desired side to create the feature. Preview of cut feature will be displayed.
• Click on the **Finish** button from the **Command Bar** to create the feature and press **ESC** to exit the tool.

## Creating Holes

• Click on the **Hole** tool from the **Hole** drop-down in the **Home** tab of **Ribbon**. The **Hole Command Bar** will be displayed and you will be asked to select a plane for creating hole.
• Select the back face of model as shown in Figure-66 and place the sketch circle of hole as shown in Figure-67.
• Click on the **Close Sketch** tool from the **Close** group in the **Home** tab of **Ribbon**. Preview of the hole will be displayed.

Figure-66. Face selected

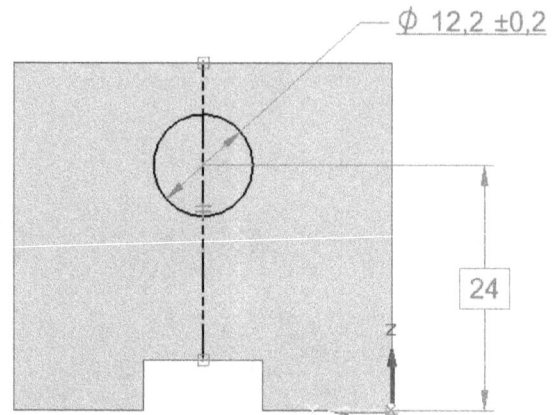

Figure-67. Sketch for hole

- By default, the holes will be created through. Click on the **Finish** button from the **Command Bar** to create the hole and press **ESC** to exit the tool.
- Similarly, create the hole of diameter **4** as per the drawing.

## Creating PMI Annotations

- Click on the **Copy to PMI** tool from the **Tools** group in the **PMI** tab of the **Ribbon**. The **Copy to PMI Command Bar** will be displayed.
- Select the model whose 2D dimensions are to be copied. Preview of dimensions will be displayed; refer to Figure-68.

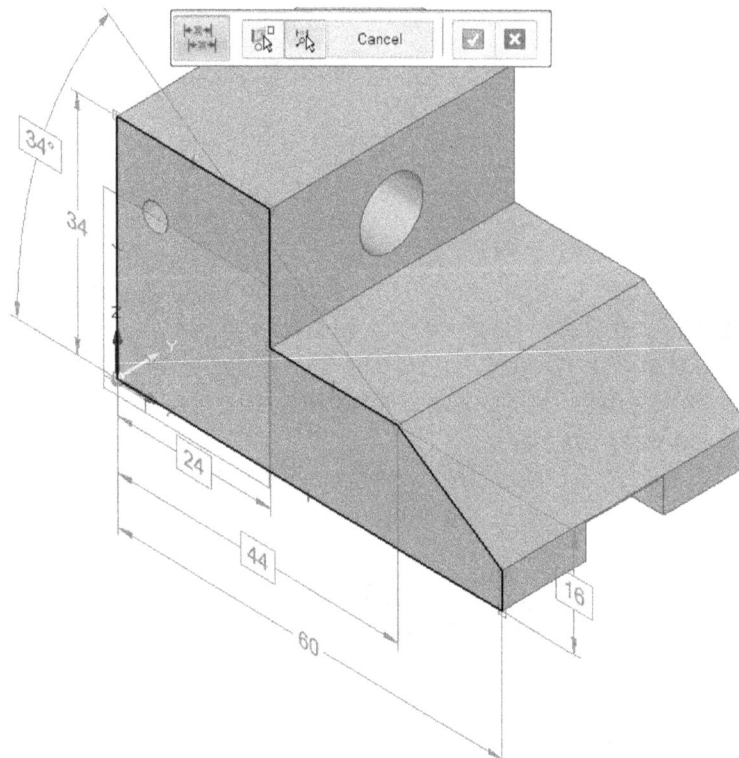

Figure-68. Dimensions displayed on selecting model

- Select all the dimensions to be copied and right-click in the drawing area.
- Repeat the steps to copy all the dimensions and then press **ESC** to exit the tool.
- Click on the **Datum Frame** tool from the **Annotation** group in the **PMI** tab of **Ribbon**. The **Datum Frame Command Bar** will be displayed.

- Click on the **Feature Control Frame** tool from the **Annotation** group in the **PMI** tab of **Ribbon** and place the control frames as shown in Figure-69 using left orientation of model.

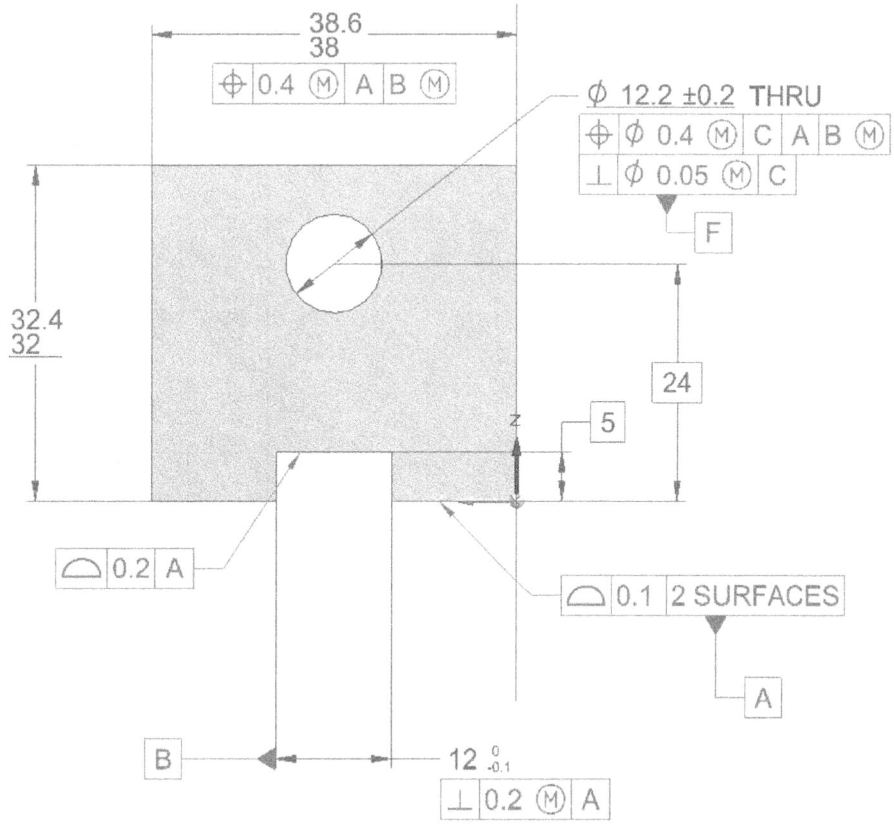

*Figure-69. Placing annotations in left orientation*

- Similarly, apply feature control frames and datum frames in Front orientation to the model; refer to Figure-70.

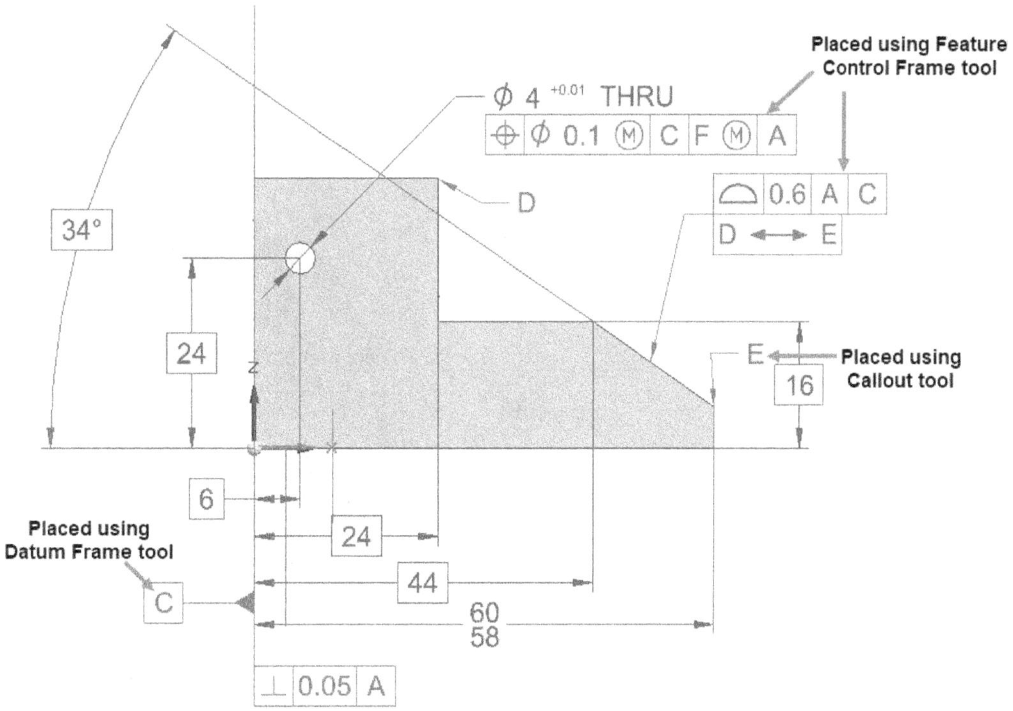

*Figure-70. Placing annotations in front orientation*

# PRACTICE 1

Create the model and apply annotations as per the drawing shown in Figure-71.

*Figure-71. Drawing for practice*

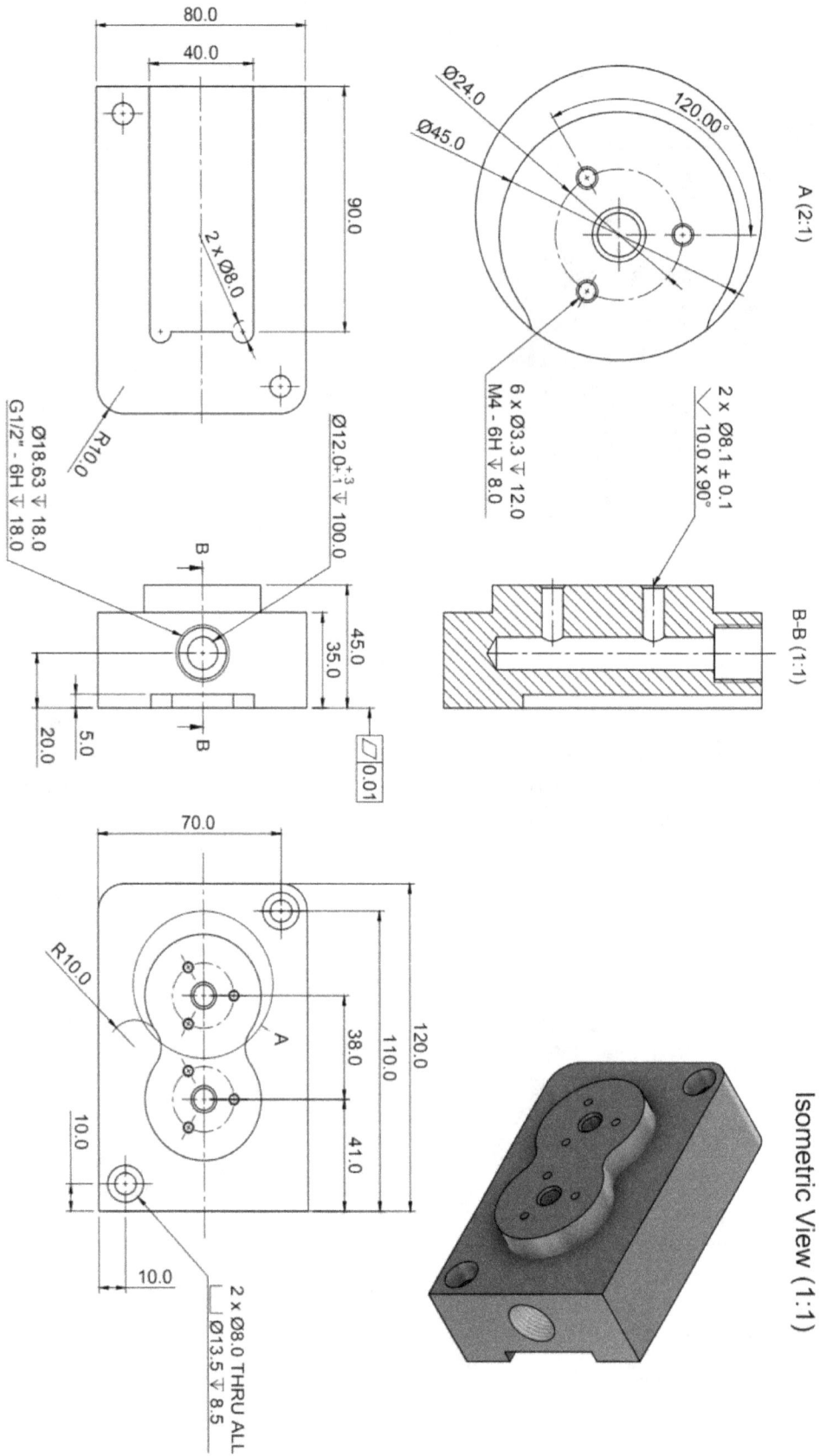

80.0

40.0

90.0

2 × Ø8.0

R10.0

Ø18.63 ⊽ 18.0
G1/2" - 6H ⊽ 18.0

Ø12.0 $^{+3}_{+1}$ ⊽ 100.0

Ø24.0

Ø45.0

120.00°

A (2:1)

6 × Ø3.3 ⊽ 12.0
M4 - 6H ⊽ 8.0

2 × Ø8.1 ± 0.1
⌵ 10.0 × 90°

B-B (1:1)

B

B

45.0

35.0

20.0

5.0

⬰ 0.01

70.0

R10.0

A

120.0

110.0

38.0

41.0

10.0

10.0

10.0

2 × Ø8.0 THRU ALL
⊔ Ø13.5 ⊽ 8.5

Isometric View (1:1)

# Chapter 8

# Assembly-I

## Topics Covered

The major topics covered in this chapter are:

- *Starting Assembly Document*
- *Inserting Component*
- *Assembly Constraint*
- *Modification Tools*
- *Applying Motors*
- *Simulation of Motors*
- *Pattern in Assembly*
- *Inspection Tools*
- *Practical*
- *Practice*

# ASSEMBLY

In Engineer's language, assembly is the combination of two or more components and these components are constrained to each other in a specified manner called assembly constraints. In Solid Edge, assembly design is a separate environment. To start the Assembly design, click on **New** button from the menu bar. The **New** dialog box will display; refer to Figure-1. Double-click on the **iso metric assembly.asm** template from the dialog box. The assembly design environment will be displayed as shown in Figure-2. The tools related to assembly are available in the **Ribbon**.

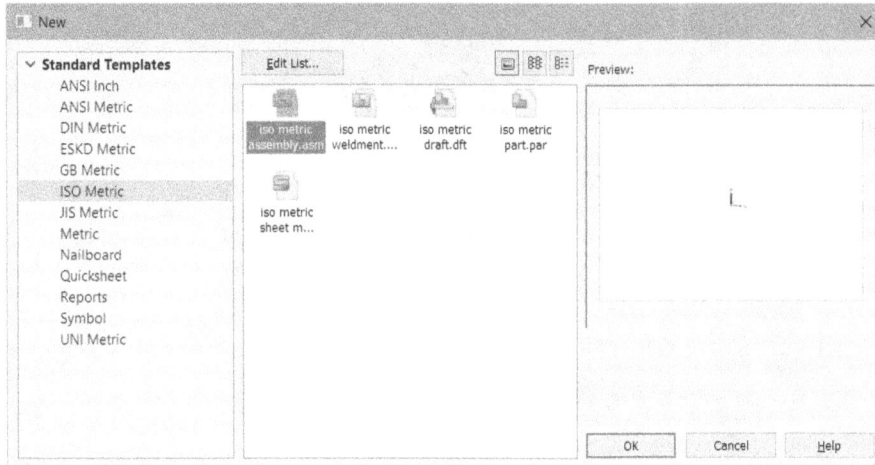

*Figure-1. New solid edge document dialog box*

*Figure-2. Assembly Design Environment*

# INSERTING GROUND COMPONENT

After starting Assembly environment, select the desired part file from the **Part Library** box which is located on the left side in application window; refer to Figure-3. If you want to fix the **Part Library** at the left in the application window then click on the **Auto Hide** (Pin) button at the top in the box; refer to Figure-3. Drag the desired part to drawing area, by default the first inserted part is grounded automatically. You can locate the newly inserted component in the **Path Finder**. When you hover the cursor on part name in the **Path Finder**, the part is highlighted in the drawing area; refer to Figure-4.

Figure-3. Part library

Figure-4. Part ground

# INSERTING COMPONENTS IN ASSEMBLY

Once, you have inserted the base component in the assembly, follow the steps given next to insert more components.

- Click on the **Insert Component** tool from the **Assemble** group in the **Home** tab of the **Ribbon**. The **Parts Library** box will be displayed.
- Browse to the location of desired part file from **Part Library** box, click on it, and drag to the drawing area. The **Assemble Command Bar** will be displayed; refer to Figure-5.

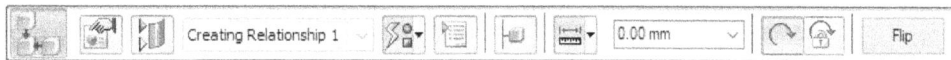

Figure-5. Assembly command bar

- Click on the **Occurrence Properties** button from the **Command Bar** to manage occurrences of inserted components. The **Occurrence Properties** dialog box will be displayed; refer to Figure-6. Set the desired parameters in the dialog box like quantity of component and other properties to be added in the assembly. After setting desired parameters, click on the **OK** button from the dialog box.

Figure-6. Occurrence Properties dialog box

- Click on the **Construction Display** button from the **Command Bar** and select the entities that you want to display in assembly; refer to Figure-7.

*Figure-7. Construction Display options*

- Select the desired option from the **Relationship Types** drop-down to define constraint applied between base component and inserted new component. Various constraints will be discussed later.

- Click on the **Options** button from the **Command Bar**. The **Options** dialog box will be displayed; refer to Figure-8.

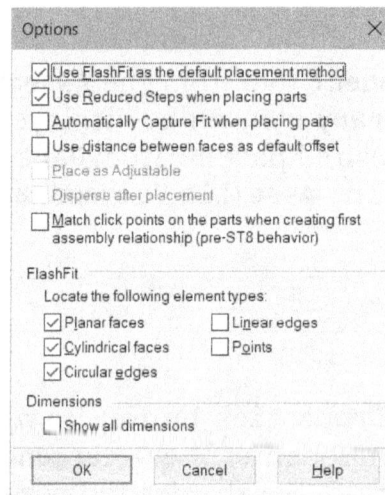

*Figure-8. Options dialog box*

- Select the desired check boxes from the dialog box to define how components will be inserted in the assembly.

  - Select the **Use FlashFit as the default placement method** check box to automatically apply FlashFit constraint when a new component is inserted in the assembly.

  - Select the **Use Reduced Steps when placing parts** check box to place parts with minimum number of steps needed like selecting the part and then applying placement constraints.

  - Select the **Automatically Capture Fit when placing parts** check box to automatically fit holes, shafts, fasteners, and so on which need to be repeatedly inserted in the assembly.

  - Select the **Use distance between faces as default offset** check box to automatically apply offset distance constraint when two faces are selected from assembly.

  - Select the **Place as Adjustable** check box to make the inserted sub-assembly adjustable.

- Select the **Disperse after placement** check box to add inserted sub-assemblies as individual parts.
- Select the **Match click points on the parts when creating first assembly relationship (pre-ST8 behavior)** check box to use match behavior used in Solid Edge ST8 and previous versions.
- Select the check boxes from the **FlashFit** area to define which elements will be used for placing component when FlashFit constraint is used.
- Select the **Show all dimensions** check box from the **Dimensions** area of dialog box to display dimensions when placing components.
- After setting desired parameters in dialog box, click on the **OK** button. Various constraints used in assembly are discussed next.

# ASSEMBLY CONSTRAINT

The assembly constraints are used to combine various parts in a single assembly. Click on the **Assemble** button from the **Assemble** group or insert the component, the **Assemble Command Bar** will be displayed. Or, you can use the respective tool from the **Assemble** group in the **Ribbon**. These constraints are discussed next.

## FlashFit

The **FlashFit** tool is active by default when you insert a new component. Using this tool, you can automatically apply mate, planar align, axial align, and connect constraints based on selected entities.

## Mate

The **Mate** tool is used to mate two parts by making their faces/planes coincident. The procedure to use this mate is given next.

- After inserting the desired components, click on the **Mate** tool from the **Assemble** group in the **Ribbon**. The **Mate Command Bar** will be displayed as shown in Figure-9.

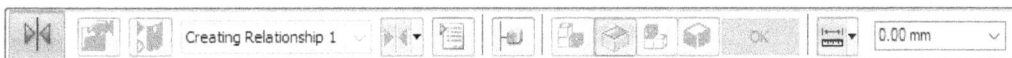

*Figure-9. Mate command bar*

- Select the face of inserted part. You will be asked to select face of target part.
- Select the desired face of target part. The constraint will be applied; refer to Figure-10.

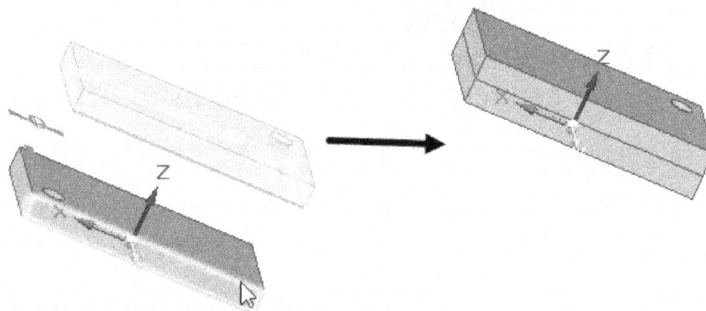

*Figure-10. Mate*

- By default, the **Fixed** option is selected in the **Offset Type** drop-down of the **Command Bar** and specify offset distance in the **Offset Value** edit box to define gap between selected faces.
- Select the **Float** option from the **Offset Type** drop-down in the **Command Bar** to allow free lateral movement of inserted part.
- Select the **Range** option from the **Offset Type** drop-down in the **Command Bar** to define minimum and maximum gap allowed between selected faces.
- Click on the **Lock Rotation** button from the **Command Bar** if you want to stop rotation of inserted part in the assembly.
- After setting the desired parameters, right-click in the drawing area. The constraint will be applied.

## Planar Align

The **Planar Align** tool is used to make faces of inserted part and base part aligned to each other. The procedure to use this tool is discuss next.

- Click on the **Planar Align** tool from the **Assemble** group in the **Home** tab of the **Ribbon**. The **Planar Align Command Bar** will be displayed; refer to Figure-11.

*Figure-11. Planar align*

- One by one select the faces of base part and inserted part. The selected faces will be aligned; refer to Figure-12. Right-click in the drawing area to create the constraint.

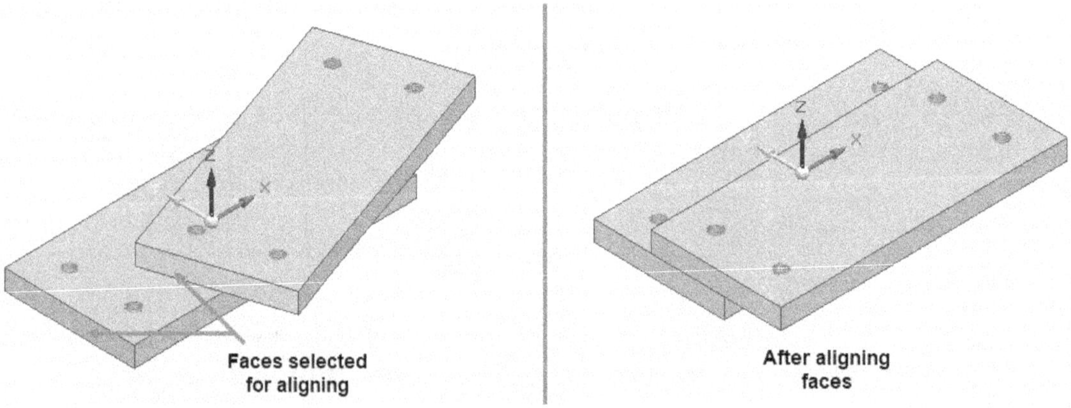

Faces selected
for aligning

After aligning
faces

*Figure-12. Planar align face*

## Axial Align

The **Axial Align** tool is used to align two cylindrical axes, a cylindrical axes and a linear element, or two linear elements by using their axes. The procedure to use this tool is given next.

- Click on the **Axial Align** tool from the **Assemble** group of **Home** tab in the **Ribbon**. The **Axial Align Command Bar** will be displayed; refer to Figure-13.

*Figure-13. Axial Align Command Bar*

- Select the axes of base component and inserted component one by one. The constraint will be applied; refer to Figure-14.
- You can lock and unlock rotation of components from the **Command Bar** as discussed earlier.

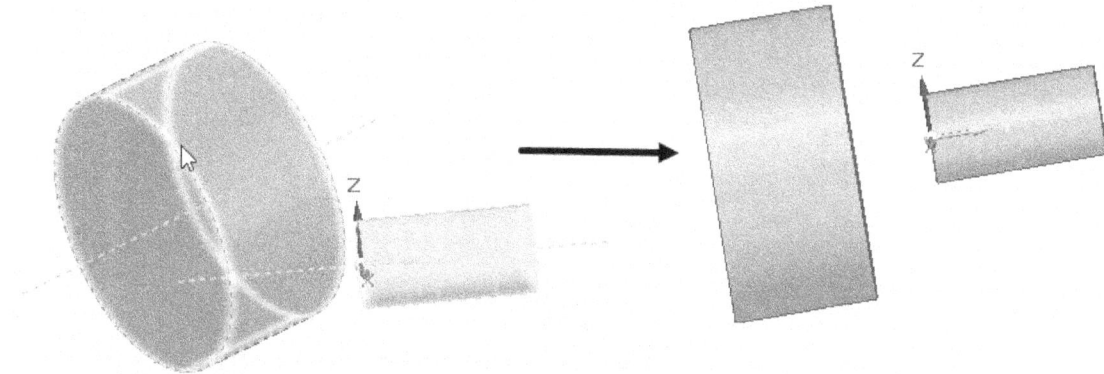

*Figure-14. Axial Align*

## Insert

The **Insert** tool is used to apply insert relationship to place axial-symmetric parts such as nuts and bolts into holes or onto cylindrical protrusions. The procedure to apply insert constraint is given next.

- Click on the **Insert** tool from the **Assemble** group in the **Home** tab of the **Ribbon**. The **Insert Command Bar** will be displayed; refer to Figure-15.

*Figure-15. Insert Command Bar*

- Select cylindrical surface of first entity and circular axis of second entity to align the axes.
- Select the face of first entity and face of second entity to make faces coincident. You can also specify offset value to put gap between selected faces; refer to Figure-16.

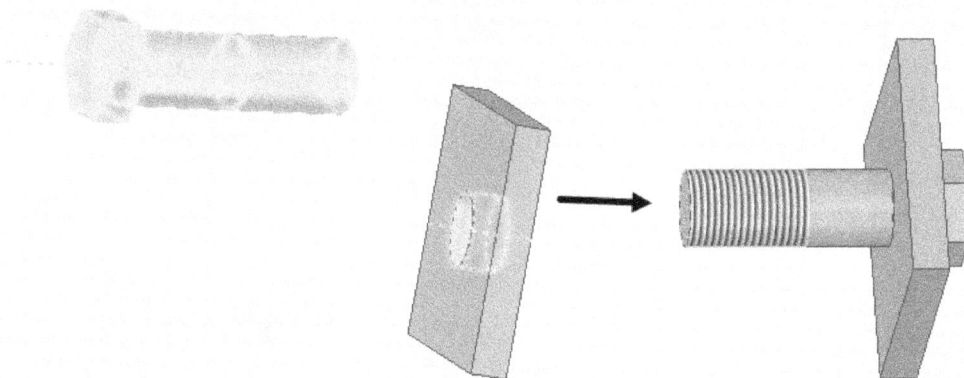

*Figure-16. Insert*

## Connect

The **Connect** tool is used to connect a keypoint of one part to another keypoint, line or plane of another part. The procedure to apply the constraint is discuss next.

- Click on the **Connect** tool from the **Assemble** group in the **Home** tab in the **Ribbon**. The **Connect Command Bar** will be displayed; refer to Figure-17.

*Figure-17. Connect Command Bar*

- Select the key points of the two components. The selected points will become coincident; refer to Figure-18.
- You can also specify offset value from the **Command Bar**.

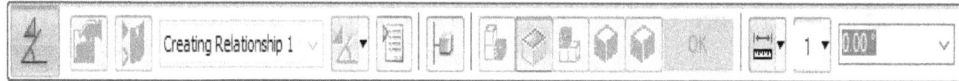

*Figure-18. Connect Constraint*

## Angle

The **Angle** tool is used to constrain angle between faces of two components. The procedure to apply this constraint is given next.

- Click on the **Angle** tool from the **Assemble** group in the **Home** tab of the **Ribbon**. The **Angle Command Bar** will be displayed as shown in Figure-19.

*Figure-19. Angle Command bar*

- Select the desired angle format from the **Angle Format** drop-down and specify the angle value in **Angle Value** edit box.
- Select the faces of first and second entity. You will be asked to select the plane along which angle will be measured.
- Select the desired plane along which angle measurement will be performed; refer to Figure-20. Right-click in the drawing area to apply constraint.

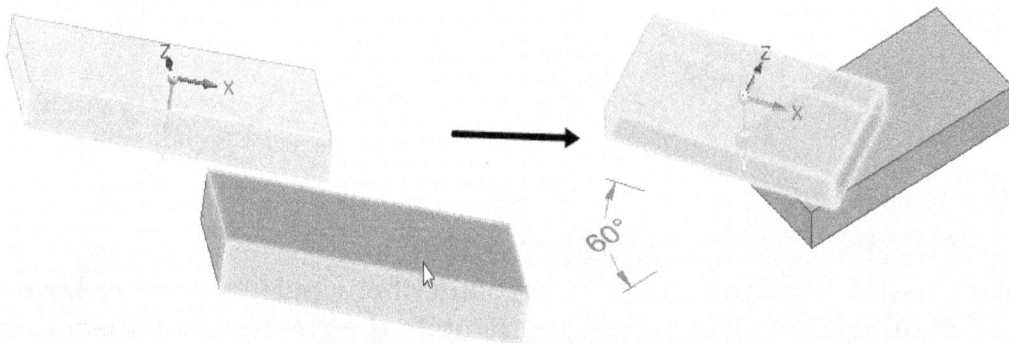

*Figure-20. Angle*

## Tangent ᐅᗶ

The **Tangent** tool is used to make two components tangent to each other. The procedure to use this tool is given next.

- Click on the **Tangent** tool from the **Assemble** group in the **Home** tab of the **Ribbon**. Select two faces that you want to make tangent. The preview of tangent constraint will be displayed; refer to Figure-21.
- Right-click in the drawing area. The constraint will be applied.

Cylinder (Protrusion 1) (Cylinder1.par:1)

*Figure-21. Tangent Constraint*

## Path ⁺

The **Path** tool is used to move part along the selected path. The procedure to use this tool is given next.

- Click on the **Path** tool from the **Path** drop-down in the **Assemble** group of the **Home** tab in the **Ribbon**. The **Path Command Bar** will be displayed; refer to Figure-22.

*Figure-22. Path Command Bar*

- Note that to use this constraint, there must be a path on the target plane so that another part can follow that path. For that you need to draw path by using sketching tools in Assembly environment. You can also use an edge of model as path for creating the constraint.
- Select the desired keypoint of inserted part and an edge or sketch path; refer to Figure-23.

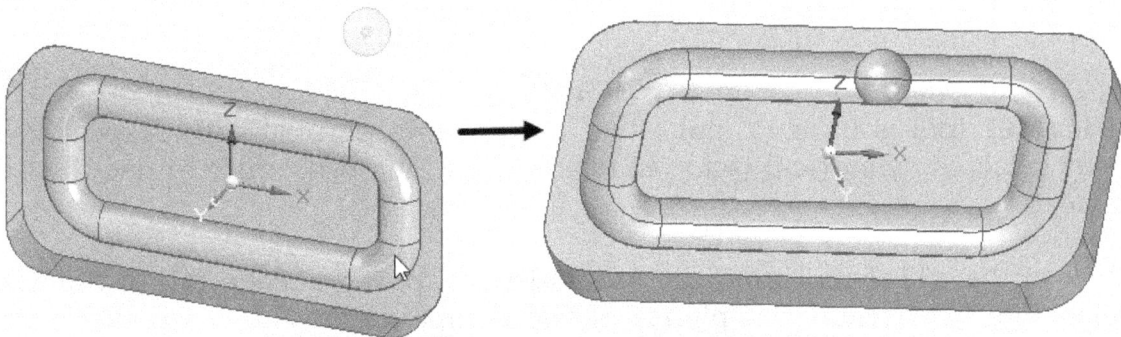

*Figure-23. Creating path constraint*

## Cam

The **Cam** tool is used to create cam-follower mate between two entities. The procedure to apply cam constraint is given next.

- Click on the **Cam** tool from the **Path** drop-down in the **Assemble** group of **Home** tab in the **Ribbon**. The **Cam Command Bar** will be displayed; refer to Figure-24.

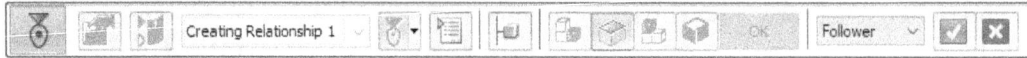

*Figure-24. Cam Command Bar*

- Select a face or point of follower which will run over cam. You will be asked to select the face chain of cam.
- Select the face chain of cam and click on the **OK** button to accept the selection.
- Right-click in the drawing area to create the constraint. Figure-25 shows an example of cam mate.

Note that you need to limit the motion of follower so that it moves only in vertical direction and do not deviate from cam path.

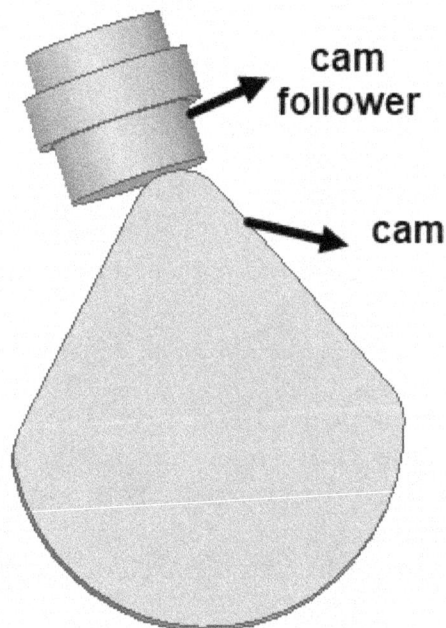

*Figure-25. Cam*

## Parallel

The **Parallel** tool is used to make two components parallel with respect to each other by selected cylindrical faces, edges, or axes. The procedure to use this tool is given next.

- Click on **Parallel** tool from the **Assemble** group in the **Home** tab of the **Ribbon**. Select the two faces/axes/planes of two components. The components will be parallel with respect to these references; refer to Figure-26.

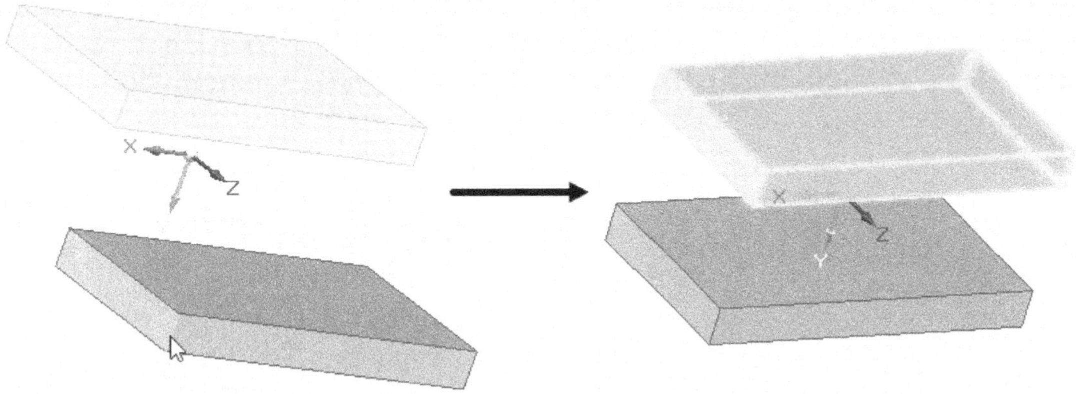

*Figure-26. Parallel Constraint*

# Gear 🗘

The **Gear** tool is used to create a joint between two gears with specified gear ratio. The procedure to use this tool is given next.

- Click on the **Gear** tool from the **Assemble** group in the **Home** tab of the **Ribbon**. The **Gear Command Bar** will be displayed; refer to Figure-27.

*Figure-27. Gear Command Bar*

- Select the cylindrical faces of two mating gears. Note that by default, the **Rotation-Rotation** option is selected in the **Gear Type** drop-down of **Command Bar** and hence two cylindrical gears are assembled. Select the **Rotation-Linear** option from the **Gear Type** drop-down of **Command Bar** if you are assembling a cylindrical gear and rack then select the cylindrical face of gear and linear edge of rack; refer to Figure-28. Select the **Linear-Linear** option from the drop-down if you want to assemble two linear gears and select the linear edges of two gears.
- Specify the desired values of gear ratio and click on the **OK** button from the **Command Bar** to accept the selection.
- Right-click in the drawing area to create the constraint.

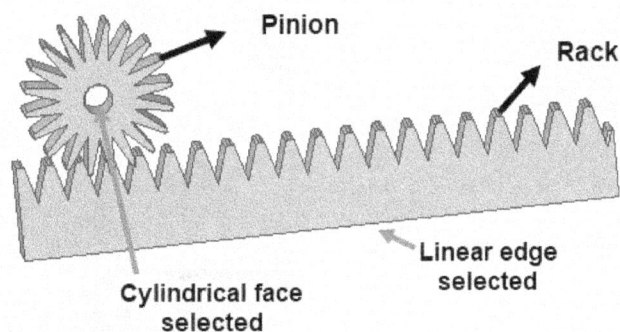

*Figure-28. Gear*

# Match Coordinate Systems ⮑

The **Match Coordinate Systems** tool is used to mate two component with their coordinate system. You can use the default coordinate systems of parts as well as user defined coordinates for applying this mate. If coordinate system of active component is not displayed by default then select the part from **Path Finder** and right-click on it. A shortcut menu will be displayed; refer to Figure-29. Click on the **Show/Hide**

**Component** option from the shortcut menu. The **Show/Hide Component** dialog box will be displayed as shown Figure-30. Select the **On** check box for **Coordinate Systems** from the dialog box and click on the **OK** button. The coordinate system will be displayed. The procedure to apply Match Coordinate Systems mate is discuss next.

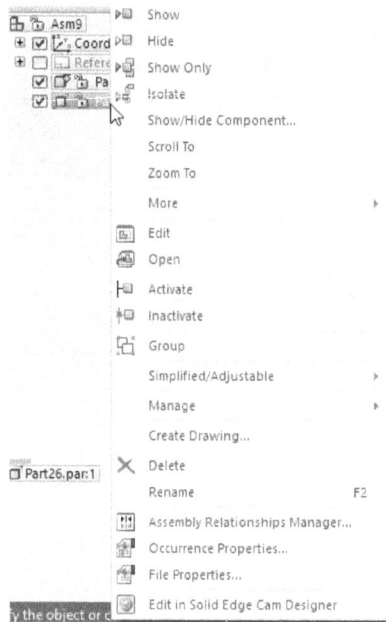

*Figure-29. Cascade menu*

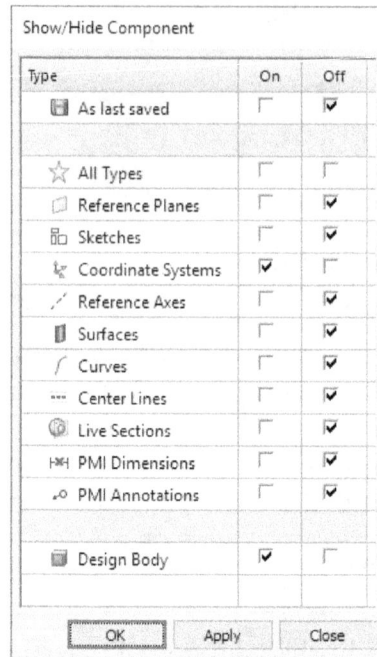

*Figure-30. Show hide component*

- Click on the **Match Coordinate Systems** tool from the **Assemble** group in the **Home** tab of the **Ribbon**. The **Match Coordinate Systems Command Bar** will be displayed; refer to Figure-31.

*Figure-31. Match Coordinate Systems Command Bar*

- Select the coordinate system of first and second component. The selected coordinate systems will become coincident; refer to Figure-32

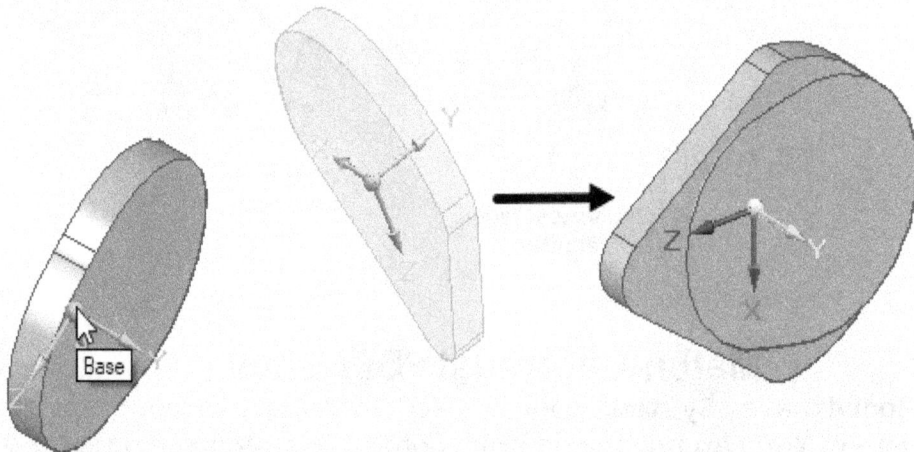

*Figure-32. Match Coordinate System*

## Centre-Plane ◄◊►

The **Center-Plane** tool is used to place a component at the center of another component. The procedure to use this tool is given next.

- Click on the **Center-Plane** tool from the **Assemble** group in the **Home** tab of the **Ribbon**. The **Center-Plane Command Bar** will be displayed and you will be asked to select the faces of components.
- Select the start and end faces of the component which is to be centered.
- Select the faces of another component to be used as boundary for component; refer to Figure-33. The component will be placed.
- Right-click in the drawing area to apply mate.

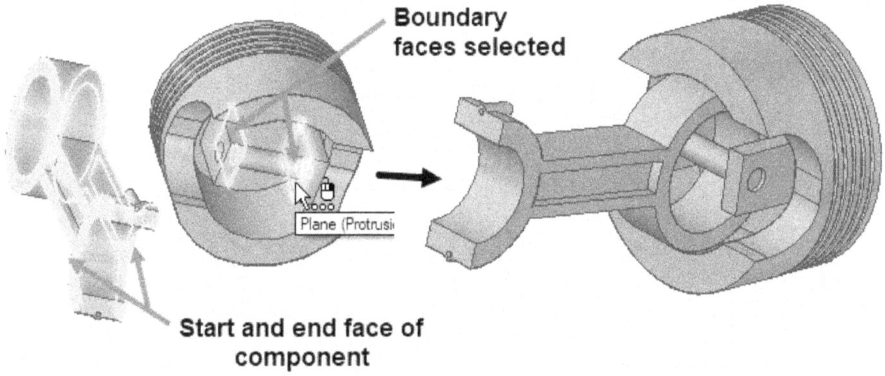

*Figure-33. Applying Center-Plane mate*

## Rigid Set ◈

The **Rigid Set** tool is used to make selected assembly components rigid or fix with respect to each other, so if we move one component then other will also move. The procedure to use this tool is given next.

- Click on the **Rigid Set** tool from the **Assemble** group in the **Home** tab of the **Ribbon**. The **Rigid Set Command Bar** will be displayed; refer to Figure-34.

*Figure-34. Rigid Set Command Bar*

- Select the desired option from the **Shared relationships** drop-down to specify whether to delete, suppress, or ignore the applied constraints when applying rigid set.
- Select the faces of components that you want to make rigid set and right-click in the drawing area. Press **ESC** to exit the tool.

## Assembly Relationships Manager ⊞

The **Assembly Relationships Manager** tool provides the list of assembly relationships applied between various components of assembly and allow you to modify/repair the relationships. The procedure to use this tool is given next.

- Click on the **Assembly Relationships Manager** tool from the **Assemble** group in the **Home** tab of the **Ribbon**. The **Assembly Relationships** dialog box will be displayed; refer to Figure-35. In this dialog box, you can locate all the relationships applied in the assembly.
- Right-click on the relationship you want to modify and select the **Edit Definition** option from the shortcut menu. The **Command Bar** will be displayed with options related to modifying the constraint.
- You can also use the **Suppress** and **Unsuppress** options in the shortcut menu to suppress or activate the selected constraint.
- After setting desired parameters, click on the **Close** button to close the dialog box.

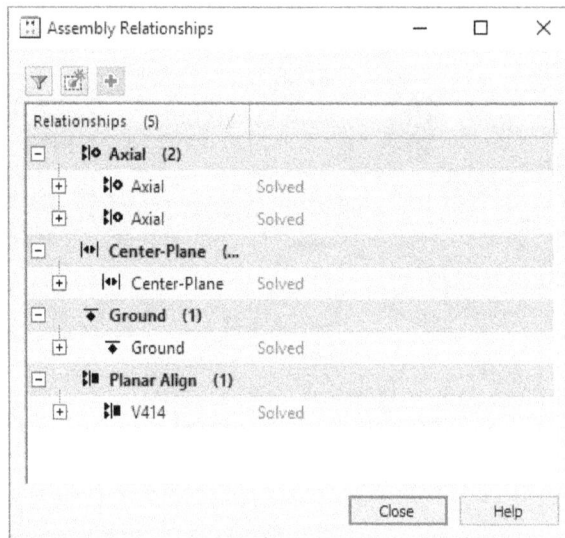

*Figure-35. Assembly Relationships Manager*

## Assembly Relationships Assistant

The **Assembly Relationship Assistant** is used to apply assembly relationships between selected parts and subassemblies based on their current geometric orientation. This command is useful when working with assemblies whose parts are oriented correctly, but do not have assembly relationships, such as assemblies that were imported into Solid Edge from another CAD system. The procedure to use this tool is given next.

- Click on the **Assembly Relationship Assistant** tool from the **Assemble** group in the **Home** tab of the **Ribbon**. The **Relationship Assistant Options** dialog box will be displayed; refer to Figure-36.

*Figure-36. Relationship Assistant Options dialog box*

- Select the desired radio button from the **Check Select Set1 Against** area to define scope of components to be checked for applying constraints automatically.
- Select the **Remove relationships on parts in Select Set1** check box to remove existing relationships when applying new one. By default, the **Remove grounded part relationships** radio button is selected which deletes the ground relationship automatically. Select the **Remove all relationships** radio button to remove all existing relationships of selected components when applying new one.
- Similarly, set the other parameters and click on the **OK** button. The **Assembly Relationship Assistant Command Bar** will be displayed; refer to Figure-37.

*Figure-37. Assembly Relationship Assistant Command Bar*

- Select the components of first set and right-click in the drawing area. You will be asked to select components of second set.
- Select the components of second set and right-click in the drawing area. The **Relationship Assistant Settings** dialog box will be displayed; refer to Figure-38.

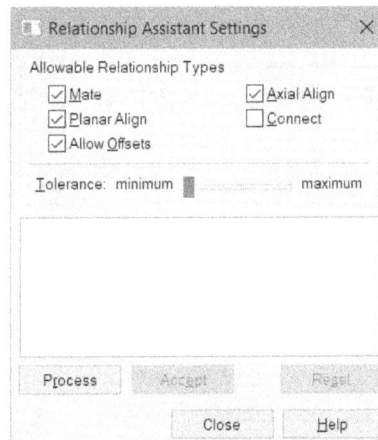

*Figure-38. Relationship Assistant Settings dialog box*

- Select the check boxes for relationships you want to apply and click on the **Process** button. The constraints that can be applied will be displayed in the list box; refer to Figure-39.

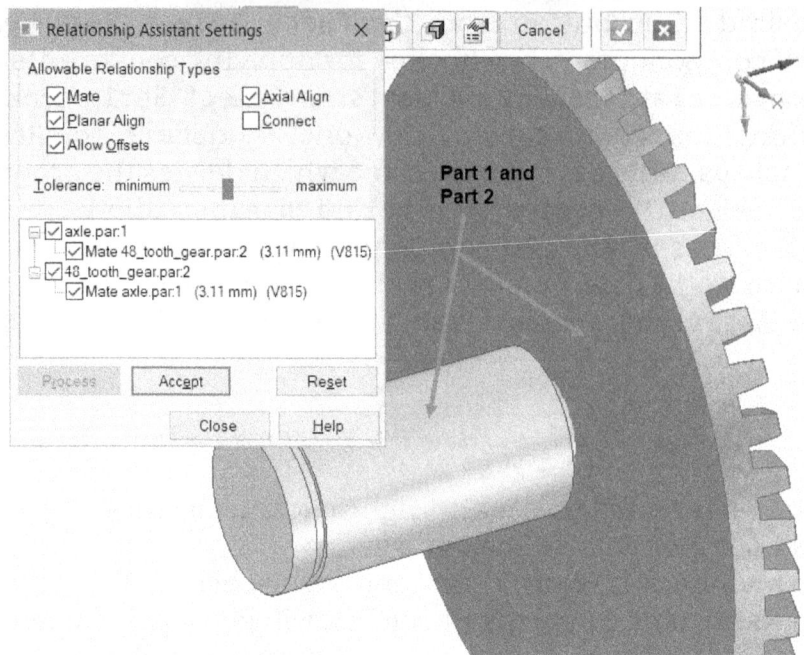

*Figure-39. List of available relationships*

- Click on the **Accept** button from the dialog box to apply new relationships.
- Click on the **Close** button from the dialog box to exit the tool.

# CAPTURE FIT

The **Capture Fit** tool is used to copy already existing relationships of a component so that when next time you insert the same component, the relationships will be applied automatically. The procedure to use this tool is given next.

- Select the assembly component whose relationships are to be copied and click on the **Capture Fit** tool from the **Assemble** group in the **Home** tab of the **Ribbon**. The **Capture Fit** dialog box will be displayed; refer to Figure-40.

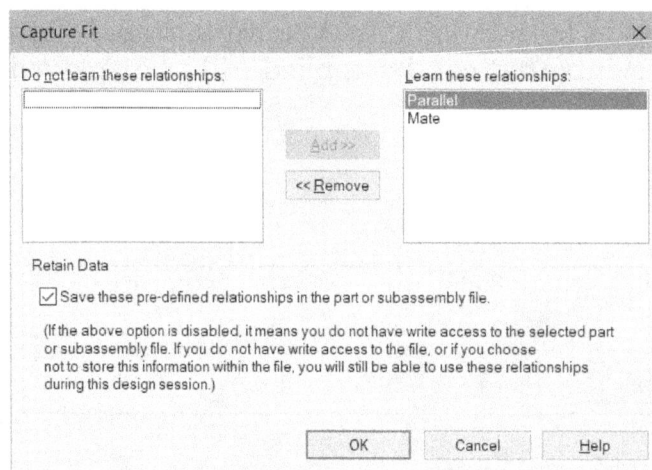

*Figure-40. Capture Fit dialog box*

- Select the relationships that you do not want to copy from the **Learn these relationships** area of the dialog box and click on the **<< Remove** button.
- Similarly, select the relationships that you want to copy from the **Do not learn these relationships** area of the dialog box and click on the **Add >>** button.

- After setting desired parameters, click on the **OK** button from the dialog box to make the software learn which relationships will be applied automatically when next time this component is inserted. Note that when next time you will insert the same component, you will be asked to select the similar reference entities on other base component as you specified while inserting the original component.

# MOVE ON SELECT

The **Move on Select** tool is used to allow moving the component in the specific direction after selecting it. When a component is selected a steering wheel appears on it; refer to Figure-41. Select the arrow of direction from steering wheel in which you want to move or rotate the component.

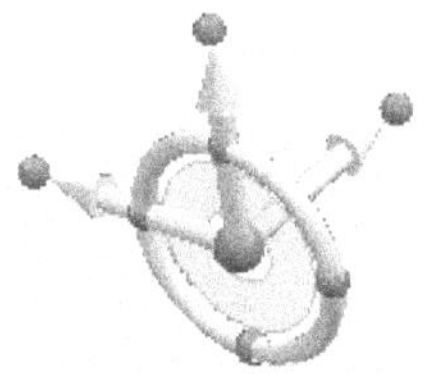

*Figure-41. Steering Wheel*

# DRAG COMPONENT

The **Drag Component** tool is used to drag component which is under constraint or grounded. The step of using this tool discuss next.

- Click on the **Drag Component** tool from the **Drag Component** drop-down in the **Modify** group of **Home** tab in the **Ribbon**. The **Analysis Options** dialog box will be displayed; refer to Figure-42.
- Select the **Locate grounded components** check box to select only the components and subassemblies which are grounded.
- Select the **Active parts only** radio button to drag only displayed parts. Select the **Both Active and Inactive parts** radio button to allow dragging both active and inactive parts of assembly.
- Select the desired radio button from the **Collision Options** area to define the scope for checking collision.
- Similarly, set the other parameters and click on the **OK** button. The **Drag Component Command Bar** will be displayed as shown in Figure-43.

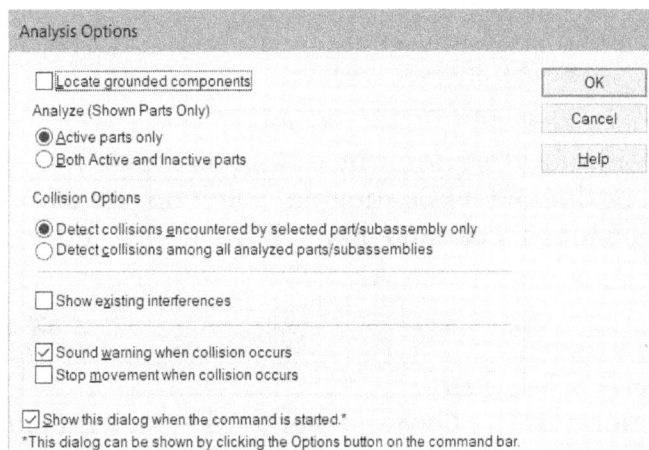

*Figure-42. Analysis Options dialog box*

*Figure-43. Drag Component*

- Select the desired button from the **Command Bar** to define motion of components. Select the **Move** button from the **Command Bar** to move components in linear directions. Select the **Rotate** button from the **Command Bar** to rotate the components about selected edge/axis. Select the **Freeform Move** button from the **Command Bar** to freely move the selected components in any direction by dragging.

- Select the desired option from the **Motion Analysis** drop-down to define how motion of components will be analyzed while dragging.

- Select object and drag component by holding Left Mouse Button; refer to Figure-44. Press **ESC** to exit the tool.

*Figure-44. Drag Component under constraint*

## MOVE COMPONENT

The **Move Components** tool is used to move or copy selected components. The procedure to use this tool is given next.

- Click on the **Move Components** tool from **Drag Component** drop-down in the **Modify** group of **Home** tab in the **Ribbon**. The **Move Options** dialog box will be displayed; refer to Figure-45.

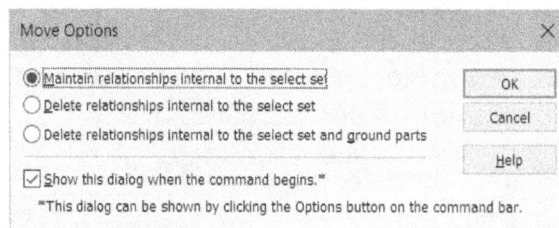

*Figure-45. Move Options dialog box*

- Select the desired radio button from the dialog box to define whether relationships will be maintained or deleted while moving. Click on the **OK** button from the dialog box. The **Move Components Command Bar** will be displayed; refer to Figure-46.

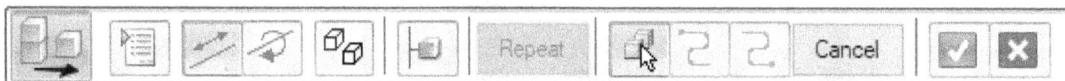

*Figure-46. Move Components Command Bar*

- Set the desired options in the **Command Bar** and select the components you want to move, rotate, or copy.

- Click on the **OK** button from the **Command Bar**. Depending on button selected in the **Command Bar** earlier, the options to specify linear or rotational movement will be displayed. (These buttons are discussed in next topic).
- Click on the desired point to specify starting reference point for the model. You will be asked to point where you want to move the selected component.
- Specify the coordinate values in the edit boxes of **Command Bar** or click at desired location.
- Press **ESC** to exit the tool.

## Linear Move

The **Move Components Option** button is used to move the component in specified direction or path. By default, this button is active in the **Move Components Command Bar** on activating the tool. The step to use this option are given next.

- Select the component(s) and click right-click in the drawing area to accept selection. The options to specify start point will be displayed; refer to Figure-47.

*Figure-47. Options for reference point of movement*

- Select the desired keypoint or specify the coordinates in **Command Bar** and press **ENTER**. You will be asked to specify the location where you want to move the components.
- Click at desired keypoint or enter the coordinates in the **Command Bar**; refer to Figure-48. Press **ESC** to exit the tool.

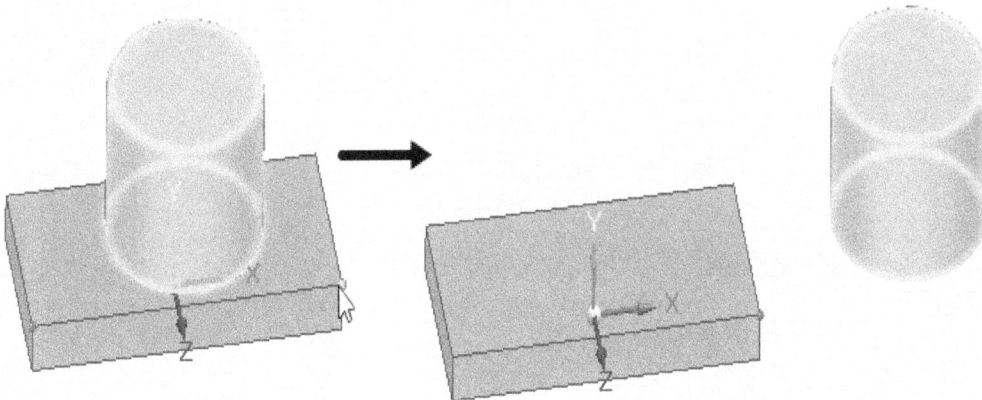

*Figure-48. Move component by pat*

## Rotational Move

The **Rotate Components Option** button is used to move component(s) along selected axis of rotation. Click on the **Rotate Components Option** button in the **Move Components Command Bar** to activate rotation mode. The steps to use this option are given next.

- Select the components that you want to rotate and right-click in the drawing area to accept selection. You will be asked to specify axis of rotation.
- Select the desired edge or axis from the model about which you want to rotate the selected components. You will be asked to specify the rotation angle.

- Specify the desired rotation angle value in the edit box of **Command Bar** and press **ENTER**. The components will rotate accordingly; refer to Figure-49. Press ESC to exit the tool.

*Figure-49. Rotating objects using Move tool*

## REPLACE PART

The tools in the **Replace Part** drop-down are used to replace existing part with another one; refer to Figure-50. Various tools in this drop-down are discussed next.

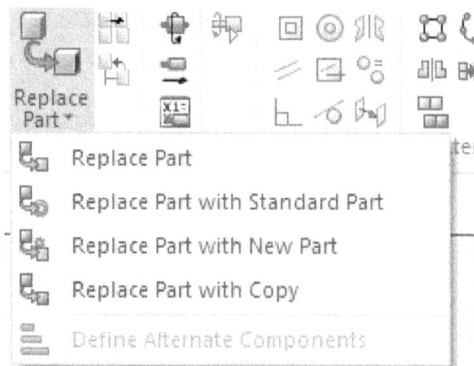

*Figure-50. Replace part drop down*

## Replace Part

The **Replace Part** tool is used to replace selected part using an existing part. The procedure to use this tool is given next.

- Click on the **Replace Part** tool from the **Replace Part** drop-down in the **Modify** group of **Home** tab in the **Ribbon**. The **Replace Part Command Bar** will be displayed; refer to Figure-51.

*Figure-51. Replace Part Command Bar*

- Select part of assembly which you want replace and right-click in the drawing area to accept the selection. The **Replacement Part** dialog box will be displayed; refer to Figure-52.
- Select the desired part from the dialog box and click on the **Open** button. The part will be replaced by new one; refer to Figure-53.

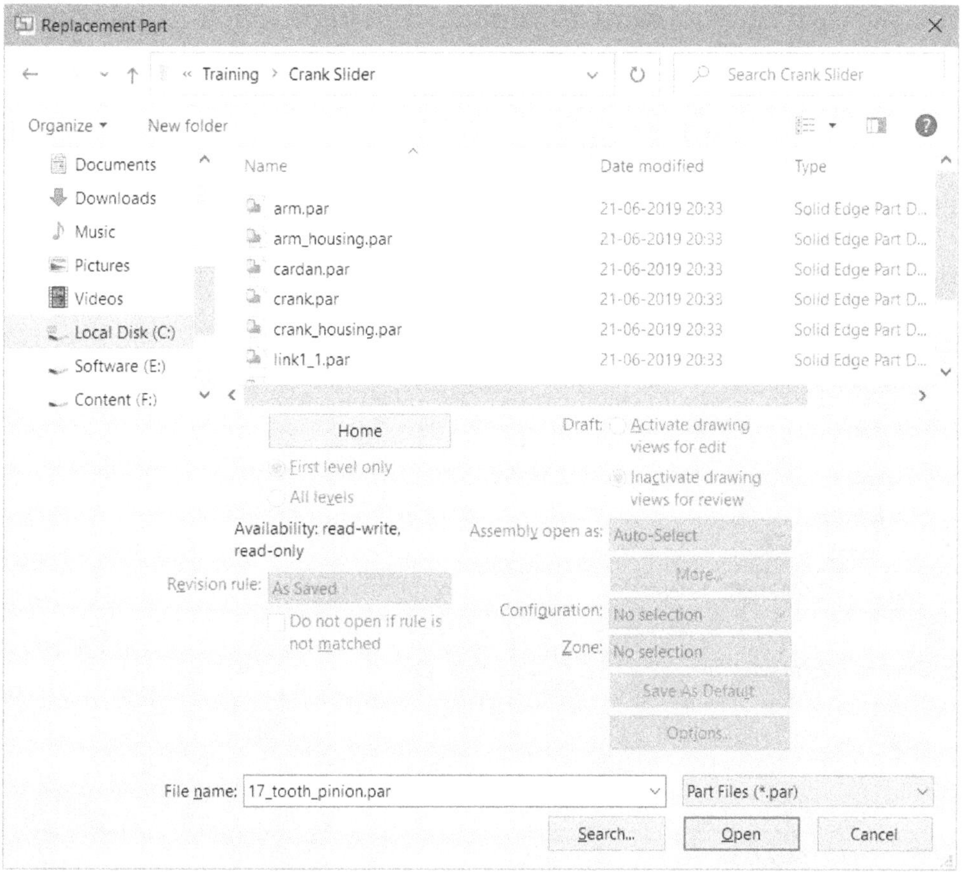

*Figure-52. Replacement part dialog box*

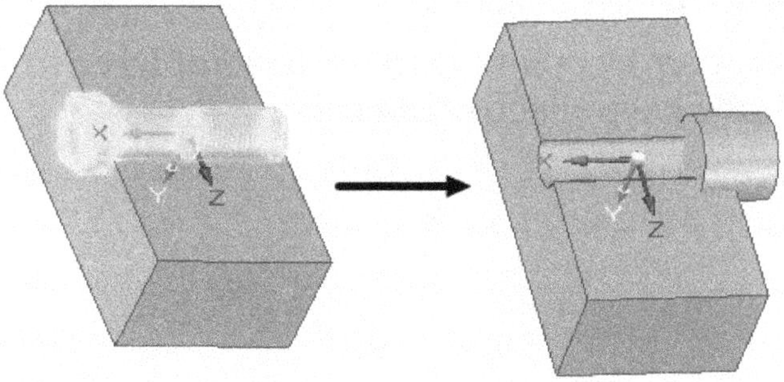

*Figure-53. Replacing part*

# Replace Part With Standard Part

The **Replace Part with Standard Part** tool is used to replace an existing part with a standard part from library. This tool is also available in the **Replace Part** drop-down. The procedure to use this tool is same as discussed for previous tool.

# Replace Part with New Part

The **Replace Part with New Part** tool is used to replace existing part by creating new part in the existing assembly. The procedure is discuss next.

- Click on the **Replace Part with New Part** tool from the **Replace Part** drop-down. The **Replace Part with New Part Command Bar** will be displayed.

- Select the part which you want to replace and right-click in the drawing area to accept the selection. The **Replacement Part - New** dialog box will be displayed; refer to Figure-54.
- Specify the name of part, template, and location of the new component. After setting desired parameters, click on the **Create Part** button. The new part will be added in the list.

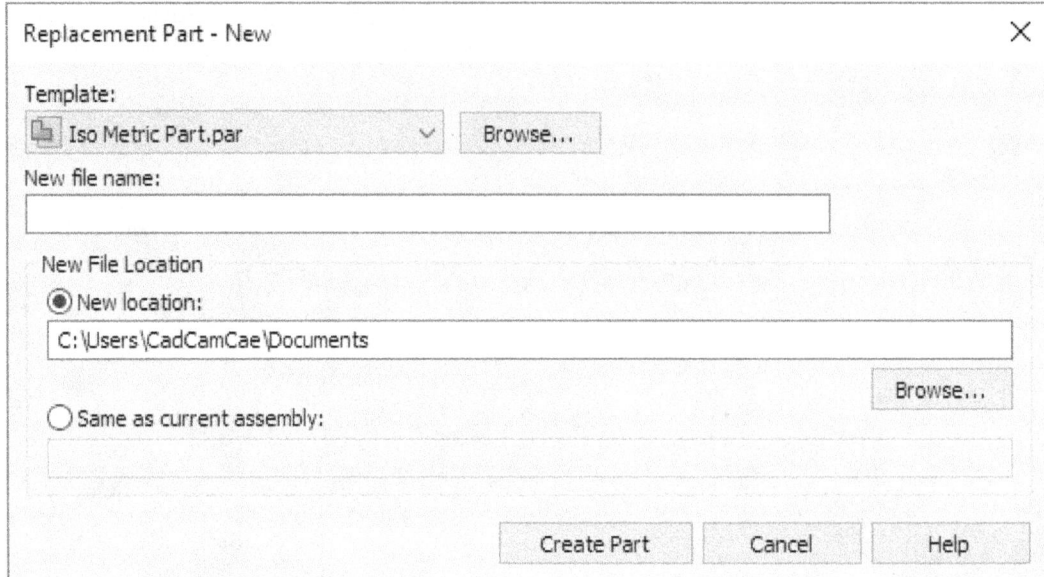

Replacement Part - New                                                    ✕

Template:

🔲 Iso Metric Part.par                    ⌄        Browse...

New file name:

```

```

New File Location

◉ New location:

```
C:\Users\CadCamCae\Documents                            
```

                                                        Browse...

◯ Same as current assembly:

```

```

                        Create Part        Cancel        Help

*Figure-54. Replace part new*

- Now, select the newly created part from the **Path Finder** in the left of application window and right-click on it. A shortcut menu will be displayed; refer to Figure-55.
- Select the **Open** option from the shortcut menu to create part in separate window or select the **Edit** option to create part in assembly environment. The tools to create part model have been discussed earlier.

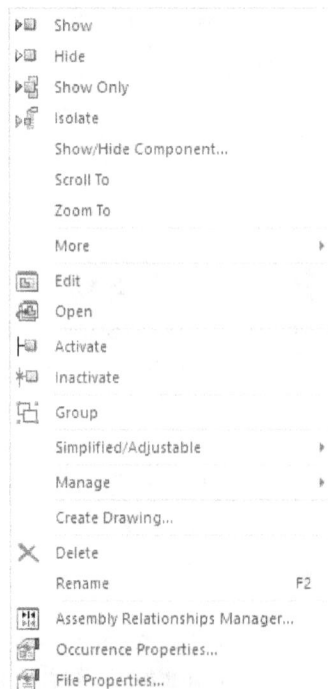

▸🔲 Show
▸🔲 Hide
▸🔲 Show Only
▸🔲 Isolate
        Show/Hide Component...
        Scroll To
        Zoom To
        More                        ▸
🔲 Edit
🔲 Open
🔲 Activate
🔲 Inactivate
🔲 Group
        Simplified/Adjustable        ▸
        Manage                        ▸
        Create Drawing...
✕ Delete
        Rename                        F2
🔲 Assembly Relationships Manager...
🔲 Occurrence Properties...
🔲 File Properties...

*Figure-55. Open dialog box*

## Replace Part with Copy

The **Replace Part with Copy** tool is used to replace a part with another copy of part. The procedure to use this tool is given next.

*   Click on the **Replace Part with Copy** tool from the **Replace Part** drop-down in the **Ribbon**. The **Replace Part with Copy Command Bar** will be displayed.
*   Select the desired part you want to replace from the assembly and right-click in the drawing area. The **Replacement Part - Copy** dialog box will be displayed.
*   Specify desired name of new copy of component and click on the **Save** button from the dialog box. Now, you can edit the new copy of component without modifying the original component.

# TRANSFERRING PART TO ANOTHER SUBASSEMBLY

The **Transfer** tool in the **Modify** group of **Ribbon** is used to transfer the selected part into another sub-assembly. The procedure to use this tool is given next.

*   Click on the **Transfer** tool from the **Modify** group in the **Home** tab of the **Ribbon** after selecting the component. The **Transfer to Assembly Level** dialog box will be displayed; refer to Figure-56.

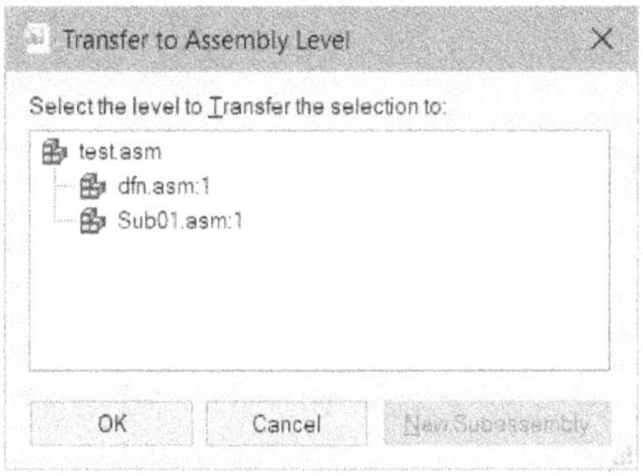

*Figure-56. Transfer to Assembly Level dialog box*

*   Select the desired sub-assembly from the dialog box and click on the **OK** button. The selected part will be transferred to new sub-assembly.

# DISPERSING SUB-ASSEMBLY

The **Disperse** tool is used to transfer all the components of selected sub-assembly into one higher level assembly. The procedure to use this tool is given next.

*   Select the sub-assembly you want to disperse and click on the **Disperse** tool from the **Modify** group in the **Home** tab of **Ribbon**. The **Disperse Assembly** message box will be displayed.
*   Click on the **Yes** button from the dialog box. The selected sub-assembly will be deleted and all its components will be added in the main assembly.

# ROTATIONAL MOTORS

The **Rotational Motor** tool is used to create a motor that rotates an under constrained component. Motors help you to observe the motion of various interrelated parts in assembly. The procedure to use this tool is given next.

- Click on the **Rotational Motor** tool from the **Motors** group in the **Home** tab of **Ribbon**. The **Rotational Motor Command Bar** will be displayed as shown in Figure-57.

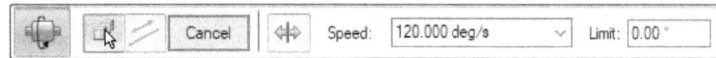

*Figure-57. Rotational Motors Command Bar*

- Select the component which you want to rotate using the motor. You will be asked to select the axis about which you want to rotate the part.
- Select the axis of component. The options to specify speed and limit will be displayed in the **Command Bar**; refer to Figure-58.

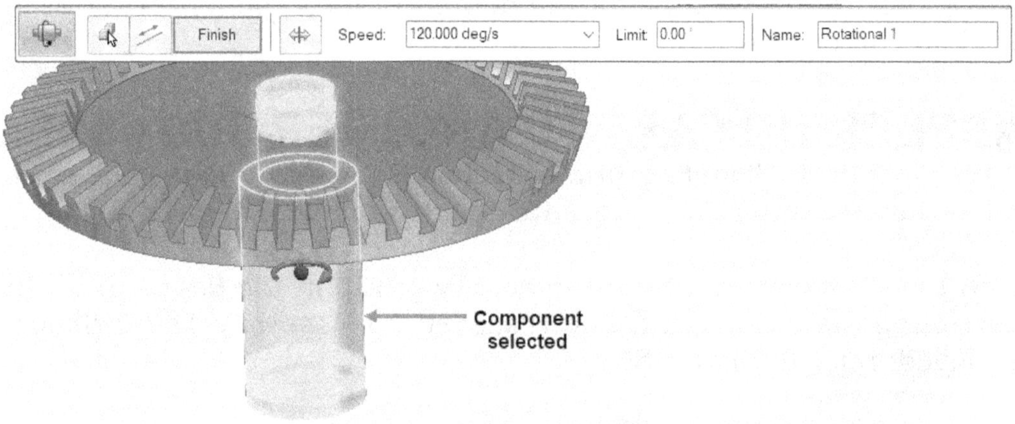

*Figure-58. Component selected for rotational motor*

- Click on the **Flip Direction** button from the **Command Bar** to reverse the rotation direction.
- Specify the desired values in **Speed** and **Limit** edit boxes of **Command Bar** to specify the rotational speed of component and total angular span of rotation, respectively.
- After setting desired parameters, click on the **Finish** button from the **Command Bar**. Press **ESC** to exit the tool.

## LINEAR MOTOR

The **Linear Motor** tool is used to move under constrained component in linear direction. The procedure to use this tool is given next.

- Click on the **Linear Motor** tool from **Motors** group in the **Home** tab of **Ribbon**. The **Linear Motor Command Bar** will be displayed; refer to Figure-59.

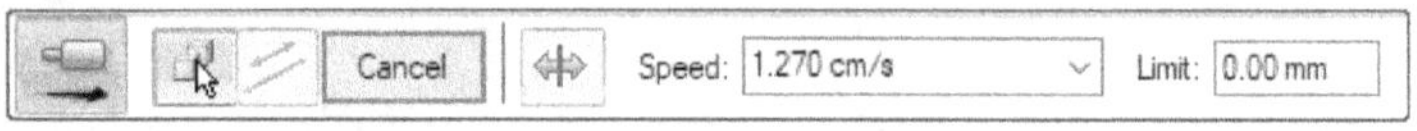

*Figure-59. Linear Motor Command Bar*

- Select the component which you want to move linearly using the motor. You will be asked to specify the direction of movement.
- Select the desired edge or axis to define direction of movement.
- Specify the speed and translation limit for component in respective edit boxes.
- After setting desired parameters, click on the **Finish** button to apply motor. Press **ESC** to exit the tool.

# SIMULATE MOTOR

The **Simulate Motor** tool is used to observe simulation of motor through animation. It is active when any kind of motor is applied to the Assembly. The procedure to use this tool is given next.

- Click on the **Simulate Motor** tool from the **Motor** group in the **Home** tab of the **Ribbon**. The **Motor Group Properties** dialog box will be displayed; refer to Figure-60.

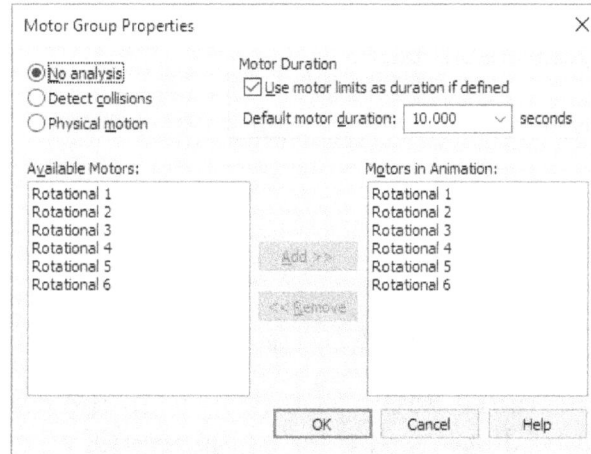

*Figure-60. Simulate Motor Dialog box*

- Select the desired radio button from the dialog box to define whether you want to perform a collision detection, physical motion, or no analysis.
- Select the **Use motor limits as duration if defined** check box to convert limits to duration during analysis. Specify the desired value of motor duration in the **Default motor** edit box.
- Select all the motors you want to include for checking motion from the **Available Motors** list box of the dialog box and click on the **Add>>** button.
- After setting desired parameters, click on the **OK** button from the dialog box.
- The animation box view in the bottom of screen will be displayed as shown in Figure-61. Click on the **Play** button to start the animation.
- Press **ESC** to exit the tool. The **Animation Editor** message box will be displayed.
- Click on the **Yes** button to save changes in animation or click **No** button to exit without saving changes.

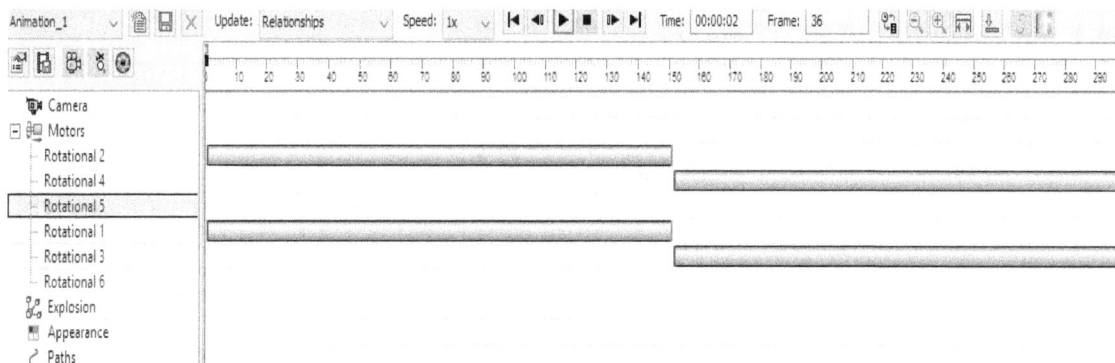

*Figure-61. Animation*

# VARIABLE TABLE MOTOR

The **Variable Table Motor** tool is used to create a motor that drives the component through the table which is created by the user. The procedure is given next.

*   Click on the **Variable Table Motor** tool from **Motor** group in the **Ribbon**. The **Variable Table Motor** dialog box will be displayed; refer to Figure-62.

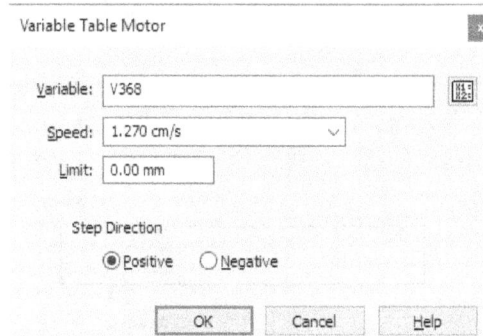

*Figure-62. Variable Table motor dialog box*

*   Click on the **Variable Table Motor** button from the dialog box. The **Variable Table** dialog box will be displayed; refer to Figure-63.

*Figure-63. Variable Table Spreadsheet*

*   Select the desired unit from the **Unit type** drop-down and set the parameters in table like name of variable, its value, and so on to create a new variable. After selecting/creating the variable, click on the **Close** button from the **Variable Table** dialog box to exit. The **Variable Table Motor** dialog box will be active again.
*   Specify the desired value in **Speed** and **Limit** edit boxes of the dialog box to specify the speed and limits of movement.
*   Click on the **OK** button from the dialog box. The motor will be applied.

# PATTERN

We have already discuss about **Pattern** tool for part modeling but in assembly, we use this tool to create copies of the component. The procedure to create assembly pattern is discuss next.

*   Click on the **Pattern** tool from the **Pattern** group in the **Home** tab of the **Ribbon**. The **Pattern Command Bar** will be displayed as shown in Figure-64.

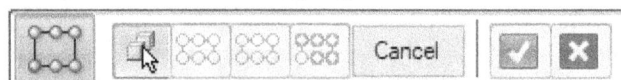

*Figure-64. Pattern Command Bar*

- Select the component which you want to pattern.
- Select the base part on which have pattern created for assembling the component.
- Select the reference feature on base part to be used as starting reference point for creating the pattern. Preview of pattern will be displayed;  refer to Figure-65.
- Click on the **Finish** button from the **Command Bar** to create the pattern.

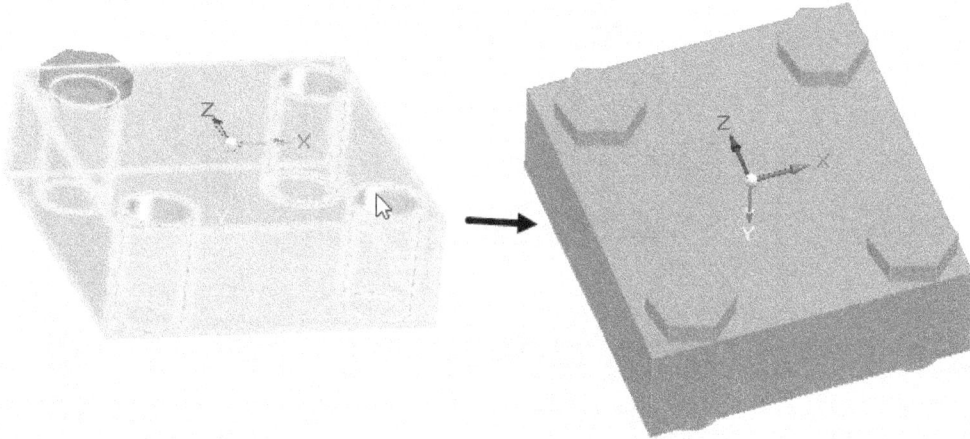

*Figure-65. Preview of pattern*

# PATTERN ALONG CURVE

The **Along Curve** tool is used to create pattern of existing components  along guided curve. To use this tool, you must have a curve on the body which will be followed while creating pattern. The procedure to use this tool is given next.

- Click on the **Along Curve** tool from the **Pattern** group in the **Home** tab of the **Ribbon**. The **Along Curve Command Bar** will be displayed; refer to Figure-66 and you will be asked to select the components you want to pattern.

*Figure-66. Along Curve Command Bar*

- Select the components that you want to pattern and right-click in the drawing area. You will be asked to select curve for 1st direction.
- Select the desired curve or edge and right-click in drawing area. You will be asked to specify start point for pattern.
- Click at desired end point of curve to specify start point of pattern.
- Click in the **Offset** edit box and enter the distance between start point and first instance of component in pattern. The options to specify number of instances and distance between two consecutive instances of pattern.
- Specify the desired parameters in the **Command Bar** and click on the **Next** button. The **Along Curve Command Bar** will be displayed with options to define orientation and rotation of part in pattern.
- Specify the desired parameters in drop-downs of **Command Bar** and click on the **Preview** button. Preview of pattern will be displayed; refer to Figure-67.
- Click on the **Finish** button from the **Command Bar** to create pattern.

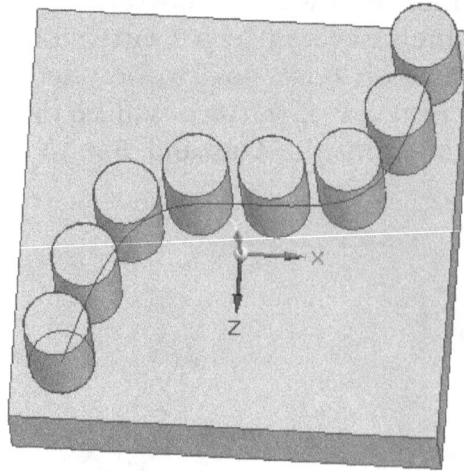

*Figure-67. Along Curve*

## MIRROR COMPONENTS

We have already discuss about the **Mirror** tool in Part modeling, here we are going to discuss about mirroring components in assembly. The procedure to use this tool is given next.

- Click on the **Mirror** tool from the **Pattern** group in the **Home** tab of **Ribbon**. The **Mirror Components Command Bar** will be displayed and you will be asked to select the components to be mirrored.
- Select the components to be mirrored and right-click in the drawing area. You will be asked to select the mirror plane.
- Select the desired mirror plane. Preview of the mirrored components will be displayed; refer to Figure-68 and the **Mirror Components** dialog box will be displayed; refer to Figure-69.
- Set the desired action and adjustment parameters for mirror copies in the dialog box and click on the **OK** button.
- Click on the **Finish** button from the **Command Bar** to create the mirror copies.

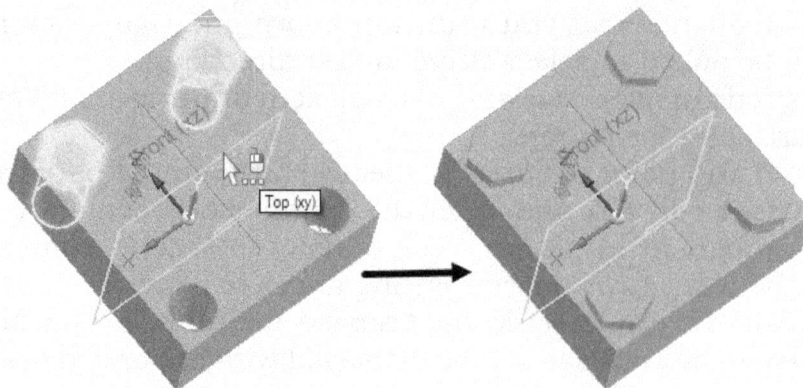

*Figure-68. Mirror Assembly*

Mirror Components

Note: Overriden actions are indicated with a shaded cell. Edit pending items manually or click the Change Pending to Mirror button to change all in a single step.

| Components | Action | Adjust | Output File | Folder |
|---|---|---|---|---|
| Nut2.par:1 | Rotate | Top (XY) ... | | |
| Nut2.par:2 | Rotate | Top (XY) ... | | |

OK   Cancel   Help

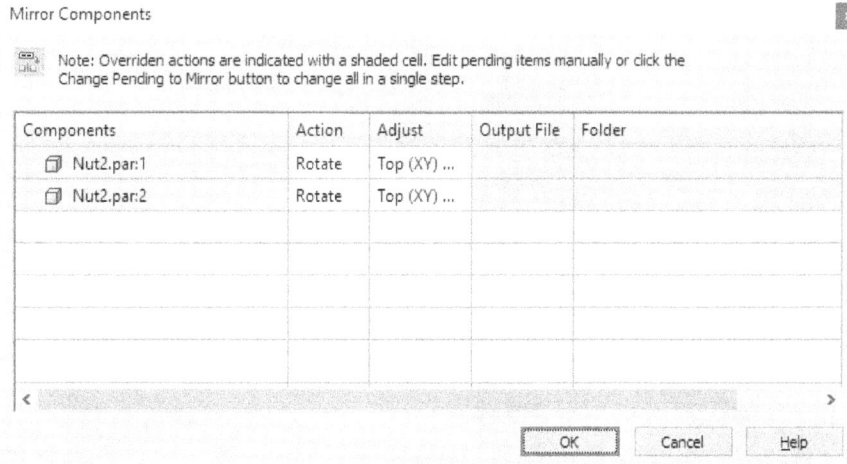

*Figure-69. Mirror Component*

# DUPLICATE PATTERN

The **Duplicate Pattern** tool in assembly is same as discussed earlier in Part modeling.

# INSPECTION TOOLS

The tools in the **Inspect** tab are used to inspect the assembly properties like material, dimension, and so on. Various tools in this tab are discuss next.

## Smart Measure

The **Smart Measure** tool is used to measure dimensions of various elements. It is useful when you are assembling components whose dimensions are not annotated. The procedure to use this tool is given next.

- Click on the **Smart Measure** tool from the **2D Measure** group in the **Inspect** tab of **Ribbon**. The **Smart Measure Command Bar** will be displayed; refer to Figure-70.

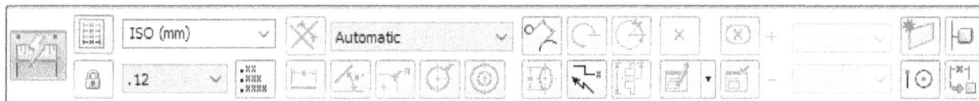

*Figure-70. Smart Measure Command Bar*

- You can measure the entities in the same way as discussed earlier for **Smart Dimension** tool; refer to Figure-71.

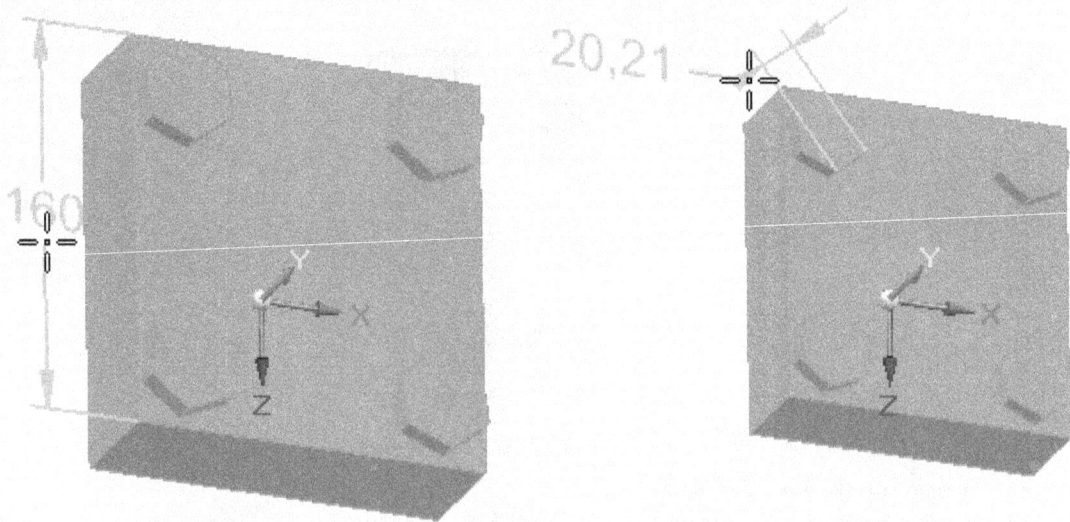

*Figure-71. Smart Measure*

## Measure

The **Measure** tool is used to measure 3D surfaces, faces, angles, lines, and others entities. The procedure to use this tool is given next.

* Click on the **Measure** tool from the **3D Measure** group in the **Inspect** tab of the **Ribbon**. The **Measure Command Bar** will be displayed; refer to Figure-72.

*Figure-72. Measure Command Bar*

* Select the entity from the part or body like face, edge, vertex, and so on; refer to Figure-73.

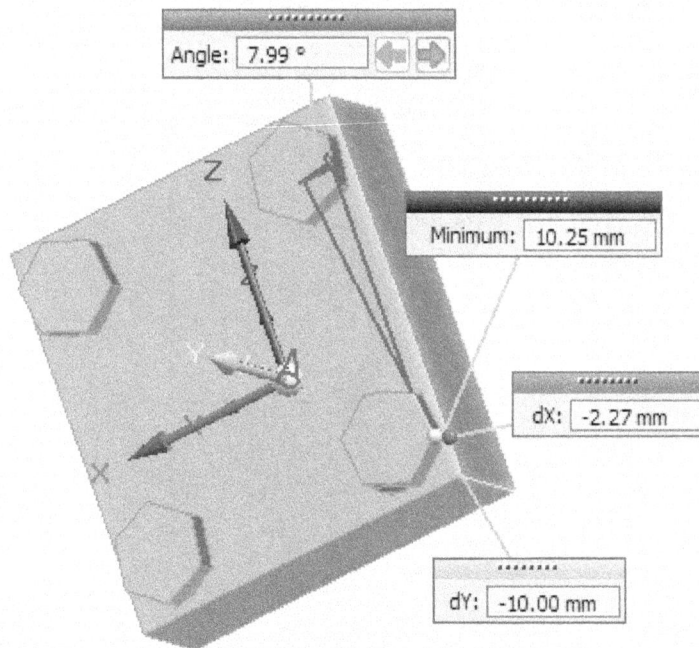

*Figure-73. Measuring entities*

## Measure Distance

The **Measure Distance** tool is used to measure the distance between two or more keypoints. The procedure to use this tool is given next.

- Click on the **Measure Distance** from the **3D Measure** group in the **Inspect** tab of **Ribbon**. The **Measure Distance Command Bar** will be displayed; refer to Figure-74.

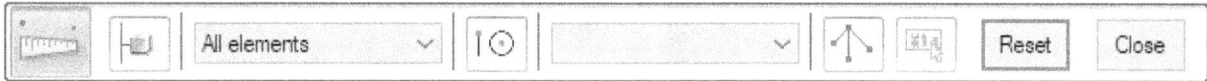

*Figure-74. Measure Distance Command Bar*

- Select the keypoints of the part or body in the order by which you want to measure them. A spreadsheet will be displayed with measurement in the **Measure Distance** dialog box; refer to Figure-75.

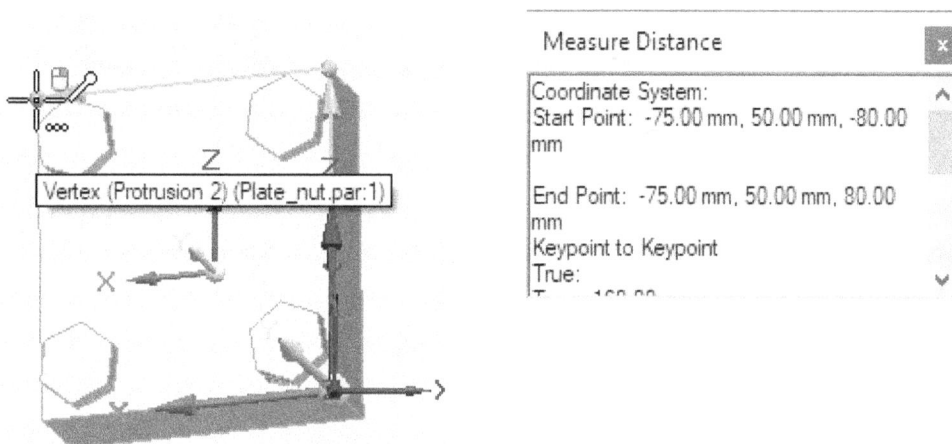

*Figure-75. Measure Distance*

## Measure Minimum Distance

The **Measure Minimum Distance** tool is used to measure minimum distance between two keypoints. The procedure to use this tool is given next.

- Click on the **Measure Minimum Distance** tool from the **3D Measure** group. The **Measure Minimum Distance Command Bar** will be displayed; refer to Figure-76.

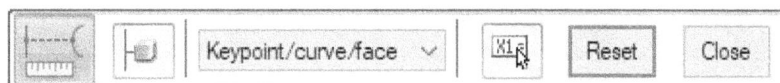

*Figure-76. Measure Minimum Distance Command Bar*

- Select the keypoints on part or body; refer to Figure-77.

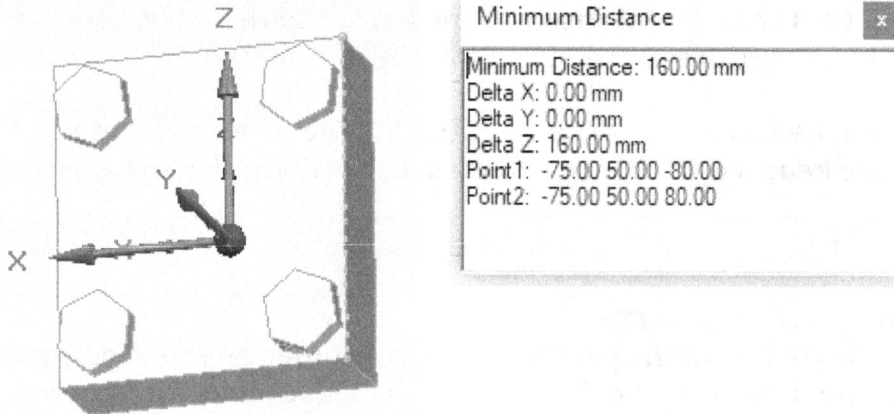

*Figure-77. Measure Minimum Distance*

## Measure Normal Distance

The **Measure Normal Distance** tool is used to measure a normal distance from a plane or line to a keypoints. The procedure is given next.

- Click on the **Measure Normal Distance** tool from the **3D Measure** group in the **Inspect** tab of **Ribbon**. The **Measure Normal Distance Command Bar** will be displayed; refer to Figure-78.

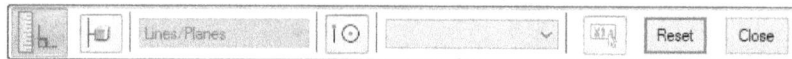

*Figure-78. Measure Normal Distance Command bar*

- Select the desired face, line, and then key point to measure distance of keypoint perpendicular to selected face/line; refer to Figure-79.

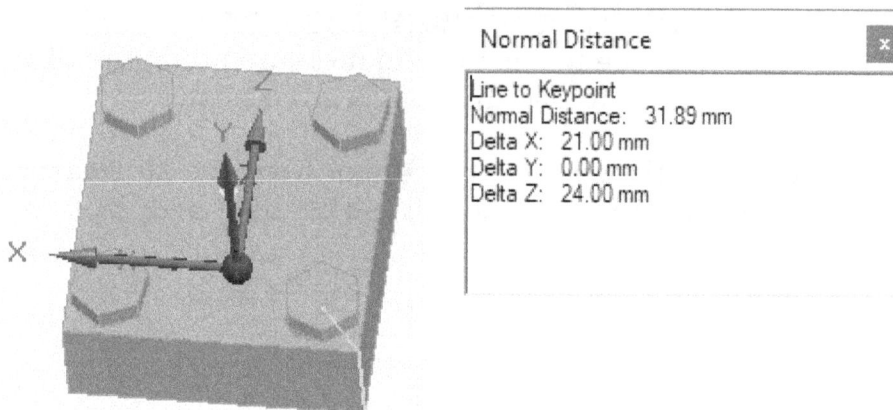

*Figure-79. Measure Normal Distance*

- Press **ESC** to exit the tool.

## Measure Angle

The **Measure Angle** tool is used to measure the angle between selected faces/planes/lines/keypoints. The procedure to use this tool is given next.

- Click on the **Measure Angle** tool from the **3D Measure** group in the **Inspect** tab of **Ribbon**. The **Measure Angle Command Bar** will be displayed; refer to Figure-80.

*Figure-80. Measure Angel*

- Select the faces/edges/keypoints. A spread sheet will be displayed with measurement; refer to Figure-81.
- Press **ESC** to exit the tool.

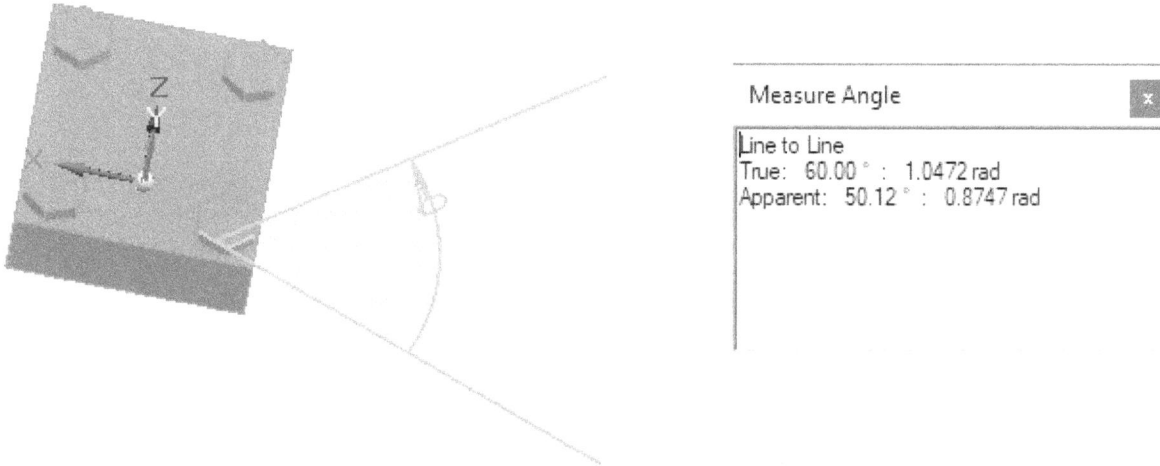

Measure Angle

Line to Line
True:   60.00 °  :   1.0472 rad
Apparent:   50.12 °  :   0.8747 rad

*Figure-81. Measuring angle*

## Inquire Element

The **Inquire Element** tool is used to get information of selected element. The procedure to use this tool is given next.

- Click on the **Inquire Element** tool from the **3D Measure** group in the **Inspect** tab of **Ribbon**. The **Inquire Element Command Bar** will be displayed; refer to Figure-82.

Locate All Elements        Close

*Figure-82. Inquire Element Command Bar*

- Select the element about which you want to get the information. The **Inquire Element** dialog box will be displayed; refer to Figure-83.

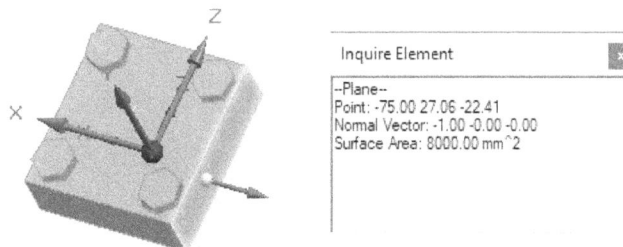

Inquire Element

--Plane--
Point: -75.00 27.06 -22.41
Normal Vector: -1.00 -0.00 -0.00
Surface Area: 8000.00 mm^2

*Figure-83. Inquire Element*

## PROPERTIES

The **Properties** tool is used to display all the physical properties of selected parts or assembly. The procedure to use this tool is given next.

- Click on the **Properties** tool from the **Physical Properties** group of the **Inspect** tab in the **Ribbon**. The **Physical Properties** dialog box will be displayed; refer to Figure-84.
- Select the desired option from the **Coordinate system** drop-down and click on the **Update** button. The mass, volume, and other properties of assembly/selected elements will be displayed in the dialog box.
- After checking the parameters, click on the **Close** button from the dialog box to exit.

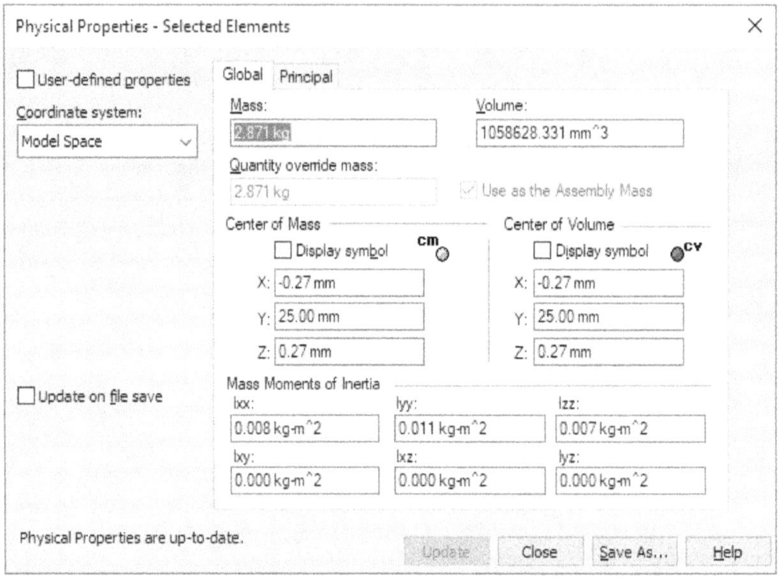

*Figure-84. Physical Properties dialog box*

# PROPERTIES MANAGER

The **Properties Manager** tool is used to modify physical properties for the top level assembly and associated component. The procedure to use this tool is given next.

- Click on the **Properties Manager** tool from the **Physical Properties** group of the **Inspect** tab in the **Ribbon**. The **Physical Properties Manager** dialog box will be displayed; refer to Figure-85.
- Select the desired material from the **Material** column of component which parameters are to be checked. After selecting the material, click on the **Update All** button from the dialog box.
- Click on the **OK** button from the dialog box to exit the dialog box.

| Document Name | Up-To-Date | Material | Density | Accuracy | Mass | Volume | User Defined | Surface Area | Bead Material | Bead Density |
|---|---|---|---|---|---|---|---|---|---|---|
| Asm2 | No | | | | 3.446 kg | 1270709.579 m... | | | | |
| Asm1.asm | No | | | | 0.000 kg | 0.000 mm^3 | | | | |
| Nut2.par | Yes | Aluminum, 60... | 2712.000 kg/m... | 0.99 | 0.144 kg | 53020.312 mm... | No | 8989.06 mm^2 | | |
| Plate_nut.par | Yes | num, 6061-T6 | 2712.000 kg/m... | 0.99 | 2.871 kg | 1058628.331 m... | No | 92194.69 mm^2 | | |

Iron
Aluminum, 6061-T6
PVC
Stainless steel
------------ Favorite Materials --------------
Stainless steel
Steel, structural
Galvanized steel
Steel
Iron
Aluminum, 1060
Bronze, 90%
Copper
Titanium, unalloyed
Zinc
Tin
Gold
Silver
Lead
ABS Plastic, high impact
Polycarbonate
Polyurethene
PVC
Cork, gasket grade
Graphite
Glass, general industrial
Silicone
Wood, Mahogany

Material Table

OK    Update All    Can...

*Figure-85. Spread Sheet of Properties Manager*

# CHECK INTERFERENCE

The **Check Interference** tool is used to find parts that physically interfere with one another. The procedure to use this tool is given next.

- Click on the **Check Interference** tool from the **Evaluate** group in the **Inspect** tab of the **Ribbon**. The **Check Interference Command Bar** will be displayed; refer to Figure-86.

*Figure-86. Check Interference Command Bar*

- Select the components of first set of parts and right-click in the drawing area. You will be asked to select parts of second set.
- Select the second set of parts used for interference checking and click on the **Process** button from the **Command Bar**. If there is an interference then it will be displayed in graphics; refer to Figure-87. If there is no interference then a message box will be displayed.
- Press **ESC** to exit the tool.

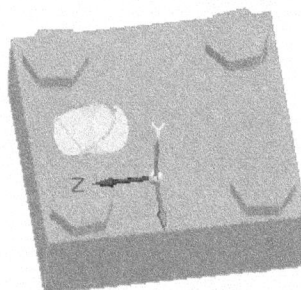

*Figure-87. Check Interference*

## ASSEMBLY STATISTICS

The **Assembly Statistics** tool is used to display the assembly statistics like;

- Assembly file name
- Total number of parts
- Number of unique parts
- Total number of subassemblies
- Number of unique subassemblies

On clicking the **Assembly Statistics** tool from the **Evaluate** group in the **Inspect** tab of the **Ribbon**. The **Assembly Statistics** dialog box will be displayed with all the parameters; refer to Figure-88.

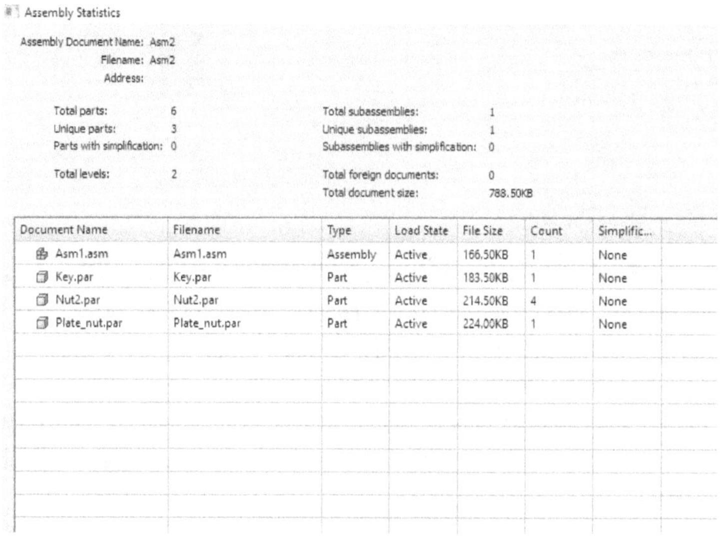

Assembly Statistics

| Assembly Document Name: | Asm2 | | | | | |
|---|---|---|---|---|---|---|
| Filename: | Asm2 | | | | | |
| Address: | | | | | | |

| | | | | | | |
|---|---|---|---|---|---|---|
| Total parts: | 6 | | Total subassemblies: | 1 | | |
| Unique parts: | 3 | | Unique subassemblies: | 1 | | |
| Parts with simplification: | 0 | | Subassemblies with simplification: | 0 | | |
| Total levels: | 2 | | Total foreign documents: | 0 | | |
| | | | Total document size: | 788.50KB | | |

| Document Name | Filename | Type | Load State | File Size | Count | Simplific... |
|---|---|---|---|---|---|---|
| Asm1.asm | Asm1.asm | Assembly | Active | 166.50KB | 1 | None |
| Key.par | Key.par | Part | Active | 183.50KB | 1 | None |
| Nut2.par | Nut2.par | Part | Active | 214.50KB | 4 | None |
| Plate_nut.par | Plate_nut.par | Part | Active | 224.00KB | 1 | None |

*Figure-88. Assembly Statistics*

## GEOMETRY INSPECTOR

The **Geometry Inspector** tool is used to check errors and small entities in the model. The procedure to use this tool is given next.

- Click on the **Geometry Inspector** tool from the **Evaluate** group in the **Inspect** tab in the **Ribbon**. The **Geometry Inspector** dialog box will be displayed; refer to Figure-89.
- Select the desired check boxes from the dialog box and parameters.
- After setting parameters, click on the **Go** button from the dialog box. The results will be displayed in the **Results** list box.

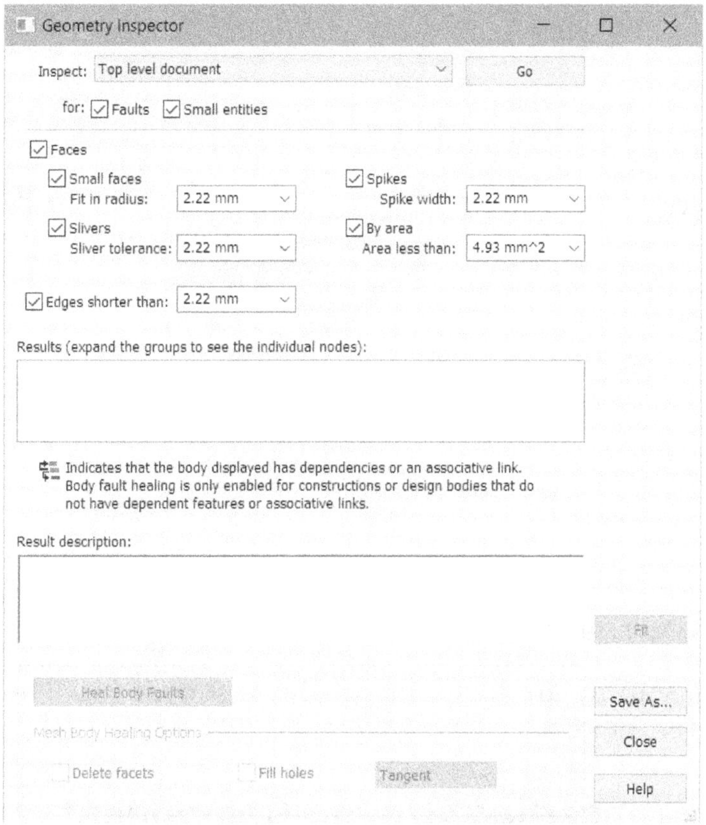

*Figure-89. Geometry Inspector dialog box*

- Click on the **Close** button from the dialog box to exit.

# PRACTICAL - 1

Assemble the parts of Knuckle Joint as shown in Figure-90. The exploded view of assembly will be displayed as shown in Figure-91.

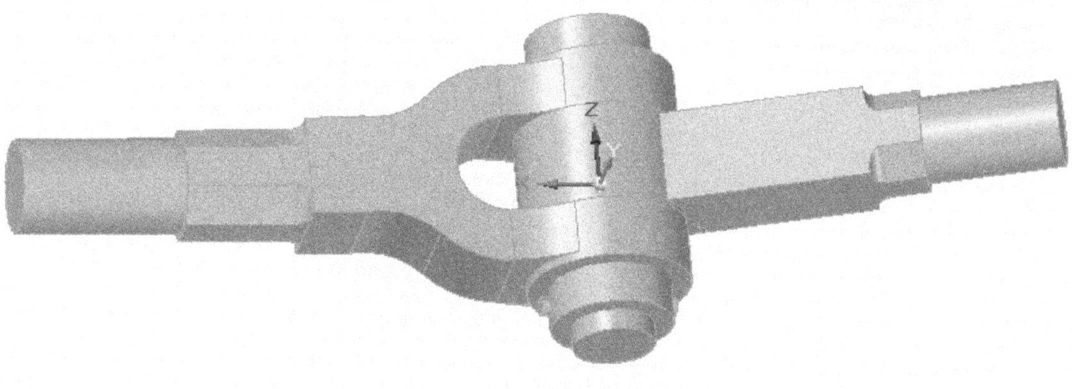

*Figure-90. Knuckle joint*

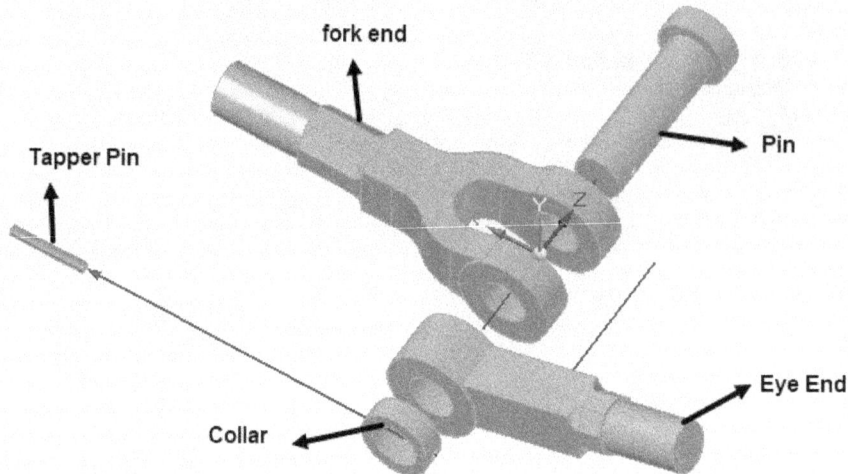

*Figure-91. Exploded View Practical 1*

Part files for this assembly are available in the resource kit. The steps to assemble components are given next.

## Inserting Ground Component

- Start Solid Edge if not started yet and click on the **New** button from **Quick Access Toolbar**. The **New** dialog box will be displayed.
- Start the assembly environment by selecting **iso metric assembly.asm** template from the **New** dialog box and click on the **OK** button. The assembly environment will be displayed.
- Click on the **Insert Component** tool from the **Assemble** group in the **Home** tab of the **Ribbon**. The **Parts Library** box will be displayed.
- Select the Fork part from the box and drag the component in drawing area; refer to Figure-92.

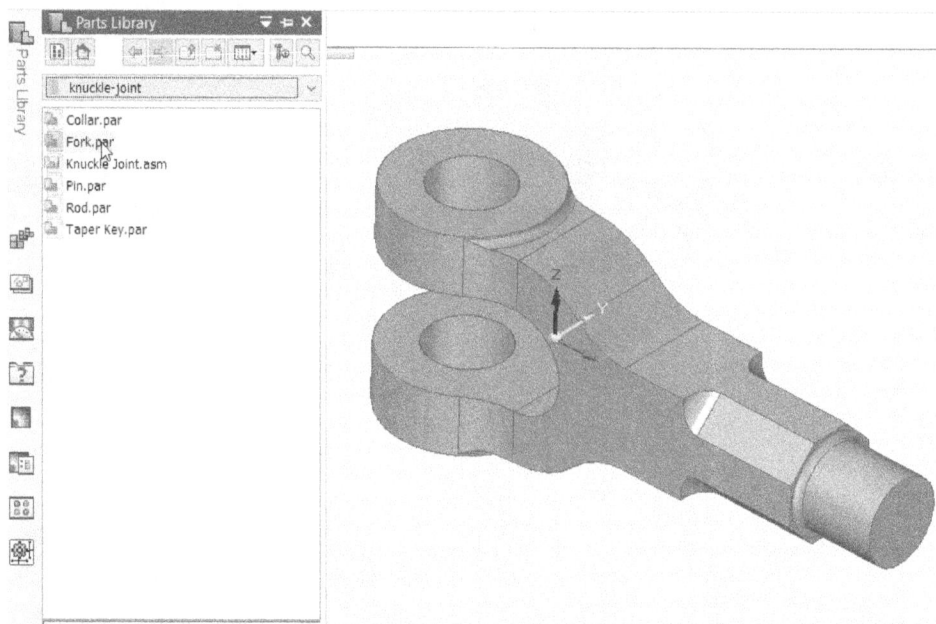

*Figure-92. Inserting Componenet*

## Inserting and Constraining Component

- Now select next component from the part library that is Rod and drag it to assembly environment.
- Select the **Planer Align** option from the **Relationship Types** drop-down in the **Command Bar** to make the face of Fork and Rod aligned. Select the inner face of fork end and outer face eye end. Refer to Figure-93. Note that you may need to flip the part.

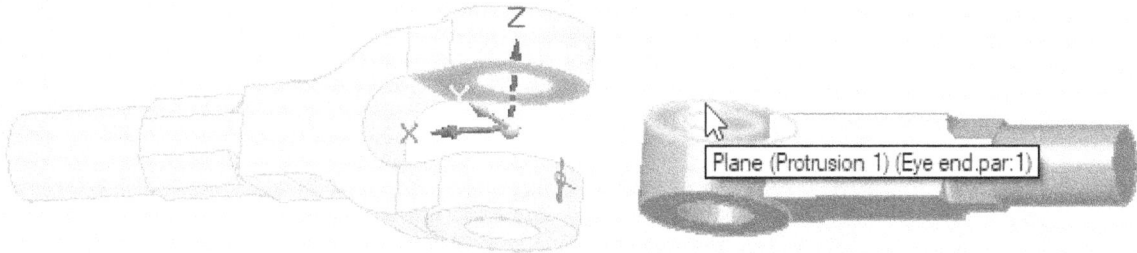

*Figure-93. Applying planar align constraint*

- Select the **Axial Align** option from the **Relationship Types** drop-down to make components aligned axially. Select the circular face of fork end and eye end; refer to Figure-94.

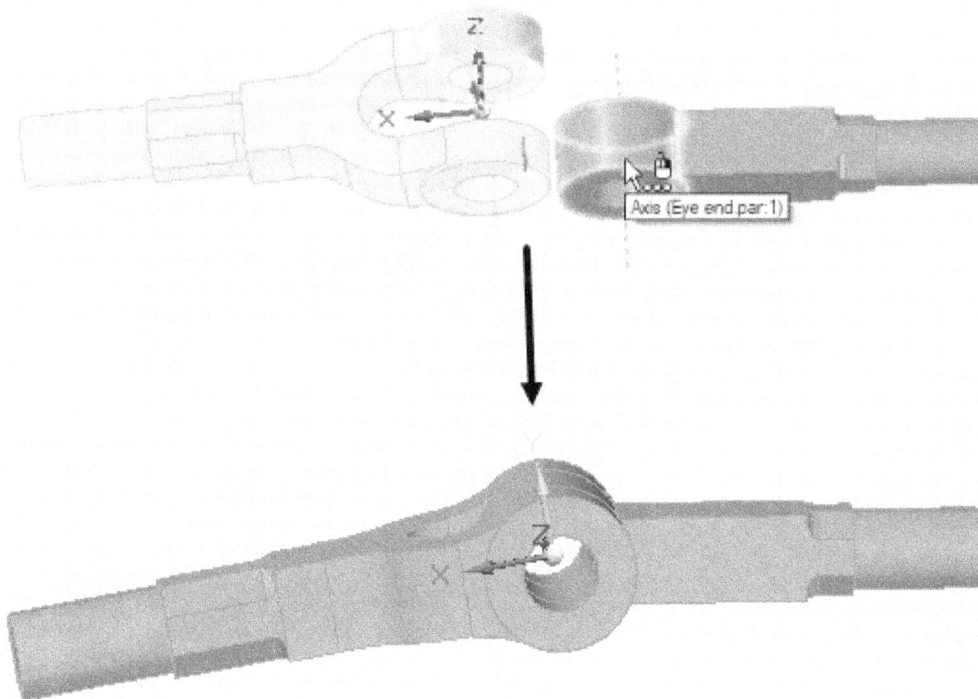

*Figure-94. Axial Align Practical 1*

- Press **ESC** to exit the **Command Bar**.

- Now, insert Pin from the Parts Library box. Select the radial axis of pin and the radial axis of Fork end; refer to Figure-95. The axes of components will be aligned.

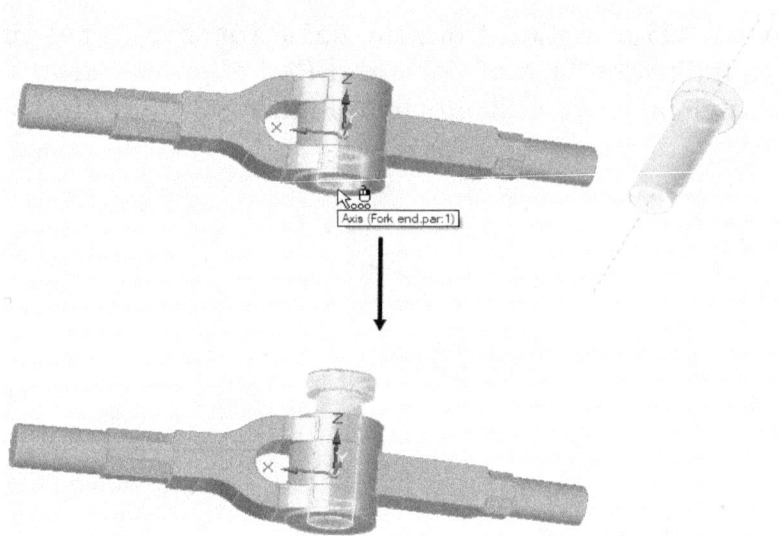

*Figure-95. Axial Align of Pin*

- Select the flat inner face of pin and top flat face of fork end; refer to Figure-96.

*Figure-96. Mate Practical 1*

- Drag the Collar part from the **Parts Library** box into the drawing area. Select the round faces of collar and pin to align their axes.
- Select the collar axial align with the hole created on the pin; refer to Figure-97.

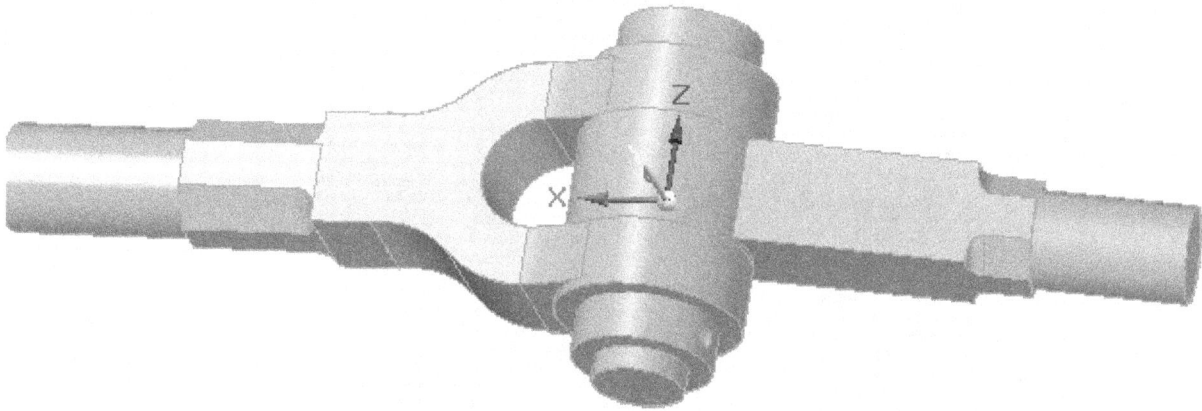

*Figure-97. Collar axial align*

- Insert Tapper pin and make the hole of collar aligned with pin. Move the taper pin by dragging inside the hole. This completes the assembly; refer to Figure-98.

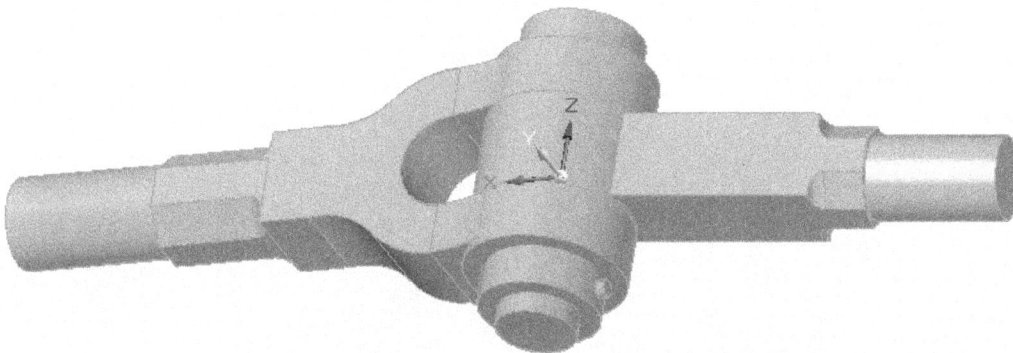

*Figure-98. Completed Assembly*

# PRACTICE - 1

Assemble the Connecting Rod parts as shown in Figure-99. The parts of assembly are available in the resource kit.

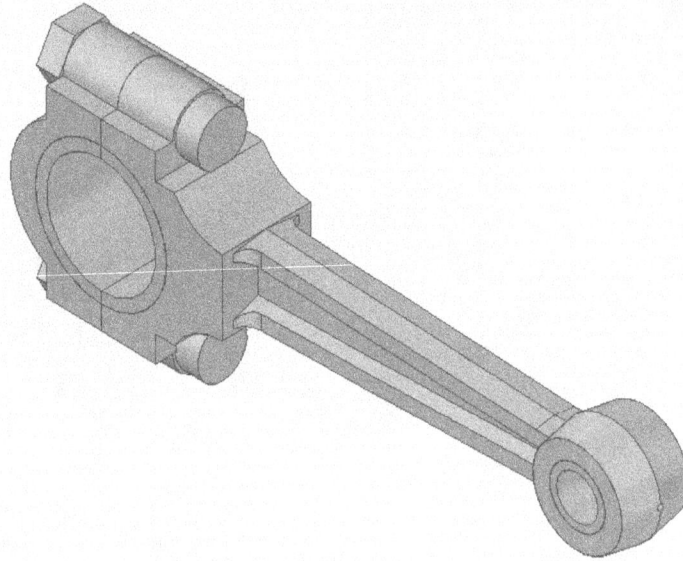

*Figure-99. Connecting Rod Assembly*

# PRACTICE - 2

Assemble the Pipe Vice as shown in Figure-100. The components of assembly are available in the resource kit.

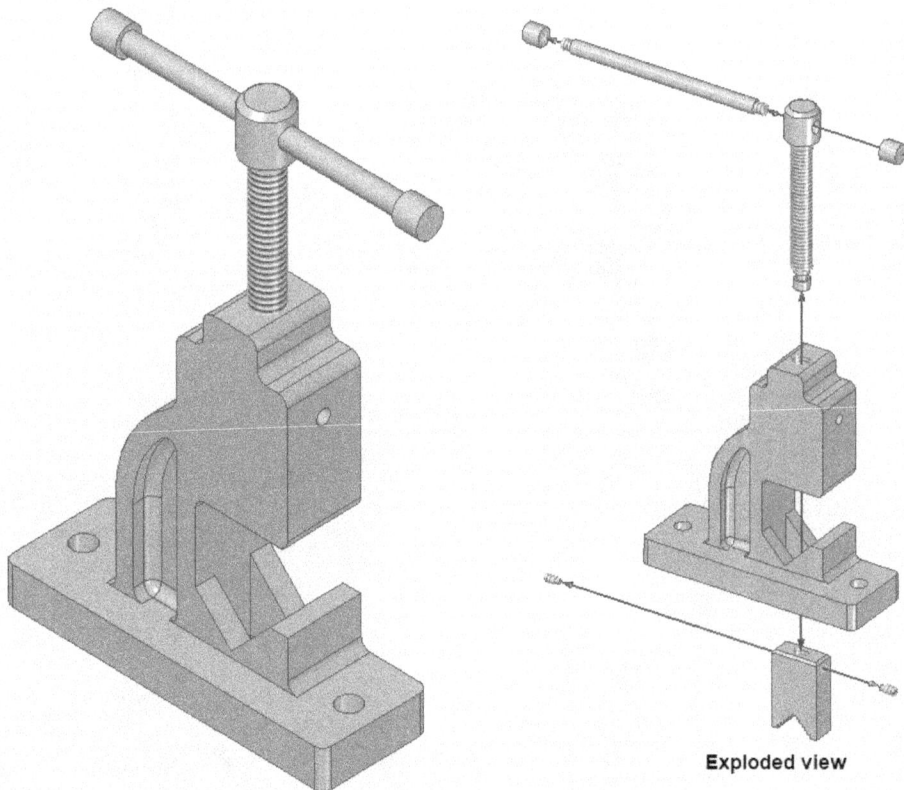

**Exploded view**

*Figure-100. Pipe vice assembly*

# Chapter 9

# Assembly Design - II

Topics Covered

The major topics covered in this chapter are:

- *Variable and Component Tracker*
- *Update Tools*
- *Reports*
- *Error Assistant, Inter-Part Manager, and Component Structure Editor*
- *Publishing Virtual Components*
- *Engineering Reference Tools*
- *Explode Render Animate Environment*
- *Modes*

In previous chapter, we have discussed about the inserting components in assembly and applying constraints. The constraints are used to arrange the parts in systematic way. Most of the time we keep some of the components under-constrained so that they can move in predefined motion. In this chapter, we are going to discuss advance assembly tools which allow motion between various components of the assembly. In previous chapter, we have discuss the tools of **Inspect** tab. Now, we will discuss the tools of **Tool** tab in the **Ribbon** and other advanced tools.

# TOOLS TAB

The tools in the **Tools** tab are used to manage variables, environments, reports, rendering, and so on. Various tools in this tab are discussed next.

## Peer Variable

The **Peer Variables** tool is used to edit variables of any attachment in the Assembly. The Steps are given next.

- Click on the **Peer Variables** tool from the **Variables** group in the **Tools** tab of the **Ribbon**. The **Peer Variable Command Bar** will be displayed; refer to Figure-1.

*Figure-1. Peer Variable Command Bar*

- Select the part in assembly whose variables are to be edited or checked. The parameters will be displayed in the dialog box; refer to Figure-2.

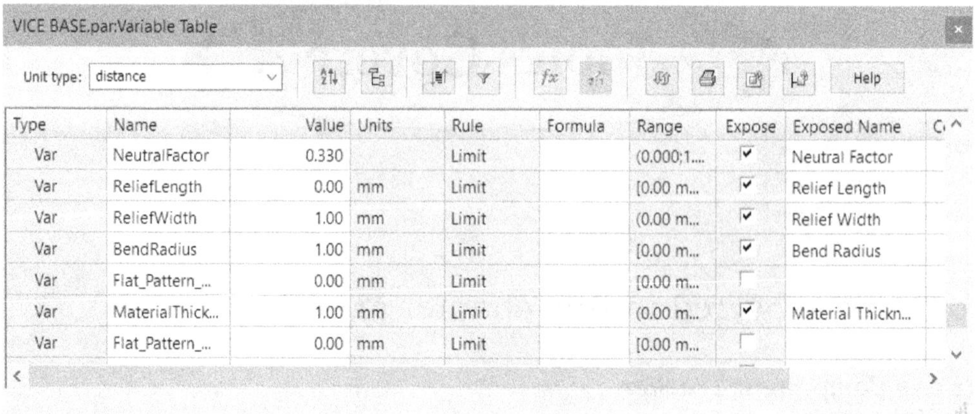

VICE BASE.par:Variable Table

Unit type: distance

| Type | Name | Value | Units | Rule | Formula | Range | Expose | Exposed Name | C |
|------|------|-------|-------|------|---------|-------|--------|--------------|---|
| Var | NeutralFactor | 0.330 | | Limit | | (0.000;1.... | ✔ | Neutral Factor | |
| Var | ReliefLength | 0.00 | mm | Limit | | [0.00 m... | ✔ | Relief Length | |
| Var | ReliefWidth | 1.00 | mm | Limit | | (0.00 m... | ✔ | Relief Width | |
| Var | BendRadius | 1.00 | mm | Limit | | [0.00 m... | ✔ | Bend Radius | |
| Var | Flat_Pattern_... | 0.00 | mm | Limit | | [0.00 m... | ☐ | | |
| Var | MaterialThick... | 1.00 | mm | Limit | | (0.00 m... | ✔ | Material Thickn... | |
| Var | Flat_Pattern_... | 0.00 | mm | Limit | | [0.00 m... | ☐ | | |

*Figure-2. Spread Sheet of Peer Variable*

- Double-click in the empty field of **Name** column and specify desired name of variable. Similarly, double-click in the empty field under **Formula** column next to new variable to create a formula for generating the value. You can use arithmetic operators (+, -, /, *) and logic rules (like greater than, less than, and so on) in the formula. Various logical rules are given next.

| Character | Meaning | Where Used | Variable Type |
|---|---|---|---|
| ( | Greater Than | Beginning Only | Limit |
| ) | Less Than | End Only | Limit |
| [ | Greater Than or Equal To | Beginning Only | Limit |
| ] | Less Than or Equal To | End Only | Limit |
| {} | Encloses a Discrete List | Use both as a set | Discrete List |
| ; | Separates values | Between values in a limit or discrete list | Limit and Discrete List |

Examples of using logic rules:
1. To define a variable that must be greater than 15 and less than 20, type the following in the Range cell: (15;20)
2. To define a variable that must be greater than or equal to 8 and less than or equal to 15, type the following: [8;15]
3. To define a variable that must be greater than or equal to 9 and less than 20, type the following: [9;20)
4. To define a variable that must be limited to the following list of values: 3, 10, 15, and 20, type the following: {3;10;15;20}

- You can also use the **Formula** button to create formula for a variable. Click in the field where you want to specify formula and then click on the **Formula** button from the dialog box. The Step 1 page of **Function Wizard** dialog box will be displayed; refer to Figure-3. Select the desired function from the list and click on the **Next** button. The Step 2 page will be displayed and you will be asked to specify value of formula; refer to Figure-4. Specify the desired value in the edit box and click on the **Finish** button.

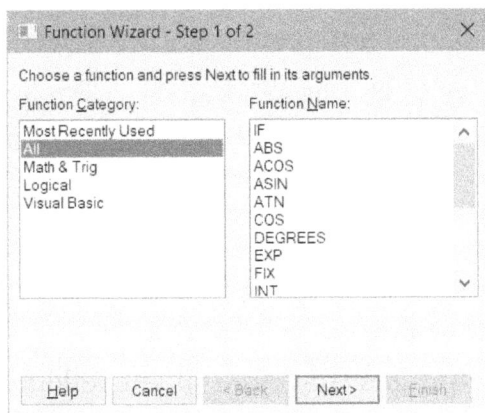

Figure-4. Step 2 of Function Wizard dialog box

Figure-3. Function Wizard dialog box

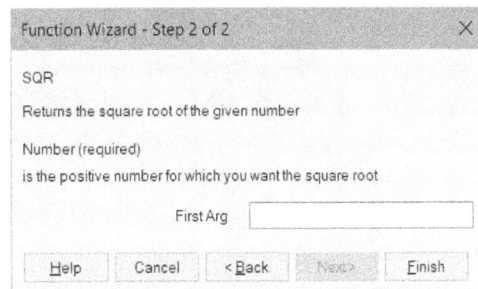

- Click on the **Variable Rule Editor** button from the dialog box to edit a parameter. The **Variable Rule Editor** dialog box will be displayed; refer to Figure-5. Select the **Value based on** check box to specify the value of variable based on formula or copied value. Select the **Limit value to** check box to specify upper and lower limits of value in the field or range of values. After setting desired parameters in the dialog box, click on the **OK** button.

*Figure-5. Variable Rule Editor dialog box*

- Click on the **Refresh** button to update the selected value.
- Click on the **Print** button to print all the parameters of table.
- Click on the **Update All** button to update all the parameters of table in the model.
- Click on the **Activate all** button to activate all parent documents.
- Click on the **Close** button to exit the dialog box.

## COMPONENT TRACKER

The **Component Tracker** tool is used to track if there is any change in the components of assembly. The **Component Tracker** dialog box will be displayed; refer to Figure-6. If all the components are updated in assembly then a tick mark will be displayed. Click on the **Close** button to exit the tool. Meaning of various marks are given next.

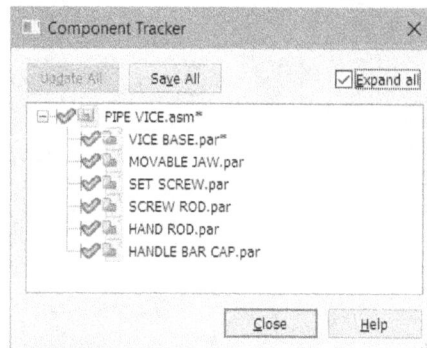

*Figure-6. Component Tracker dialog box*

 One or more documents have geometric changes that are not yet reflected at this level. An update is required.

 Document is up to date in memory. May need to be updated to disk.

 Document has changes and is causing higher level documents to be out of date.

? Document is not found.

 Document has changes in memory that have not been saved.

## UPDATE TOOLS

The tools in the **Update** drop-down of **Update** group are used to update components, documents, and relationships in assembly. Click on the **Update Active Level** tool from the **Update** drop-down in the **Tools** tab of the **Ribbon** to update components of current model's active assembly. Click on the **Update All Open Documents** tool from the drop-down to update all the documents currently open in Solid Edge. Click on the **Update Relationships** tool from the **Update** drop-down in the **Ribbon** to update all the relationships applied in the assembly. Toggle on the **Automatic Update** tool from the drop-down to automatically update model and other parameters while working.

## LIMITED UPDATE AND LIMITED SAVE

The **Limited Update** and **Limited Save** tools are active only when these options are selected in the **Assembly** page of the **Solid Edge Options** dialog box; refer to Figure-7.

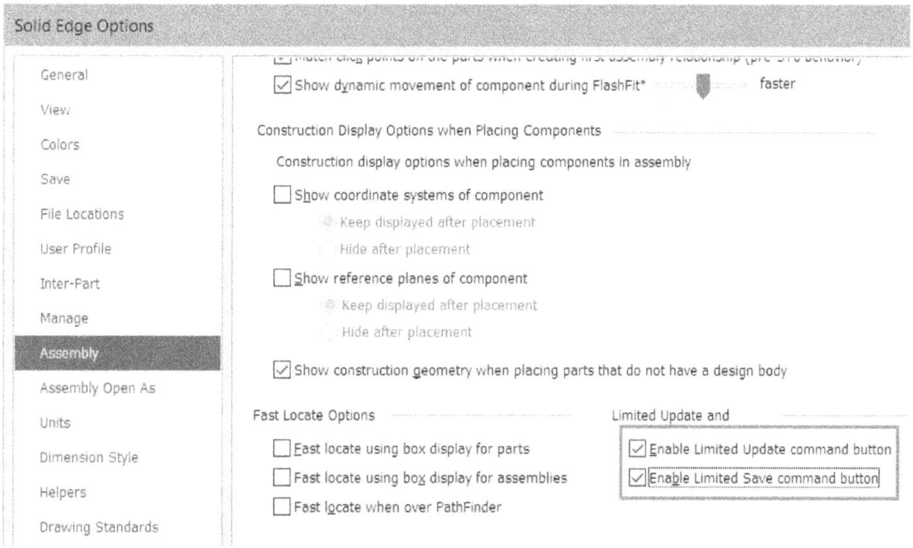

*Figure-7. Limited update options*

Toggle 'ON' the **Limited Update** and **Limited Save** buttons to let Solid Edge decide when to update and save the model while using minimal resources and keeping balance in performance.

## REPORTS

The **Reports** tool is used to generate common assembly reports like Bill of materials, part list, and so on. The Bill of materials is used to list the components of the assembly in the form of a table. The procedure to use this tool is discussed next.

- Click on the **Reports** tool from the **Assistants** group in the **Tools** panel of the **Ribbon**. The **Report** dialog box will be displayed; refer to Figure-8.

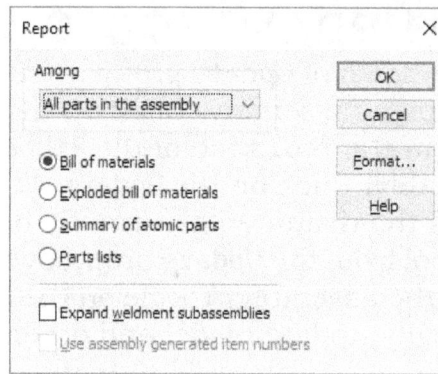

*Figure-8. Report dialog box*

- Select the desire radio button (Bill of materials in our case) from the dialog box and click on the **OK** button. The Bill of Materials spread sheet will be displayed with description of assembly parts; refer to Figure-9.
- Click on the **Save As** button to save the bill of materials in the form of a text file. Click on the **Print** button from the dialog box to get hard copy of the bill of materials.
- Using the **Copy** button, you can copy parameters of the bill of materials.
- Click on the **New Report** button from the dialog box to generate different report.
- Click on the **Close** button from the dialog box to exit.

*Figure-9. Bill of Material*

# ERRORS

The **Errors** tool is used to track the errors in parts and assembly that have failed to recompute properly. Using this tool makes easier to fix the errors in parts and assembly. On clicking this tool, the **Error Assistant** dialog box will be displayed; refer to Figure-10. Select the component which has error and click on the **Zoom** button from the dialog box. The model will be displayed highlighted in the drawing area. Select the error object and click on the **Edit Definition** button to modify the part.

*Figure-10. Error Assistant*

# INTER-PART MANAGER

The **Inter-Part Manager** tool is used to check the components of model in the form of children or parent hierarchy. Click on the **Inter-Part Manager** tool from the **Inter-Part Links** dialog box will be displayed; refer to Figure-11. Select the desired radio button from the dialog box and click on the **Update All** button. The updated list of components and inter-part links will be displayed. Click on the **Close** button to exit from dialog box.

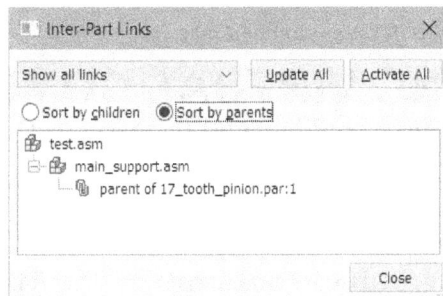

*Figure-11. Inter-Part Links dialog box*

# COMPONENT STRUCTURE EDITOR

The **Component Structure Editor** tool is used to add virtual components in the assembly structure. Virtual components are used to represent the objects which are not in assembly but you want them in bill of materials and other reports. The procedure to use this tool is given next.

- Click on the **Component Structure Editor** tool from the **Assistants** group of the **Tools** tab in the **Ribbon**. The **Virtual Component Structure Editor** dialog box will be displayed; refer to Figure-12.
- Select the desired radio button from the dialog box to define what type of object you want to add in the structure.
- Specify the desired name of object in the **Name** edit box and click on the **Add Virtual Component** button from the dialog box. The object will be added in the assembly.
- You can also add a real component in the assembly by selecting the Add as real component check box and then dragging the component from the left box to right box.
- After setting desired parameters, click on the **OK** button. The component will be added in the assembly.

*Figure-12. Virtual Component Structure Editor dialog box*

## PUBLISHING VIRTUAL COMPONENTS

The **Publish Virtual Components** tool is used to publish the list of virtual components. The procedure to use this tool is given next.

• Click on the **Publish Virtual Components** tool from the **Publish Virtual** drop-down in the **Assistants** group of **Tools** tab in the **Ribbon**. The **Publish Virtual Components** dialog box will be displayed; refer to Figure-13.

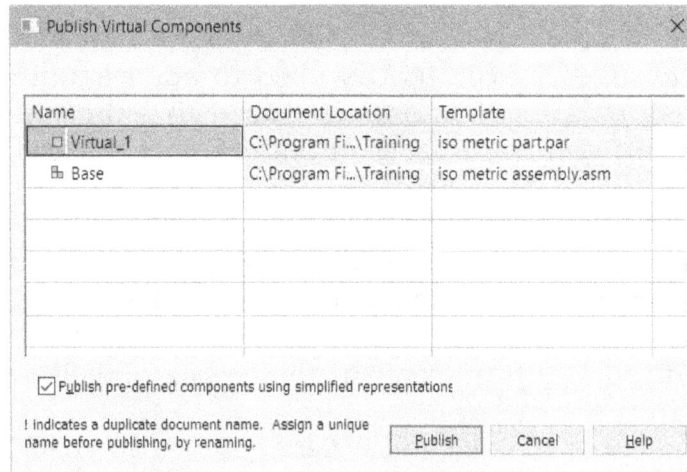

*Figure-13. Publish Virtual Components dialog box*

• Select the **Publish pre-defined components using simplified representations** check box to publish components in simplified format.
• Click on the **Publish** button from the dialog box. The virtual components will be added in the **Path Finder**.

# PUBLISHING TERRAIN MODELS

The **Publish Terrain Models** tool is used to publish virtual objects used to define terrain of environment. The terrain model is imported in the form of an STL file generated by a GIS (Geographic Information System) software. The procedure to import and publish a terrain model is given next.

- Import the STL file of terrain in a new part file and save it in the default location of assembly.
- Create a virtual component using the new created terrain part using the **Component Structure Editor** tool.
- Click on the **Publish Terrain Models** tool from the **Publish Virtual** drop-down in the **Assistants** tab of **Ribbon** and generate the model as discussed for **Publish Virtual Components** tool.

# UPDATE STRUCTURE

The **Update Structure** tool in **Assistants** group of **Tools** tab in **Ribbon** is used to update the model after modifications have been performed on the model.

# ENGINEERING REFERENCE

The tools in **Engineering Reference** drop-down are used to design a standard part using engineering reference; refer to Figure-14. Note that you need to first save the assembly model before using the tools in this drop-down. Various tools of this drop-down are discussed next.

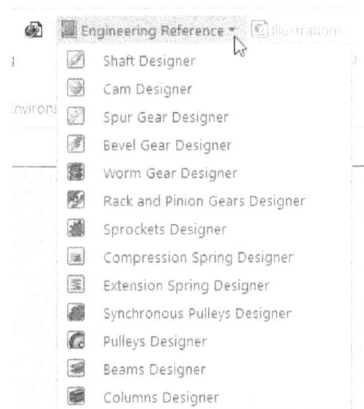

*Figure-14. Engineering Reference drop-down*

## Shaft Designer

A shaft is a rotating machine element used to transmit power from one location to other. Power is delivered to the shaft by some tangential force and resultant torque generated in the shaft makes the transfer of power to other linked elements. The standard size of transmission shafts are given as:

25 mm to 60 mm diameter with 5 mm steps
60 mm to 110 mm diameter with 10 mm steps
110 mm to 140 mm diameter with 15 mm steps
140 mm to 500 mm diameter with 20 mm steps

Standard lengths of shafts are 5 m, 6 m, and 7 m.

While designing a shaft, we need to take care of following stresses:

1. Shear stresses due to the transmission of torque (i.e. due to torsional load).
2. Bending stresses (tensile or compressive) due to the forces acting upon machine elements like gears, pulleys etc. as well as due to the weight of the shaft itself.
3. Stresses due to combined torsional and bending loads.

According to American Society of Mechanical Engineers (ASME) code for the design of transmission shafts, the maximum permissible working stresses in tension or compression may be taken as,

(a) 112 MPa for shafts without allowance for keyways.
(b) 84 MPa for shafts with allowance for keyways.

The maximum permissible shear stress may be taken as
(a) 56 MPa for shafts without allowance for key ways.
(b) 42 MPa for shafts with allowance for keyways.

The **Shaft Designer** tool is used to create a standard shaft based on specified engineering parameters. The procedure to use this tool is given next.

- Click on the **Shaft Designer** tool from the **Engineering Reference** drop-down in the **Environs** group of **Tools** tab in the **Ribbon**. The **Solid Edge Shaft Designer** dialog box will be displayed; refer to Figure-15.

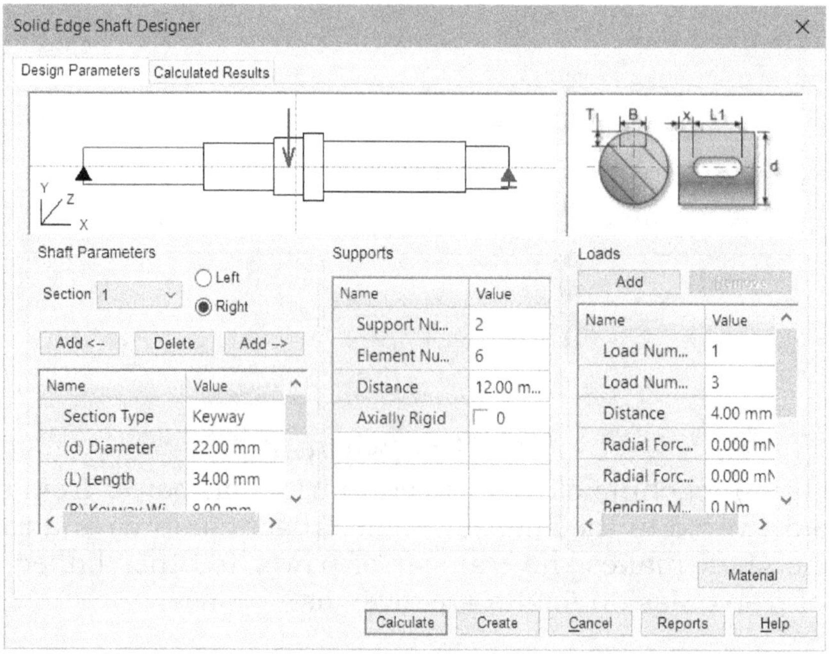

*Figure-15. Solid Edge Shaft Designer dialog box*

- There are three sections of the dialog box to define parameters of shaft in the **Design Parameters** tab of dialog box; **Shaft Parameters**, **Supports**, and **Loads**. These options are discussed next.

## Shaft Parameters

The options in the Shaft Parameters section of the dialog box are used to add or remove sections of the shaft. These options are discussed next.

*   The **Section** drop-down in the **Shaft Parameters** section is used to select the section of shaft to be modified.
*   Select the **Left** or **Right** radio button to define start point of the shaft.
*   Click on the **Add <--** button to add a new section before selected section. Click on the **Add-->** button to add a new section after selected section in **Section** drop-down.
*   Select the desired option from the **Section Type** drop-down to specify what type of shaft section you want to create. Select the **Simple** option from the drop-down to create a simple shaft of specified diameter and length. Select the **Ring Groove** option to create a shaft with ring groove at specified distance from starting of section. Select the **Keyway** option from the drop-down to create a shaft section with cut create for key installation. Select the **Locknut Groove** option from the drop-down to create grooves for installing locknut. Select the **Cone** option from the drop-down to create the shaft section in the form of a cone which has different diameters at start and end point of section. Figure-16 show different shapes of shaft sections.

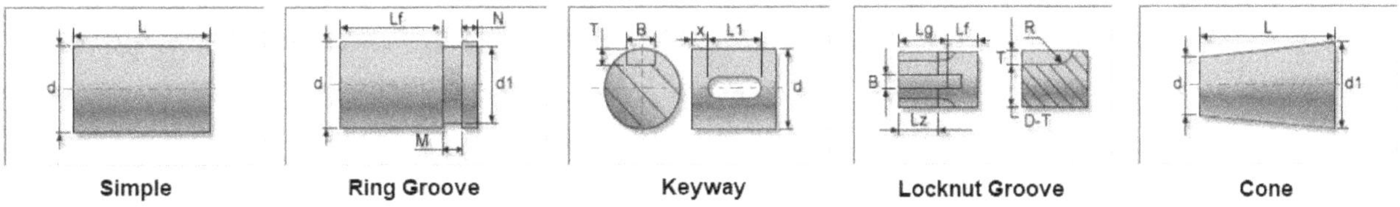

|  Simple  |  Ring Groove  |  Keyway  |  Locknut Groove  |  Cone  |

*Figure-16. Shaft section shapes*

*   After selecting the desired type of shaft, specify the related parameters in the table.

## Support Parameters

The options in the **Supports** section are used to how shaft will be supported in the assembly for load distribution. The options in this section are discussed next.

*   Select the desired number of support whose position is to be modified from the **Support Number** drop-down. There are two supports applied to shaft by default.
*   Select the desired option from the **Element Number** drop-down to define at which section of shaft will the support be applied.
*   Specify the desired value in the **Distance** edit box to define the distance of support from starting location of the section.
*   Select the **Axially Rigid** check box to make the shaft rigidly held at intersection.

## Load Parameters

The options in the **Loads** section are used to specify number of loads, location of load, and type of loads applied on the shaft. These options are discussed next.

*   Click on the **Add** button from the **Loads** section of dialog box to add a new load on the shaft. A new load will be applied and preview will be displayed in the preview area of the dialog box.
*   Select the desired option from the **Load Number** edit box to define which load you want to modify.
*   Select the desired option from the next drop-down to specify on which section of shaft, the load will be applied.

- Click in the **Distance** edit box and specify the distance from start of section for applying load.
- Specify the desired values of radial force, bending moment, axial force, and torque in the respective edit boxes of the **Loads** section.
- Click on the **Material** button from the dialog box and select the desired material from the **Material Values** dialog box displayed; refer to Figure-17. Click on the **OK** button from the dialog box to apply the material.

*Figure-17. Material Values dialog box*

- After setting desired parameters, click on the **Calculate** button from the **Solid Edge Shaft Designer** dialog box. The **Calculated Results** tab will be displayed in the dialog box with a graph showing distribution of load on shaft; refer to Figure-18.
- Select the desired radio buttons from the **Graph Options** area to check results.
- Click on the **Reports** button from the dialog box to generate a report file based on results.

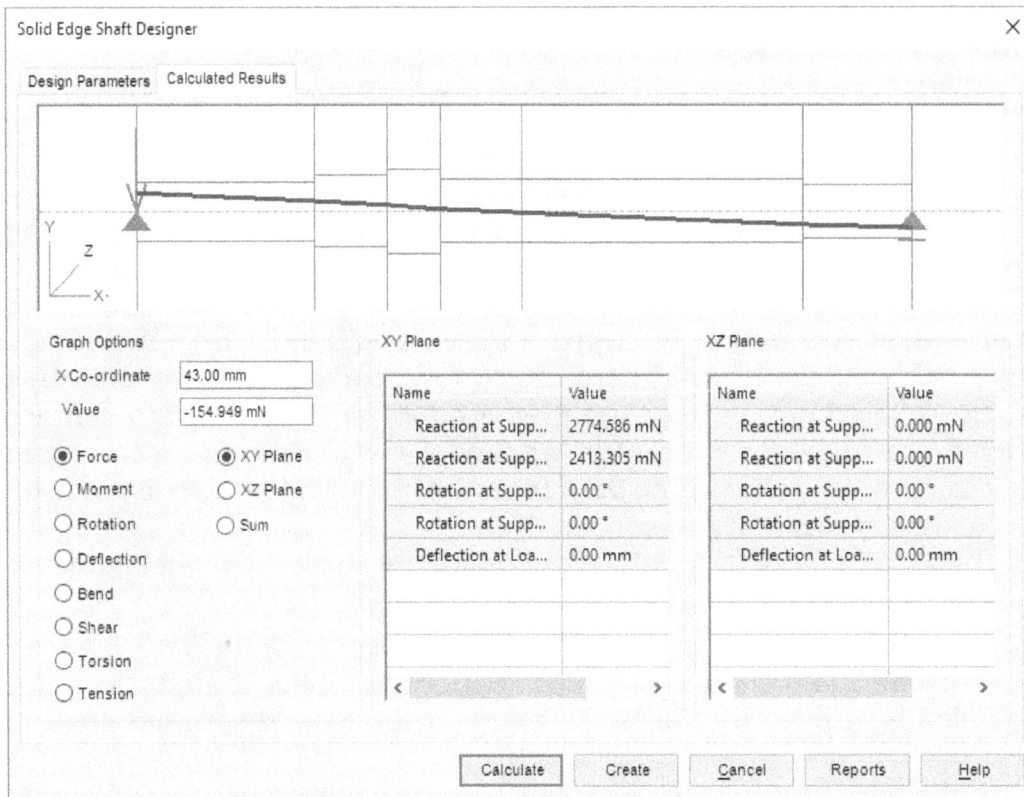

*Figure-18. Calculated Results tab*

- Click on the **Create** button from the dialog box. The **Save As** dialog box will be displayed; refer to Figure-19.

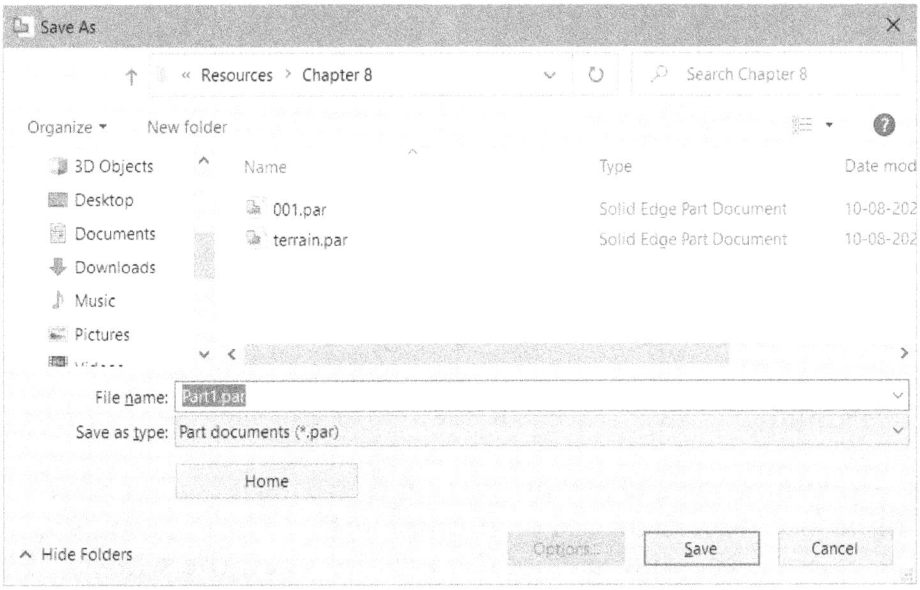

*Figure-19. Save As dialog box*

- Specify desired name of the shaft file in the **File name** edit box and click on the Save button. The file will be saved and added in the assembly; refer to Figure-20.

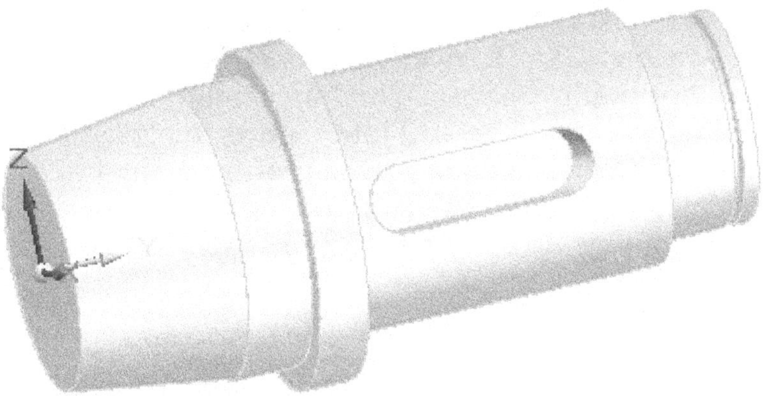

*Figure-20. Shaft created*

### Finding Torque

Most of the time in real world problems, you will get the load in Watt and RPM like, a shaft need to transfer 20KW load at 200 RPM. In such cases, you need to convert this value into torque. The formula is given next.

$$T = \frac{P \times 60}{2\pi N}$$

Here, T is torque, P is the power to be transferred, and N is the RPM.

After getting the torque value, you can apply it on the model to test the failure of shaft.

Example, A line shaft rotating at 200 r.p.m. is to transmit 20 kW. The shaft may be assumed to be made of mild steel with an allowable shear stress of 42 MPa. Determine the diameter of the shaft, neglecting the bending moment on the shaft.

### Finding Bending Moment

Sometimes, you will get the problems like a load of 50kN is working at a distance of 100 mm outside the wheelbase. In such cases, you need to find out the bending moment by using the formula,

$$M = W \times L$$

Here, M is the bending moment, W is the load, and L is the distance from wheelbase.

Example, A pair of wheels of a railway wagon carries a load of 50 kN on each axle box, acting at a distance of 100 mm outside the wheel base. The gauge of the rails is 1.4 m. Find the diameter of the axle between the wheels, if the stress is not to exceed 100 MPa.

## Cam Designer

The transformation of one of the simple motions, such as rotation, into any other motions is often conveniently accomplished by means of a cam mechanism. A cam mechanism usually consists of two moving elements, the cam and the follower, mounted on a fixed frame. Cam devices are versatile, and almost any arbitrarily-specified motion can be obtained. In some instances, they offer the simplest and most compact way to transform motions. In Plate cam or disk cam, the follower moves in a plane perpendicular to the axis of rotation of the camshaft. A translating or a swing arm follower must be constrained to maintain contact with the cam profile. Figure-21 shows the nomenclature of a disc cam.

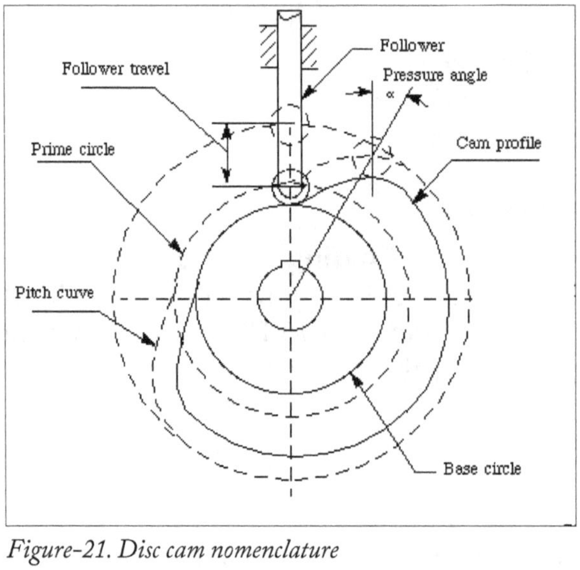

*Figure-21. Disc cam nomenclature*

The **Cam Designer** tool is used to create standard cam based on specified engineering parameters. The procedure to use this tool is given next.

*   Click on the **Cam Designer** tool from the **Engineering Reference** drop-down in the **Environs** group of **Tools** tab in the **Ribbon**. The **Solid Edge Cam Designer** dialog box will be displayed; refer to Figure-22.

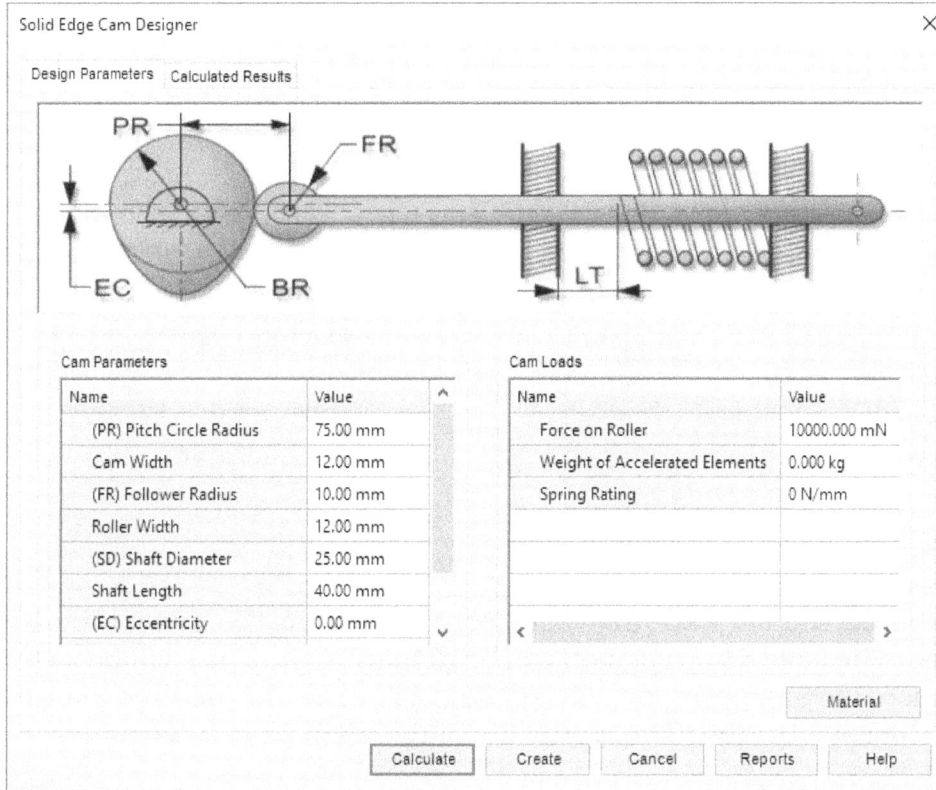

*Figure-22. Solid Edge Cam Designer dialog box*

*   Specify the cam parameters and load parameters in respective sections of the dialog box as discussed earlier.
*   After setting desired parameters, click on the **Calculate** button to check design and loading graph of Cam in calculated result; refer to Figure-23.

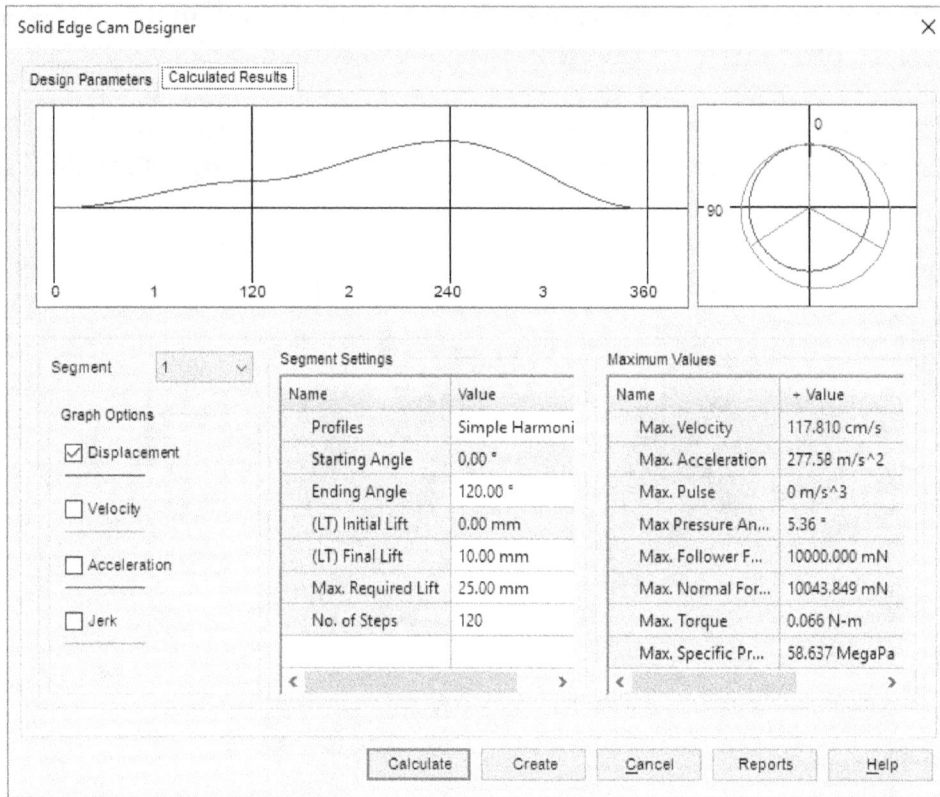

*Figure-23. Calculated Result Cam*

- Once you find the cam suitable for design, click on the **Create** button from the dialog box. The **Save As** dialog box will be displayed and you will be asked to specify the location for saving model file of Cam. Specify the desired name of file and click on the **Save** button. The model file of cam will be created; refer to Figure-24.

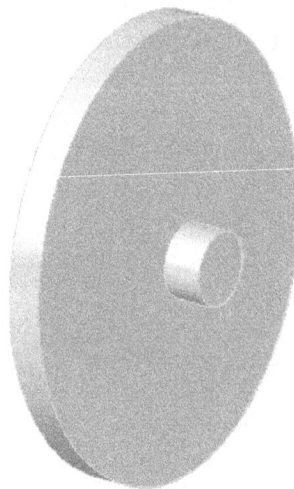

*Figure-24. Cam Desiner*

## Spur Gear Designer

For any kind of gear design, there are a few requirements of designers. In designing a gear drive, following data is usually given :

1. The power to be transmitted,
2. The speed of the driving gear,
3. The speed of the driven gear or the velocity ratio, and
4. The centre distance.

The following requirements must be met in the design of a gear drive :

(a) The gear teeth should have sufficient strength so that they will not fail under static loading or dynamic loading during normal running conditions.
(b) The gear teeth should have wear characteristics so that their life is satisfactory.
(c) The use of space and material should be economical.
(d) The alignment of the gears and deflections of the shafts must be considered because they effect on the performance of the gears.
(e) The lubrication of the gears must be satisfactory.

The different modes of failure of gear teeth and their possible remedies to avoid the failure, are as follows :

1. **Bending failure**. Every gear tooth acts as a cantilever. If the total repetitive dynamic load acting on the gear tooth is greater than the beam strength of the gear tooth, then the gear tooth will fail in bending, i.e. the gear tooth will break. In order to avoid such failure, the module and face width of the gear is adjusted so that the beam strength is greater than the dynamic load.

2. **Pitting**. It is the surface fatigue failure which occurs due to many repetition of Hertz contact stresses. The failure occurs when the surface contact stresses are higher than the endurance limit of the material. The failure starts with the formation of pits which continue to grow resulting in the rupture of the tooth surface. In order to avoid the pitting, the dynamic load between the gear tooth should be less than the wear strength of the gear tooth.

3. **Scoring**. The excessive heat is generated when there is an excessive surface pressure, high speed or supply of lubricant fails. It is a stick-slip phenomenon in which alternate shearing and welding takes place rapidly at high spots. This type of failure can be avoided by properly designing the parameters such as speed, pressure and proper flow of the lubricant, so that the temperature at the rubbing faces is within the permissible limits.

4. **Abrasive wear**. The foreign particles in the lubricants such as dirt, dust or burr enter between the tooth and damage the form of tooth. This type of failure can be avoided by providing filters for the lubricating oil or by using high viscosity lubricant oil which enables the formation of thicker oil film and hence permits easy passage of such particles without damaging the gear surface.

5. **Corrosive wear**. The corrosion of the tooth surfaces is mainly caused due to the presence of corrosive elements such as additives present in the lubricating oils. In order to avoid this type of wear, proper anti-corrosive additives should be used.

## Designing Spur Gear

In order to design spur gears, the following procedure may be followed :

The design tangential tooth load is obtained from the power transmitted and the pitch line velocity by using the following relation :

$$W_T = \frac{P}{v} \times C_S$$

where $W_T$ = Permissible tangential tooth load in newtons,

$\quad\quad$ P = Power transmitted in watts,

$\quad\quad$ v = Pitch line velocity in m / s = $(\pi DN)/60$

$\quad\quad$ D = Pitch circle diameter in metres,

We know that circular pitch,

$\quad\quad p_c = \pi\ D\ /\ T = \pi\ m$
$\quad\quad D = m.T$

Thus, the pitch line velocity may also be obtained by using the following relation, i.e.

$$v = \frac{\pi D.N}{60} = \frac{\pi\ m.T.N}{60} = \frac{p_c.T.N}{60}$$

where,$\quad\quad$ m = Module in metres, and

$\quad\quad\quad\quad$ T = Number of teeth.

The **Spur Gear Designer** tool is used to design standard spur gear based on specified geometry parameters and loads. The procedure to use this tool is given next.

- Click on the **Spur Gear Designer** tool from the **Engineering Reference** drop-down in the **Environs** group of **Tools** tab in the **Ribbon**. The **Solid Edge Spur Gear Designer** dialog box will be displayed. Specify the design parameter of spur gear in dialog box; refer to Figure-25.
- You can also specify the material of spur gear by clicking on **Material** button as discussed earlier.

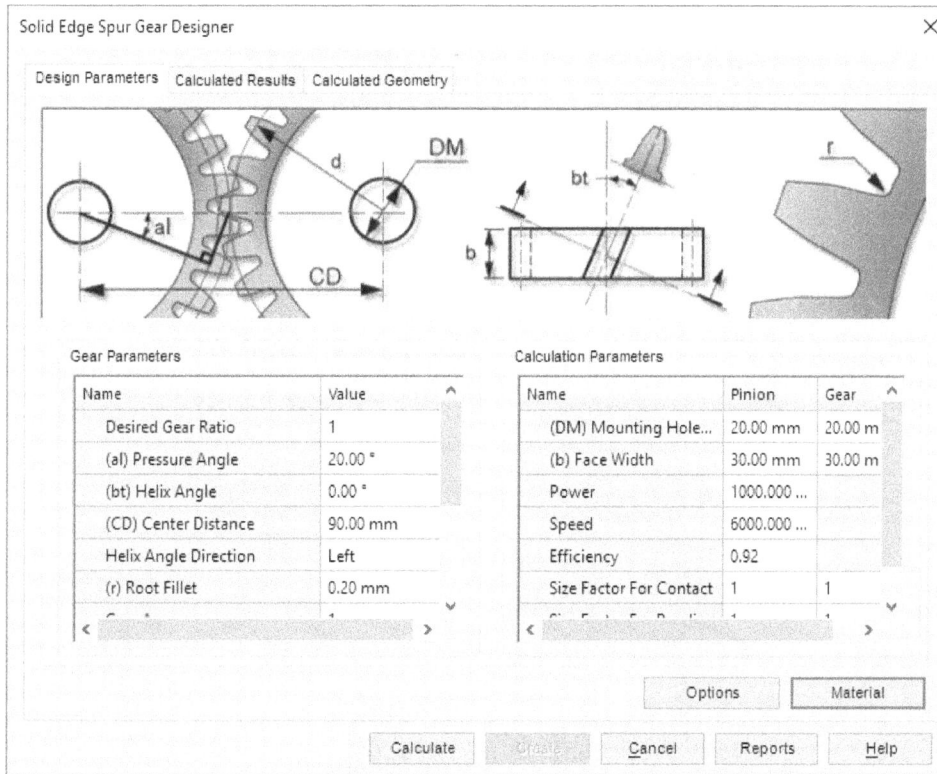

Figure-25. Spur Gear Designer dialog box

- Click on the **Options** button from the dialog box to define the type of spur gear. The **Design Parameters - Input Conditions** dialog box will be displayed; refer to Figure-26. Select the desired radio button from the **Type of Gearing** area to define whether you want to create an internal gear or external gear. Select the desired radio button from the **Output Geometrical Parameters** area to specify what type of parameters you want to output for gear. Similarly, select the desired radio buttons to define the input conditions and click on the **OK** button.

Figure-26. Design Parameters-Input Conditions dialog box

- After setting parameters, click on the **Calculate** button from the dialog box. The **Finalize Output Results** dialog box will be displayed; refer to Figure-27.
- Specify the desired parameters in the dialog box and click on the **Apply** button. The **Correction Methods** dialog box will be displayed; refer to Figure-28.

*Figure-27. Finalize Output Results dialog box*

*Figure-28. Correction Methods dialog box*

- Select the desired option from the **Correction Methods** drop-down to apply size correction of gears and click on the **Apply** button. The **Calculated Results** tab will be displayed in the dialog box; refer to Figure-29.

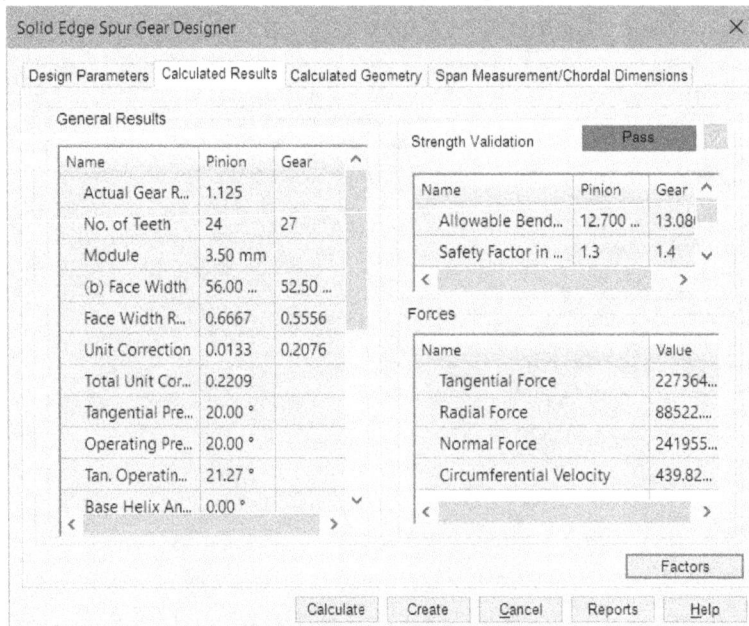

*Figure-29. Calculated Results tab of Spur gear design*

- You can also check the calculated geometry of spur gear, by clicking on the **Calculated Geometry** tab; refer to Figure-30.

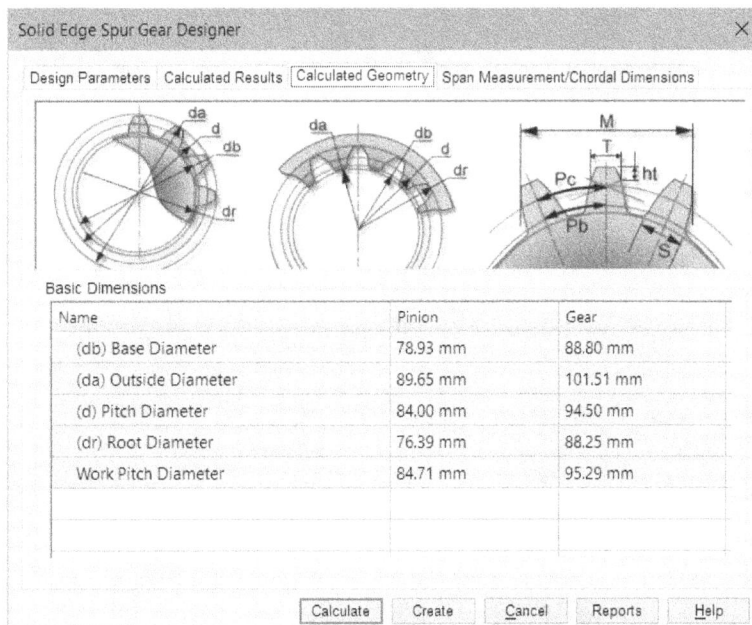

Figure-30. Calculated geometry Spur gear

- Click on the **Span Measurement/Chordal Dimensions** tab in the dialog box to check chordal dimensions.
- Click on the **Create** button from the dialog box. The **Save As** dialog box will be displayed. Specify the name of gear model and click on the **Save** button. The spur gears will be created; refer to Figure-31.

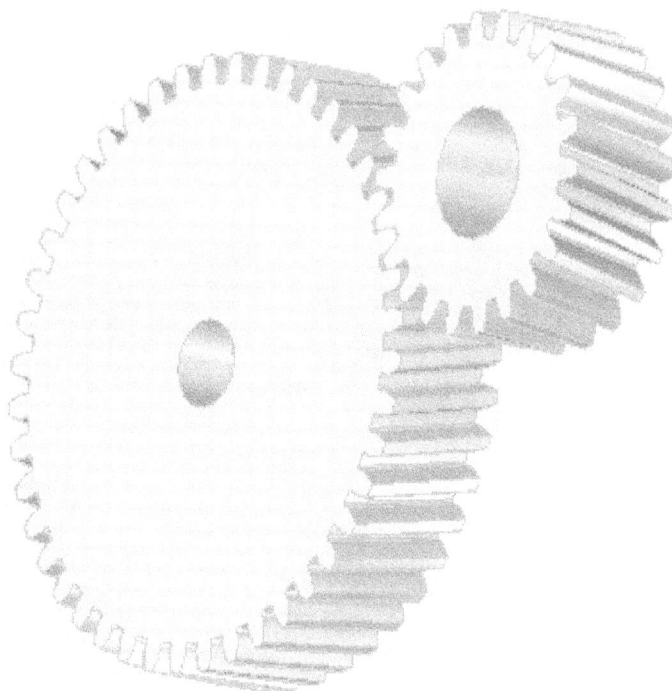

Figure-31. Spur gears created

## Bevel Gear Designer

The bevel gears are used for transmitting power at a constant velocity ratio between two shafts whose axes intersect at a certain angle. The pitch surfaces for the bevel gear are frustums of cones; refer to Figure-32.

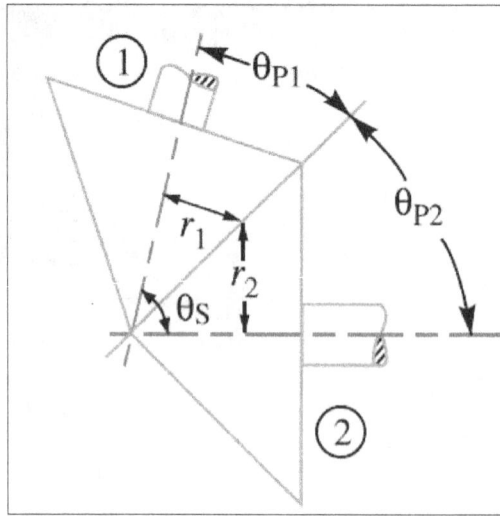

*Figure-32. Bevel gear arrangement*

Here, $r_1$ and $r_2$ are radii of gear 1 and gear 2 respectively.
$\Theta_{P1}$ and $\Theta_{P2}$ are cone angle for gears
and, $\Theta_S$ is shaft angle.

The **Bevel Gear Designer** tool is used to create bevel gears based on specified engineering parameters. The procedure to use this tool is given next.

- Click on the **Bevel Gear Designer** tool from the **Engineering Reference** drop-down in the **Environs** group of **Tools** tab in the **Ribbon**. The **Solid Edge Bevel Gear Designer** dialog box will be displayed; refer to Figure-33.
- Specify the design parameters and material as discussed earlier and click on the **Calculate** button. The **Calculated Results** tab of dialog box will be displayed; refer to Figure-34.

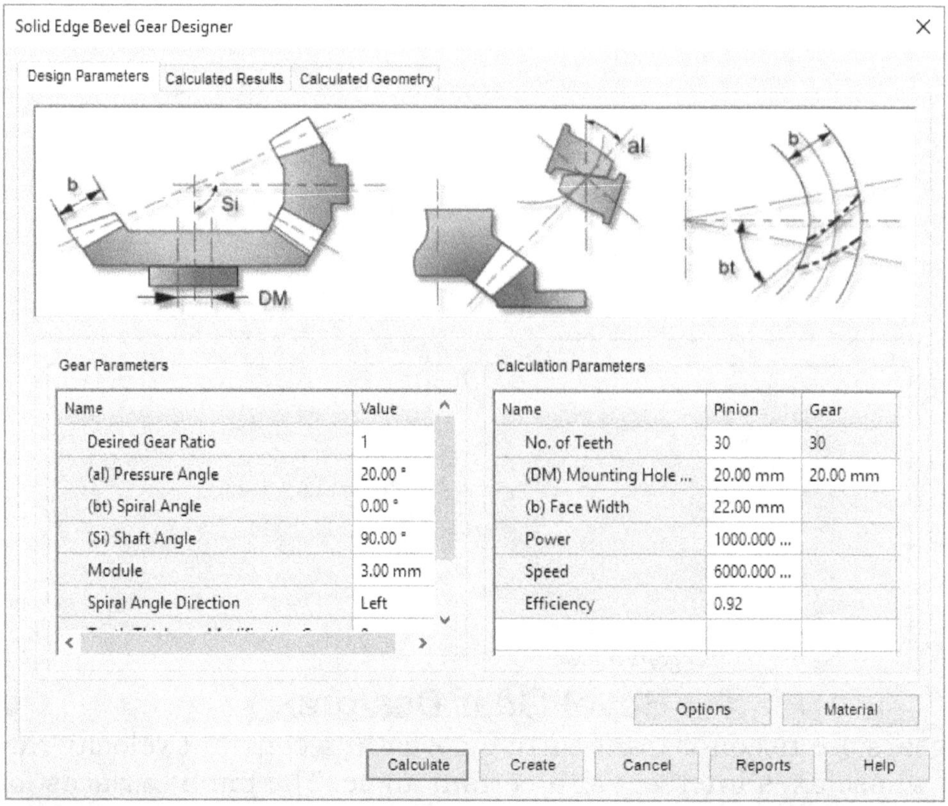

*Figure-33. Design paramters of Bevel gear*

*Figure-34. Calculated Results tab for Bevel Gear*

Rest of the tools are same as we discuss earlier.

# EXPLODE RENDER ANIMATE ENVIRONMENT

The **ERA** stands for **Explode- Render- Animate** environment. Click on the **ERA** tool from the **Environs** group of **Tools** tab in the **Ribbon**. The **Explode - Render - Animate** environment will be displayed and ERA tools will be available in the **Ribbon**; refer to Figure-35.

*Figure-35. ERA Ribbon*

# ANIMATION EDITOR

The **Animation Editor** tool is used to play and produce the animation of assembly. On clicking this tool, the **Animation Editor** window will be displayed at the bottom in the application window; refer to Figure-36. Various tools and options of the **Animation Editor** are discussed next.

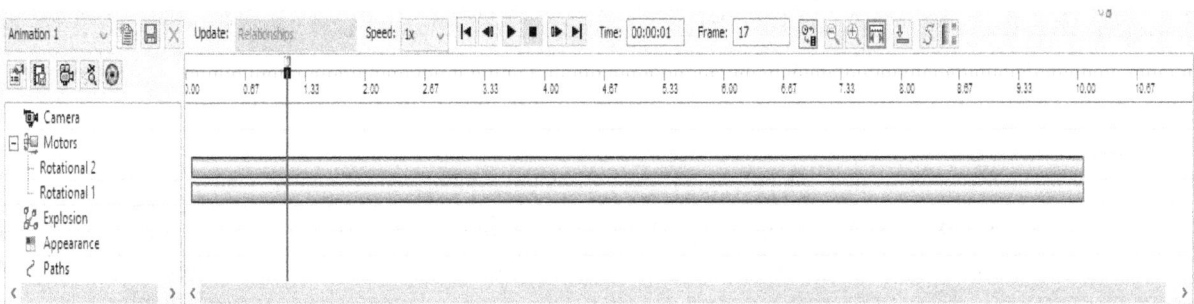

*Figure-36. Animation Editor*

## Animation Drop-down

If you have created multiple animations then a list of animation will be displayed in this drop-down. Select the desired animation from the list to modify it.

## Creating New Animation

The **New Animation** button in the **Animation Editor** is used to create a new animation. The new animation will be listed in the **Animation** drop-down. On clicking this button, the **Animation Properties** dialog box will be displayed; refer to Figure-37. Specify the name of new animation in the **Animation** edit box of dialog box. Select the desired radio button from the **Frames per second** section of the dialog box to specify the number of frames generated per second in new animation. You can select the **NTSC** and **PAL** radio buttons to use standard frame rates or select the **Custom** radio button and specify the desired value in the edit box. Specify the desired value in the **Animation length** edit box to define total length of animation. Select the **Stop playback after last event** check box to automatically stop animation when duration has passed. After setting desired parameters, click on the **OK** button to create the animation.

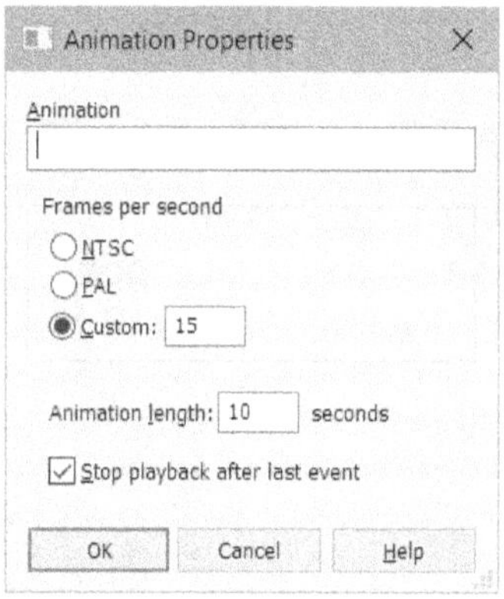

*Figure-37. Animation Properties*
*dialog box*

## Saving Animation

Click on the **Save Animation** button from the **Animation Editor** to save the modifications in animation.

## Deleting Animation

Click on the **Delete Animation** button from the **Animation Editor** to delete the animation selected in the Animation drop-down. A confirmation box will be displayed. Select the **Yes** button from the confirmation box to delete the animation.

## Animation Properties

The **Animation Properties** button in **Animation Editor** is used to modify the properties of the selected animation. On clicking this button, the **Animation Properties** dialog box is displayed. The options in this dialog box have been discussed earlier.

## Save as Movie

The **Save as Movie** button in the **Animation Editor** is used to save the current animation in movie clip format (WMV, AVI, and BIP). On clicking this button, the **Save As Movie** dialog box will be displayed; refer to Figure-38.

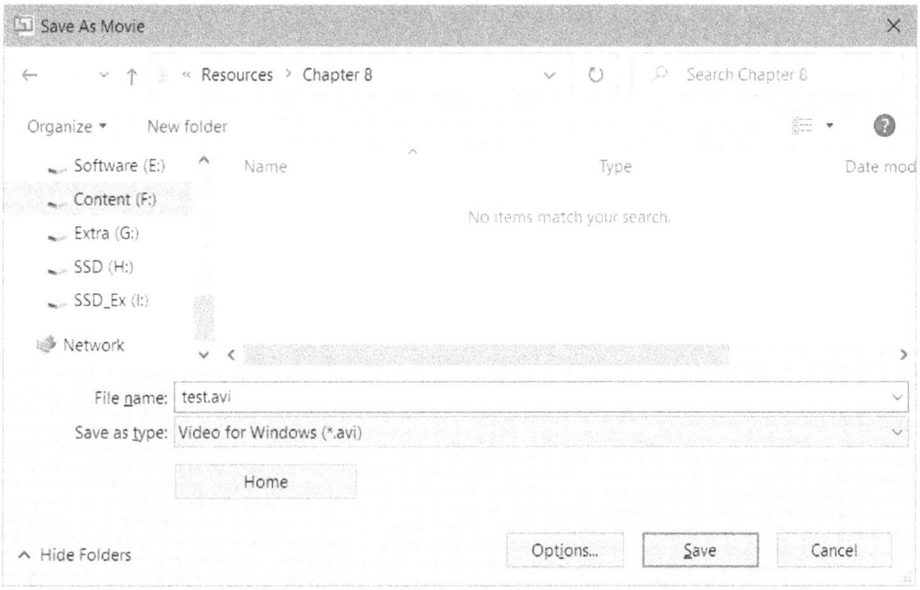

*Figure-38. Save As Movie dialog box*

Specify the desired name for animation video in the **File name** edit box. Specify the desired format for video file in the **Save as type** drop-down. Click on the **Options** button from the dialog box to specify parameters of video. The **Save As Movie Options** dialog box will be displayed; refer to Figure-39. Specify the desired parameters like output size, time line, codec in the dialog box and click on the **OK** button. After setting all the parameters, click on the **Save** button from the dialog box to save the animation video file.

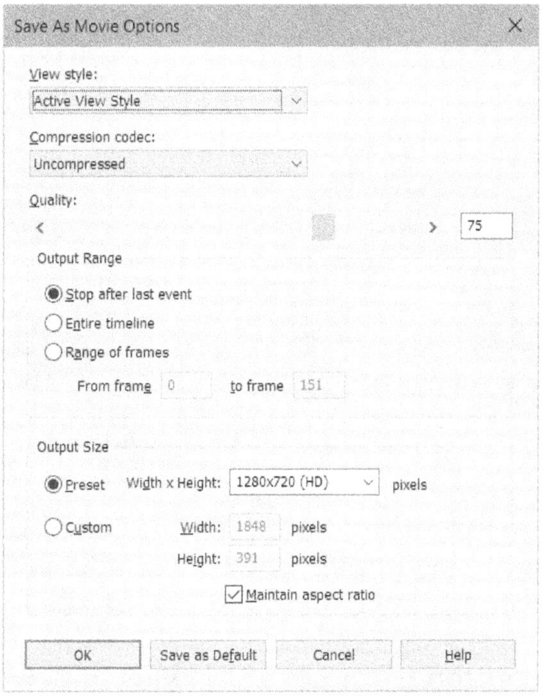

*Figure-39. Save As Movie Options dialog box*

## Creating Camera Path and Enabling Camera Movement

The **Camera Path** button in the **Animation Editor** is used to enable or disable the camera movement in animation. After enabling camera movement in animation, now we will create camera path. The procedure to create camera path is given next.

- Right-click on the **Camera** option from the **Animation PathFinder**; refer to Figure-40 and select the **Edit Definition** option from the shortcut menu. The **Camera Path Wizard** dialog box will be displayed; refer to Figure-41.

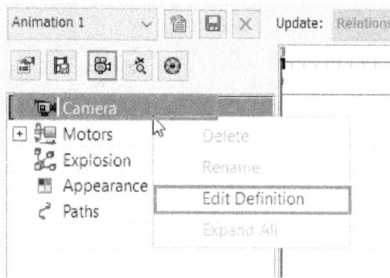

*Figure-40. Edit Definition option for Camera*

*Figure-41. Camera Path Wizard dialog box*

- Select the desired option from the dialog box like if you want to fly around the model using **Clockwise** or **Counterclockwise** radio button. If you want to use named views for camera rotation then select the **Build using Named Views** radio button. If you want to copy camera path from an existing camera path then select the **Copy existing Camera Path from** radio button and then select the desired path from the drop-down. After selecting the desired radio button (in our case, it is **Build using Named Views** radio button) and click on the **Next** button. The options in the dialog box will be displayed as shown in Figure-42.

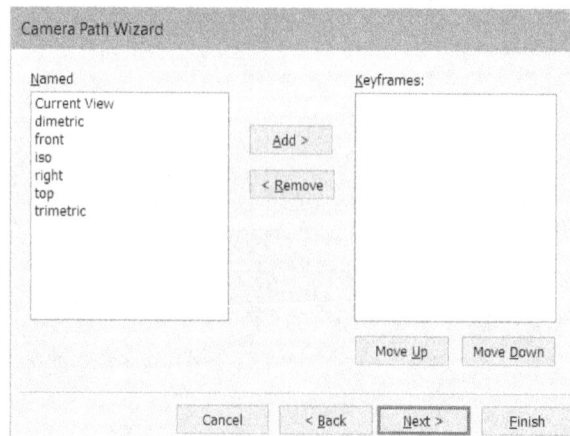

*Figure-42. Views for camera path*

- Select the desired plane from the **Named** list and click on the **Add >** button. The plane will be added in the keyframe. One by one add the desired planes and then click on the **Next >** button. Preview page of the camera path will be displayed in the dialog box.
- Click on the **Preview** button to check preview of camera path. Click on the **Play** button to check the camera path.
- Click on the **Finish** button from the dialog box to create the camera path.

## SHOW CAMERA PATH

The **Show Camera Path** button from the **Animation PathFinder** to display path of camera movement during animation.

## WORKING WITH ANIMATION

The buttons at the top in the **Animation Editor** are used to play, stop, restart, and switch frames in the animation.

## CREATING KEYSHOT ANIMATION

Once you have checked the animation, click on the **KeyShot Animate** button from the **Animation Editor** box if you want to create animation using the KeyShot software. The procedure to use this button is given next.

* Click on the **KeyShot Animate** button from the **Animation Editor**. The **KeyShot Options** dialog box will be displayed; refer to Figure-43.

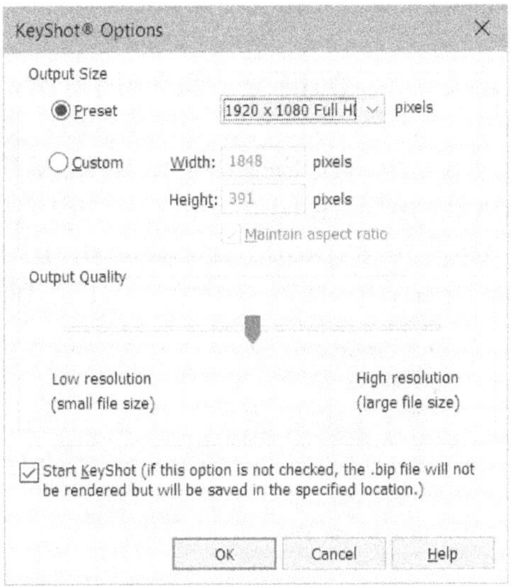

*Figure-43. KeyShot Options dialog box*

* Select the **Preset** radio button and then select the desired resolution for animation video from the drop-down. You can also specify custom resolution of animation video by selecting the **Custom** radio button.
* Set the desired output quality using the **Output Quality** slider and click on the **OK** button. The **KeyShot For Solid Edge** application window will be displayed with rendering of model.
* Click on the **Animation** button at the bottom in the application window to check animation and save the animation using the **Save** tool from **File** menu.
* Close the application by **Close** button at the top right corner.

## KEYSHOT RENDER

The **KeyShot Render** tool is used to render an assembly or part by applying realistic material and environment. On clicking this tool from the **KeyShot** group of **Home** tab in **Ribbon**, the KeyShot application will be displayed. We will discuss about the options of this application later in this book.

## AUTO EXPLODE

The **Auto Explode** tool is used to explode the assembly automatically. The procedure to use this tool is given next.

- Click on the **Auto Explode** tool from the **Explode** group in the **Home** tab of the **Ribbon**. The **Auto Explode Command Bar** will be displayed; refer to Figure-44.

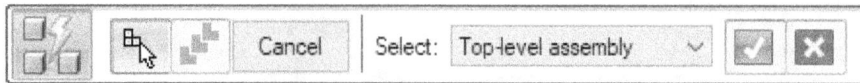

*Figure-44. Auto Explode Command Bar*

- Select the desired option from the **Select** drop-down. If you want to explode a subassembly then select the **Subassembly** option from the drop-down and select the desired subassembly from the model. After selecting assembly/subassembly, right-click in the drawing area to accept the selection.
- Select the **Automatic Spread Distance** button from the **Command Bar** to automatically set the gap between parts. If you want to manually specify gap between components then toggle **OFF** the **Automatic Spread Distance** button and specify desired value in the **Distance** edit box.
- Click on the **Automatic Explode Options** button from the **Command Bar** to specify how subassemblies will be exploded. The **Automatic Explode Options** dialog box will be displayed; refer to Figure-45. Select the **Bind all subassemblies** check box to keep the exploded subassemblies together. Select the desired radio button from the **Explode Technique** area to explode parts individually or using subassembly level. After setting desired parameters, click on the **OK** button.

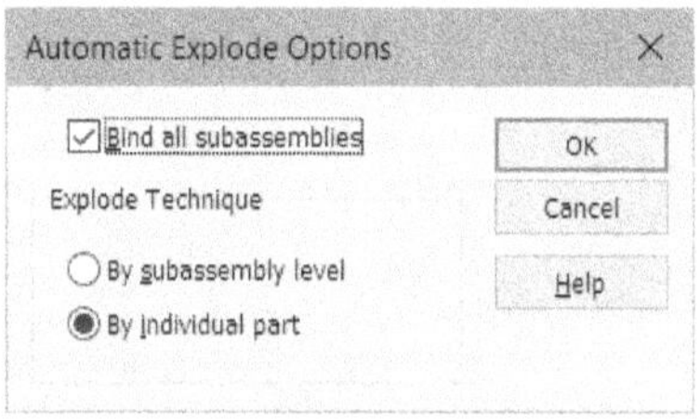

*Figure-45. Automatic Explode Options dialog box*

- Click on the **Explode** button from the **Command Bar** to apply automatic explode view of assembly. The exploded view of assembly will be displayed; refer to Figure-46.

*Figure-46. Auto Exploded view*

- Click on the **Finish** button from the **Command Bar** and press **ESC** to exit.

# EXPLODE

The **Explode** tool is used to explode the assembly manually. The procedure to use this tool is given next.

*   Click on the **Explode** button from the **Explode** group in the **Home** tab of **Ribbon**. The **Explode Command Bar** will displayed; refer to Figure-47.

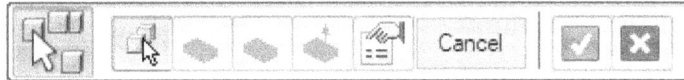

*Figure-47. Explode Commnad bar*

*   Select the part you want to move and right-click in the drawing area. You will be asked to select a base part.
*   Select the desired base part to use as reference for explode. You will be asked to select a face of base part.
*   Select the desired face of base part. You will be asked to specify direction in which you want to move exploded part.
*   Click to specify the direction and specify desired value in the **Distance** edit box. After specifying value, click on the **Explode** button from the **Command Bar**. The selected part will move at specified distance; refer to Figure-48. Click on the **Finish** button to complete exploding the component. Repeat the procedure to explode other components and then press **ESC** to exit the tool.

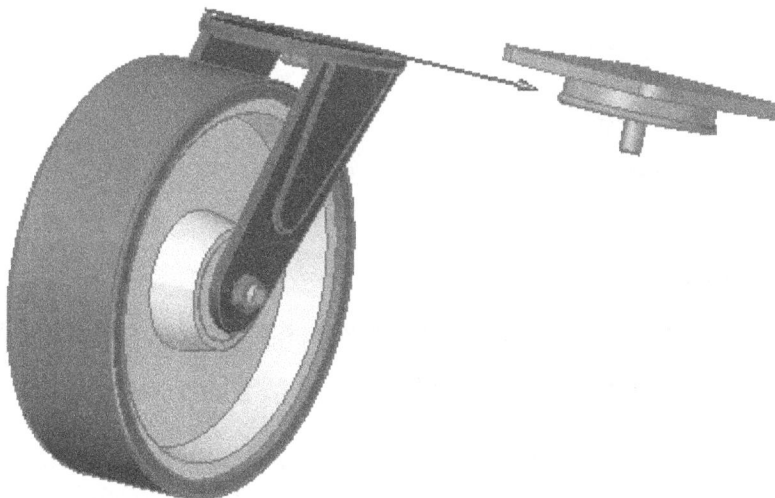

*Figure-48. Explode manually*

# REPOSITION

The **Reposition** tool is used to reorder an exploded part. The procedure to use this tool is given next.

*   Click on the **Reposition** tool from the **Modify** group in the **Ribbon**. The **Reposition Command Bar** will be displayed; refer to Figure-49.

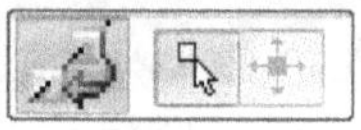

*Figure-49. Reposition Command Bar*

- Select the part to be repositioned. You will be asked to select an earlier exploded component to position the new component.
- Select the desired component. You will be asked to specify the direction of movement.
- Click in the desired direction. The component will move accordingly; refer to Figure-50. Press **ESC** to exit the tool.

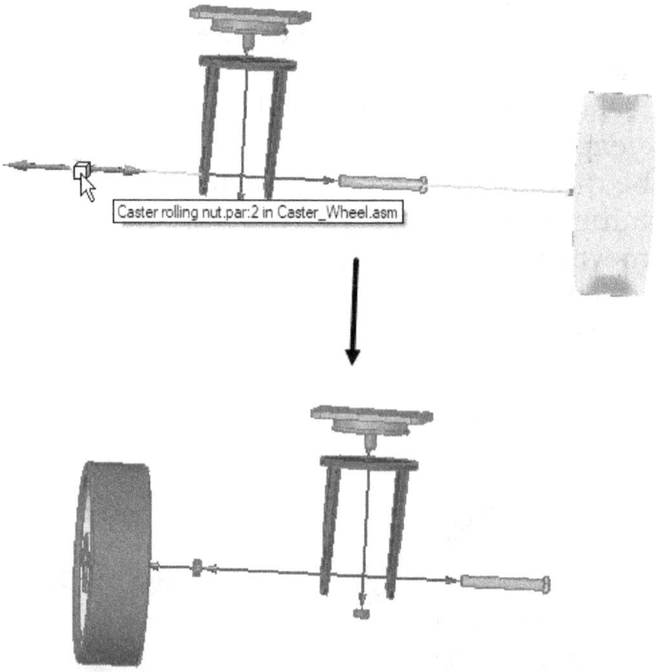

*Figure-50. Reposition View*

## REMOVING COMPONENT FROM EXPLODED STATE

The **Remove** tool in **Modify** group of **Ribbon** is used to remove a component from exploded state and return it to assembled position. To do so, select the component to be removed from the exploded state and click on the **Remove** tool from **Modify** group in **Ribbon**. The component will return to original assembled position and will be hidden.

## COLLAPSING EXPLODED COMPONENT

The **Collapse** tool is used to return selected exploded component to its assembled position. To do so, select the component to be collapsed and click on the **Collapse** tool from the **Modify** group in the **Home** tab of **Ribbon**. The component will return to assembled position.

The **Drag Component** and **Move on Select** tools work in the same way as discussed earlier.

# UNEXPLODE

The **Unexplode** tool is used to return the components of assembly to their native assembled position. The procedure to use this tool is given next.

- Click on the **Unexplode** tool from the **Modify** group of **Home** tab in the **Ribbon**. The Solid Edge message box will displayed as shown in Figure-51. You are asked to either save the exploded view file or the data will erase. If you do not want to save the exploded view then click on **Yes** button to continue. The unexploded view of model will be displayed. Click on the **No** button and save the exploded view as a new configuration if you do not want to see the message box next time.

*Figure-51. Unexplode dialog box*

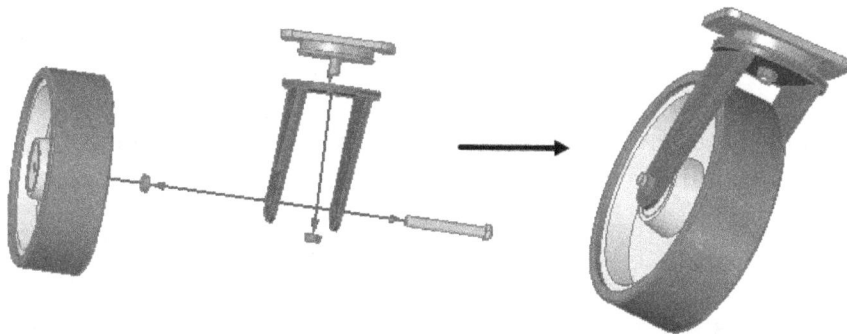

*Figure-52. Unexplode*

# BINDING SUB-ASSEMBLY

The **Bind** tool in **Modify** group is used to keep all the components of a sub-assembly together. To use this tool, select the sub-assembly from the **Path Finder** that you want to bind together and click on the **Bind** tool from the **Modify** group of **Home** tab in the **Ribbon**. The sub-assembly components will be exploded as a single unit.

# UNBINDING SUB-ASSEMBLY

The **Unbind** tool is used to make sub-assemblies explode when exploded of assembly is generated. To do so, select the sub-assembly to be unbound from the **Path Finder** and click on the **Unbind** tool from the **Modify** group in the **Home** tab of the **Ribbon**.

# FLOW LINES

The tools in the **Flow Lines** group is used to create and display annotation lines showing how components are assembled in exploded view. Various tools in this group are discussed next.

## Drop

The **Drop** tool is used to create an individual entry in the **Explode PathFinder** for each flow line created automatically while exploding; refer to Figure-53. You can select a flow line to see its path and what its connects to. It converts the event flow lines in the exploded assembly into annotation flow lines. You can use annotation flow lines of exploded views in draft documents.

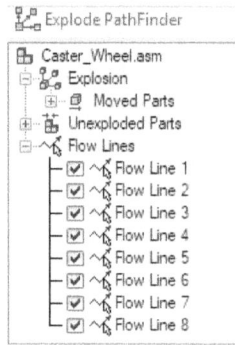

*Figure-53. Explode path finder*

## Draw

The **Draw** button is used to draw annotation flow lines for components of exploded view. You can use annotation flow lines to create an exploded view of assembly model on a drawing sheet. The procedure to use this tool is given next.

*   Click on the **Draw** tool from the **Flow Lines** group in the **Home** tab of **Ribbon**. The **Draw Command Bar** will be displayed; refer to Figure-54.

*Figure-54. Draw Command Bar*

*   Select the desired start key point and end key point to define flow line.
*   Specify the desired value of length of the segments which terminate at ports in the **Port segment** edit box. Preview of the line will be displayed; refer to Figure-55.
*   Click on the **Next** and **Previous** buttons to switch between variations of flow lines between same points.
*   Click on the **Change Orientation** button from **Command Bar** and select the desired face to orient middle section of flow line.
*   Click on the **Finish** button to create flow line and press **ESC** to exit the tool.

*Figure-55. Preview of flow lines*

## Modify

The **Modify** tool is used to modify flow lines. Modification an event flow line or annotation flow line between two exploded parts in an assembly. You can:

- Change the shape of an event flow line.
- Split annotation flow line into multiple segments.
- Change the location of the entire flow line.
- Edit the length of an end segment or the position of a joggle segment.

The procedure to use this tool is given next.

- Click on the **Modify** tool from the **Flow Lines** group of the **Home** tab in the **Ribbon**. The **Modify Command Bar** will be displayed; refer to Figure-56.

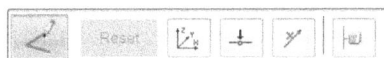

*Figure-56. Modify Command Bar*

- To change orientation, select the **Change Orientation** button from the **Command Bar** and select the face/plane which you want to use for orientation and then select the flow line to be oriented. After selecting flow line, right-click in the drawing area.
- To split a flow line, select the **Split Flow Line** button from the **Command Bar** and then click at desired location on flow line to split it. Right-click in drawing area to exit the button.
- To delete an annotation flow line, select the **Delete Annotation Flow Line** button from the **Command Bar** and select the desired flow line to be deleted. Right-click in the drawing area to delete the flow line.
- Press **ESC** to exit the tool.

Select the **Flow Lines** and **Flow Line Terminators** buttons to display flow lines and flow line terminators, respectively.

# CONFIGURATIONS

Configurations are variations of assembly. A display configuration captures the display status of the parts, assemblies, assembly sketches, weld beads, and reference planes in an assembly. The tools to create and manage configurations are available in the **Configurations** group of the **Home** tab in **Ribbon**; refer to Figure-57. Various tools in this group are discussed next.

*Figure-57. Configurations group*

## Saving and Retrieving Display Configurations

The **Display Configurations** tool is used to save the current displaying configurations of assembly components and retrieve them later. The procedure to use this tool is given next.

- After performing desired modifications of model like exploded view, changed position of assembly components, and so on; click on the **Display Configurations** tool from the **Configurations** group of **Home** tab in the **Ribbon**. The **Display Configurations** dialog box will be displayed; refer to Figure-58.

*Figure-58. Display Configurations dialog box*

- Click on the **New** button from the dialog box. The **New Configuration** dialog box will be displayed.
- Specify desired name in the edit box and click on the **OK** button. A new configuration will be created.
- Select the desired configuration and click on the **Rename** button to change its name.
- Select the configuration and click on the **Update** button after making changes in the model to update the configuration.
- If you want to delete a configuration then select it from dialog box and click on the **Delete** button.
- Click on the **Close** button from the dialog box to exit the tool.

You can apply a configuration by selecting it from the **Current Configuration** drop-down in the **Configurations** group of the **Ribbon**; refer to Figure-59.

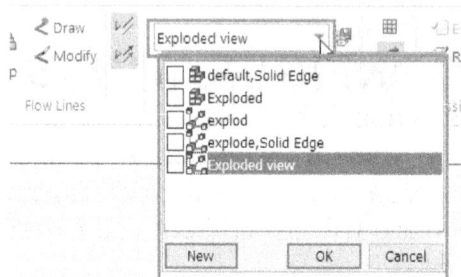

*Figure-59. Current Configuration drop-down*

## Configuration Manager

The **Configuration Manager** tool is used to display/hide components in an assembly configuration. The procedure to use this tool is given next.

- Click on the **Configuration Manager** tool from the **Configurations** group of **Home** tab in the **Ribbon**. The **Configuration Manager** dialog box will be displayed; refer to Figure-60.

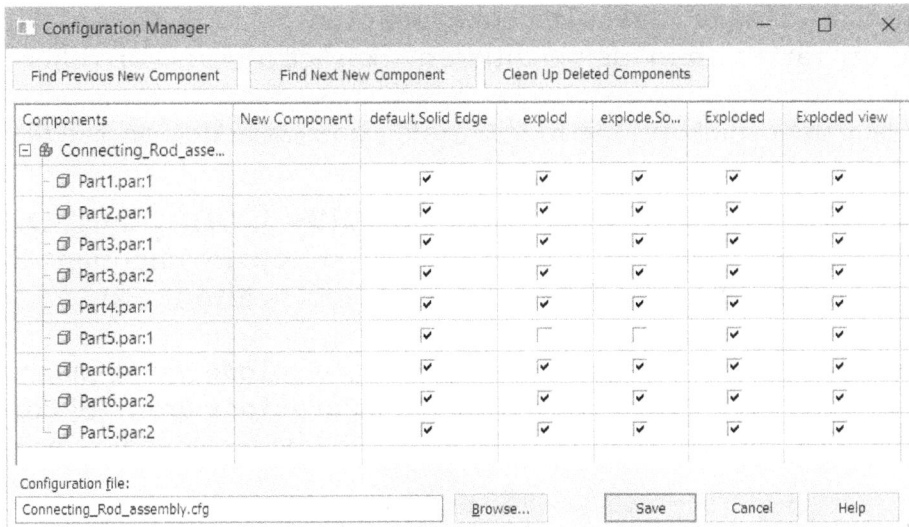

*Figure-60. Configuration Manager dialog box*

- Select or clear check boxes from the configurations to add or remove respective components from them.
- Click on the **Save** button from the dialog box to save the configuration.

## Configuration Options

The **Configuration Options** tool is used to specify parameters for configurations. The procedure to use this tool is given next.

- Click on the **Configuration Options** tool from the **Configurations** group in the **Home** tab of **Ribbon**. The **Display Configuration Options** dialog box will be displayed; refer to Figure-61.

*Figure-61. Display Configuration Options dialog box*

- Select the **Fit view after apply** check box to automatically fit view when a display configuration is applied.
- Select the **List exploded configuration in modeling mode** check box to display exploded view configurations in the modeling mode after you exit ERA environment. After selecting the check box, select the desired radio button from the When section to specify when the configuration will be displayed.

- Select the **Use single default configuration** check box to a single default configuration rather than using different configurations for different users of software.
- Similarly, set the other options in the dialog box and click on the **OK** button.

## Copying Components to Display Configuration

The **Copy Current Display** tool is used to copy selected component to a different display configuration. The procedure to use this tool is given next.

- Select the component you want to copy from the **Explode PathFinder** and click on the **Copy Current Display** tool from the **Configurations** group of **Home** tab in the **Ribbon**. The **Copy Current Display** dialog box will be displayed; refer to Figure-62.

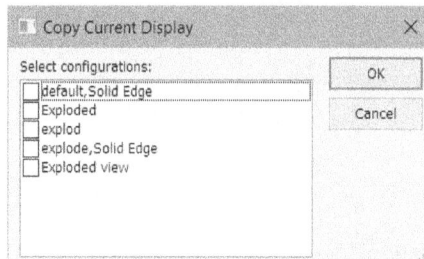

*Figure-62. Copy Current Display dialog box*

- Select the desired check boxes from the dialog box to specify the configurations in which the component will be copied.
- Click on the **OK** button to apply the changes.

## Taking Snapshot and Restoring

The **Take Snapshot** tool is used to keep a record of shown and hidden components in the current configuration of assembly. After applying the show/hide settings to components, click on the **Take Snapshot** tool from the **Configurations** group of the **Home** tab in the **Ribbon**. A snapshot of settings will be saved.

Open another configuration of assembly with different show/hide settings and change the settings of components. Click on the **Restore Snapshot** button to apply the settings earlier saved using **Take Snapshot** tool.

## Unloading Hidden Parts

The **Unload Hidden Parts** tool is used to remove hidden parts from the temporary memory of assembly. This tool is available in the **Configurations** group of **Home** tab in the **Ribbon**.

## MODES

The tools in the **Modes** group are used to active different assembly modes for getting better performance. Select the **Large Assembly Performance** toggle button to increase the performance of manipulating a large assembly at the cost of better visualization. Select the **Limited Update** toggle button to reduce the frequency of automatic model updation when you have made changes in the assembly. Select the **Limited Save** toggle button to reduce the load on system when you save the assembly. When this button is ON, only the directly modified parts will be saved with assembly.

# FOR STUDENT NOTES

FOR STUDENT NOTES

# Chapter 10

# Assembly - III

## Topics Covered

The major topics covered in this chapter are:

- *Introduction to XpressRoute*
- *Introduction to Electrical Routing*
- *Introduction to Frame Design*
- *Publishing Model to HTML Web Page*

# INTRODUCTION TO XPRESROUTE

The tools in the Xpresroute environment are used to create tube and pipe routes in the assembly. To activate this environment, click on the **Xpresroute** tool from the **Environs** group in the **Tools** tab of **Ribbon**. The Xpresroute environment will be displayed; refer to Figure-1.

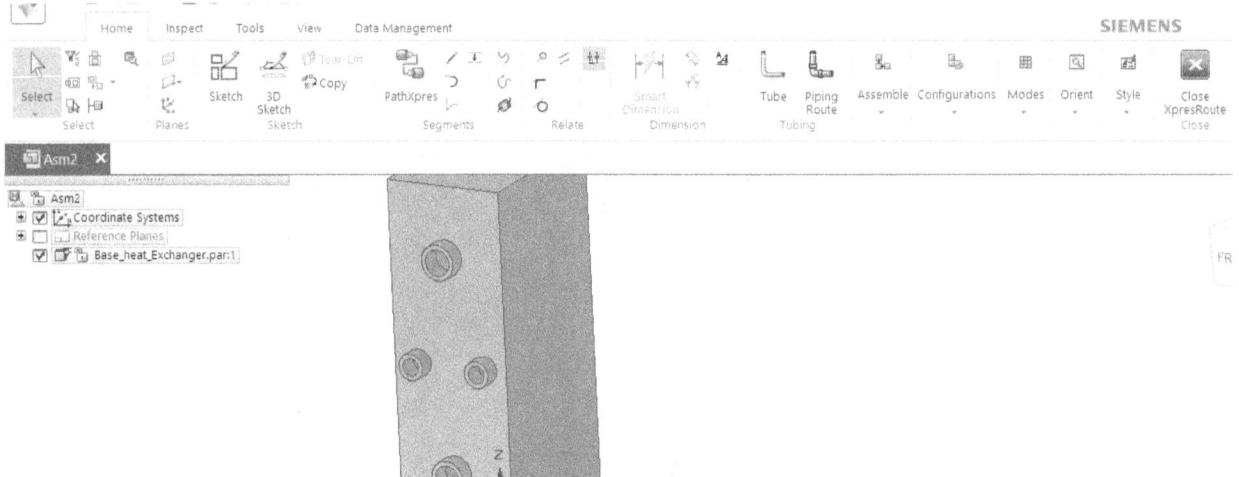

*Figure-1. Xpresroute environment*

# CREATING PATH FOR ROUTE

The **PathXpres** tool is used to automatically create path for routing pipes and tubes. The procedure to use this tool is given next.

- Click on the **PathXpres** tool from the **Segments** group in the **Home** tab of **Ribbon**. The **PathXpres Command Bar** will be displayed; refer to Figure-2 and you will be asked to specify start point of route path.

*Figure-2. PathXpres Command Bar*

- Select the desired points (centers of circles) to specify start point and end point of route. Preview of route path will be displayed; refer to Figure-3.

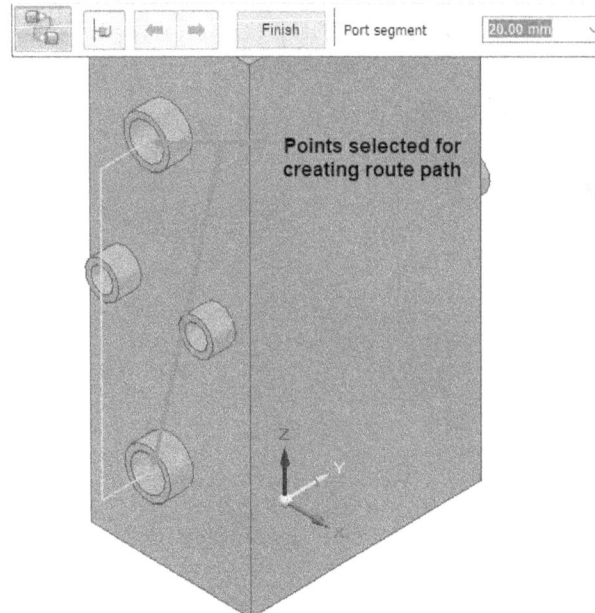

*Figure-3. Preview of route path*

- Specify the desired value in the **Port segment** edit box to specify length of route at the start and end points of route path. Press **TAB** to check the preview of route path.
- Click on the **Next** and **Preview** buttons to switch between different shapes of path curves.
- Click on the **Finish** button from the **Command Bar** to create the route path; refer to Figure-4.

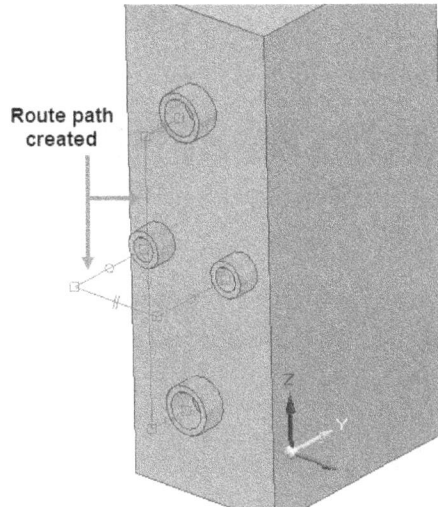

Route path
created

*Figure-4. Route path created*

- Press **ESC** to exit the tool.

## CREATING LINE SEGMENT OF ROUTE PATH

The **Line Segment** tool is used to create a line segment of route path. The procedure to use this tool is given next.

- Click on the **Line Segment** tool from the **Segments** group in the **Home** tab of **Ribbon**. The **Line Segment Command Bar** will be displayed with tips; refer to Figure-5.

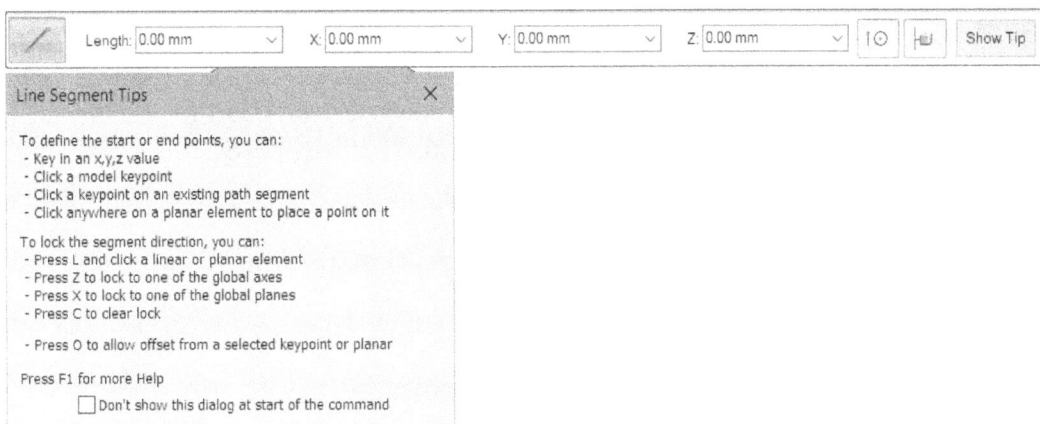

| Length: 0.00 mm | X: 0.00 mm | Y: 0.00 mm | Z: 0.00 mm | Show Tip |

Line Segment Tips                                              ✕

To define the start or end points, you can:
  - Key in an x,y,z value
  - Click a model keypoint
  - Click a keypoint on an existing path segment
  - Click anywhere on a planar element to place a point on it

To lock the segment direction, you can:
  - Press L and click a linear or planar element
  - Press Z to lock to one of the global axes
  - Press X to lock to one of the global planes
  - Press C to clear lock

  - Press O to allow offset from a selected keypoint or planar

Press F1 for more Help
  ☐ Don't show this dialog at start of the command

*Figure-5. Line Segment Command Bar*

- Click on the desired point or specify coordinate values in the **X**, **Y**, and **Z** edit boxes of the **Command Bar** to specify start point of line segment. The other end of line segment will get attached to cursor; refer to Figure-6.
- Click at desired location or enter desired coordinates in the **Command Bar** to specify end point.

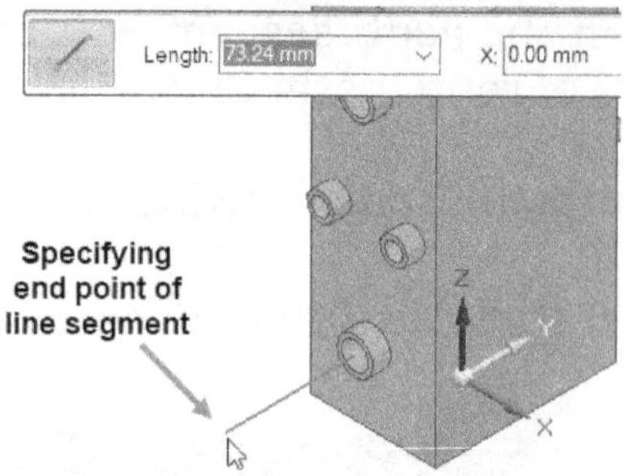

Length: 73.24 mm    X: 0.00 mm

Specifying
end point of
line segment

*Figure-6. Specifying end point of line segment*

# CREATING ARC SEGMENT

The **Arc Segment** tool is used to create segment in the form of an arc. The procedure to use this tool is given next.

- Click on the **Arc Segment** tool from the **Segments** group in the **Home** tab of **Ribbon**. The **Arc Segment Command Bar** will be displayed; refer to Figure-7.

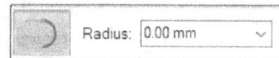

Radius: 0.00 mm

*Figure-7. Arc Segment Command Bar*

- Click to specify the start point and end point of arc segment. You will be asked to specify radius of arc segment.
- Specify the desired value of radius in the **Radius** edit box of **Command Bar** and press **ENTER**. You will be asked to specify the desired side to create the arc.
- Click on the desired side. The arc segment will be created.
- Press **ESC** to exit the tool.

# MOVING LINE SEGMENT

The **Move Segment** tool is used to move selected line segment to desired location. The procedure to use this tool is given next.

- Click on the **Move Segment** tool from the **Segments** group in the **Home** tab of **Ribbon**. You will be asked to select and drag desired segments to be moved.
- Drag the line segment to desired location; refer to Figure-8.

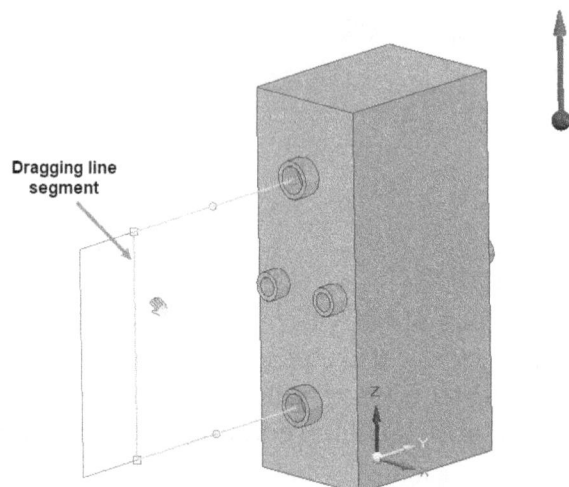

Dragging line
segment

*Figure-8. Dragging line segment*

*   Press **ESC** to exit the tool.

# SPLITTING PATH SEGMENT

The **Split Path** tool is use to break the selected path segment at specified point. The procedure to use this tool is given next.

*   Click on the **Split Path** tool from the **Segments** group in the **Home** tab of **Ribbon**. You will be asked to specify point on line at which you want to split the line.
*   Click at desired location on line segment to split the path segment; refer to Figure-9.

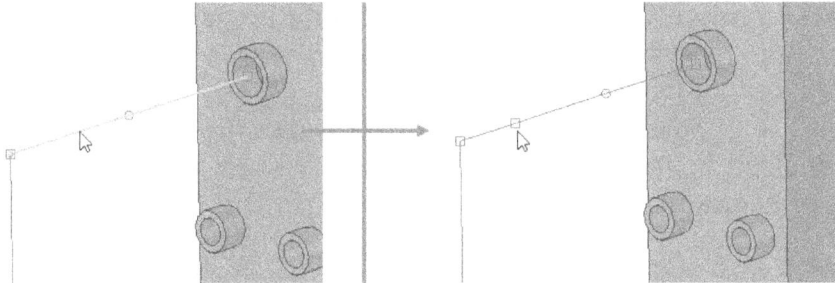

*Figure-9. Creating segment on line*

*   Press **ESC** to exit the tool.

# CREATING CURVE SEGMENT

The **Curve Segment** tool is used to create curved segment for route path. The procedure to use this tool is given next.

*   Click on the **Curve Segment** tool from the **Segments** group in the **Home** tab of the **Ribbon**. The **Curve Segment Command Bar** will be displayed; refer to Figure-10.

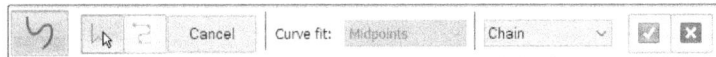

*Figure-10. Curve Segment Command Bar*

*   Select the desired segment to be used as reference for creating curve segment.
*   Select the desired option from the **Curve fit** drop-down in the **Command Bar** to fit curve on respective key points.
*   Click on the **Preview** button from the **Command Bar** to check preview of curves; refer to Figure-11.

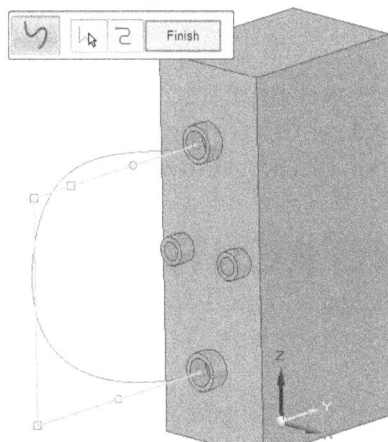

*Figure-11. Preview of curve segment*

- Click on the **Finish** button from the **Command Bar** to create the segment.
- Press **ESC** to exit the tool.

## CREATING KEYPOINT CURVE SEGMENT

The **Keypoint Curve Segment** tool is used to create curve segment through selected keypoints. The procedure to use this tool is given next.

- Click on the **Keypoint Curve Segment** tool from the **Segments** group in the **Home** tab of **Ribbon**. The **Keypoint Curve Segment Command Bar** will be displayed; refer to Figure-12 and you will be asked to specify keypoints.

*Figure-12. Keypoint Curve Segment Command Bar*

- Click at desired keypoint of model to specify start point of curve segment. If the direction of segment is opposite then click on the **Flip** button from the **Command Bar**.
- Click on the next points to create the curve segment; refer to Figure-13.

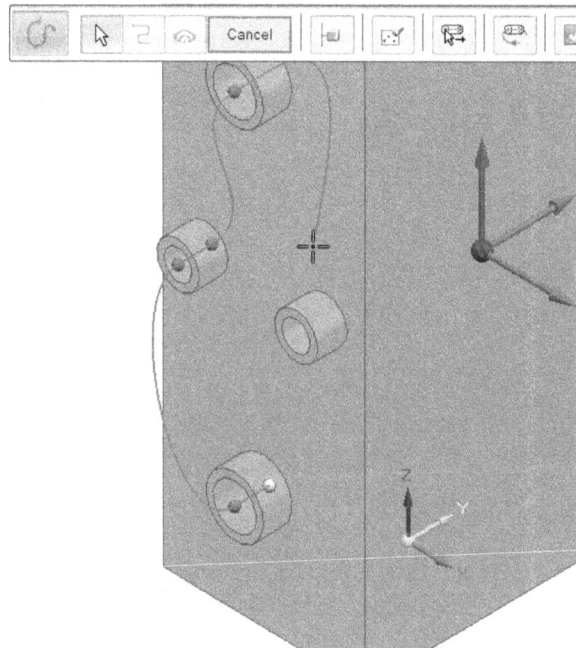

*Figure-13. Preview of keypoint curve segment*

- After you have specified desired points, right-click in the drawing area to accept selection. Click on the **Preview** button to check preview of the segment.
- Click on the **Finish** button from the **Command Bar** to create the segment.

## CREATING ROUTE USING PATHS

The **Route** tool in **Segments** group is used to create route using a path segment and then continuing to next selected openings. The procedure to use this tool is given next.

- Click on the **Route** tool from the **Segments** group in the **Home** tab of **Ribbon**. You will be asked to select the path segments to be used for creating the route.
- Select the desired path segment to be continued. You will be asked to select next circular opening.

- Select the desired openings one by one to continue route path. After specifying desired path, right-click in the drawing area to create the path.

# CREATING TUBE ALONG PATH

The **Tube** tool is used to create tube of desired material along selected route path. The procedure to use this tool is given next.

- Click on the **Tube** tool from the **Tubing** group in the **Home** tab of **Ribbon**. The **Tube Options** dialog box will be displayed; refer to Figure-14.

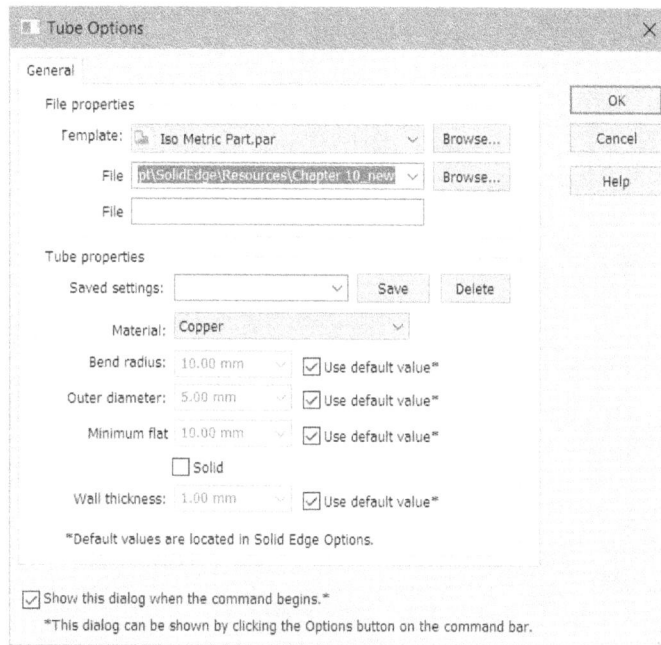

*Figure-14. Tube Options dialog box*

- Specify the desired template, file location, and other properties for tube in the File properties area of the dialog box.
- Select the desired material for tube from the **Material** drop-down in dialog box.
- Clear the **Use default value** check boxes and specify the desired values for bend radius, outer diameter, minimum flat, wall thickness etc. in the **Tube properties** area of the dialog box.
- After setting desired parameters, click on the **OK** button. The **Tube Command Bar** will be displayed; refer to Figure-15.

*Figure-15. Tube Command Bar*

- If you want to change the tube options again then click on the **Tube Options** button from the **Command Bar**.
- Select the path curves for creating pipe and specify the desired name for pipe parts in the **Name** edit box of the **Command Bar**. The **Preview** button will be displayed in the **Command Bar**.
- Click on the **Preview** button to check preview of pipe.
- Click on the **Finish** button from the **Command Bar** to create the tube. The **Tube Options** dialog box will be displayed again for next pipe segment. Specify the parameters for next segment and create the pipes or press **ESC** to exit the tool.

## CREATING PIPING ROUTE

The **Piping Route** tool is used to create pipes and connectors for piping system while following the route path. Note that to use this tool, you need to install Standard Parts and Standard Parts Administrator application separately using installation media provided by Siemens. The procedure to use this tool is given next.

- After creating the route path, click on the **Piping Route** tool from the **Tubing** group in the **Home** tab of **Ribbon**. The **Piping Options** dialog box will be displayed; refer to Figure-16.

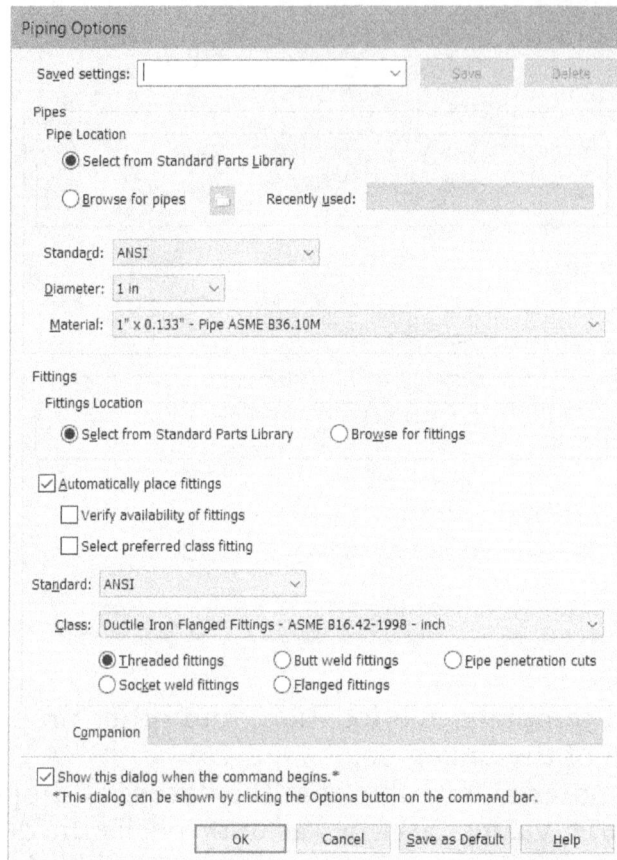

*Figure-16. Piping Options dialog box*

- Select the desired option from the **Pipe Location** area of the dialog box to define source part files for creating pipe. By default, the **Select from Standard Parts Library** radio button is selected so you can use the standard parts for piping using the library of Solid Edge. Select the **Browse for pipes** radio button and click on the Browse button next to it if you want to select the part file of pipe manually. You can also select desired pipe from the **Recently used** drop-down if you have earlier created pipes in the system.
- If you are using standard parts library then set the desired parameters in the **Standard**, **Diameter**, and **Material** drop-downs to define shape and size of pipes.
- Similarly, you can set the desired parameters in **Fittings** area to define pipe fittings.
- After setting desired parameters, click on the **OK** button. The **Piping Route Command Bar** will be displayed; refer to Figure-17 and you will be asked to select path segment.

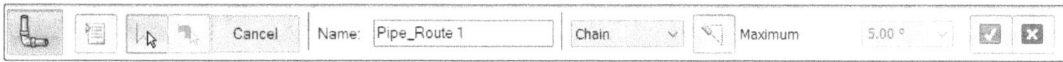

Figure-17. Piping Route Command Bar

- Select the desired path segments and right-click in the drawing area. The nodes which need pipe fittings will be highlighted on path; refer to Figure-18.

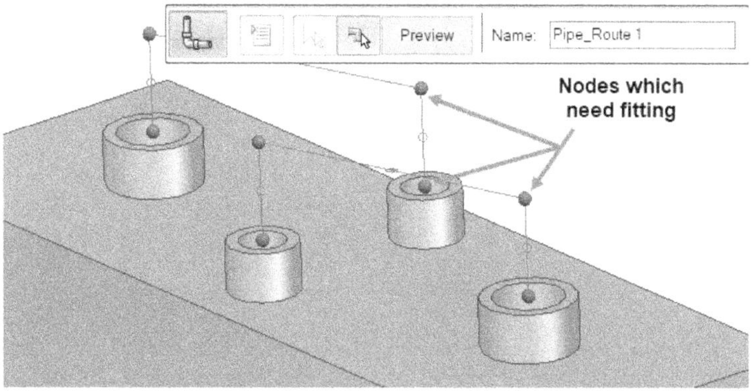

Figure-18. Nodes which need pipe fittings

- Select the desired node on which you want to apply pipe fitting. The **Standard Parts** dialog box will be displayed; refer to Figure-19.

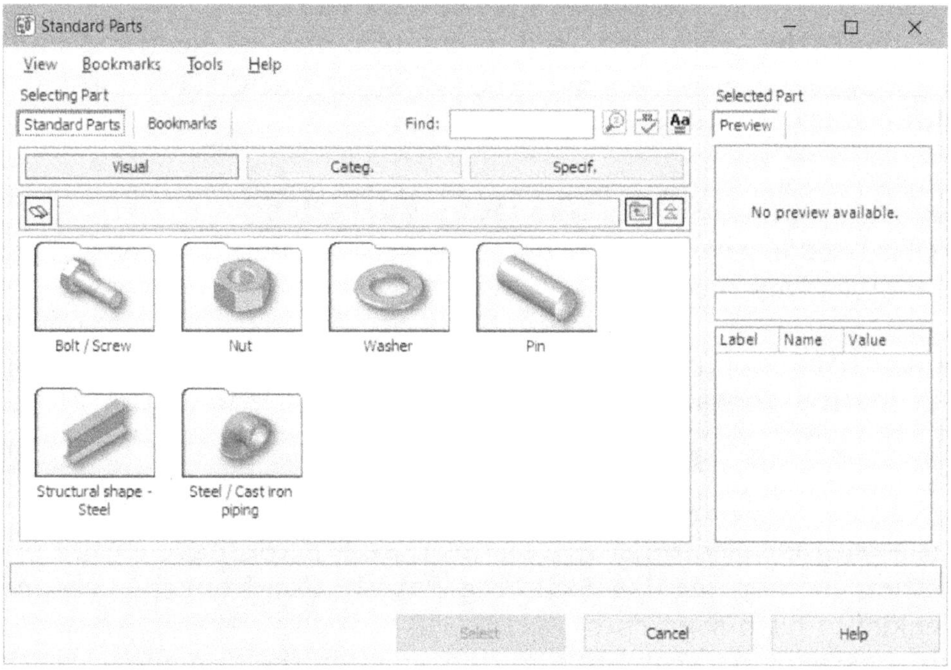

Figure-19. Standard Parts dialog box

- Select the desired pipe fitting from the desired category and click on the **Select** button. Preview of the fitting will be displayed.
- After applying all the desired fittings, click on the **Preview** button to check preview of piping.
- Click on the **Finish** button to create the piping route and press **ESC** to exit the tool.

To exit the XpresRoute environment, click on the **Close XpresRoute** tool from the **Close** group in the **Home** tab of **Ribbon**.

# INTRODUCTION TO ELECTRICAL ROUTING

The tools in the Electrical Routing environment are used to create wiring harness and cable for electrical system of component. To activate the **Electrical Routing** environment, click on the **Electrical Routing** tool from the **Environs** group in the **Tools** tab of **Ribbon**. The Electrical Routing environment will be displayed; refer to Figure-20.

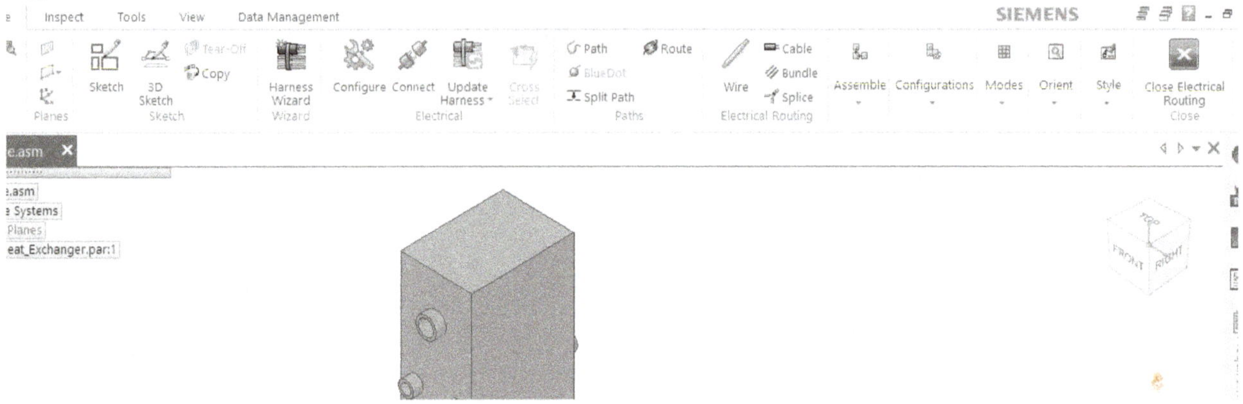

Figure-20. Electrical Routing environment

# CREATING PATHS

The tools in the **Paths** group are used to create paths for electrical routing. These tools work in the same way as discussed for XpresRoute. Create the desired path segments for electrical route.

# CREATING WIRES

The **Wire** tool is used to generate wire using the selected path segments. The procedure to use this tool is given next.

- Click on the **Wire** tool from the **Electrical Routing** group in the **Home** tab of the **Ribbon**. The **Wire Command Bar** will be displayed; refer to Figure-21.

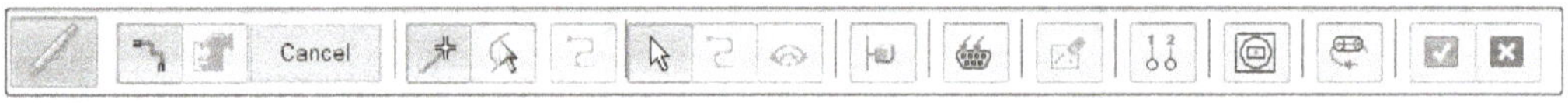

Figure-21. Wire Command Bar

- Create the path for electrical wire as discussed. If you have earlier created path for wire then click on the **Use Existing Path** button from the **Command Bar**. You will be asked to path segments.
- Select the desired path segment and right-click in the drawing area. The **Command Bar** will be displayed as shown in Figure-22.

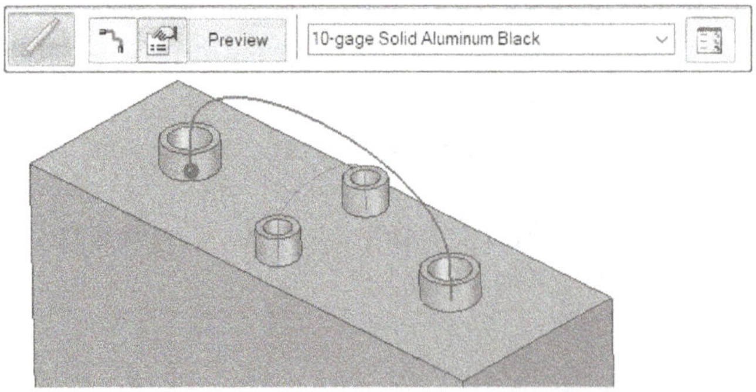

Figure-22. Command Bar with Wire Type options

- Select the desired wire type from the **Material** drop-down in the **Command Bar**. After selecting the wire type, click on the **Properties** button from **Command Bar** if you want to modify wire properties. The **Properties - Wire** dialog box will be displayed; refer to Figure-23.

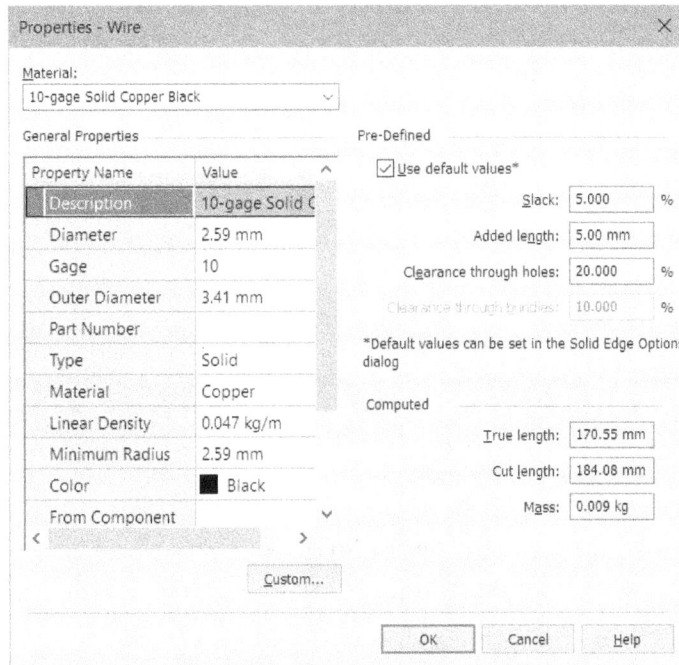

*Figure-23. Properties-Wire dialog box*

- Specify the desired parameters in the dialog box to modify properties of wire and click on the **OK** button. The **Wire Command Bar** will be displayed again.
- Click on the **Preview** button from the **Command Bar** to check preview of wire.
- Specify desired name of wire in the **Name** edit box of **Command Bar** and click on **Finish** button to create the wire. Press **ESC** to exit the tool.

# CREATING CABLE

The **Cable** tool is used to combine two or more wires for making a cable. The procedure to use this tool is given next.

- Click on the **Cable** tool from the **Electrical Routing** group in the **Home** tab of **Ribbon**. The **Cable Command Bar** will be displayed; refer to Figure-24.

*Figure-24. Cable Command Bar*

- Click at desired point of wire to define start point of cable and right-click. You will be asked to select a path for cable.
- Select the desired path curve and right-click in the drawing area. The option to select cable type will be displayed in the **Command Bar**.
- Select the desired cable type and click on the **Finish** button to create the cable.

Similarly, you can use the **Bundle** and **Splice** tools. Click on the **Close Electrical Routing** tool from the **Close** group in the **Home** tab of **Ribbon**.

# INTRODUCTION TO FRAME DESIGN

The Frame Design environment is used to create steel frame members along selected paths and edges. To activate the Frame Design environment, click on the **Frame** tool from the **Environs** group in the **Tools** tab of **Ribbon**. The tools to create frame design will be displayed; refer to Figure-25. Most of the tools in this environment are same as discussed for XpresRoute and Electrical Routing. Here, we will discuss the Frame tool used for creating frame members.

Figure-25. Tools for frame design

# CREATING FRAME MEMBERS

The **Frame** tool is used to insert frame members along selected paths and edges. The procedure to use this tool is given next.

*   Click on the **Frame** tool from the **Frame** group in the **Home** tab of **Ribbon**. The **Frame Options** dialog box will be displayed; refer to Figure-26.

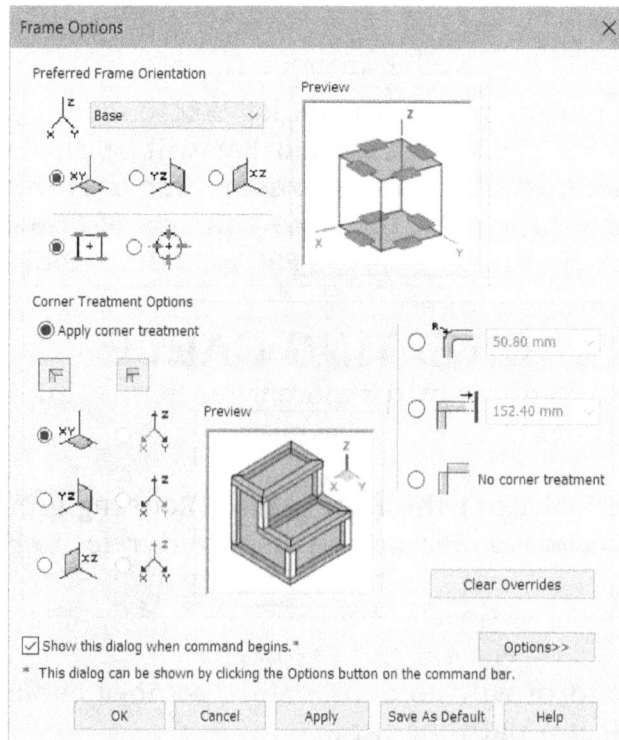

Figure-26. Frame Options dialog box

*   Select the desired radio buttons from the **Preferred Frame Orientation** area of the dialog box to define orientation of frame members.
*   Set the desired options in the **Corner Treatment Options** area of the dialog box to define how corners will be created at intersections of two frame members.
*   Click on the **Options** button from the dialog box to check advanced options of the frame; refer to Figure-27.

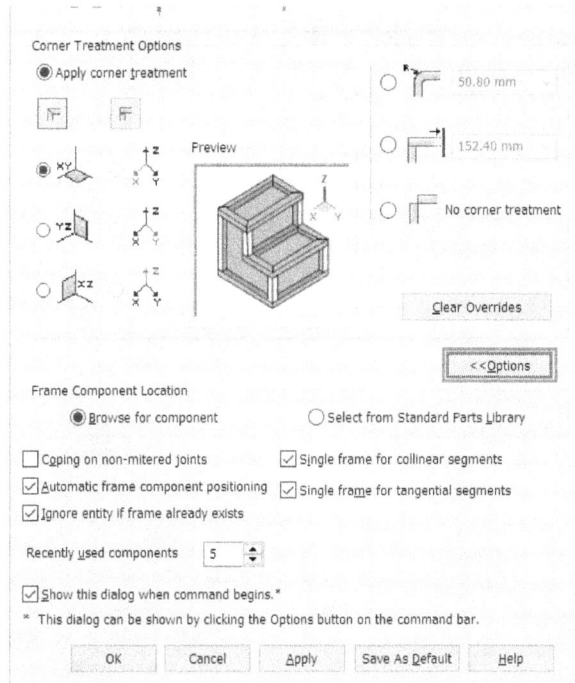

*Figure-27. Advanced options*

- Set the desired check boxes in the dialog box to define the placement of frame members and click on the **OK** button. The **Frame Command Bar** will be displayed; refer to Figure-28.

*Figure-28. Frame Command Bar*

- Click on the **Select Cross Section Component** button from the **Command Bar** to select desired frame member. The **Frame** dialog box will be displayed; refer to Figure-29.

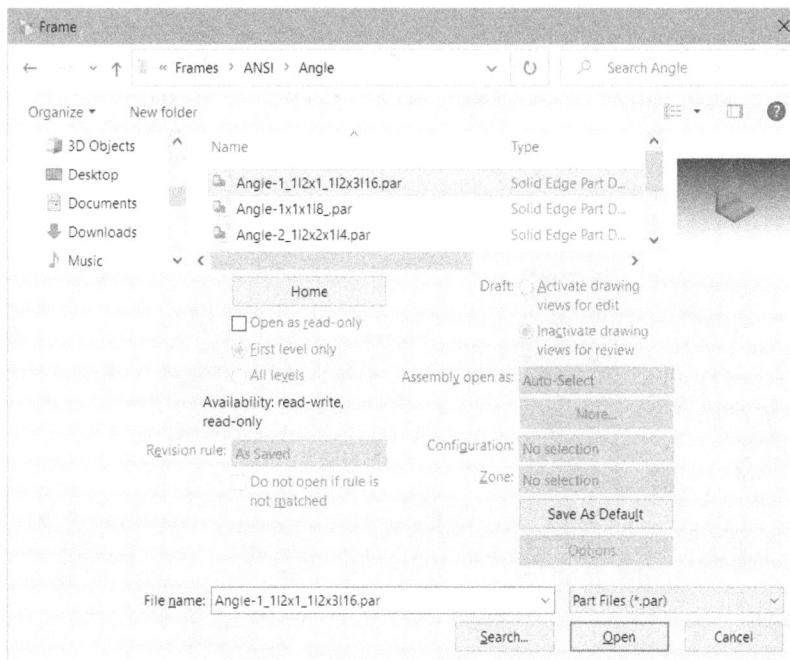

*Figure-29. Frame dialog box*

- Select the desired frame member part file from the default library folder (C:\ Program Files\Siemens\Solid Edge 2020\Frames in our case) and click on the **Open** button. You will be asked to select the edges or path curves from the model.
- Select the desired entities and right-click in the drawing area. Preview of the frame will be displayed; refer to Figure-30.

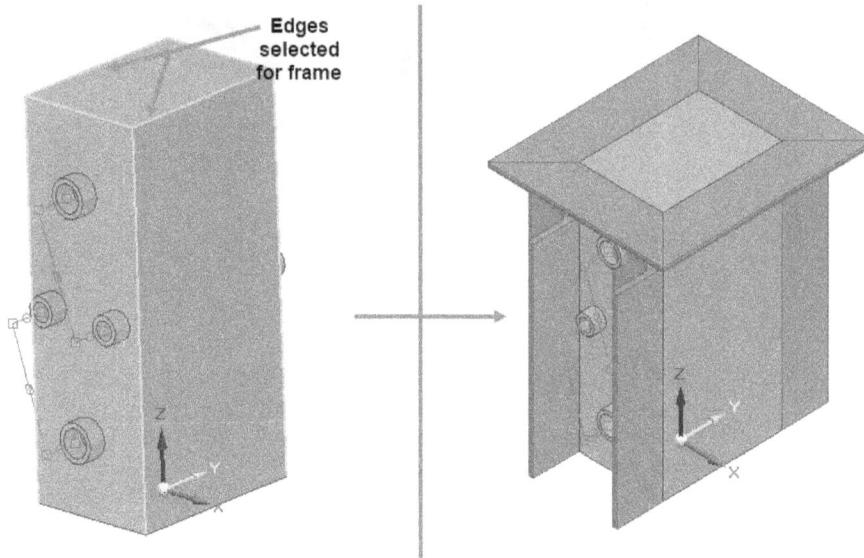

*Figure-30. Creating frame*

- Click on the **Finish** button from the **Command Bar** to create the frame. Press **ESC** to exit the tool.

Click on the **Close Frame** tool from the **Close** group in the **Home** tab of **Ribbon** to exit the environment.

## PUBLISHING MODEL TO HTML WEB PAGE

The **Publish** tool in the **Environs** group is used to generate an HTML file of current active model. The procedure to use this tool is given next.

- Click on the **Publish** tool from the **Environs** group in the **Tools** tab of the **Ribbon**. The **Solid Edge Web Page Publisher Wizard** will be displayed; refer to Figure-31.

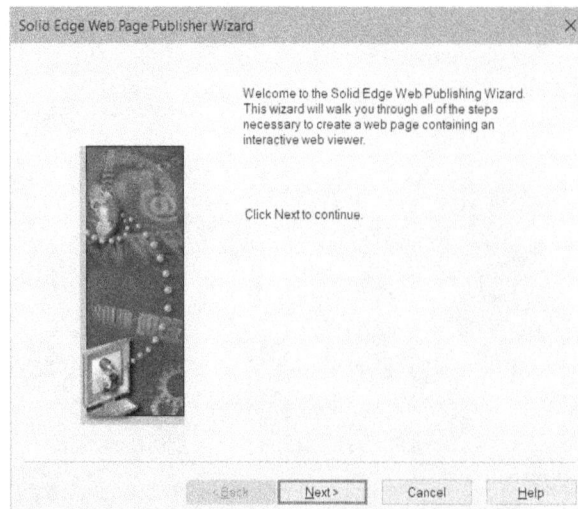

*Figure-31. Solid Edge Web Page Publisher Wizard*

- Click on the **Next** button from the dialog box. The options to select template for web page will be displayed; refer to Figure-32.

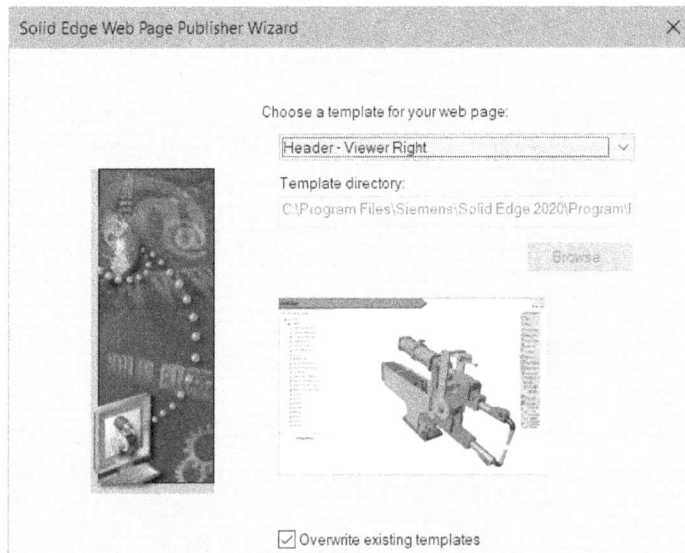

*Figure-32. Template page*

- Select the desired template for web page from the drop-down and click on the **Next** button. The options to specify web page title, author, company name, and date will be displayed; refer to Figure-33.

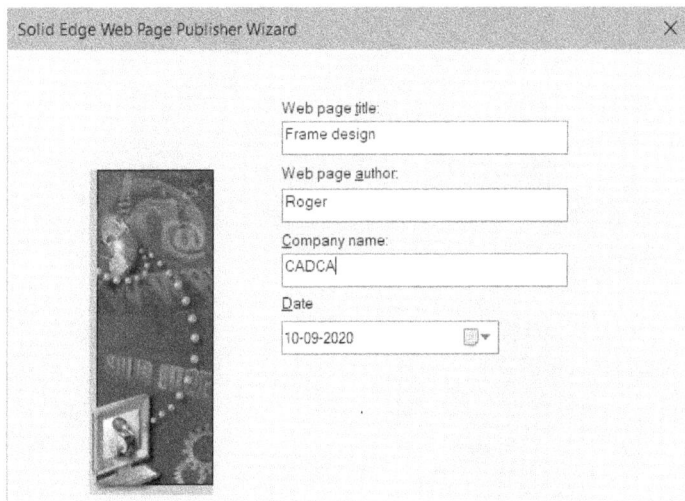

*Figure-33. Web page user details page*

- Click on the **Next** button from the dialog box. The options to save format for web page will be displayed; refer to Figure-34.

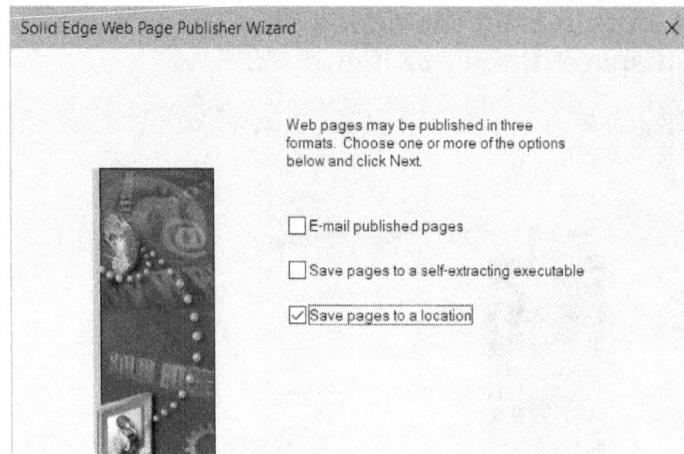

*Figure-34. Options for setting format*

- Select the desired check boxes from the dialog box and click on the **Next** button to save/share the web pages. The options to specify location of web page will be displayed.
- Specify the page location and click on the **Next** buttons from subsequent pages. After setting all the desired parameters, click on the **Finish** button. The web page will be saved at specified location.

# Chapter 11

# Drawing

## Topics Covered

The major topics covered in this chapter are:

- *Introduction*
- *Starting Drawing Environment*
- *Placing Views*
- *Creating Section View using Cutting Plane*
- *Creating Parts List*
- *Creating Hole Table and Bend Table*
- *Dimensioning Tools*

# INTRODUCTION

Drawing is the engineering representation of a model on the paper. For manufacturing a model in real world, we need some means by which we can tell the manufacturer what to manufacture. For this purpose, we create drawings from the models. These drawings have information like dimensions, material, tolerances, objective, precautions and so on. In Solid Edge, we create drawings by using the Drawing environment. There are two methods to start drawing environment; creating drawing from assembly/part environment and create a new drawing. The procedure to start drawing environment is given next.

# STARTING DRAWING ENVIRONMENT

There are two methods to start drawing environment in Solid Edge. These methods are discussed next.

## Starting a New Drawing

The procedure to use **New** tool for starting a new drawing is given next.

* Click on the **New** tool from the **New** cascading menu of the **Application** menu. The New dialog box will be displayed as discussed earlier.
* Select the desired **.dft** template from the dialog box and click on the **OK** button from the dialog box. The drawing environment will be displayed without a model selected for placing views.

## Starting a Drawing using Model or Assembly

The procedure to start drawing environment using a model/assembly is given next.

* Open the part file or assembly file in Solid Edge for which you want to generate the views.
* Click on the **Drawing of Active Model** option from the **New** drop-down in the **Quick Access Toolbar**; refer to Figure-1. The **Create Drawing** dialog box will be displayed; refer to Figure-2.

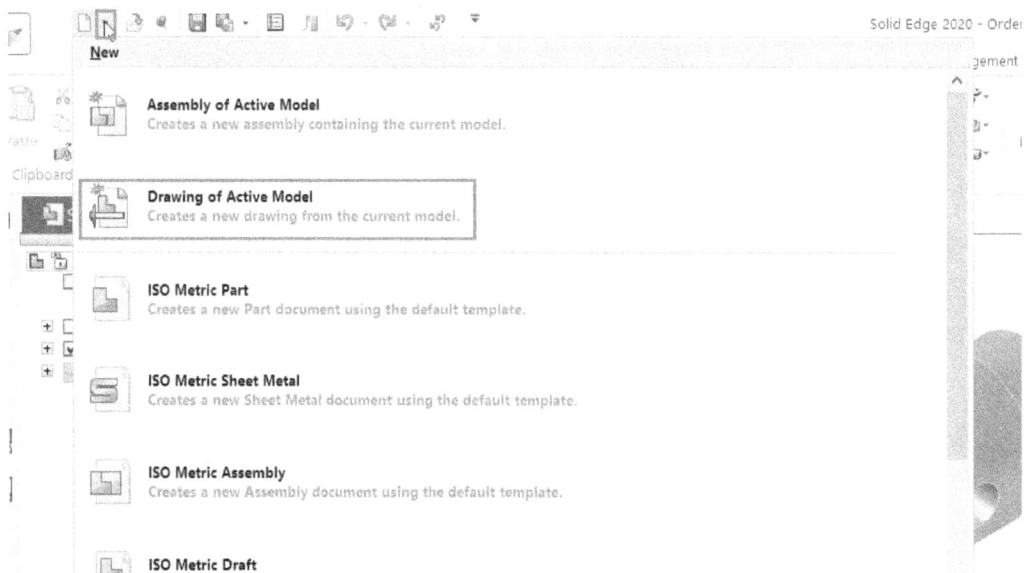

*Figure-1. Drawing of Active Model tool*

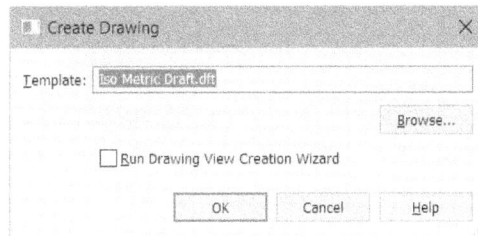

*Figure-2. Create Drawing dialog box*

- Click on the **Browse** button from the dialog box and select desired template for creating drawing.
- Select the **Run Drawing View Creation Wizard** check box if you want to create primary views using Drawing View Creation Wizard.
- After setting parameters, click on the **OK** button. If you have not selected the **Run Drawing View Creation Wizard** check box then three standard views will be placed automatically in the drawing. If you have selected the **Run Drawing View Creation Wizard** check box then **View Wizard Command Bar** will be displayed and you will be asked to place the primary view. Click at desired locations to place the views and press **ESC** to exit the tool. You will learn more about the **View Wizard Command Bar** in next topic.

## PLACING VIEWS WITH VIEW WIZARD

The **View Wizard** tool is used to select the model for creating drawing views and place the drawing views. The procedure to use this tool is given next.

- Click on the **View Wizard** tool from the **Drawing Views** group in the **Home** tab of **Ribbon**. The **Select Attachment** dialog box will be displayed; refer to Figure-3.

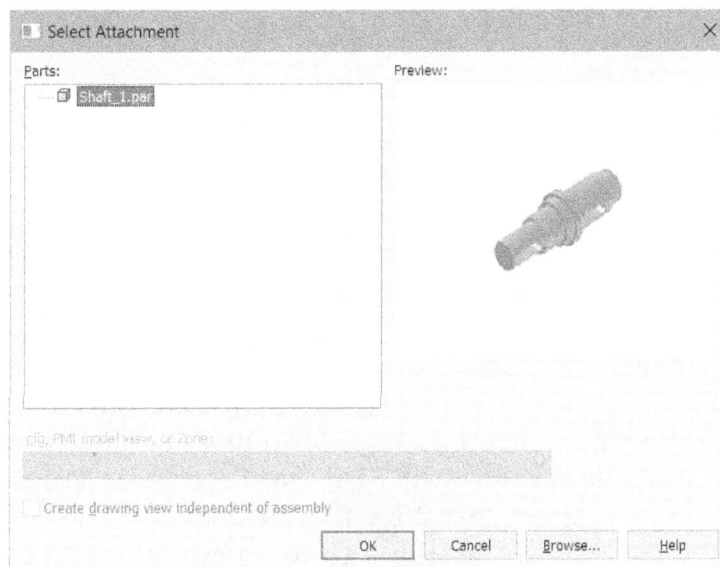

*Figure-3. Select Attachment dialog box*

- Click on the **Browse** button from the dialog box to select desired model. The **Select Model** dialog box will be displayed.
- Select the desired model and click on the **Open** button. The front view of model will get attached to cursor and the **View Wizard Command Bar** will be displayed; refer to Figure-4.

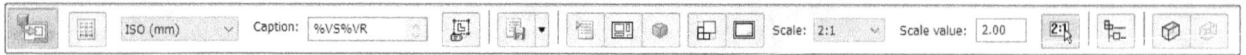

*Figure-4. View Wizard Command Bar*

- Select the desired option from the **Drawing View Style** drop-down in the **Command Bar** to define view placement style and unit for view.
- Click in the **Caption** edit box and specify desired text or symbol to be displayed below the view.
- Click on the **Show Caption** button from the **Command Bar**. A list of options will be displayed; refer to Figure-5. Select the desired options in the list to include them with view.

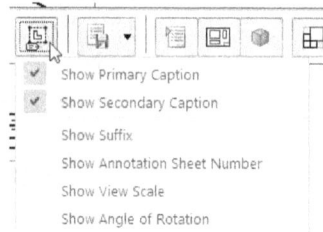

*Figure-5. Show Captions list*

- Click on the **Saved Settings** button from the **Command Bar** to save current settings of the **Command Bar** for using them later. The **Drawing View Wizard** dialog box will be displayed; refer to Figure-6. Specify the desired name for the setting in the **Saved settings** edit box and click on the **Save** button. After that click on the **OK** button from the dialog box.

*Figure-6. Drawing View Wizard dialog box*

- Click on the **Drawing View Wizard Options** button from the **Command Bar**. The **Drawing View Wizard** dialog box will be displayed; refer to Figure-7.
  - Select the **Designed part** radio button from the **Part and Sheet Metal Drawing View** section to use designed model as it looks while creating. Select the **Simplified part** radio button to use simplified version of model with lesser details for creating view. Select the **Flat pattern** radio button from the section to use flat pattern of sheet metal part while creating the view.
  - If you have applied PMI annotations to the model then select the desired PMI view created in the Part/Assembly environment from the **PMI model view** drop-down. The check boxes below the drop-down will become active. Select the **Include PMI dimensions from model views** check box to automatically import PMI dimension in the view. Similarly, select the **Include PMI dimensions from model views** check box to include the annotations specified in PMI.
  - Select the **Show tube centerlines** check box to display center lines in the view.
  - Similarly, you can set the other options in the dialog box for displaying various entities in the view.
  - After setting desired parameters, click on the **OK** button from the dialog box.

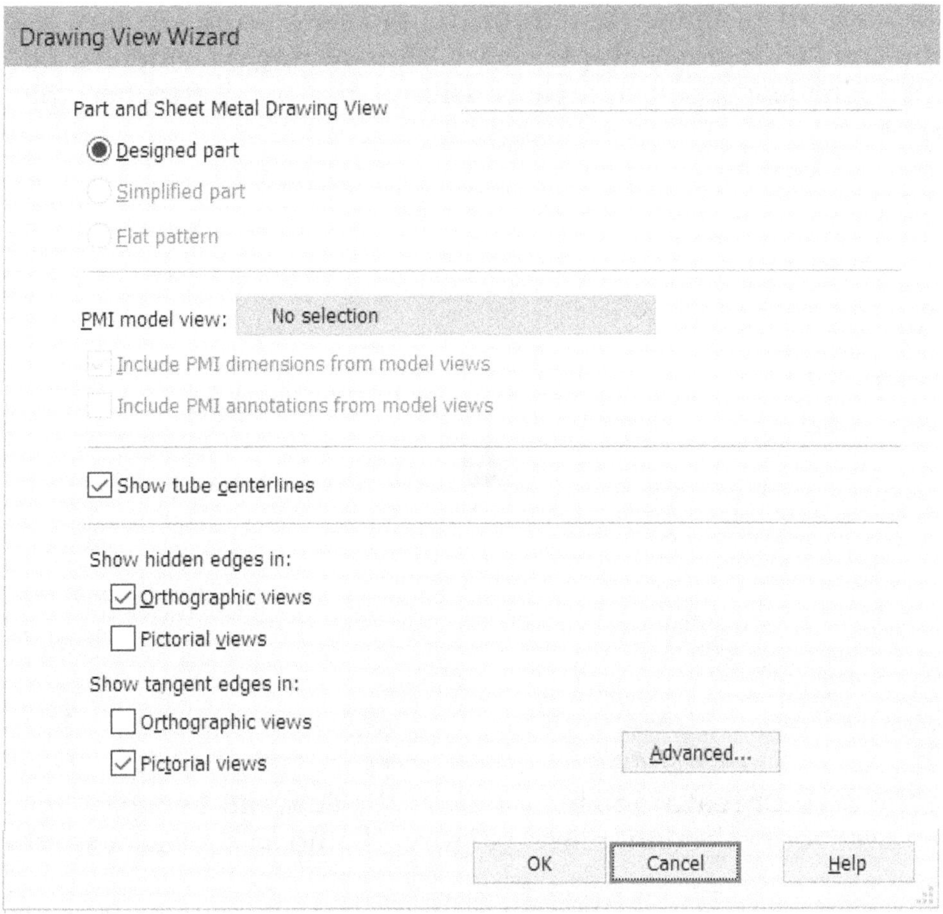

*Figure-7. Drawing View Wizard dialog box*

- Click on the **Drawing View Layout** button from the **Command Bar** to set desired standard view of model. The **Drawing View Wizard** dialog box will be displayed as shown in Figure-8.

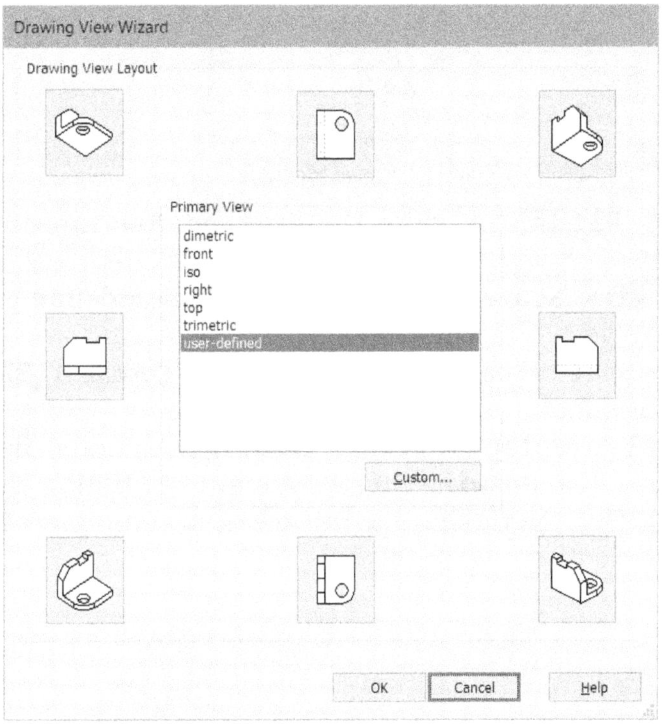

*Figure-8. Drawing View Wizard dialog box*

- Select the desired primary view from the **Primary View** list box and select the desired buttons to create multiple views. After setting parameters, click on the **OK** button. The preview of selected views will be displayed attached to cursor; refer to Figure-9.

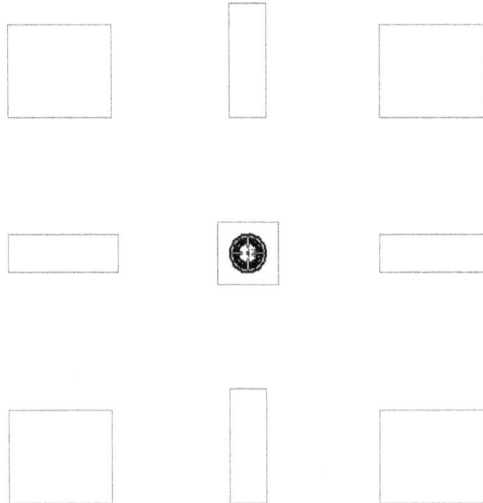

*Figure-9. Multiple views attached to cursor*

- Click on the **View Orientation** button from the **Command Bar** to select the primary view to be used for projection; refer to Figure-10 and select the desired view option.

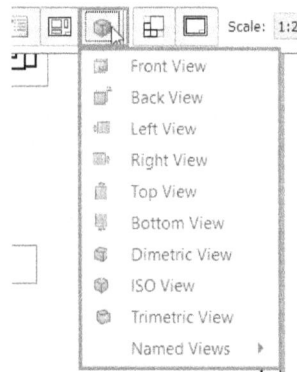

*Figure-10. View Orientation options*

- Click on the **Best Fit** button from the **Command Bar** to automatically fit all the views in the sheet.
- Click on the **Set View Scale** button from the **Command Bar** to use specified scale value for size of views.
- Click on the **Set Sheet Scale** button from the **Command Bar** to match scale of views with scale of sheet.
- Click on the **Model Display Settings** button to specify display settings of the model in the views. The **Drawing View Properties** dialog box will be displayed; refer to Figure-11. Set the desired parameters and click on the **OK** button.

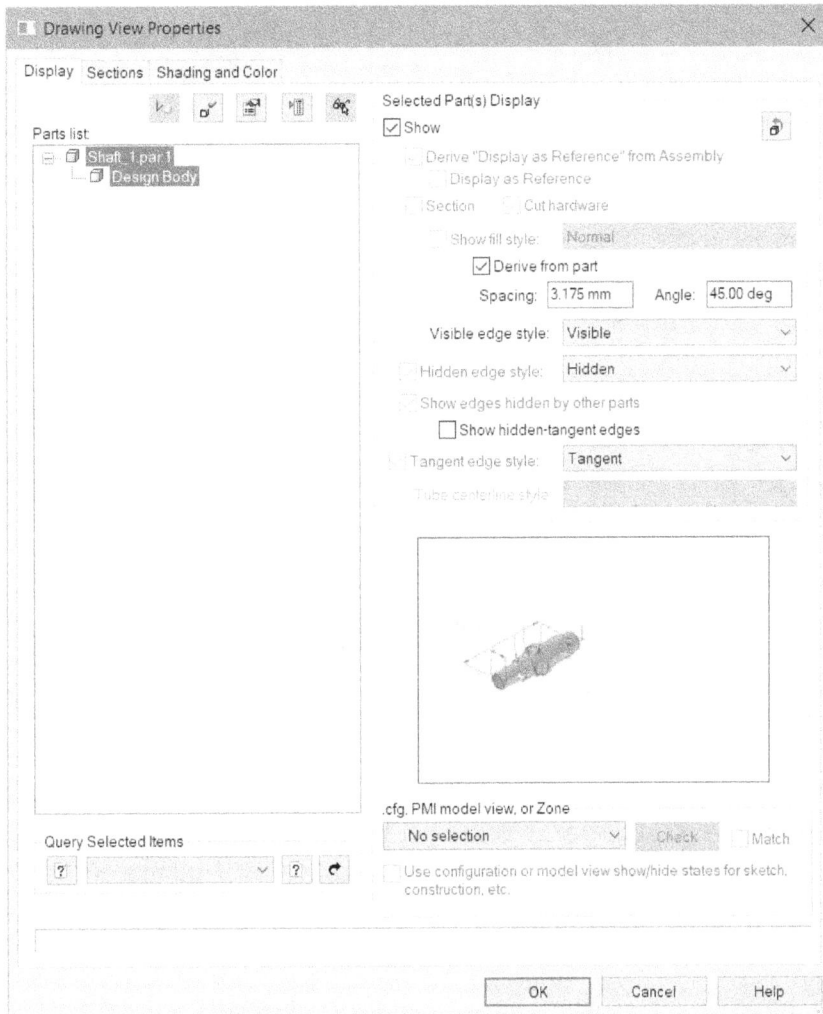

*Figure-11. Drawing View Properties dialog box*

- Click on the **Shading Options** button to define how 3D models will be displayed in the view; refer to Figure-12. Select the desired option from the list to specify shading style.

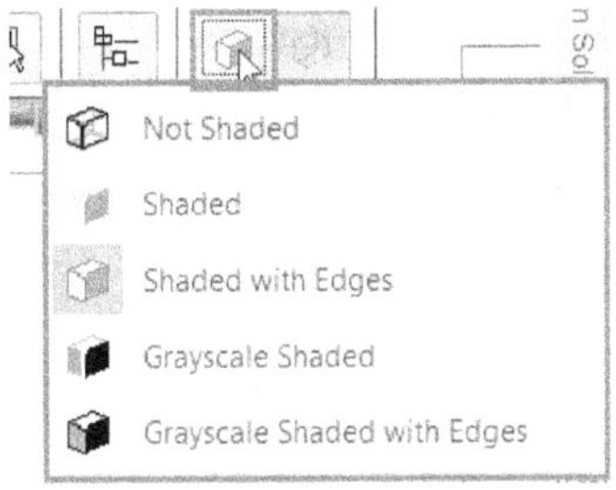

*Figure-12. Shading Options*

- Select the **Use Model Colors** button from the **Command Bar** to use colors specified for model in the view.
- After setting desired parameters, click at desired location in the sheet. The views will be created; refer to Figure-13.

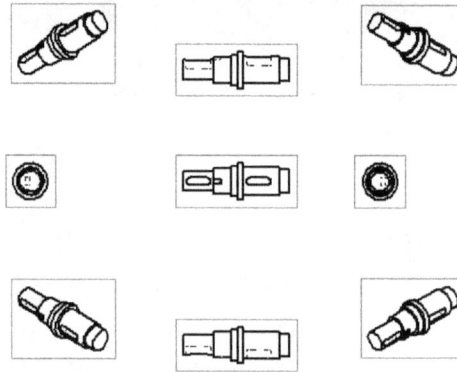

*Figure-13. Views created*

# UPDATING VIEWS

The **Update Views** tool in the **Drawing Views** group is used to update views in the drawing based on changes applied in the model.

# CREATING PRINCIPAL VIEWS

The **Principal** tool in the **Drawing Views** group is used to create principal views based on selected view in the drawing. The procedure to use this tool is given next.

*   Click on the **Principal** tool from the **Drawing Views** group in the **Home** tab of the **Ribbon**. The **Principal Command Bar** will be displayed; refer to Figure-14 and you will be asked to select the view to be used for creating projected principal views.

*Figure-14. Principal Command Bar*

*   Select the desired base view from the drawing. The projected principal view will get attached to cursor.
*   Click at desired locations. The views will be created based on their angle from base view.
*   Press **ESC** to exit the tool.

# CREATING AUXILIARY VIEW

The auxiliary view is a projected view created by making selected edge of view parallel to screen. The procedure to create auxiliary view is given next.

*   Click on the **Auxiliary** tool from the **Drawing Views** group in the **Home** tab of **Ribbon**. The **Auxiliary Command Bar** will be displayed; refer to Figure-15.

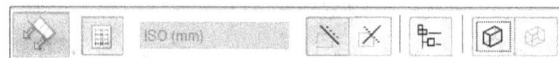

*Figure-15. Auxiliary Command Bar*

*   Select the **Parallel** or **Perpendicular** button from the **Command Bar** to create a view parallel or perpendicular to selected edge of model.
*   Select two key points of model or select the desired edge of model. The view will get attached to the cursor; refer to Figure-16.

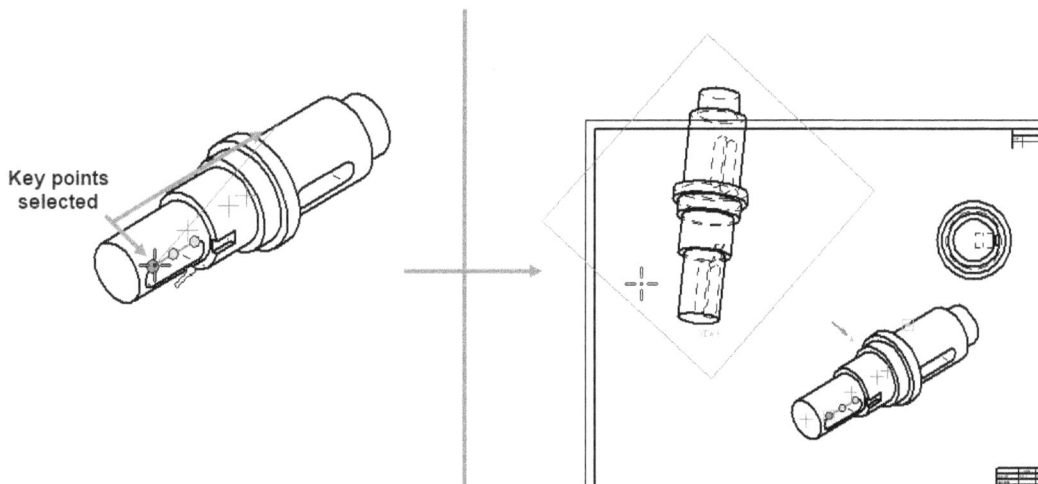

*Figure-16. Creating auxiliary view*

• Click at desired location to place the view. The auxiliary view will be created.

## CREATING DETAIL VIEW

The **Detail** tool is used to create a scaled up view of local region of the model. The procedure to use this tool is given next.

• Click on the **Detail** tool from the **Drawing Views** group in the **Home** tab of the **Ribbon**. The **Detail Command Bar** will be displayed; refer to Figure-17.

*Figure-17. Detail Command Bar*

• By default, the **Circular Detail View** button is used to create a circular boundary around the feature to create detailed circular view. If you want to use sketched curves for defining boundary of detail view then click on the **Define Profile** button from the **Command Bar** and click on the view to be used for creating detail view. The sketching environment will be displayed. Create the desired closed loop sketch and click on the **Close Detail Envelope** button from the **Close** group of the **Ribbon**. The detail view will get attached to cursor.
• Click at desired location to place the view.

## CREATING BROKEN VIEW

The **Broken** tool is used to represent very long objects in the drawing by breaking them at specific span. The procedure to use this tool is given next.

• Select the desired view to be used for broken view and click on the **Broken** tool from the **Drawing Views** group in the **Home** tab of the **Ribbon**. You will be asked to specify location of first break line.
• Select the desired option from the **Style** drop-down to define style of break lines.
• Click on the **Vertical Break** or **Horizontal Break** button from the **Command Bar** to define direction of break lines.
• Click on the **Break Line Type** button from the **Command Bar** and select the desired option from the list displayed to define type of break lines.
• Specify the desired value of distance between two break lines in the **Break gap** edit box of the **Command Bar**.

- Similarly, set the other parameters in the
- Click at desired location to specify first break line and then click to specify location of end break line; refer to Figure-18.
- Click on the **Finish** button from the **Command Bar** to create the view.

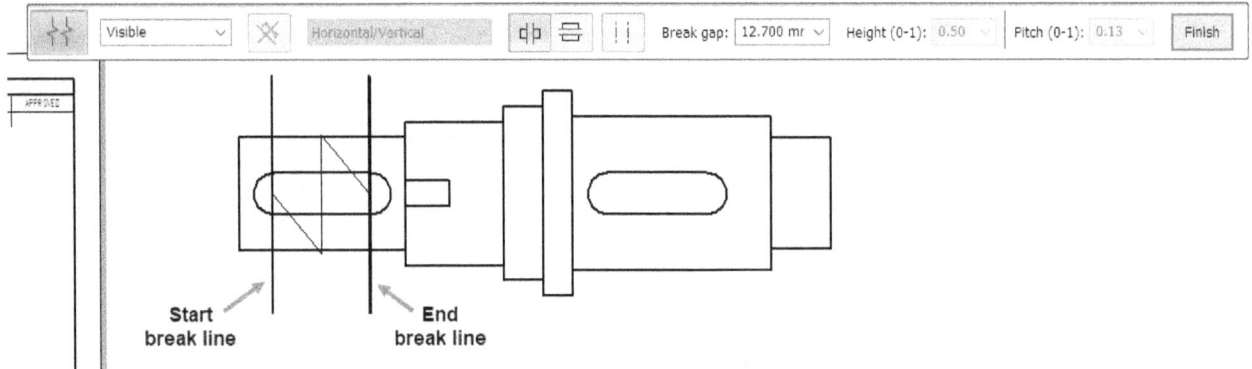

*Figure-18. Break view lines*

- Press **ESC** to exit the tool.

## CREATING SECTION VIEW CUTTING PLANE

The **Cutting Plane** tool is used to create cutting plane for generating section view of the model. The procedure to use this tool is given next.

- Click on the **Cutting Plane** tool from the **Drawing Views** group in the **Home** tab of **Ribbon**. The **Cutting Plane Command Bar** will be displayed; refer to Figure-19 and you will be asked to select the view for which you want to create the cutting plane.

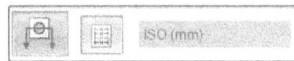

*Figure-19. Cutting Plane Command Bar*

- Select the desired view from the drawing. You will be asked to draw shape of cutting plane.
- Draw the shape of cutting plane using line or arc. Note that the sketch for cutting plane should be a linear curve which is not self-intersecting; refer to Figure-20.

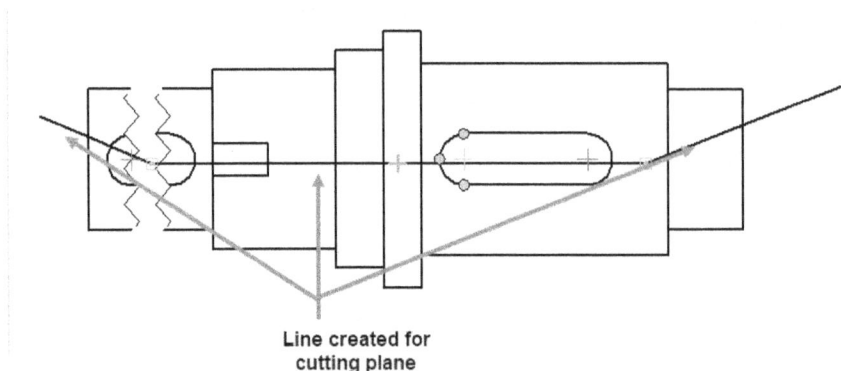

Line created for cutting plane

*Figure-20. Creating lines for cutting plane*

- After creating sketch, click on the **Close Cutting Plane** tool from the **Close** group in the **Ribbon**. The preview lines/curves for cutting plane will be displayed; refer to Figure-21.

*Figure-21. Cutting plane lines*

- Click above or below the line to define direction of cutting plane. The plane will be created.

## CREATING SECTION USING CUTTING PLANE LINE

The **Section** tool is used to create section plane using cutting plane line. The procedure to use this tool is given next.

- Click on the **Section** tool from the **Drawing Views** group in the **Home** tab of the **Ribbon**. The **Section Command Bar** will be displayed.
- Select the desired cutting plane line. The section view will get attached to cursor and the **Section Command Bar** will be displayed; refer to Figure-22.

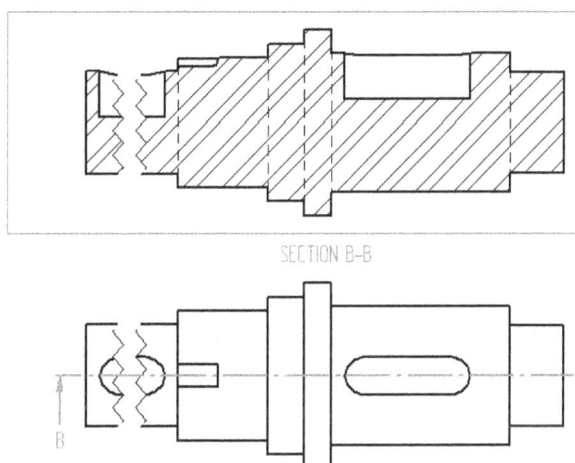

*Figure-22. Section Command Bar and section view*

- Set the desired values in the **Fill**, **Angle**, and **Spacing** edit boxes.
- Select the **Section Only** button from the **Command Bar** to display section only and do not display hidden edges.
- Click at desired location to place the section view after setting desired parameters. The section view will be created. Press **ESC** to exit the tool.

## CREATING BROKEN-OUT DRAWING VIEW

The **Broken-Out** tool is used to create broken out view for checking inside region of part. The procedure to use this tool is given next.

- Click on the **Broken-Out** tool from the **Drawing Views** group in the **Home** tab of **Ribbon**. You will be asked to select the drawing view to be used for creating broken-out view.
- Select the desired drawing view. You will be asked to create sketch for broken out section.

- Create the desired closed loop sketch for section and click on the **Close Broken Out Section** button from the **Close** group in the **Ribbon**. The **Broken-Out Command Bar** will be displayed and you will be asked to specify depth of broken-out section view.
- Specify the desired value in the **Depth** edit box and press **ENTER**. You will be asked to select the view which will be sectioned.
- Select the desired drawing view. The broken out section view will be displayed; refer to Figure-23. Press **ESC** to exit the tool.

*Figure-23. Broken-out section view*

## CREATING PARTS LIST

The **Parts List** tool is used to create a list of parts with balloons. The procedure to use this tool is given next.

- Click on the **Parts List** tool from the **Tables** group in the **Home** tab of the **Ribbon**. The **Parts List Command Bar** will be displayed and you will be asked to select a drawing view.
- Select the desired view from the drawing. The table of parts list will get attached to cursor.
- Click at desired location to place the table. The balloons will be applied to parts with parts list; refer to Figure-24.

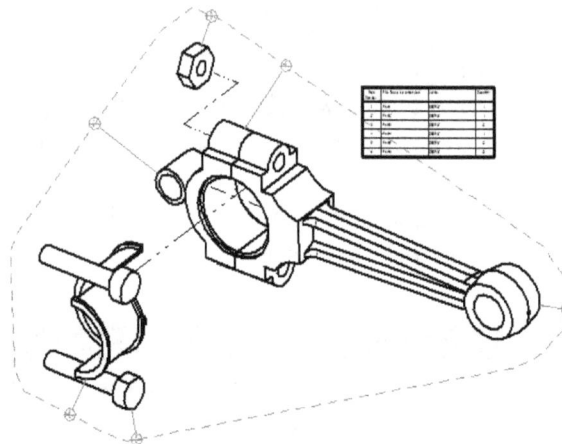

*Figure-24. Parts list with balloons*

## CREATING HOLE TABLE

The **Hole Table** tool is used to create a table of holes in the drawing. The procedure to use this tool is given next.

- Click on the **Hole Table** tool from the **Table** group in the **Home** tab of **Ribbon**. The **Hole Table Command Bar** will be displayed and you will be asked to specify location of origin point.
- Click at desired point to specify x and y location of origin for table; refer to Figure-25. You will be asked to select the view which contains holes.

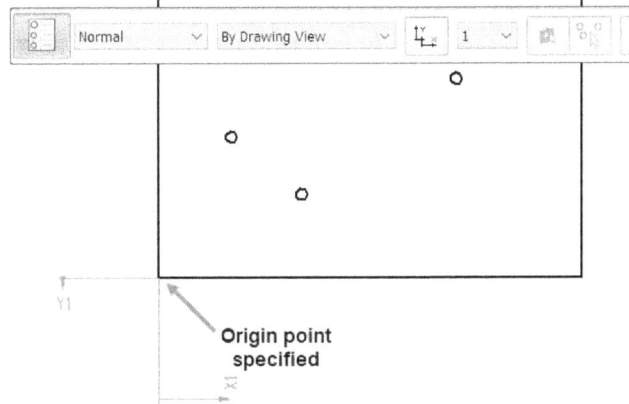

*Figure-25. Origin point specified*

- Select the desired view with holes. Note that the holes should have been created by Hole tool in modeling environment. After selecting view, right-click in the drawing to accept the selection. The table will get attached to cursor.
- Click at desired location to place the table. The holes will be annotated with hole table; refer to Figure-26.

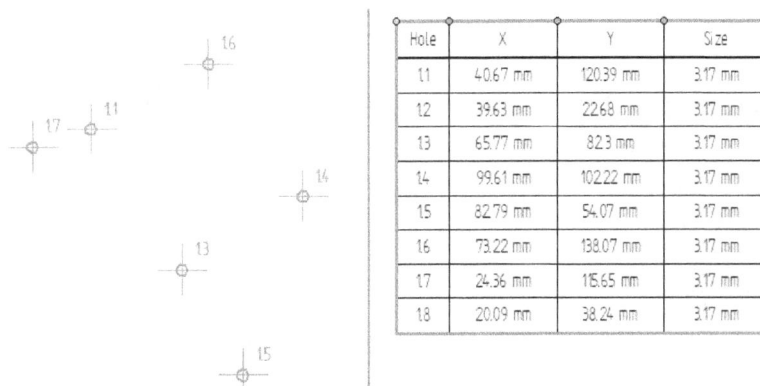

| Hole | X | Y | Size |
|------|-----|-----|------|
| 1.1 | 40.67 mm | 120.39 mm | 3.17 mm |
| 1.2 | 39.63 mm | 22.68 mm | 3.17 mm |
| 1.3 | 65.77 mm | 82.3 mm | 3.17 mm |
| 1.4 | 99.61 mm | 102.22 mm | 3.17 mm |
| 1.5 | 82.79 mm | 54.07 mm | 3.17 mm |
| 1.6 | 73.22 mm | 138.07 mm | 3.17 mm |
| 1.7 | 24.36 mm | 15.65 mm | 3.17 mm |
| 1.8 | 20.09 mm | 38.24 mm | 3.17 mm |

*Figure-26. Holes with table*

- Press **ESC** to exit the tool.

## CREATING BEND TABLE

The **Bend Table** tool is used to create table of bends in the sheetmetal part. Note that there should be a sheetmetal part with bends in the drawing to use this tool. The procedure to use this tool is given next.

- Click on the **Bend Table** tool from the **Tables** group in the **Home** tab of the **Ribbon**. The **Bend Table Command Bar** will be displayed and you will be asked to select the drawing view which contains flat pattern of sheet metal part.
- Select the desired drawing view. The bend table will get attached to cursor.
- Select the **Auto-Callout** button from the **Command Bar** to automatically add callouts to bends in the drawing view.

- Select the **Place Table** button from the **Command Bar** to create bend table and place it in drawing sheet.
- Click at the desired location in the drawing to place the table. The table will be created; refer to Figure-27.

| Sequence | Feature | Radius | Angle | Direction | Included Angle |
|---|---|---|---|---|---|
| 1 | Bend 1 | 0.002 mm | 180.00 deg | Up | 0.00 deg |
| 2 | Bend 2 | 8.000 mm | 25.18 deg | Down | 154.82 deg |
| 3 | Bend 3 | 1.000 mm | 90.00 deg | Up | 90.00 deg |

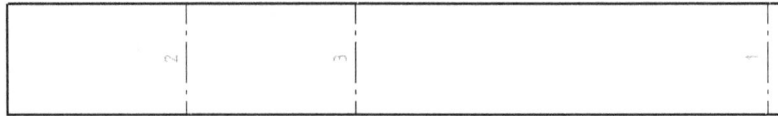

*Figure-27. Bend table*

- Press **ESC** to exit the tool.

Similarly, you can use the other tools in the **Tables** group to generate tables.

# DIMENSIONING TOOLS

Most of the tools used for creating dimensions have been discussed earlier in chapters related to modeling and PMI. We will now discuss rest of the tools for dimensioning and annotation.

## Creating Symmetric Diameter Dimension

The **Symmetric Diameter** tool is used to create a diameter dimension which is double of the distance between two selected points to create diameter dimension. The procedure to use this tool is given next.

- Click on the **Symmetric Diameter** tool from the **Dimension** group in the **Home** tab of **Ribbon**. The **Symmetric Diameter Command Bar** will be displayed and you will be asked to select the points.
- One by one click on the two points to be dimensioned. The dimension will get attached to cursor; refer to Figure-28.

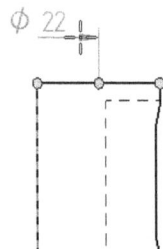

*Figure-28. Diameter dimension*

- Click at desired location to place the dimension and press **ESC** to exit the tool.

## Applying Chamfer Dimension

The **Chamfer Dimension** tool is used to dimension chamfers in the drawing. The procedure to apply chamfer dimension is given next.

- Click on the **Chamfer Dimension** tool from the **Dimension** group in the **Home** tab of the **Ribbon**. The **Chamfer Dimension Command Bar** will be displayed and you will be asked to select a chamfer for dimensioning.
- Select the desired option from the **Orientation** drop-down in the **Command Bar** to specify how dimension will be placed. Select the **Along Axis** option from the drop-down to create dimension aligned to selected chamfer edge. Select the **Callout Perpendicular** option from the drop-down to create chamfer dimension callout perpendicular to selected edge. Select the **Callout Parallel** to place the dimension parallel to selected edge.
- Select the chamfer line to be dimensioned and then select the base line to be used for aligning dimension. The dimension get attached to cursor; refer to Figure-29.

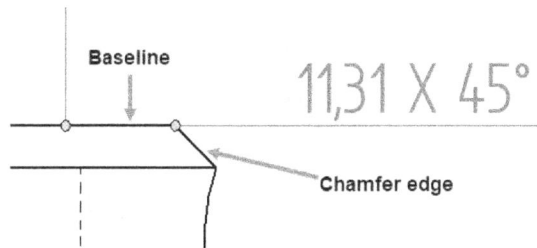

*Figure-29. Chamfer dimension created*

- Click at desired location to place the dimension.

## Retrieving Dimension

The **Retrieve Dimensions** tool is used to copy PMI dimensions and annotations of the model in to a drawing view. The procedure to use this tool is given next.

- Click on the **Retrieve Dimensions** tool from the **Dimension** group in the **Home** tab of the **Ribbon**. The **Retrieve Dimensions Command Bar** will be displayed; refer to Figure-30.

*Figure-30. Retrieve Dimensions Command Bar*

- Select the desired buttons from the **Command Bar** to define which type of dimensions and annotations will be retrieved in the drawing views.
- Select the desired drawing views to copy dimensions and press **ESC** to exit the tool.

## Line Up Text

The **Line Up Text** tool is used to align dimension and annotation texts. The procedure to use this tool is given next.

- Click on the **Line Up Text** tool from the **Dimension** group in the **Home** tab of the **Ribbon**. The **Line Up Text Command Bar** will be displayed; refer to Figure-31.

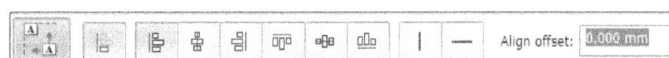

*Figure-31. Line Up Text Command Bar*

- Select the desired button from the **Command Bar** to define alignment style and then select the dimensions & annotations to align them.
- Press **ESC** to exit the tool.

## Copying Attributes

The **Copy Attributes** tool is used to copy attributes from one annotation to other. The procedure to use this tool is given next.

- Click on the **Copy Attributes** tool from the **Dimension** group in the **Home** tab of the **Ribbon**. The **Copy Attributes Command Bar** will be displayed; refer to Figure-32.

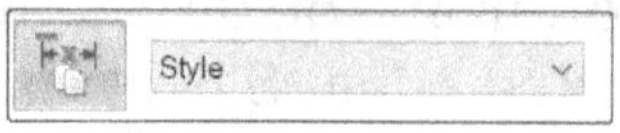

*Figure-32. Copy Attributes Command Bar*

- Select the desired option from the drop-down to define which attributes are to be copied.
- Select the desired dimension/annotation you want to copy from the drawing and then select the other dimensions to which you want to copy the attributes.
- After setting parameters, press **ESC** to exit the tool.

## Removing Dimensions from Alignment Set

Click on the **Remove from Alignment Set** tool from the **Dimension** group in the **Home** tab of **Ribbon** to remove selected dimensions from alignment set. After removing a dimension from alignment set, you can individually move the selected dimension.

## Stacking Dimensions

The **Arrange Dimensions** tool in **Dimension** group of **Ribbon** is used to stack dimensions one over the other at fixed distance between them. To use this tool, select it from the **Dimension** group in the **Home** tab of the **Ribbon**. The **Arrange Dimensions Command Bar** will be displayed. Specify the desired gap multiplier in the **Command Bar** and then select the dimensions to be arranged.

Similarly, the **Update Retrieved Dimensions** tool in **Dimension** group of **Ribbon** is used to update the changes made in PMI of model in drawing.

The tools in the **Annotation** group have been discussed earlier in the book.

# PRACTICAL 1

Create engineering drawing of model as shown in Figure-33. The procedure to create this model has been discussed earlier in Chapter 7 of this book.

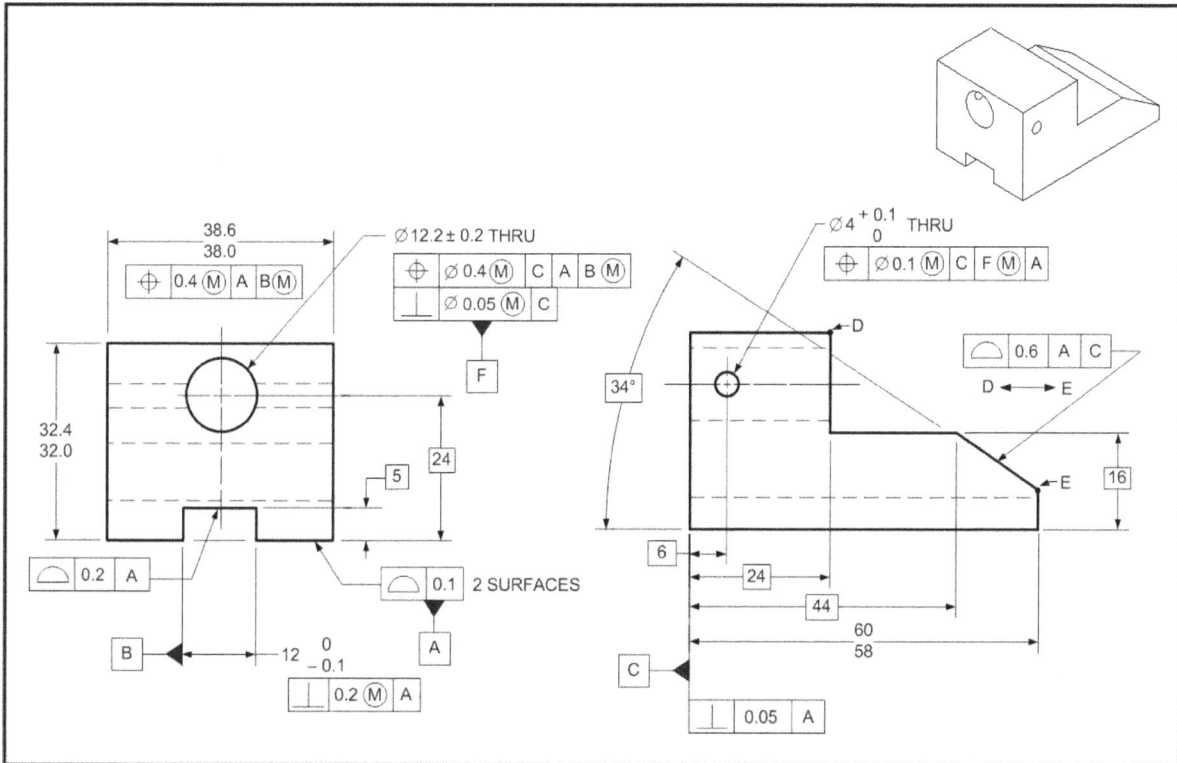

*Figure-33. Practical*

## Starting A Drawing File

• Start Solid Edge using Start menu or desktop icon, if not started yet.
• Click on the **ISO Metric Draft** tool from the **New** cascading menu of the **Application** menu. The drawing creation interface will be displayed with a new draft document opened.

## Placing Views

• Click on the **View Wizard** tool from the **Drawing Views** group in the **Home** tab of **Ribbon**. The Select Model dialog box will be displayed.
• Select the model file of PMI Practical in Chapter 7 of resource kit. The Front view of model will get attached to cursor.
• Make sure the scale value in **Scale value** edit box of **Command Bar** is specified as **4** and then click at the right side in the drawing area to place the view; refer to Figure-34.
• Move the cursor to left side of model. The right view of model will get attached to cursor.
• Click on the left side of front view to place the view; refer to Figure-35 and then press **ESC** to exit the tool.

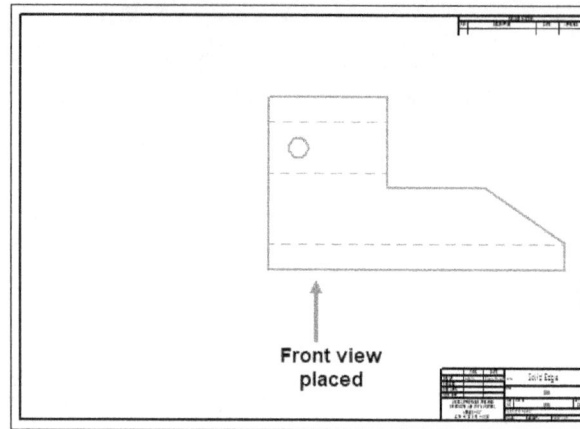

*Figure-34. Placing front view*

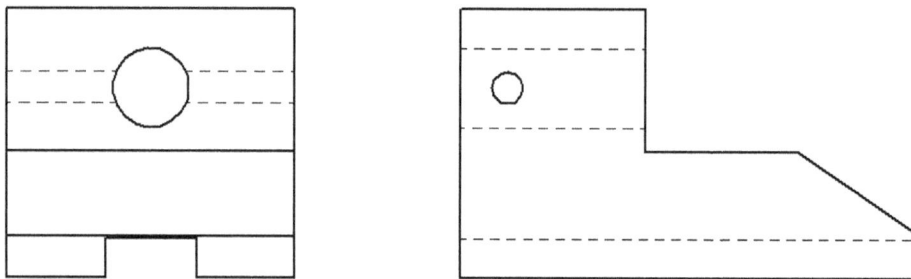

*Figure-35. Views placed in drawing*

## Retrieving and Applying Dimensions

- Click on the **Retrieve Dimensions** tool from the **Dimension** group in the **Home** tab of **Ribbon**. The **Retrieve Dimensions Command Bar** will be displayed.
- Select the two views one by one. All the dimensions from PMI data will be retrieved in the drawing views. Press **ESC** to exit the tool.
- Move the dimensions and annotations to desired locations. Delete the extra dimensions and generate a neat drawing as shown in Figure-36.

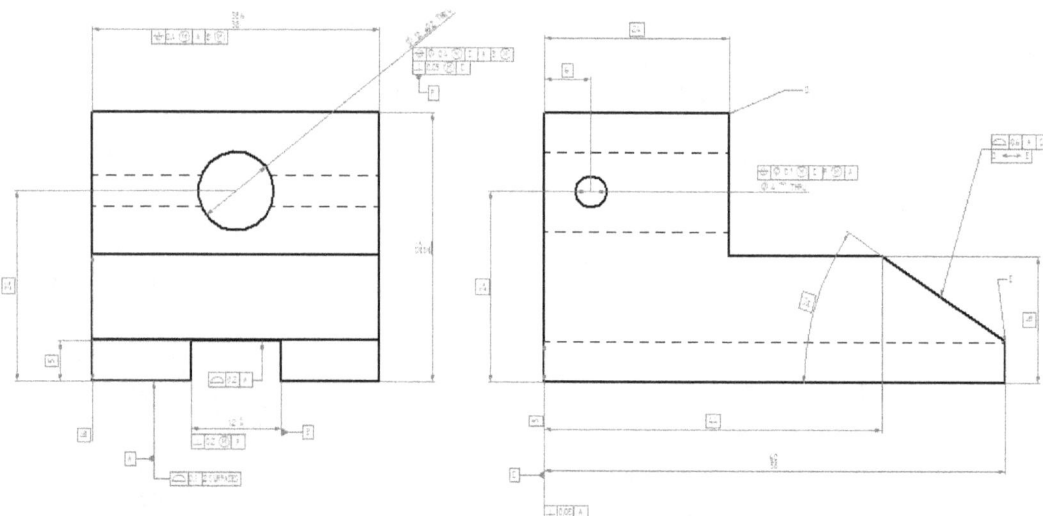

*Figure-36. Drawing after applying dimensions*

# PRACTICE 1

Create the model and generate drawing as shown in Figure-37.

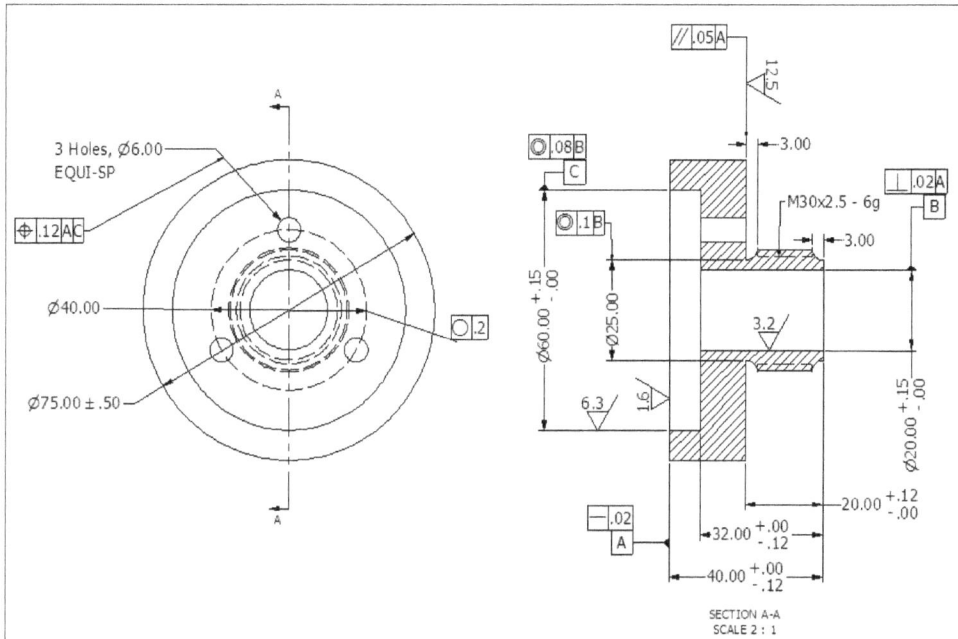

*Figure-37. Production drawing for Practice 1*

# PRACTICE 2

Create the model and drawing per the views shown in Figure-38.

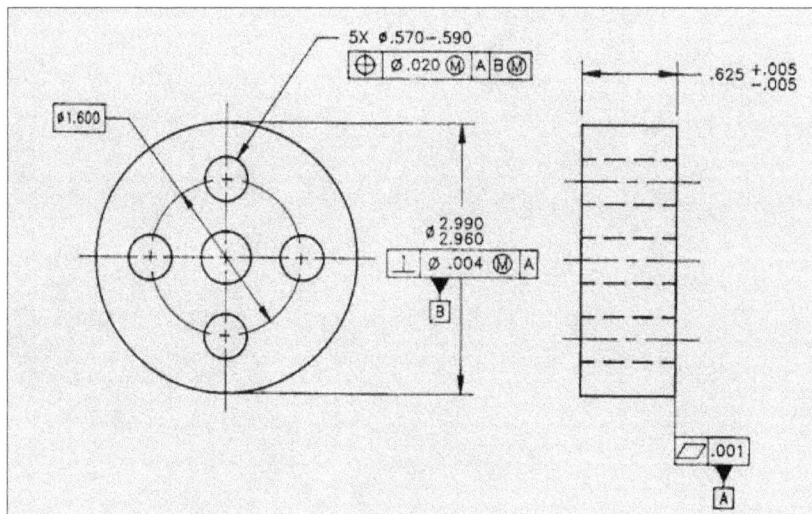

*Figure-38. Practice 2*

FOR STUDENT NOTES

# Chapter 12

# Sheetmetal Design

## Topics Covered

The major topics covered in this chapter are:

- *Starting Sheetmetal Design*
- *Creating Tab Feature*
- *Creating Different types of Flanges*
- *Creating Hem, Dimple, and Louver Features*
- *Creating Drawn Cutouts, Bead, Gusset, Cross Brake, Emboss, and Etch Features*
- *Modifying Bend Corners*
- *Bending, Unbending, and Rebending*
- *Creating Jog features,*
- *Sheetmetal Modification Tools*

# INTRODUCTION

Sheet metal work is an important aspect of Mechanical engineering. Many parts around us are manufactured via sheetmetal processes. For example, car body, vents in houses, Air-conditioner ducts, spoon, metal bowls, and so on. The sheetmetal parts generally have thickness ranging from fraction of millimeter to 12.5 millimeters i.e. up to half inch. Like welding and machining, sheetmetal also has its own processes like, bending, punching, stamping, spinning, rolling, and so on. In this chapter, we will discuss about the tools available in Solid Edge related to sheetmetal designing. Note that we will be working in Ordered modeling environment in this chapter. You can use the same tools in Synchronous modeling environment as well.

# STARTING SHEETMETAL PART DOCUMENT

In Solid Edge, Sheetmetal parts are designed in a different environment. The procedure to create sheetmetal part document is given next.

*   Click on the **New** tool from the **New** cascading menu in the **Application** menu. The **New** dialog box will be displayed.
*   Select the desired .psm template from the dialog box (**iso metric sheet metal. psm** in our case) and click on the **OK** button. The Sheetmetal environment interface will be displayed in Synchronous style; refer to Figure-1.

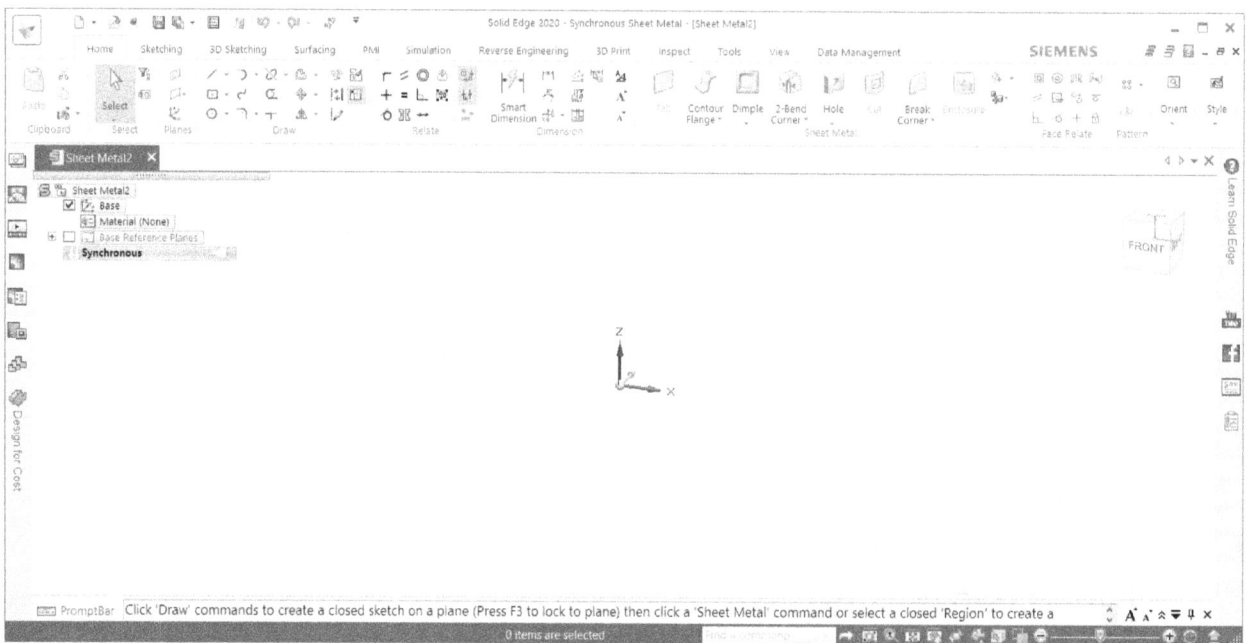

*Figure-1. Sheet metal environment interface*

*   Select the **Ordered** radio button from the **Model** group in the **Tools** tab of the **Ribbon**. The interface will be displayed in Ordered environment.

# CREATING TAB FEATURE

Tab is a flat plate of sheetmetal generally used as base for flanges and other features. The procedure to create a tab feature is given next.

*   Create a closed loop sketch to be used for defining boundary of tab feature.

- Click on the **Tab** tool from the **Sheet Metal** group in the **Home** tab of **Ribbon**. The **Tab Command Bar** will be displayed; refer to Figure-2 and you will be asked to select a sketching plane.

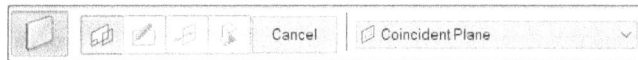

*Figure-2. Tab Command Bar*

- If you have created a sketch already for tab then select the **Select from Sketch** option from the **Create-From Options** drop-down in the **Command Bar**, select the desired sketch, right-click in the drawing area. Otherwise, select the desired plane/face and create the sketch. The **Command Bar** will be displayed as shown in Figure-3 to specify thickness of tab.

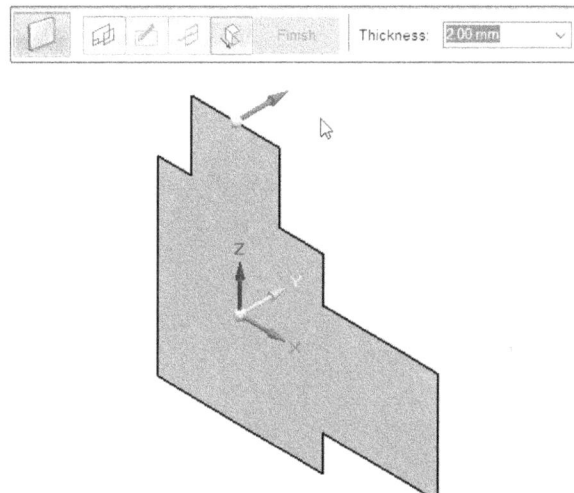

*Figure-3. Specifying thickness of tab feature*

- Specify the desired value of thickness in the **Thickness** edit box and click at desired location to define side of thickness.
- Click on the **Finish** button from the **Command Bar** to create tab feature. Press **ESC** to exit the tool.

## CREATING FLANGE FEATURE

The flange feature is used to create sheet metal walls on selected edges of tab. The procedure to create the feature is given next.

- Click on the **Flange** tool from the **Sheet Metal** group in the **Home** tab of **Ribbon**. The **Flange Command Bar** will be displayed; refer to Figure-4.

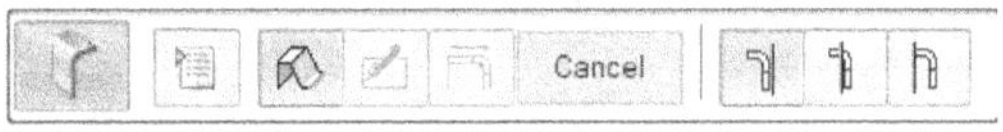

*Figure-4. Flange Command Bar*

- Click on the **Flange Options** button from the **Command Bar**. The **Flange Options** dialog box will be displayed; refer to Figure-5. Clear the **Use default value** check box to specify desired bend radius value in the **Bend radius** edit box. Select the **Bend relief**, **Extend relief**, and **Corner relief** check boxes to apply relief cuts at bends and corners of the flange sheet.
- Specify the desired values in the edit box and click on the **OK** button to apply to flange.

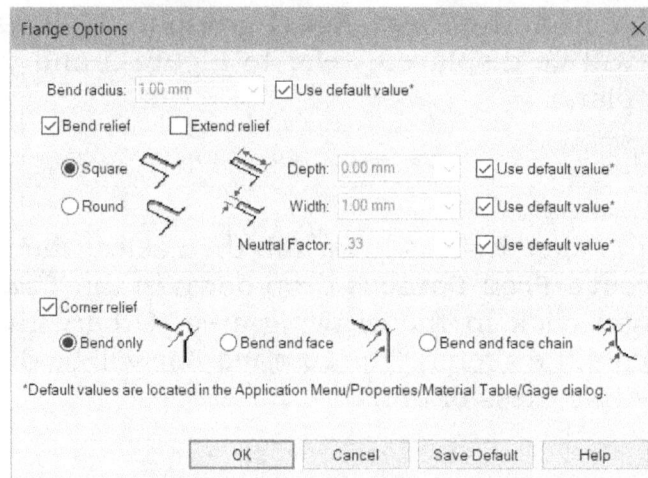

*Figure-5. Flange Options dialog box*

- Select the **Material Inside** button from the **Command Bar** to create flange inside the selected edge. Similarly, select the **Material Outside** and **Bend Outside** buttons from the **Command Bar** to create flange wall and bend of flange outside selected edge.
- Click on the desired edge to create the flange. The flange will get attached to cursor; refer to Figure-6.

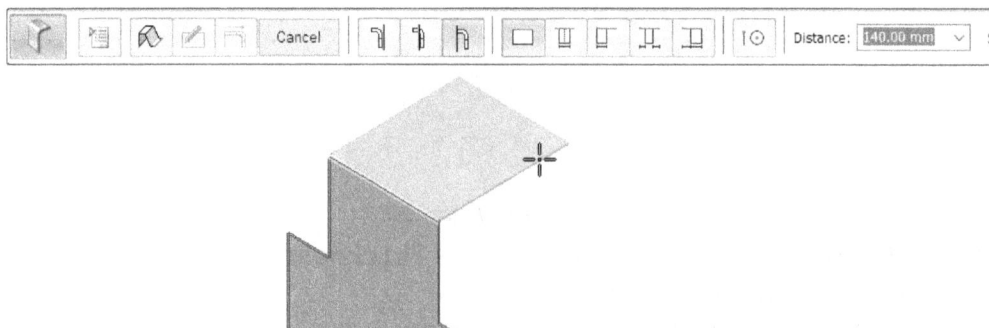

*Figure-6. Flange attached to cursor*

- Select the desired button from the **Command Bar** to define width of flange. Select the **Full Width** button from the **Command Bar** to create flange with width equal to length of selected edge. Select the **Centered** button from the **Command Bar** to create flange of specified width at center of selected edge. Select the **At End** button to create flange of specified width at selected end point and then select the end point from where you want to start the flange; refer to Figure-7. Select the **From Both Ends** button to create flange at specified distance from the end points. Select the **From End** button to create flange at specified distance from the selected end point.
- Specify the desired distance value in **Distance** edit box to define length of flange and press **ENTER**. Preview of the flange will be displayed; refer to Figure-8.

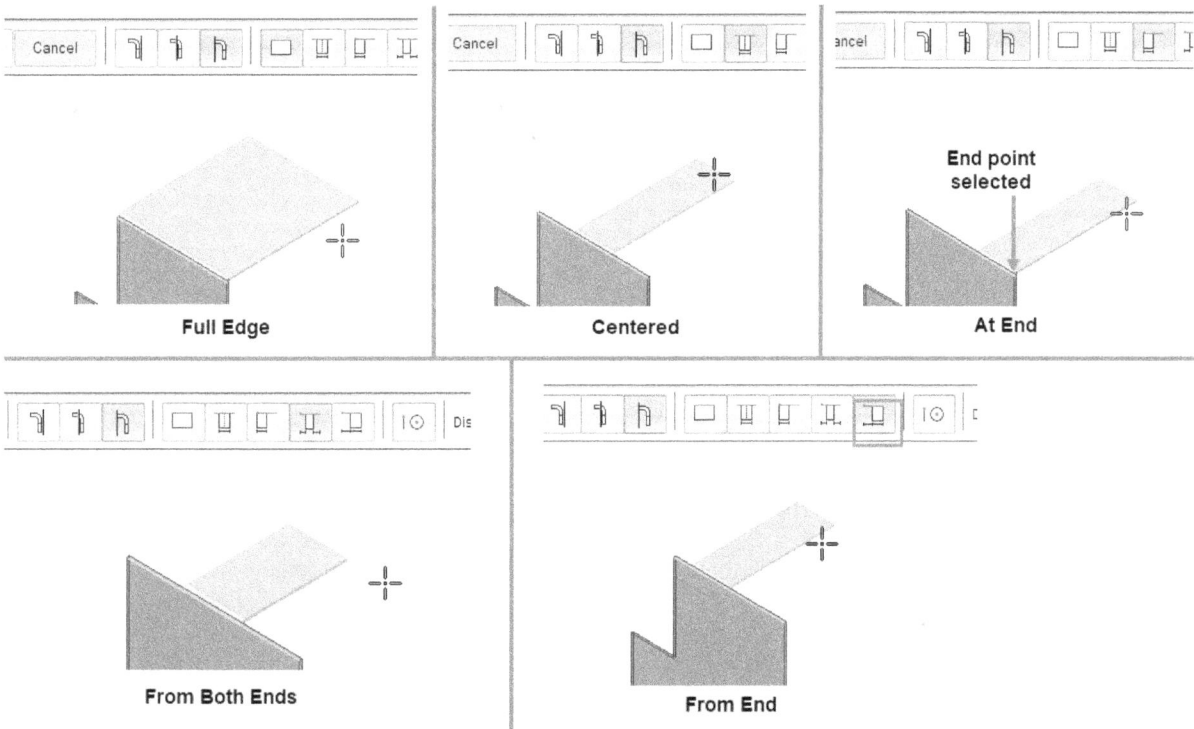

Figure-7. Specifying width of flange

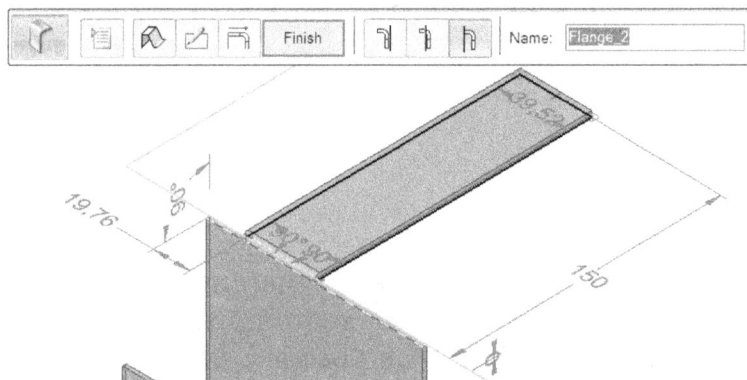

Figure-8. Sketch for flange

- Double-click on desired dimensions to modify the flange and click on the Finish button. The flange will be created. Press **ESC** to exit the tool.

## CREATING CONTOUR FLANGE

The **Contour Flange** tool is used to create flange at desired edge with create shape. The procedure to use this tool is given next.

- Click on the **Contour Flange** tool from the **Contour Flange** drop-down in the **Sheet Metal** group of the **Home** tab in the **Ribbon**. The **Contour Flange Command Bar** will be displayed; refer to Figure-9.

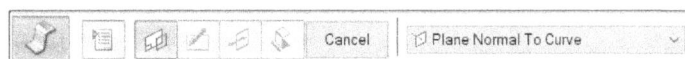

Figure-9. Contour Flange Command Bar

- Select the desired edge to be used for creating flange. You will be asked to specify location of plane for creating sketch of flange shape.
- Click at desired location to specify location of plane and create desired sketch for flange; refer to Figure-10.

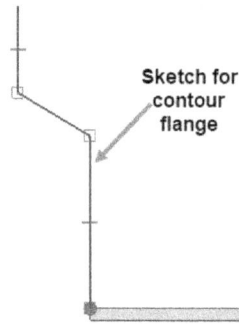

*Figure-10. Sketch created for contour flange*

- After creating sketch, click on the **Close Sketch** button from the **Close** group in the **Home** tab of the **Ribbon**. The flange will get attached to cursor; refer to Figure-11.

*Figure-11. Contour flange attached to cursor*

- Select the **Finite Extent** button to specify distance up to which the flange will be created. Select the **To End** button to create flange up to end point of selected edge. Select the **Chain** button from the **Command Bar** to create the flanges along multiple edges of base feature in a chain; refer to Figure-12.

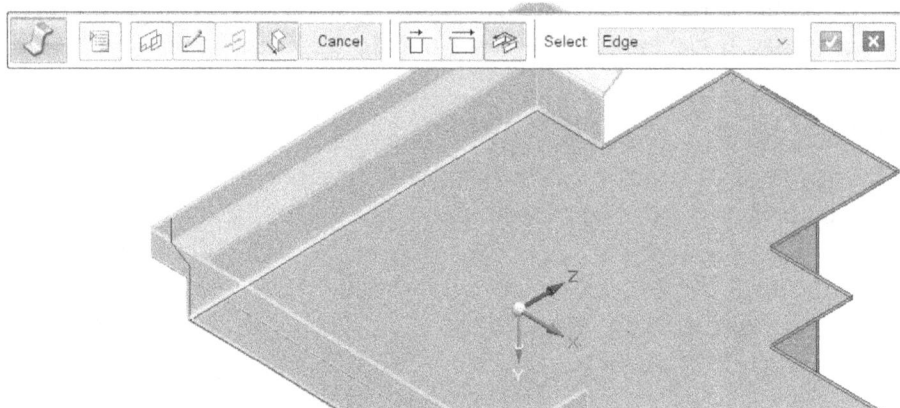

*Figure-12. Contour flanges using chain selection*

- Click on the **OK** button from the **Command Bar** to create the flange and click on the **Finish** button. Press **ESC** to exit the tool.

## CREATING LOFTED FLANGE

The **Lofted Flange** tool is used to create flange using sketched sections. The procedure to use this tool is given next.

- Click on the **Lofted Flange** tool from the **Contour Flange** drop-down in the **Sheet Metal** group of the **Home** tab in the **Ribbon**. The **Lofted Flange Command Bar** will be displayed; refer to Figure-13.

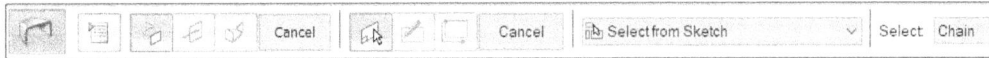

*Figure-13. Lofted Flange Command Bar*

- Select the desired plane/face to create the first sketch of lofted flange and create the desired open sketch; refer to Figure-14. You will be asked to specify the start point of section.

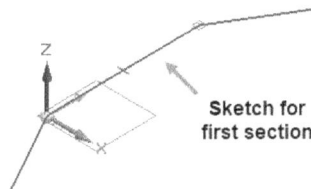

*Figure-14. Sketch 1 for lofted section*

- Select the desired vertex of sketch and right-click in the drawing area. You will be asked to select a plane for creating second section of sketch.
- Select the **Parallel Plane** or other option from the **Create-From Options** drop-down in the **Command Bar** and select the desired sketching plane. You will be asked to create sketch for second section.
- Create the desired open sketch and click on the **Close Sketch** button from **Ribbon**. You will be asked to select starting vertex of second section.
- Select the desired vertex and right-click in the drawing area. You will be asked to specify direction of applying thickness to feature.
- Click at desired side of profile. The preview of flange will be displayed; refer to Figure-15.

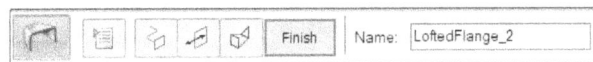

*Figure-15. Preview of lofted flange feature*

- Click on the **Finish** button from the **Command Bar** to create the feature and press **ESC** to exit the tool.

# CREATING HEM FEATURE

The **Hem** tool is used to create hem bend in the sheet metal at selected edges. The procedure to use this tool is given next.

- Click on the **Hem** tool from the **Contour Flange** drop-down in the **Sheet Metal** group in the **Home** tab of **Ribbon**. The **Hem Command Bar** will be displayed; refer to Figure-16.

*Figure-16. Hem Command Bar*

- Click on the **Hem Options** button from the **Command Bar**. The **Hem Options** dialog box will be displayed; refer to Figure-17.

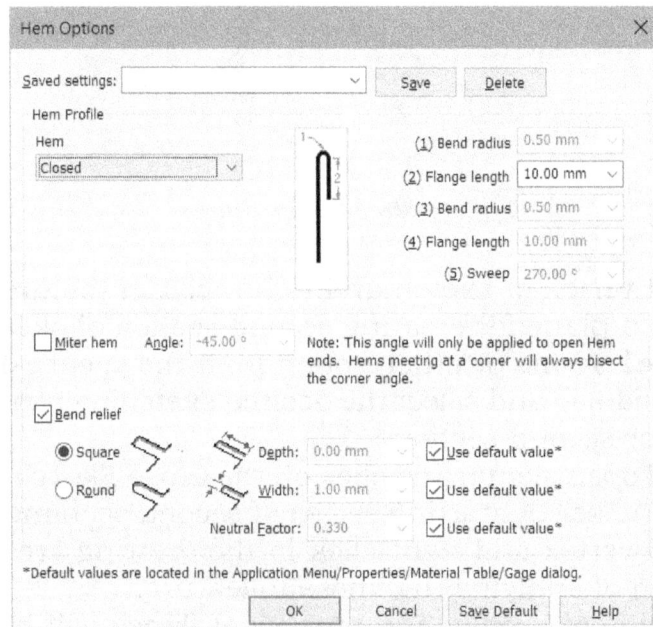

*Figure-17. Hem Options dialog box*

- Select the desired option from the **Hem** drop-down in the **Hem Profile** section and set the desired parameters in the dialog box.
- Select the **Miter hem** check box to apply miter cut at the end sides of the hem feature. Click on the **OK** button from the dialog box to apply parameters.
- Select the desired edge to apply the feature.
- Click on the **Material Inside**, **Material Outside**, or **Bend Outside** button to define location of hem.
- After setting parameters, click on the **OK** button and then **Finish** button from the **Command Bar** to create the hem feature.
- Press **ESC** to exit the tool.

# CREATING DIMPLE FEATURE

The **Dimple** tool is used to create dent in sheet metal with specified boundary curves. The procedure to use this tool is given next.

- Click on the **Dimple** tool from the **Dimple** drop-down in the **Sheet Metal** group of the **Home** tab in the **Ribbon**. The **Dimple Command Bar** will be displayed; refer to Figure-18 and you will be asked to select a planar face of sheet metal.

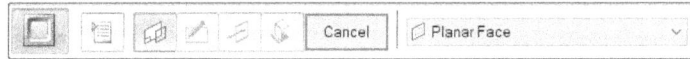

*Figure-18. Dimple Command Bar*

- Select the desired face of sheet metal. The sketching environment will be displayed and you will be asked to create a closed loop sketch of dimple.
- Create the sketch as desired for dimple; refer to Figure-19. After creating sketch, click on the **Close Sketch** button from the **Command Bar**. You will be asked to specify depth of dimple.

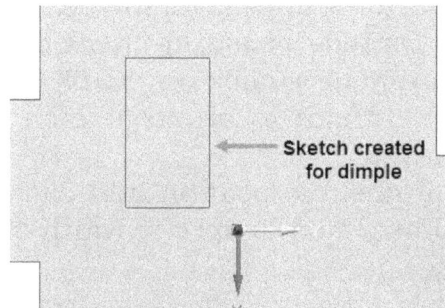

*Figure-19. Sketch created for dimple*

- Click at desired depth level to specify depth of dimple; refer to Figure-20. You can double-click on the depth dimension to change it.

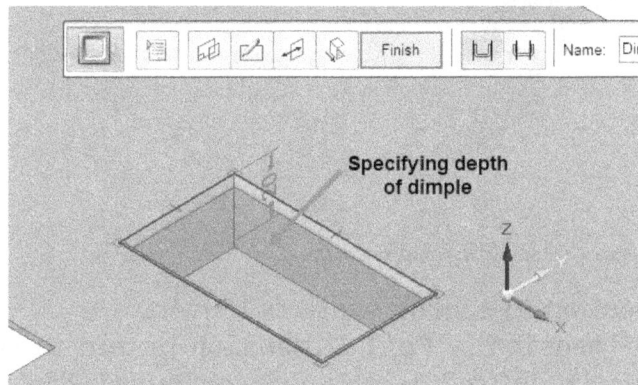

*Figure-20. Specifying depth of dimple*

- Click on the **Finish** button from the **Command Bar** and then press **ESC** to exit the tool.

## CREATING LOUVER FEATURE

The **Louver** tool is used to create louver cuts in the sheet. The procedure to use this tool is given next.

- Click on the **Louver** tool from the **Dimple** drop-down in the **Sheet Metal** group of the **Home** tab in the **Ribbon**. The **Louver Command Bar** will be displayed; refer to Figure-21 and you will be asked to select a face of sheet metal.

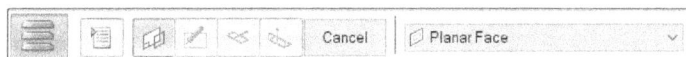

*Figure-21. Louver Command Bar*

- Click on the **Louver Options** button from the **Command Bar**. The **Louver Options** dialog box will be displayed; refer to Figure-22.

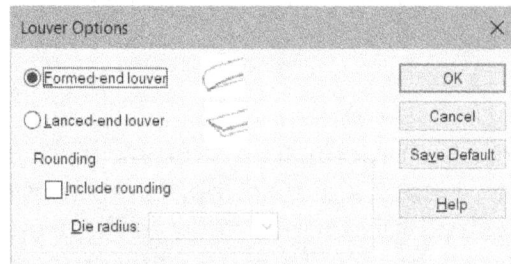

*Figure-22. Louver Options dialog box*

- Select the desired radio button from the dialog box to define what type of louver will be created. Select the **Include rounding** check box to specify radius at louver edges. After specifying parameters, click on the **OK** button.
- Click on the face of sheet metal to create sketch for louver. The sketching environment will be displayed.
- Create a line for louver at desired location and click on the **Close Sketch** tool from the **Ribbon**. You will be asked to specify width of louver; refer to Figure-23.

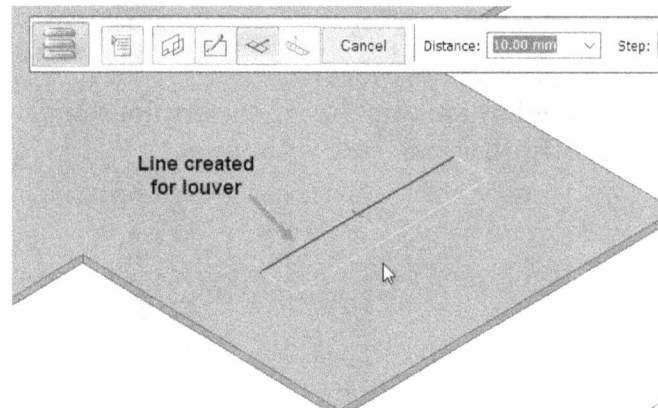

*Figure-23. Specifying width of louver*

- Click at desired location to specify width of louver.
- Select the **Offset Dimension** or **Full Dimension** button to specify depth style of louver and then specify desired depth value for louver. The preview of louver will be created; refer to Figure-24.

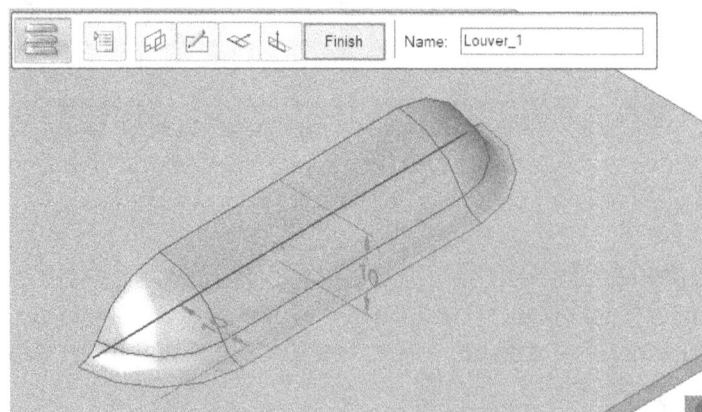

*Figure-24. Preview of louver*

- Click on the **Finish** button to create the louver. Press **ESC** to exit the tool.

## CREATING DRAWN CUTOUT

The **Drawn Cutout** tool is used to cut material from the sheet using closed sketch boundary and use the boundary to draw flange walls. The procedure to use this tool is given next.

- Click on the **Drawn Cutout** tool from the **Dimple** drop-down in the **Sheet Metal** group of the **Home** tab in the **Ribbon**. The **Drawn Cutout Command Bar** will be displayed; refer to Figure-25.

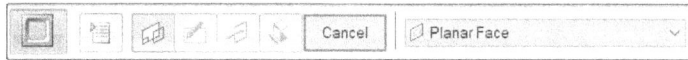

*Figure-25. Drawn Cutout Command Bar*

- Click on the **Drawn Cutout Options** button from the **Command Bar**. The **Drawn Cutout Options** dialog box will be displayed; refer to Figure-26. Set the desired parameters in the dialog box like taper angle, round radius at edges, corner radius, and so on. After setting desired parameters, click on the **OK** button.

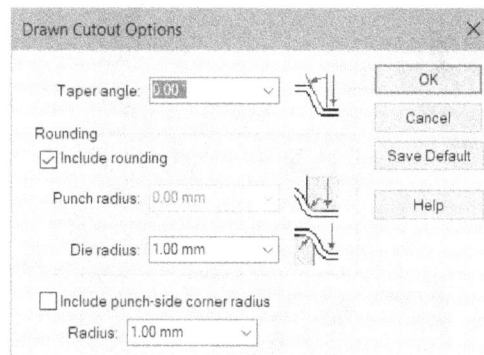

*Figure-26. Drawn Cutout Options dialog box*

- Click at desired face of sheet metal. The sketching environment will be displayed.
- Create the desired closed loop sketch for cutout and click on the **Close Sketch** tool from the **Ribbon**. You will be asked to specify height/depth of cutout.
- Click at desired level to create the cutout. The preview of cutout will be displayed; refer to Figure-27.

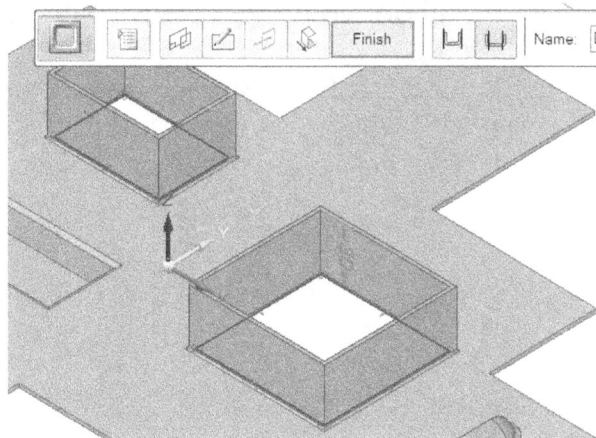

*Figure-27. Preview of cutout*

- Select the **Profile Represents Punch** or **Profile Represents Die** button from the **Command Bar** to define how boundaries will be used for creating cutout.
- After setting desired parameters, click on the **Finish** button from the **Command Bar**. The cutout will be created. Press **ESC** to exit the tool.

## CREATING BEAD FEATURE

The **Bead** tool is used to create dent in sheet metal using line curves as reference. The procedure to use this tool is given next.

- Click on the **Bead** tool from the **Dimple** drop-down in the **Sheet Metal** group of the **Home** tab in the **Ribbon**. The **Bead Command Bar** will be displayed as shown in Figure-28 and you will be asked to select a plane for creating base sketch.

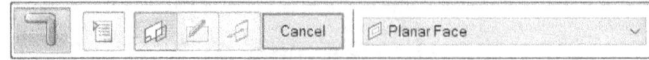

*Figure-28. Bead Command Bar*

- Click on the **Bead Options** button from the **Command Bar**. The **Bead Options** dialog box will be displayed; refer to Figure-29.

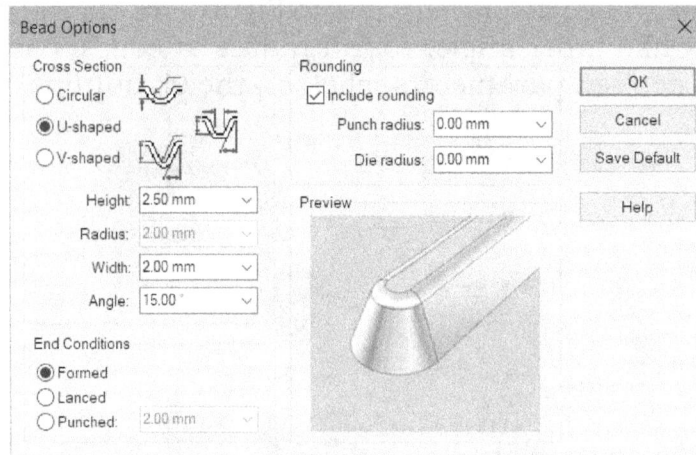

*Figure-29. Bead Options dialog box*

- Select the desired radio button from the **Cross Section** area of the dialog box to define shape of bead and specify the parameters of beads in the edit boxes of this area.
- Select the desired option from the **End Conditions** area to define how ends of beads will be created.
- Set the other parameters as desired in the dialog box and click on the **OK** button.
- Click on the face of sheet metal on which you want to create bead. The sketching environment will be displayed.
- Create the desired open line sketch (note that lines should be in tangent chain) and click on the **Close Sketch** button. Preview of bead will be displayed; refer to Figure-30.

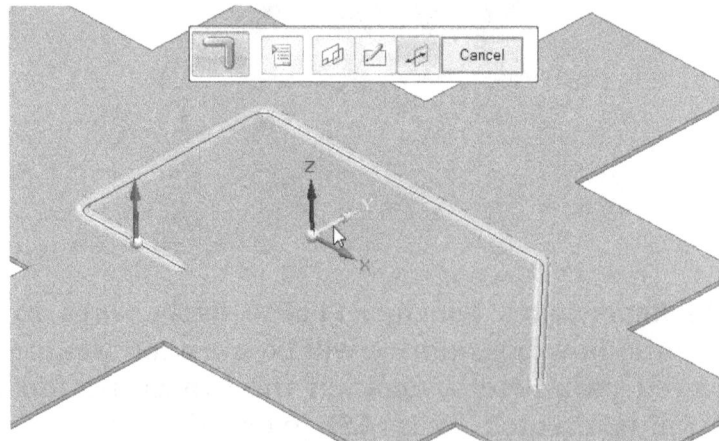

*Figure-30. Preview of bead*

- Click at desired side of sheet metal to create the bead and click on the **Finish** button from the **Command Bar**. Press **ESC** to exit the tool.

## CREATING GUSSET FEATURE

The **Gusset** tool is used to create gussets in sheet metal for supporting flange walls. The procedure to use this tool is given next.

- Click on the **Gusset** tool from the **Dimple** drop-down in the **Sheet Metal** group of the **Home** tab in **Ribbon**. The **Gusset Command Bar** will be displayed; refer to Figure-31.

*Figure-31. Gusset Command Bar*

- Click on the **Gusset Options** button from the **Command Bar**. The **Gusset Options** dialog box will be displayed; refer to Figure-32.

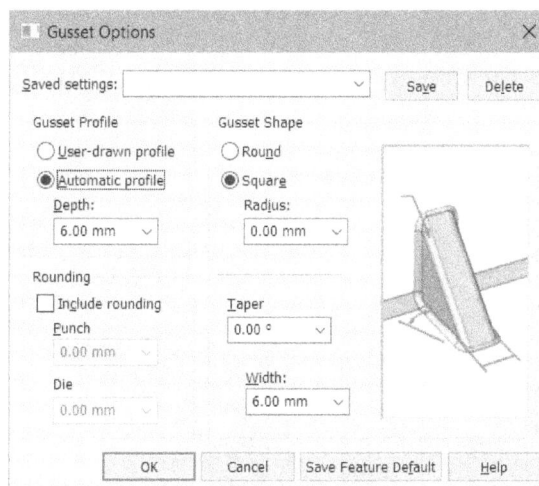

*Figure-32. Gusset Options dialog box*

- Set the desired profile and thickness of gusset in the dialog box and click on the **OK** button.
- Select the desired bend from the model to apply gusset. The gusset will get attached to cursor; refer to Figure-33.

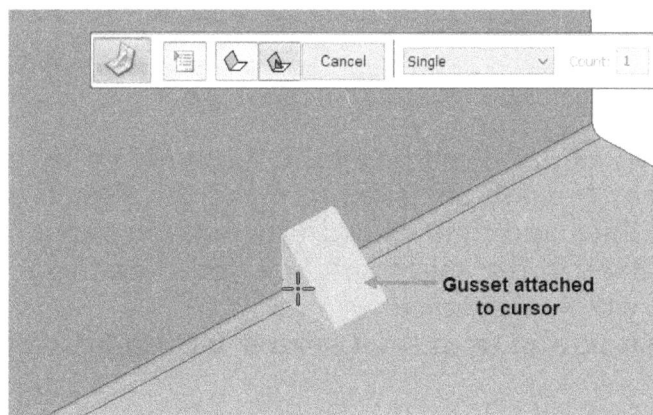

*Figure-33. Gusset attached to cursor*

- Select the **Single** option from the **Pattern Type** drop-down in the **Command Bar** to create single gusset at bend.

- Select the **Fit** option from the **Pattern Type** drop-down and specify total number of gussets to be created in the **Count** edit box. Press **TAB** after specifying value to check preview.
- Select the **Fill** option from the **Pattern Type** drop-down to create as many gussets as possible on a bend with specified gap. Specify the desired distance in the **Spacing** edit box and click on the **OK** button to create the gussets.
- Select the **Fixed** option from the **Pattern Type** drop-down and specify gap as well as number of instances of gussets in respective edit boxes of the **Command Bar**.
- After setting desired parameters, click on the **OK** button from the **Command Bar** to create the gussets.
- Click on the **Finish** button from **Command Bar** and then press **ESC** to exit the tool.

## CREATING CROSS BRAKE

The **Cross Brake** tool is used to create cross-brake feature on the sheet metal using sketch. The procedure to use this tool is given next.

- Create the line sketch for cross brake on the sheet metal; refer to Figure-34.

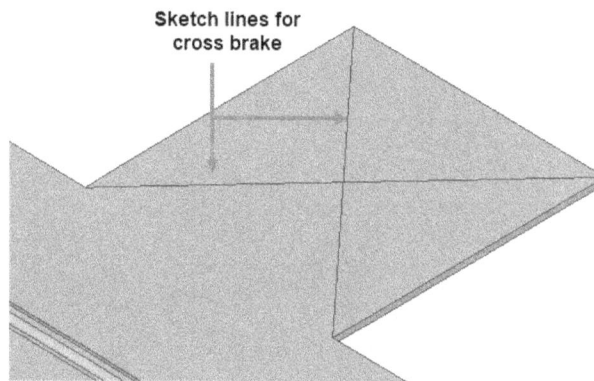

*Figure-34. Sketch for cross brake*

- Click on the **Cross Brake** tool from the **Dimple** drop-down in the **Sheet Metal** group of the **Home** tab in the **Ribbon**. The **Cross Brake Command Bar** will be displayed; refer to Figure-35.

*Figure-35. Cross Brake Command Bar*

- Select the face of sheet on which you want to create cross brake and right-click in the drawing area. You will be asked to select the sketch lines.
- Select the desired lines and right-click in the drawing area.
- Specify the bend angle in the **Angle** edit box and click to specify direction of cut. Preview of feature will be displayed.
- Click on the **Finish** button from the **Command Bar** and then press **ESC** to exit the tool.

## CREATING ETCH FEATURE

The **Etch** tool is used to create etch feature using selected sketch. The procedure to use this tool is given next.

- Click on the **Etch** tool from the **Dimple** drop-down in the **Sheet Metal** group of the **Ribbon**. The **Etch Command Bar** will be displayed; refer to Figure-36.

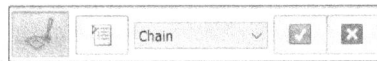

*Figure-36. Etch Command Bar*

- Click on the **Etch Options** dialog box will be displayed; refer to Figure-37. Set the desired color, width, and line type for etching. Click on the **OK** button from the dialog box to apply changes.

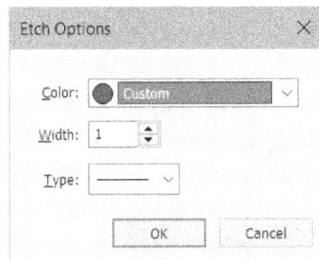

*Figure-37. Etch Options dialog box*

- Select the desired curves to be used for creating etch feature and click on the **OK** button. The etch feature will be created; refer to Figure-38.

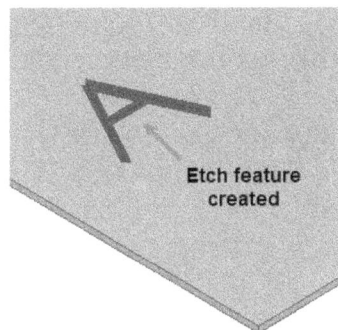

*Figure-38. Etch feature created*

## CREATING EMBOSS FEATURE

The **Emboss** tool is used to emboss a solid body on sheet metal. Note that to use this tool, you will need a solid body along with sheet metal body. The procedure to use this tool is given next.

- Click on the **Emboss** tool from the **Dimple** drop-down in the **Sheet Metal** group of **Home** tab in the **Ribbon**. The **Emboss Command Bar** will be displayed; refer to Figure-39.

*Figure-39. Emboss*
*Command Bar*

- Select the sheet metal part on which the solid body will be embossed. You will be asked to select the tool body (solid body) to be used for embossing.
- Select the desired solid body; refer to Figure-40.
- Select the **Thicken** button from the **Command Bar** to create embossed sheet of thickness equal to sheet metal thickness. If this button is not selected then a cut is made equal to boundary of tool body.

- Click on the **Direction** button from the **Command Bar** to flip direction of embossing.

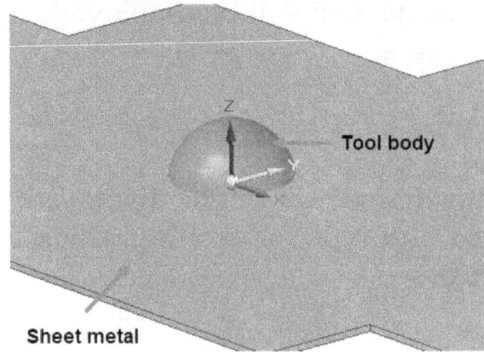

*Figure-40. Objects selected for embossing*

- Specify the desired value in the **Clearance** edit box to generate gap of specified value between tool body and target body.
- You can also press **t** to toggle between thicken On and Off. Similarly, press **f** to flip the direction of embossing.
- After setting desired parameters, click on the **OK** button from the **Command Bar** to create the feature. Press **ESC** to exit the tool.

## CLOSING 2-BEND CORNER

The **Close 2-Bend Corner** tool is used to close or overlap flange walls at 2-bend corners. The procedure to use this tool is given next.

- Click on the **Close 2-Bend Corner** tool from the **2-Bend Corner** drop-down in the **Sheet Metal** group of the **Home** tab in the **Ribbon**. The **Close 2-Bend Corner Command Bar** will be displayed; refer to Figure-41.

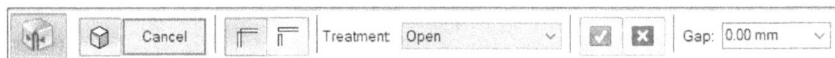

*Figure-41. Close 2-Bend Corner Command Bar*

- Select the desired option from the **Treatment** drop-down in the **Command Bar** and specify respective parameters in the edit boxes of the **Command Bar**.
- Select the **Close** button or **Overlap** button to create closed bends or overlapping bends at flange walls. If you have selected the **Overlap** button then specify desired value in **Overlap** ratio edit box to define the extent of two overlapping flange walls. For example if you have specified **0.5** in the edit box then half of thickness of flange walls will be overlapping.
- After specifying the parameters, select two consecutive bends. The bends will be made closed or overlapping and preview will be displayed; refer to Figure-42.

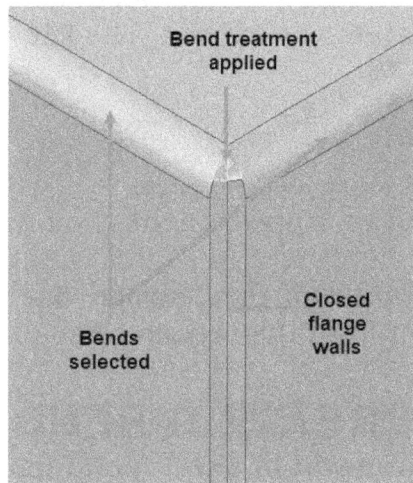

*Figure-42. Preview of closed bends*

- After checking preview, click on the **OK** button from **Command Bar** to create feature.
- Click on the **Finish** button and then press **ESC** key to exit the tool.

## CLOSING 3-BEND CORNERS

The **Close 3-Bend Corner** tool is used to modify corners created at intersection of 3 bends. The procedure to use this tool is given next.

- Click on the **Close 3-Bend Corner** tool from the **2-Bend Corner** drop-down in the **Sheet Metal** group of **Home** tab in the **Ribbon**. The **Close 3-Bend Corner Command Bar** will be displayed; refer to Figure-43.

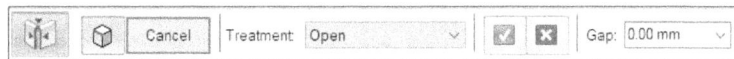

*Figure-43. Close 3-Bend Corner Command Bar*

- Select the desired option from **Treatment** drop-down to define corner condition to be applied and set the desired parameters in the edit boxes of the **Command Bar**.
- Select two bends of three bends set to be used for modifying the cut; refer to Figure-44.

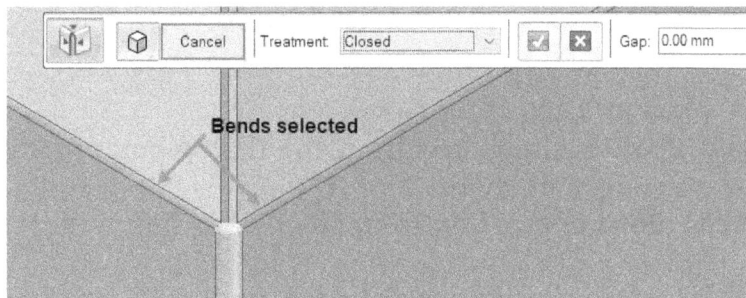

*Figure-44. Bends selected for 3-Bends*

- Click on the **OK** button from the **Command Bar** to create the feature.
- Click on the **Finish** button from the **Command Bar** and press **ESC** to exit the tool.

## CREATING RIPPED CORNERS

The **Rip Corner** tool is used to create ripped corners by replacing the bends. The procedure to use this tool is given next.

- Click on the **Rip Corner** tool from the **2-Bend Corner** drop-down in the **Sheet Metal** group of the **Home** tab in the **Ribbon**. The **Rip Corner Command Bar** will be displayed; refer to Figure-45.

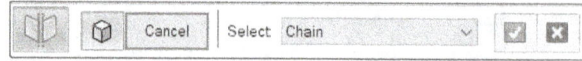

*Figure-45. Rip Corner Command Bar*

- Select the desired edges to be ripped. Right-click in the drawing area to accept the selection.
- Click on the **Preview** button from the **Command Bar** to check preview of ripping. Click on the **Finish** button from the **Command Bar** to create the feature.

## CREATING BEND

The **Bend** tool is used to create bend in sheet metal using specified bend lines. The procedure to use this tool is given next.

- Click on the **Bend** tool from the **Bend** drop-down in the **Sheet Metal** group of **Home** tab in the **Ribbon**. The **Bend Command Bar** will be displayed; refer to Figure-46.

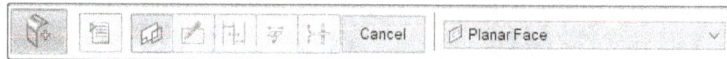

*Figure-46. Bend Command Bar*

- Click on the **Bend Options** button from the **Command Bar**. The **Bend Options** dialog box will be displayed; refer to Figure-47.

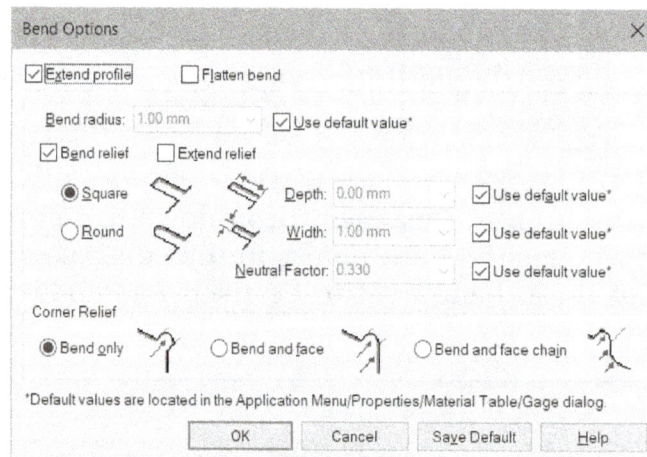

*Figure-47. Bend Options dialog box*

- Select the **Extend Profile** check box from the dialog box to extend the bend line automatically to the edges of sheet.
- Select the **Flatten Bend** check box from the dialog box to flatten the bend after creating it.
- Set the other parameters as discussed earlier and click on the **OK** button.
- Select the desired planar face of sheet metal to draw the bend line. The sketching environment will be displayed.
- Create a straight line to define bend line; refer to Figure-48 and click on the **Close Sketch** button from **Ribbon**. You will be asked to specify direction of bend.
- Click on the left or right side of bend line to define direction of bend. You will be asked to specify which side of profile will move.
- Click on the desired side to define which side will be folded by bend line. You will be asked to specify whether you want to bend the sheet inward or outward.

- Move the cursor to desired side and click to specify the bend direction. Preview of the bend will be displayed; refer to Figure-49.

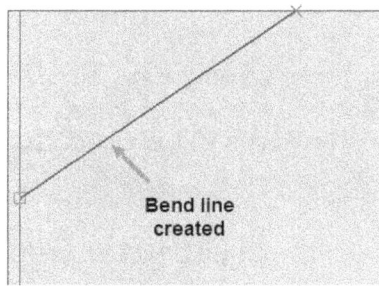

Figure-48. Bend line

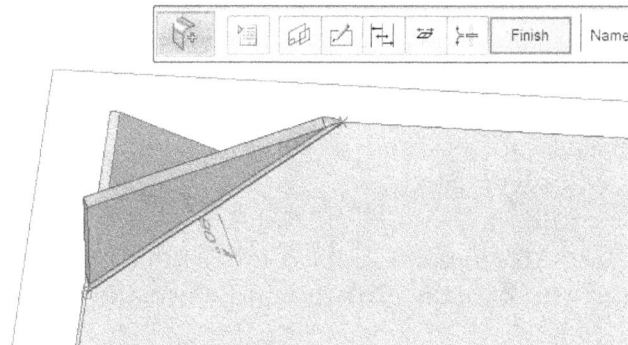

Figure-49. Preview of the bend

- Click on the **Finish** button from the **Command Bar** to create bend and press **ESC** to exit the tool.

## UNBENDING SHEET METAL

The **Unbend** tool is used to unbend selected portion of sheet metal. The procedure to use this tool is given next.

- Click on the **Unbend** tool from the **Bend** drop-down in the **Sheet Metal** group of **Home** tab in the **Ribbon**. The **Unbend Command Bar** will be displayed; refer to Figure-50 and you will be asked to select fixed face.

Figure-50. Unbend Command Bar

- Select the desired face of sheet metal to be fixed. You will be asked to select the bends to be unbend.
- Select the desired bends from the model. The preview of unbend feature will be displayed; refer to Figure-51.

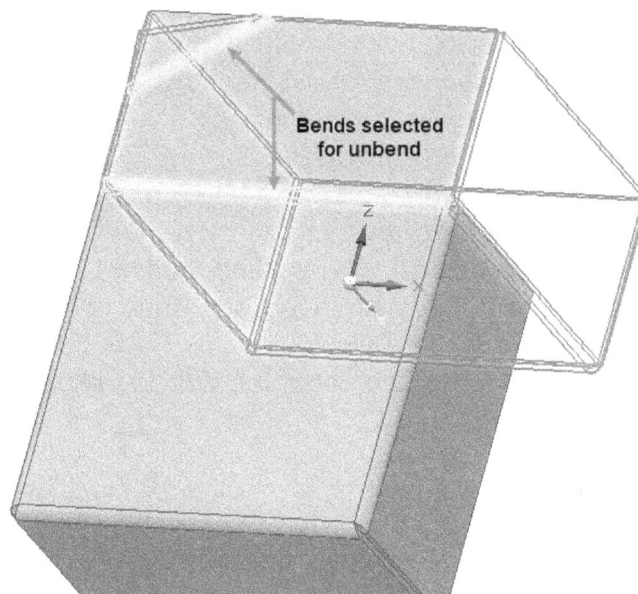

Figure-51. Preview of unbend

- Select the **All Bends** option from the **Select** drop-down if you want to select all the bends in model for unbend. The preview of unbend will be displayed.
- Click on the **Finish** button from the **Command Bar** to create the feature and press **ESC** to exit the tool.

# REBENDING

The **Rebend** tool is used to rebend selected unbend features. The procedure to use this tool is given next.

- Click on the **Rebend** tool from the **Bend** drop-down in the **Sheet Metal** group of **Home** tab in the **Ribbon**. The **Rebend Command Bar** will be displayed; refer to Figure-52.

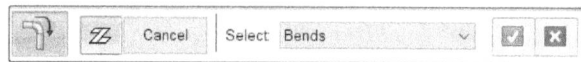

*Figure-52. Rebend Command Bar*

- Select the **Bends** option from the **Select** drop-down in the **Command Bar** to select the bends which were unbend earlier.
- Select the **All Bends** option from the drop-down if you want to rebend all the unbends created in the model.
- After selecting bends, click on the **Finish** button from the **Command Bar** to create features.

# CREATING JOG FEATURE

The **Jog** tool is used to jog features in the sheet metal. The procedure to use this tool is given next.

- Click on the **Jog** tool from the **Bend** drop-down in the **Sheet Metal** group of **Home** tab in the **Ribbon**. The **Jog Command Bar** will be displayed; refer to Figure-53.

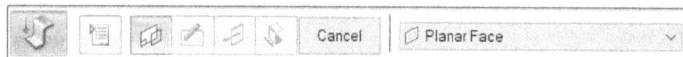

*Figure-53. Jog Command Bar*

- Select the desired option from the **Create-From Options** drop-down in the **Command Bar** to define plane for creating jog line or select a sketch of jog line created on sheetmetal face.
- Select/create the sketch for jog line and then right-click in the drawing area. You will be asked to specify side to be moved by jog.
- Click on the desired side of jog line to specify which side will be moved.
- The jogged face will be attached to cursor; refer to Figure-54. Select the desired buttons from the **Command Bar** to define how bends will be applied to jog.
- Specify the desired radius of bend in the **Bend radius** edit box.
- Click at desired location to specify length of jog base or click in the **Distance** edit box of **Command Bar** and enter the value.
- Click on the **Finish** button from the **Command Bar** to create the feature. Press **ESC** to exit the tool.

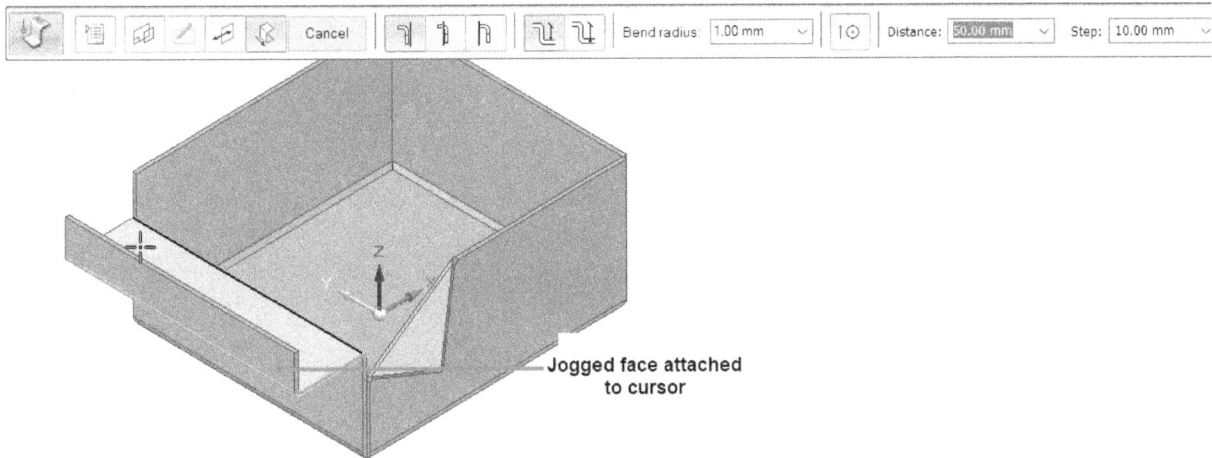

*Figure-54. Jogged face attached to cursor*

## APPLYING BEND BULGE RELIEF

The **Bend Bulge Relief** tool is used to provide relief at the bends which cause bulge in the sheet. The procedure to use this tool is given next.

- Click on the **Bend Bulge Relief** tool from the **Bend** drop-down in the **Sheet Metal** group of the **Home** tab in the **Ribbon**. The **Bend Bulge Relief Command Bar** will be displayed; refer to Figure-55.

*Figure-55. Bend Bulge Relief Command Bar*

- Click on the **Bend Bulge Relief Options** button from the **Command Bar**. The **Bulge Relief Options** dialog box will be displayed; refer to Figure-56.

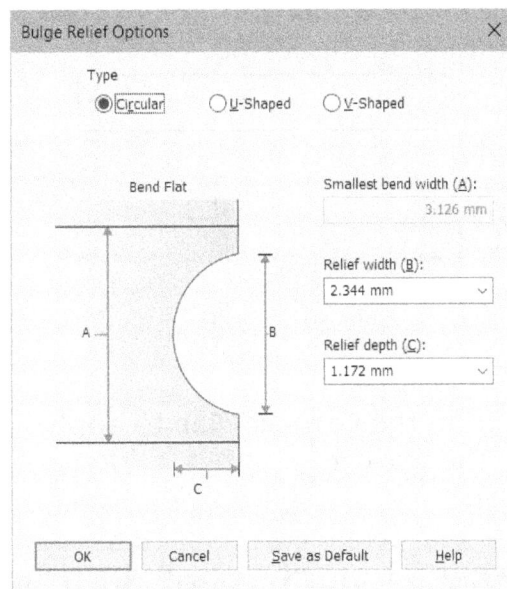

*Figure-56. Bulge Relief Options dialog box*

- Select the desired radio button from the dialog box and specify the desired parameters. Click on the **OK** button to apply settings.

*   Select the desired end point of bend which is deformed. The preview of relief will be displayed; refer to Figure-57.

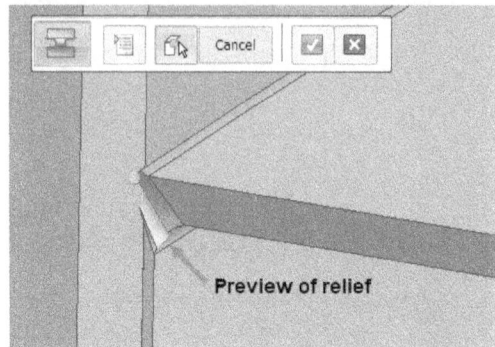

*Figure-57. Preview of bulge relief*

*   Click on the **OK** button from the **Command Bar** to apply the relief.
*   Click on the **Finish** button from the **Command Bar** and then press **ESC** to exit.

## APPLYING BREAK CORNER

The **Break Corner** tool is used to apply round or chamfer to selected edge of sheet metal. The procedure to use this tool is given next.

*   Click on the **Break Corner** tool from the **Break Corner** drop-down in the **Sheet Metal** group of **Home** tab in the **Ribbon**. The **Break Corner Command Bar** will be displayed; refer to Figure-58.

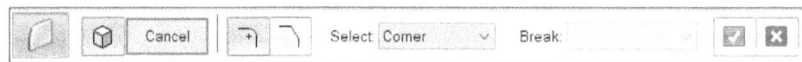

*Figure-58. Break Corner Command Bar*

*   Select the **Radius Corner** or **Chamfer Corner** button from the **Command Bar** to create round or chamfer respectively at the selected corner edge.
*   Select the desired edge to be modified from the model and specify the desired value of radius/setback in the **Break** edit box of **Command Bar**. Preview of corner break will be displayed; refer to Figure-59.

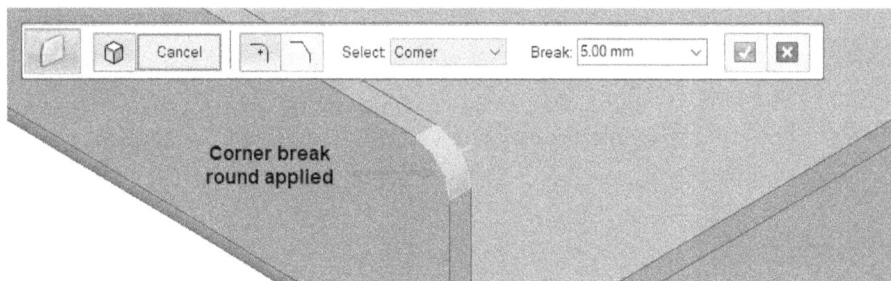

*Figure-59. Preview of corner break*

*   Click on the **OK** button from the **Command Bar** to apply corner break. Click on the Finish button to create the feature.
*   Press **ESC** to exit the tool.

## SHEET METAL MODIFICATION TOOLS

The tools in the **Modify** group are used to modify the shape and size of sheet metal. Various tools in this **Modify** group which have not been discussed are discussed next.

## Modifying Bend Angle

The **Bend Angle** tool in **Modify** group is used to modify angle of selected bend. The procedure to use this tool is given next.

- Click on the **Bend Angle** tool from the **Modify** group in the **Home** tab of the **Ribbon**. The **Bend Angle Command Bar** will be displayed; refer to Figure-60.

*Figure-60. Bend Angle Command Bar*

- Select the bend whose angle is to be modified. You will be asked to select a planar face to be fixed.
- Select the desired face to be used for angle reference. The **Angle** edit box will be displayed in the **Command Bar**.
- Specify the desired value in the edit box and press **ENTER**. Preview of the modified bend will be displayed; refer to Figure-61.

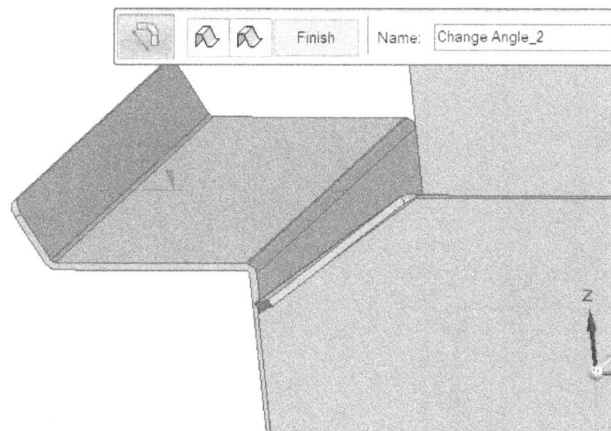

*Figure-61. Modified bend angle*

- Click on the **Finish** button from the **Command Bar** to apply modification and press **ESC** to exit the tool.

## Modifying Bend Radius

The **Bend Radius** tool is used modify radius of bend. The procedure to use this tool is given next.

- Click on the **Bend Radius** tool from the **Modify** group of the **Home** tab in the **Ribbon**. The **Bend Radius Command Bar** will be displayed; refer to Figure-62.

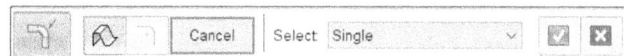

*Figure-62. Bend Radius Command Bar*

- Select the bends to be modified and press **ENTER**. The New radius edit box will be displayed in the **Command Bar**.
- Specify the desired value in the edit box and press **ENTER**. The **Preview** button will be displayed.
- Click on the **Preview** button to check preview of modified bend. Click on the **Finish** button to apply modifications.
- Press **ESC** to exit the tool.

The other tools in this group have been discussed earlier.

# PRACTICE 1

Create the sheet metal model as shown in Figure-63. Dimensions are given in Figure-64.

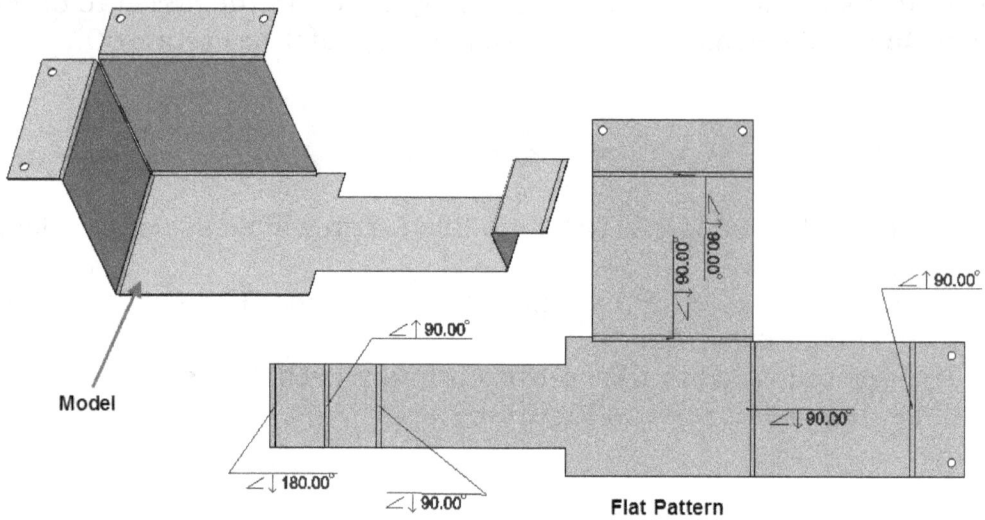

Figure-63. Sheet metal model

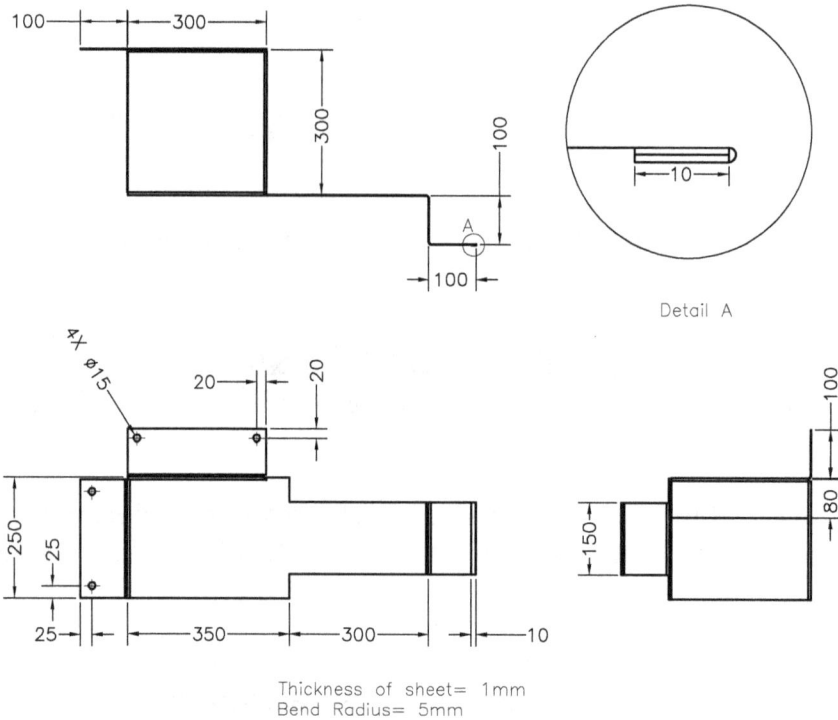

Detail A

Thickness of sheet= 1mm
Bend Radius= 5mm

**Figure-64.** *Drawing views*

# PRACTICE 2

Create the sheet metal model as shown in Figure-65. Dimensions are given in Figure-66.

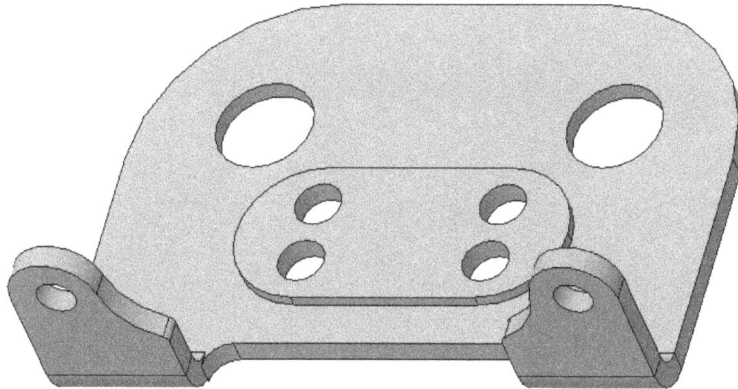

*Figure-65. Practice 2 Model*

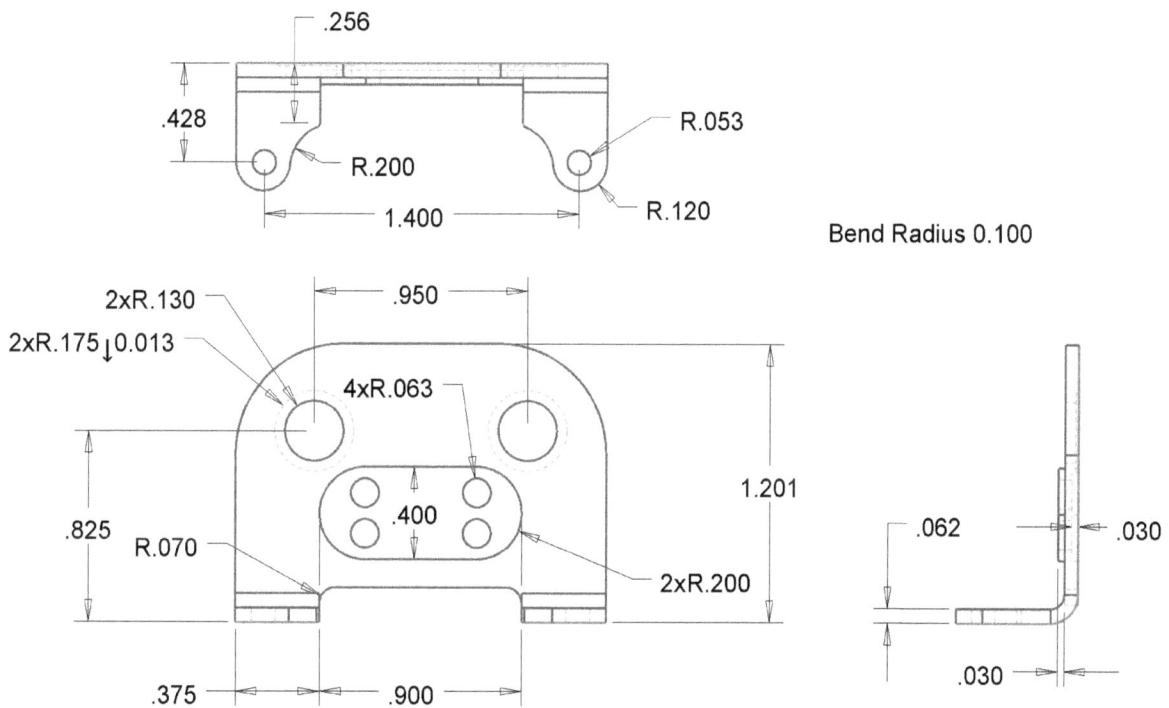

Bend Radius 0.100

*Figure-66. Practice 2 Drawing Views*

FOR STUDENT NOTES

# Chapter 13

# Simulation Study

Topics Covered

The major topics covered in this chapter are:

- *Introduction*
- *Simulation Studies in Solid Edge*
- *Performing Linear Static Analysis*
- *Performing Normal Modes Analysis*
- *Performing Linear Buckling Analysis*
- *Analyzing results*

# INTRODUCTION TO SIMULATION

Simulation is the study of effects caused on an object due to real-world loading conditions. Computer Simulation is a type of simulation which uses CAD models to represent real objects and it applies various load conditions on the model to study the real-world effects. Solid Edge Simulation is one of the Computer Simulation programs available in the market. In Solid Edge Simulation, we apply loads on a constrained model under predefined environmental conditions and check the result(visually and/or in the form of tabular data). The types of analyses that can be performed in Solid Edge are given next.

# TYPES OF ANALYSES PERFORMED IN SOLID EDGE SIMULATION

Solid Edge Simulation allows to perform various analyses that are generally performed in Industries. These analyses and their uses are given next.

## Linear Static Analysis

This is the most common type of analysis we perform. In this analysis, loads are applied to a body due to which the body deforms and the effects of the loads are transmitted throughout the body. To absorb the effect of loads, the body generates internal forces and reactions at the supports to balance the applied external loads. These internal forces and reactions cause stress and strain in the body. Static analysis refers to the calculation of displacements, strains, and stresses under the effect of external loads, based on some assumptions. The assumptions are as follows.

1. All loads are applied slowly and gradually until they reach their full magnitudes. After reaching their full magnitudes, load will remain constant (i.e. load will not vary against time).
2. Linearity assumption: The relationship between loads and resulting responses is linear. For example, if you double the magnitude of loads, the response of the model (displacements, strains and stresses) will also double. You can make linearity assumption if:

- All materials in the model comply with Hooke's Law that is stress is directly proportional to strain.
- The induced displacements are small enough to ignore the change is stiffness caused by loading.
- Boundary conditions do not vary during the application of loads. Loads must be constant in magnitude, direction and distribution. They should not change while the model is deforming.

If the above assumptions are valid for your analysis, then you can perform **Linear Static Analysis**. For example, a cantilever beam fixed at one end and force applied on other end; refer to Figure-1.

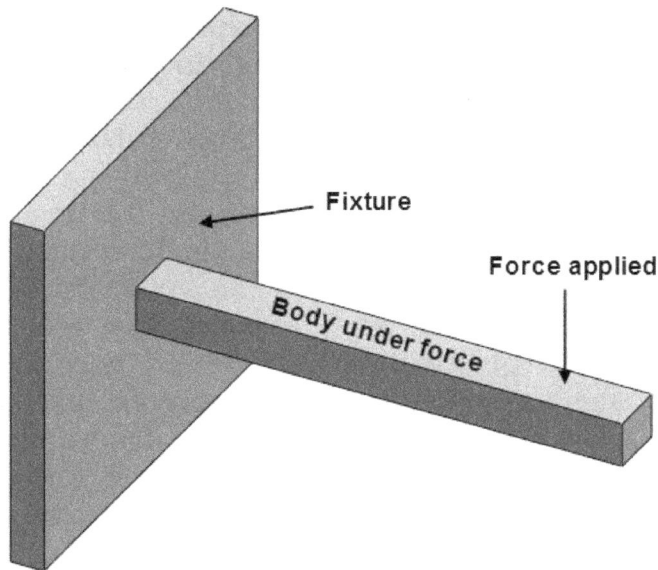

*Figure-1. Linear static analysis example*

## Normal Modes Analysis

The Normal Modes analysis is used to check natural frequency of the model. This study is also called Vibration Analysis. By its very nature, vibration involves repetitive motion. Each occurrence of a complete motion sequence is called a "cycle." Frequency is defined as so many cycles in a given time period. "Cycles per seconds" or "Hertz". Individual parts have what engineers call "natural" frequencies. For example, a violin string at a certain tension will vibrate only at a set number of frequencies, which is why you can produce specific musical tones. There is a base frequency in which the entire string is going back and forth in a simple bow shape.

Harmonics and overtones occur because individual sections of the string can vibrate independently within the larger vibration. These various shapes are called "modes". The base frequency is said to vibrate in the first mode, and so on up the ladder. Each mode shape will have an associated frequency. Higher mode shapes have higher frequencies. The most disastrous kinds of consequences occur when a power-driven device such as a motor for example, produces a frequency at which an attached structure naturally vibrates. This event is called "resonance." If sufficient power is applied, the attached structure will be destroyed. Note that ancient armies, which normally marched "in step," were taken out of step when crossing bridges. Should the beat of the marching feet align with a natural frequency of the bridge, it could fall down. Engineers must design so that resonance does not occur during regular operation of machines. This is a major purpose of Modal Analysis. Ideally, the first mode has a frequency higher than any potential driving frequency. Frequently, resonance cannot be avoided, especially for short periods of time. For example, when a motor comes up to speed it produces a variety of frequencies. So it may pass through a resonant frequency.

## Linear Buckling Analysis

If you press down on an empty soft drink can with your hand, not much will seem to happen. If you put the can on the floor and gradually increase the force by stepping down on it with your foot, at some point it will suddenly squash. This sudden scrunching is known as "buckling."

Models with thin parts tend to buckle under axial loading. Buckling can be defined as the sudden deformation, which occurs when the stored membrane(axial) energy is converted into bending energy with no change in the externally applied loads. Mathematically, when buckling occurs, the total stiffness matrix becomes singular. In the normal use of most products, buckling can be catastrophic if it occurs. The failure is not one because of stress but geometric stability. Once the geometry of the part starts to deform, it can no longer support even a fraction of the force initially applied. The worst part about buckling for engineers is that buckling usually occurs at relatively low stress values for what the material can withstand. So they have to make a separate check to see if a product or part thereof is okay with respect to buckling. Slender structures and structures with slender parts loaded in the axial direction buckle under relatively small axial loads. Such structures may fail in buckling while their stresses are far below critical levels. For such structures, the buckling load becomes a critical design factor. Stocky structures, on the other hand, require large loads to buckle, therefore buckling analysis is usually not required.

Buckling almost always involves compression; refer to Figure-2. In mechanical engineering, designs involving thin parts in flexible structures like airplanes and automobiles are susceptible to buckling. Even though stress can be very low, buckling of local areas can cause the whole structure to collapse by a rapid series of 'propagating buckling'. Buckling analysis calculates the smallest (critical) loading required buckling a model. Buckling loads are associated with buckling modes. Designers are usually interested in the lowest mode because it is associated with the lowest critical load. When buckling is the critical design factor, calculating multiple buckling modes helps in locating the weak areas of the model. This may prevent the occurrence of lower buckling modes by simple modifications.

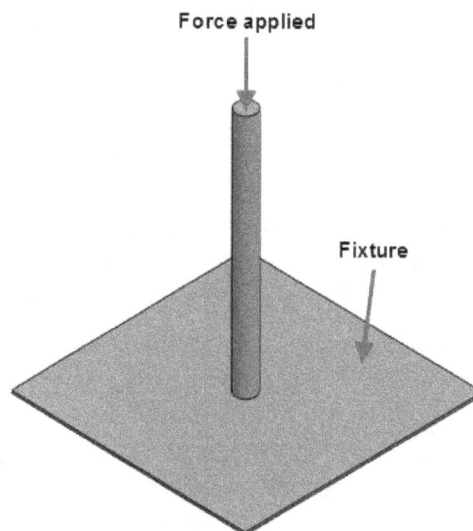

*Figure-2. Buckling example*

You will learn about performing various studies in Solid Edge later. Various tools in **Simulation** tab are discussed next.

# STARTING A STUDY

The **New Study** tool is used to start a new study in the Solid Edge. The procedure to start a new study is given next.

- Click on the **New Study** tool from the **Study** group in the **Simulation** tab of the **Ribbon**. The **Create Study** dialog box will be displayed; refer to Figure-3.

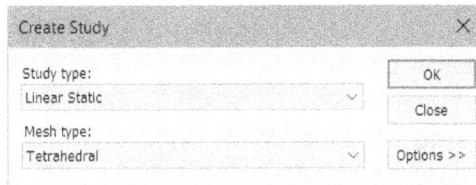

*Figure-3. Create Study dialog box*

- Select the desired option from the **Study Type** drop-down to define which type of study you want to create.
- Select the desired option from the **Mesh type** drop-down to define type of elements to be generated in mesh for the study. The **Tetrahedral** element type is used for solid body. Use the **Surface** element type for thin or surface bodies. Select the **Mixed and General Bodies** option from the drop-down if want to mesh a model which has solid as well as thin bodies.
- Click on the **Options>>** button from the dialog box to display parameters related to study. The options will be displayed as shown in Figure-4.

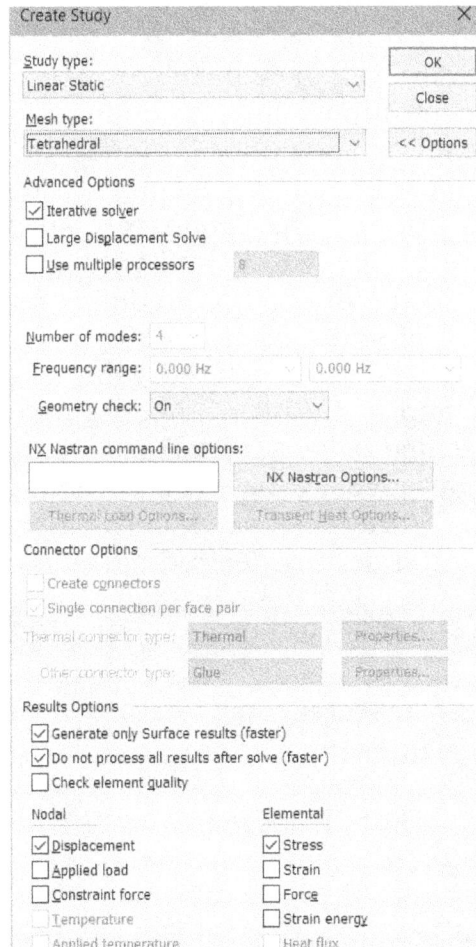

*Figure-4. Options for creating studies*

- Select the **Iterative solver** check box to use NX Nastran iterative solver for perform analysis.
- Select the **Large Displacement Solve** check box to perform study on model in which large deformation will be caused.

- Select the **Use multiple processors** check box and then select the desired number of processors to be used for running analysis.
- Select the **Use Inertia Relief** check box to make model stable using load equilibrium rather than using constraints while performing analysis.
- Specify desired number of modes to be checked for Normal Modes or Linear Buckling analyses.
- Specify desired value of frequencies to be checked for finding natural frequencies in the edit boxes of **Frequency range**.
- Select the **On** option from the **Geometry check** drop-down to use default threshold for checking if mesh has desired type and number of element shapes for running the analysis. If there are less number of elements then analysis will fail to run. Select the **Warning Only** option from the drop-down if you want to show warnings in output file when number of elements are less than desired rather than causing the solution to fail.
- Specify desired NX Nastran command in the edit box of **NX Nastran command line options** area. Click on the **NX Nastran Options** button to modify options related to NX Nastran solver.
- The options in the **Connector Options** area are active when you are simulating analysis on an assembly. Select the **Create connectors** check box to automatically create connectors when running analysis.
- Select the **Single connection per face pair** check box to create single connection for each face rather than multiple connections for a face.
- Select the **Glue** or **No penetration** option from the **Other connector type** drop-down to apply respective connector in assembly. After selecting desired option, click on the **Properties** button. The **Connector Properties** dialog box will be displayed; refer to Figure-5. Set the desired values of friction, search distance, penalty value and so on in the dialog box to define properties of No penetration connector. The coefficient of static friction defines friction between two faces of connector during study. Specify the desired value of minimum search distance to define the minimum gap to be used for creating automatic no penetration connector. The value specified in **Penalty value** edit box is used to define stiffness factor for displacement transmission between element pairs. If you are performing thermal study then specify the desired value of thermal conductance between elements of connectors. After setting desired parameters, click on the **OK** button.

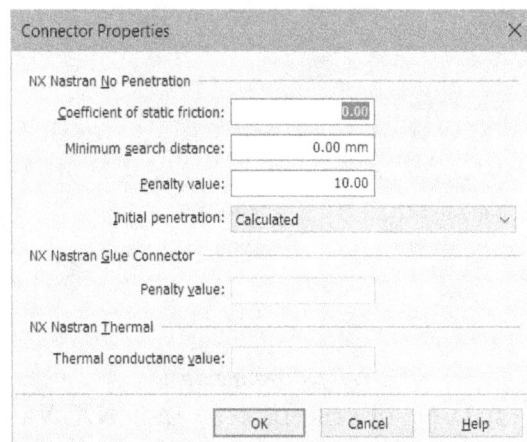

*Figure-5. Connector Properties dialog box*

- Select the **Generate only Surface results (faster)** check box to mesh only surfaces of solid and run the analysis as a thin part.

- Select the **Do not process all results after solve (faster)** check box to load only standard results. You can later use **Simulation** pane or tools in **Ribbon** to generate additional results.
- Select the **Check element quality** check box to ignore the elements which are of poor quality due to fine meshing in tight spots of model.
- Select the desired check boxes from the **Nodal** section and **Elemental** section to define the results to be generated by the study.
- After setting desired parameters, click on the **OK** button from the dialog box to create a new study.

## SWITCHING BETWEEN STUDIES

Select the desired option from the **Study List** drop-down in the **Study** group of **Simulation** tab in the **Ribbon** to activate respective simulation study; refer to Figure-6.

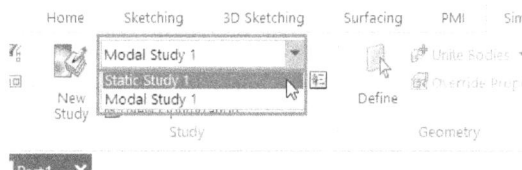

*Figure-6. Drop-down for switching study*

## SELECTING MATERIAL OF MODEL

Click in the **Material List** drop-down and select the desired material from list; refer to Figure-7. The selected material will be applied to the part or assembly as default material.

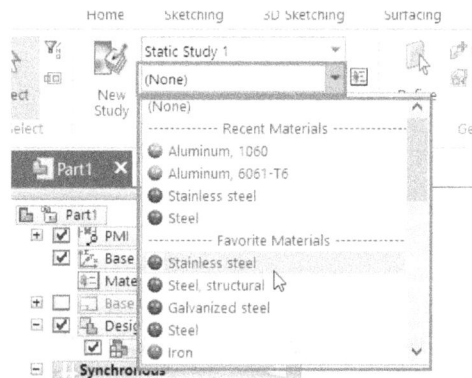

*Figure-7. Selecting material for simulation*

## SELECTING GEOMETRY FOR ANALYSIS

The **Define** tool is used to select bodies to be used for performing analyses. The procedure to use this tool is given next.

- Click on the **Define** tool from the **Geometry** panel in the **Simulation** tab of the **Ribbon**. The **Define Command Bar** will be displayed; refer to Figure-8.

*Figure-8. Define Command Bar*

- Select the desired body to be used for simulation study and right-click in the drawing area.

# UNITING BODIES FOR SIMULATION

The **Unite Bodies** tool is used to combine two or more bodies for performing analysis. The procedure to use this tool is given next.

*   Click on the **Unite Bodies** tool from the **Unite Bodies** drop-down in the **Geometry** group of the **Ribbon**. You will be asked to select surfaces and solids to be united.
*   Select the desired surfaces and solids to be united, and right-click in the drawing area. The selected bodies and surfaces will be combined to form single body.

# RECOVERING BODIES

The **Recover Bodies** tool is used to extract surfaces and solids from the selected united body. Note that this tool is active only when surface and solid bodies are united in a single body. The procedure to use this tool is given next.

*   Click on the **Recover Bodies** tool from the **Unite Bodies** drop-down in the **Geometry** group of **Simulation** tab in the **Ribbon**. You will be asked to select the united body to be split into surface and solids.
*   Select the desired body and right-click in the drawing area. The bodies will be separated; refer to Figure-9.

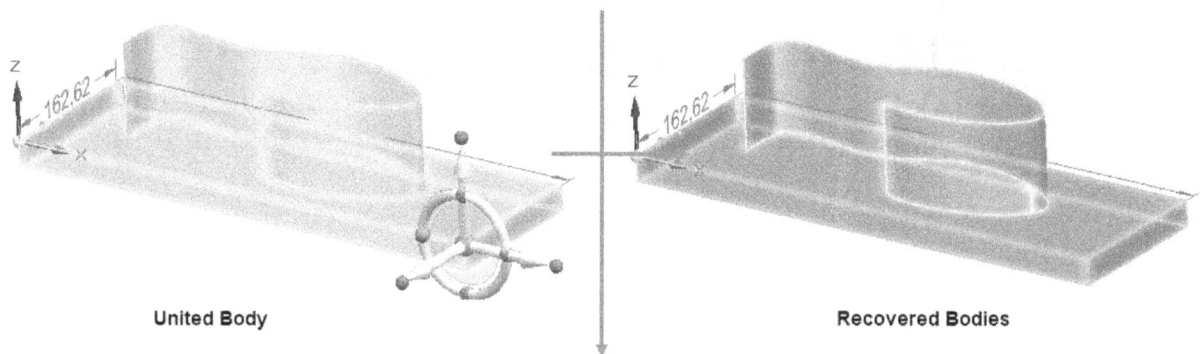

United Body                                          Recovered Bodies

*Figure-9. Recovering bodies*

*   Press **ESC** to exit the tool.

Note that once you have defined bodies for simulation then tools for applying loads will be active in the **Ribbon**. These tools are discussed next.

# APPLYING STRUCTURAL LOADS

The tools in the **Structural Loads** group of **Simulation** tab in **Ribbon** are used to apply different type of structural loads like force, pressure, torque, and so on. Various tools in this group are discussed next.

## Applying Force Load

The **Force** tool is used to apply force of specified value on selected faces/edges/ points/features in the model. The procedure to use this tool is given next.

*   Click on the **Force** tool from the **Structural Loads** group in the **Simulation** tab of the **Ribbon**. The **Force Command Bar** will be displayed; refer to Figure-10.

*Figure-10. Force Command Bar*

- Select the desired option from the **Select** drop-down to define which type of objects are to be selected for applying force. (In our case, it is **Face** option.)
- Click on the desired faces/edges/points/features on which you want to apply force. Preview of the force will be displayed; refer to Figure-11.

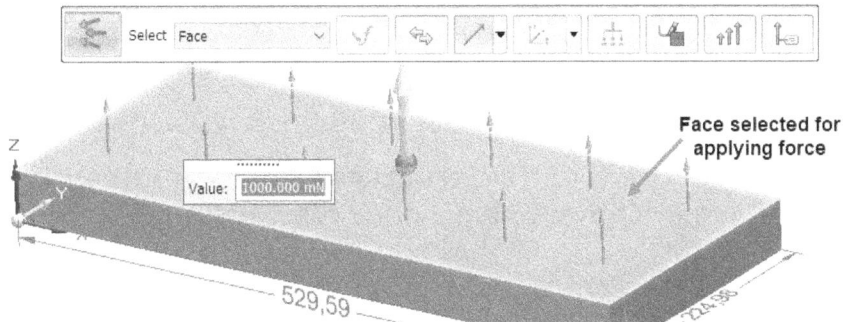
*Figure-11. Applying force on face*

- Click on the **Flip Direction** button from the Command Bar to reverse the direction of force application.
- Select the **Along a Vector** option from the **Direction Type** drop-down to define direction of load along direction vector of selected face. Select the **Normal to Face** option from the drop-down to apply force perpendicular to selected face. Select the **Components** option from the drop-down to define each component of load separately in respective edit box; refer to Figure-12.

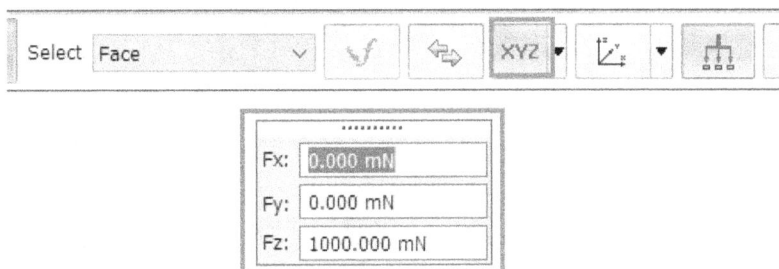
*Figure-12. Components of force*

- Specify the desired load values in the edit box(es).
- Click on the **Select Coordinate System** button to select a different coordinate system for defining direction reference of force.
- Click on the **Total Load** button from the **Command Bar** if you want to specify combined value of force on all the selected faces. If this button is not selected then specified force value will be individually applied to all the selected faces.
- After setting the desired parameters, right-click in the drawing area to apply force.
- Press **ESC** to exit the tool.

## Applying Pressure Load

The **Pressure** tool is used to apply the pressure load on selected entity. The procedure to use this tool is given next.

- Click on the **Pressure** tool from the **Structural Loads** group in the **Simulation** tab of the **Ribbon**. The **Pressure Command Bar** will be displayed; refer to Figure-13.

*Figure-13. Pressure Command Bar*

- Select the desired faces on which you want to apply the pressure load and specify the desired value in the **Value** edit box.
- The buttons in **Command Bar** are same as discussed earlier. Right-click in the drawing area to apply the load.

## Applying Torque Load

The **Torque** tool is used to apply turning force on selected faces. The procedure to use this tool is given next.

- Click on the **Torque** tool from the **Structural Loads** group in the **Simulation** tab of **Ribbon**. The **Torque Command Bar** will be displayed; refer to Figure-14.

*Figure-14. Torque Command Bar*

- Select the face on which you want to apply the load and drag the axis of rotation to desired location; refer to Figure-15.

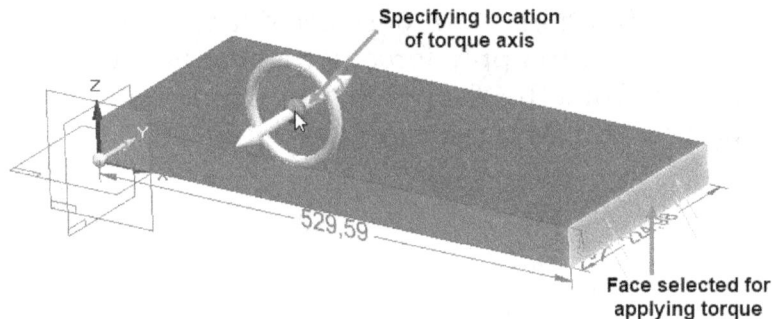

Specifying location of torque axis

Face selected for applying torque

529,59

*Figure-15. Applying torque on selected face*

- Specify desired value of torque in the **Value** edit box and press **ENTER** to apply torque.

## Applying Displacement

The **Displacement** tool is used to apply specified value of displacement to selected faces/edge/point/features. Note that there must be matching constraint where selected faces/edge/point/features can move by specified value. The procedure to use this tool is given next.

- Click on the **Displacement** tool from the **Structural Loads** group in the **Simulation** tab of **Ribbon**. The **Displacement Command Bar** will be displayed; refer to Figure-16.

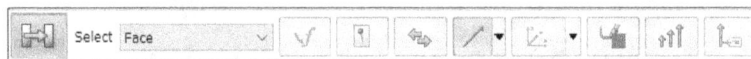
*Figure-16. Displacement Command Bar*

- Select the desired entity of model to be displaced and specify the desired value of displacement in the **Value** edit box.
- Set the other parameters as discussed earlier and right-click in the drawing area to apply displacement.

## Applying Bearing Load

The **Bearing** tool is used to apply physical loading of bearing on cylindrical and non cylindrical parts. The procedure to use this tool is given next.

- Click on the **Bearing** tool from the **Structural Loads** group in the **Simulation** tab of **Ribbon**. The **Bearing Command Bar** will be displayed; refer to Figure-17.

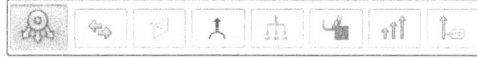

*Figure-17. Bearing Command Bar*

- Select the desired face of model on which you want to apply bearing load and drag the axis center at desired location.
- Select the **Traction** button from the **Command Bar** if you want to pull the faces opposite to default push motion of the force.
- Specify the desired values in the dynamic edit boxes. Right-click in the drawing area to apply the load.

## Applying Body Temperature

The **Body Temperature** tool is used to apply initial temperature to the model. Note that this tool is useful for transient thermal analysis and radiation analysis. The procedure to use this tool is given next.

- Click on the **Body Temperature** tool from the **Body Loads** group in the **Simulation** tab of **Ribbon**. The **Body Temperature Command Bar** will be displayed with **Value** edit box; refer to Figure-18.

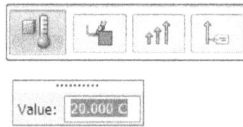

Value: 20.000 C

*Figure-18. Body Temperature Command Bar*

- Enter the desired value of temperature in the **Value** edit box to specify initial temperature of model. The temperature will be applied to system.

## Applying Centrifugal Load

The **Centrifugal** tool is used to apply centrifugal load on whole model. Centrifugal force is generated when a part rotates about its axis. The procedure to apply this load is given next.

- Click on the **Centrifugal** tool from the **Body Loads** group in the **Simulation** tab of **Ribbon**. The preview of load will be displayed with **Centrifugal Command Bar** and edit boxes; refer to Figure-19.
- Drag the center point of axis of rotation to desired location in model. The direction of centrifugal load will be modified accordingly.
- Specify the desired values in **Angular Velocity** and **Angular Acceleration** edit boxes.
- Using the **Flip** buttons in **Command Bar**, you can flip direction of angular velocity and angular acceleration.

- Right-click in the drawing area to apply the loads.

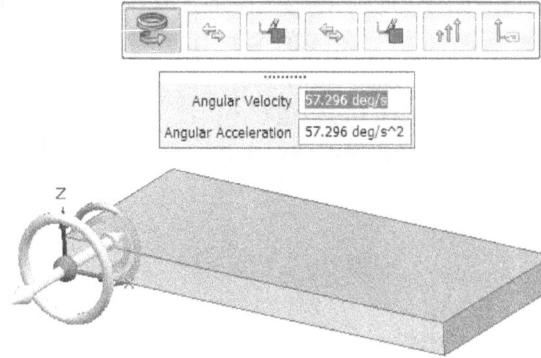

*Figure-19. Centrifugal Command Bar*

## Applying Gravity Load

The **Gravity** tool is used to represent effect of gravity on the model. The procedure to use this tool is given next.

- Click on the **Gravity** tool from the **Body Loads** group in the **Simulation** tab of **Ribbon**. The **Gravity Command Bar** will be displayed; refer to Figure-20.

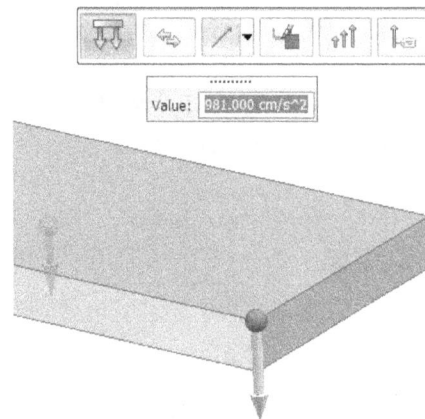

*Figure-20. Gravity Command Bar*

- Drag the center point of direction vector on desired edge or face to define direction of gravity.
- Specify the desired value in the **Value** edit box to define value of gravity.
- Right-click in the drawing area to apply the load.

## APPLYING CONSTRAINTS

The tools in the **Constraints** group are used to restrict the free motion of part(s) during analysis; refer to Figure-21. Various tools in this group are discussed next.

*Figure-21. Constraints group*

## Applying Fixed Constraint

The Fixed constraint is used to stop movement of selected entities in all 6 directions (3 translation and 3 rotation). The procedure to apply this constraint is given next.

- Click on the **Fixed** tool from the **Constraints** group in the **Simulation** tab of **Ribbon**. The **Fixed Command Bar** will be displayed; refer to Figure-22.

*Figure-22. Fixed Command Bar*

- Select the desired option from the **Select** drop-down in **Command Bar** to define what type of entities are to be selected and then select the desired faces/edges/ points/features to be fixed.
- Right-click in the drawing area to apply constraint.

## Applying Pinned Constraint

The Pinned constraint is used to allow free rotation of selected entity while restricting translation in all three directions. The procedure to apply this constraint is given next.

- Click on the **Pinned** tool from the **Constraints** group in the **Simulation** tab of **Ribbon**. The **Pinned Command Bar** will be displayed; refer to Figure-23.

*Figure-23. Pinned Command Bar*

- Select the desired option from the **Select** drop-down and then select the respective entities from model to be constrained as pinned.

## Applying No Rotation Constraint

The No Rotation constraint is used to stop rotation of model while restricting the translation movement of model. The procedure to use this tool is given next.

- Click on the **No Rotation** tool from the **Constraints** group in the **Simulation** tab of **Ribbon**. The **No Rotation Command Bar** will be displayed.
- Select the desired entities and right-click in the drawing area to apply the constraints.

## Applying Slide Along Surface Constraint

The **Sliding Along Surface** tool is used to allow sliding of selected faces along the selected surfaces. The procedure to use this tool is given next.

- Click on the **Sliding Along Surface** tool from the **Constraints** group in the **Simulation** tab of the **Ribbon**. The **Sliding Along Surface Command Bar** will be displayed; refer to Figure-24.

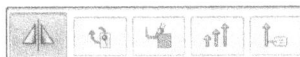

*Figure-24. Sliding Along Surface Command Bar*

- Select the planar faces of model that you want to use for sliding and right-click in the drawing area. The constraint will be applied; refer to Figure-25.

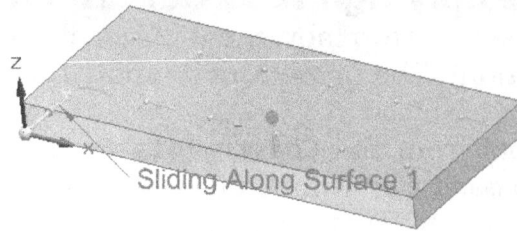

*Figure-25. Sliding along surface constraint applied*

## Applying Cylindrical Constraint

The Cylindrical constraint is applied to the model for allowing rotation about an axis and translation along that same axis. The procedure to apply this constraint is given next.

* Click on the **Cylindrical** tool from the **Constraints** group in the **Simulation** tab of **Ribbon**. The **Cylindrical Command Bar** will be displayed; refer to Figure-26 and you will be asked to select cylindrical faces for apply constraint.

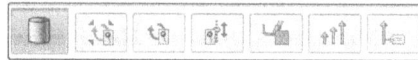

*Figure-26. Cylindrical Command Bar*

* Select the cylindrical faces of model to which you want to apply the constraint. The buttons in the **Command Bar** will become active; refer to Figure-27.

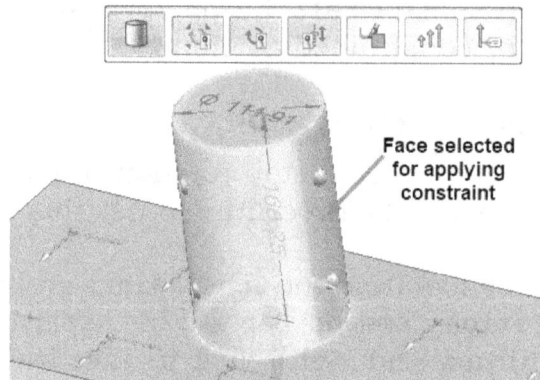

*Figure-27. Face selected for cylindrical constraint*

* Select the **Constrain Radial Growth** button to restrict radial changes in the model.
* Select the **Constrain Rotation** button to stop rotation of cylindrical part about its axis.
* Select the **Constrain Sliding Along Axis** button to restrict translation movement of part along its center axis.
* After setting desired parameters, right-click in the drawing area to apply constraint.

## Applying User-Defined Constraint

The **User-Defined** tool is used to restrict selected degrees of freedom of part. The procedure to use this tool is given next.

* Click on the **User-Defined** tool from the **Constraints** group in the **Simulation** tab of **Ribbon**. The **User-Defined Command Bar** will be displayed; refer to Figure-28.

*Figure-28. User-Defined Command Bar*

- Select the desired option from the **Select** drop-down and then select the desired entity of model to which you want to apply the constraint. Preview of constraint will be displayed; refer to Figure-29.

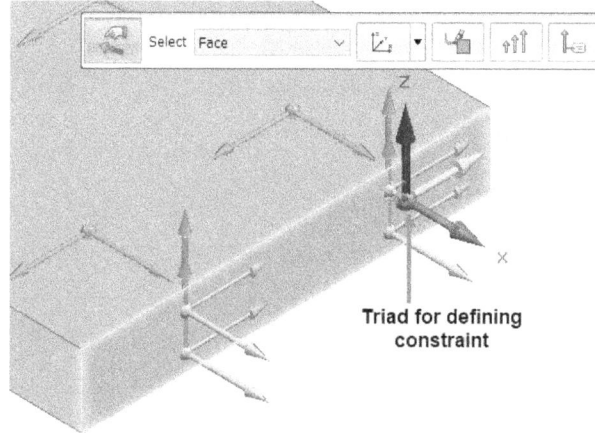

*Figure-29. Preview of constraint*

- Click on the desired axis of triad to restrict movement in respective direction.
- After setting desired parameters, right-click in the drawing area. The constraint will be applied.

# MESHING

The tools to create mesh are available in the **Mesh** group of the **Ribbon**; refer to Figure-30. Various tools of this group are discussed next.

*Figure-30. Mesh group*

## Creating Mesh

The **Mesh** tool is used to create mesh of specified element size. The procedure to use this tool is given next.

- Click on the **Mesh** tool from the **Mesh** group in the **Simulation** tab of **Ribbon**. The **Mesh** dialog box will be displayed related to selected element type in the **Create Study** dialog box; refer to Figure-31.

*Figure-31. Tetrahedral Mesh dialog box*

- Move the slider to desired side (Coarser or Finer) for creating coarse or fine mesh.
- Click on the **Options** button from the dialog box to specify meshing options for different elements. The **Mesh Options** dialog box will be displayed; refer to Figure-32.

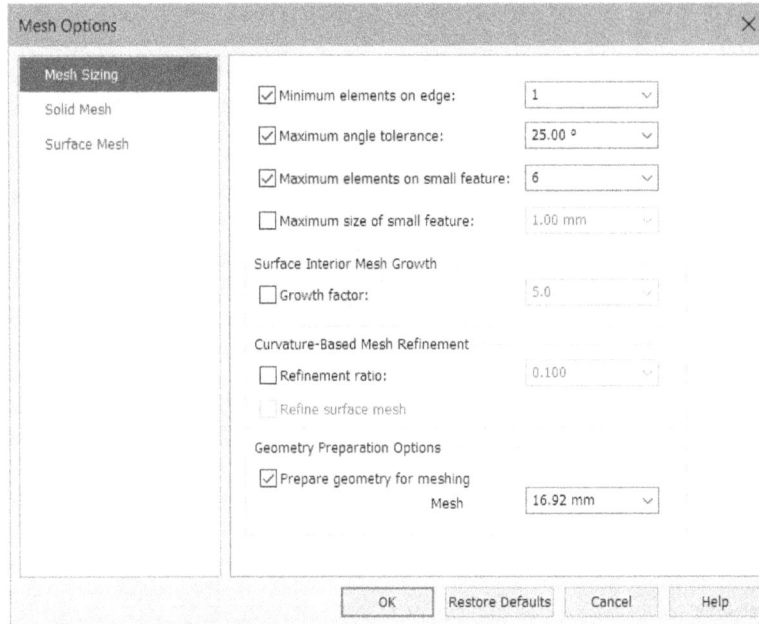

*Figure-32. Mesh Options dialog box*

- Set the desired parameters in the **Mesh Sizing**, **Solid Mesh**, and **Surface Mesh** pages of the dialog box and click on the **OK** button.
- Click on the **Mesh** button from the **Tetrahedral Mesh** dialog box to create the mesh; refer to Figure-33.

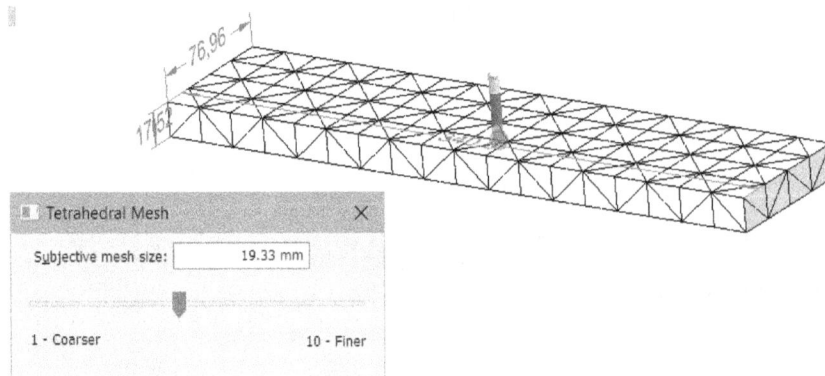

*Figure-33. Preview of mesh*

- Click on the **Close** button from the dialog box to exit the dialog box.

## Setting Mesh Size for Edges

The **Edge Size** tool is used to set size of mesh elements along selected edges. The procedure to set mesh size of edges is given next.

- Click on the **Edge Size** tool from the **Mesh** group in the **Simulation** tab of **Ribbon**. The **Edge Size** dialog box will be displayed; refer to Figure-34.
- Select the desired radio button from the dialog box and specify the parameters for mesh elements.

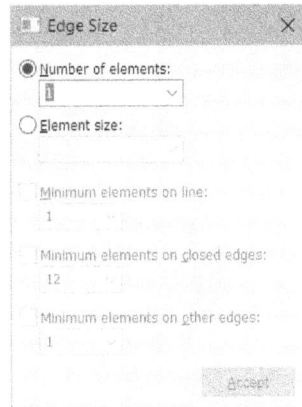

*Figure-34. Edge Size dialog box*

- Select the edges for which you want to specify edge mesh elements' sizes. Preview of the mesh will be displayed; refer to Figure-35.

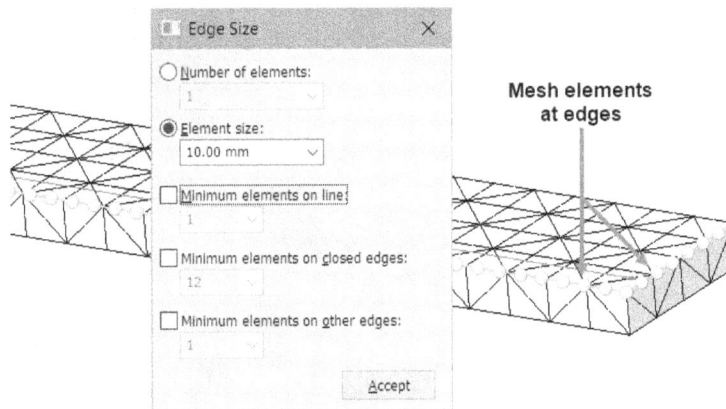

*Figure-35. Preview of edge mesh*

- Click on the **Accept** button from the dialog box to apply changes. Press **ESC** to exit the tool.

## Specifying Surface Mesh Sizes

The **Surface Size** tool is used to modify size of mesh at surface of model. The procedure to use this tool is given next.

- Click on the **Surface Size** tool from the **Mesh** group in the **Simulation** tab of **Ribbon**. The **Surface Size** dialog box will be displayed; refer to Figure-36.

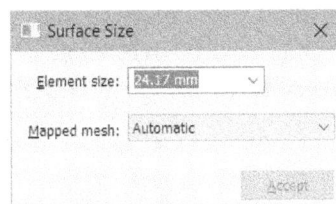

*Figure-36. Surface Size dialog box*

- Specify the desired element size in the edit box and select the desired option from the **Mapped mesh** drop-down to define which type of mesh elements will be created.
- Select the surfaces whose meshing is to be modified. Preview of mesh will be displayed; refer to Figure-37.

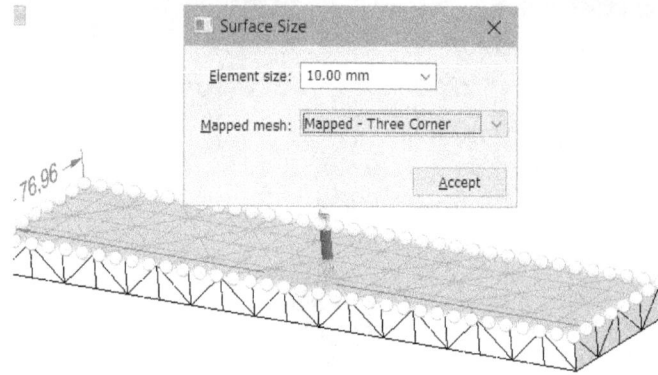

*Figure-37. Preview of surface mesh*

- Click on the **Accept** button from the dialog box to apply changes. Press **ESC** to exit the tool.

## Specifying Body Mesh Size

The **Body Size** tool is used set size of body mesh in the model. The procedure to use this tool is given next.

- Click on the **Body Size** tool from the **Mesh** group in the **Simulation** tab of **Ribbon**. The **Body Size** dialog box will be displayed; refer to Figure-38.

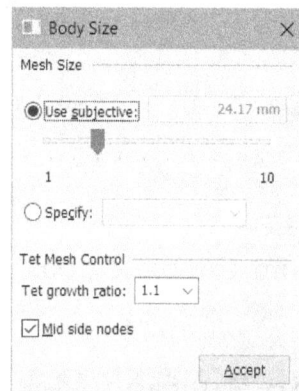

*Figure-38. Body Size dialog box*

- Set the desired parameters in the dialog box and click on the **Accept** button.
- Press **ESC** to exit the tool.

## SOLVING ANALYSIS

The **Solve** tool in **Ribbon** is active when material, load, and constraints have been applied to the model. After setting the parameters, click on the **Solve** tool from the **Solve** group in the **Simulation** tab of **Ribbon**. The results of analysis will be displayed; refer to Figure-39.

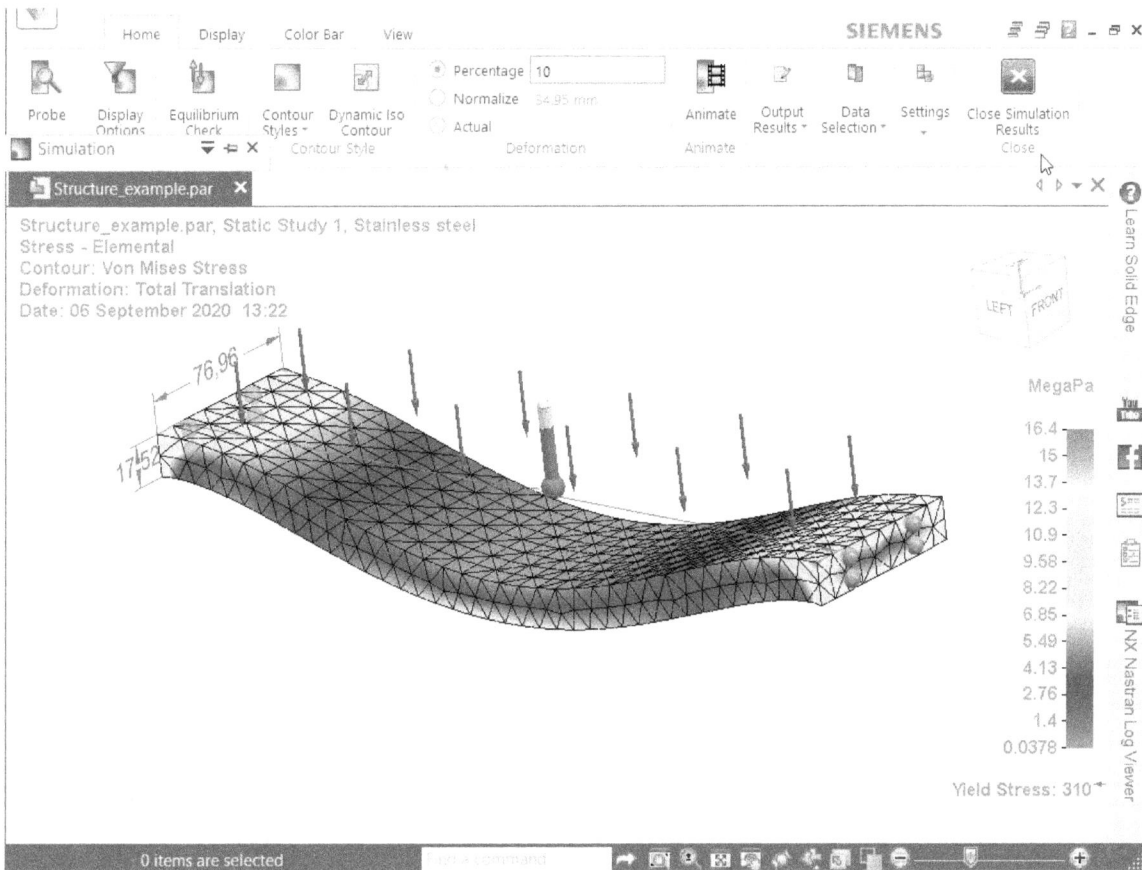

*Figure-39. Results of structural analysis*

# SIMULATION RESULTS

Once you have performed the analysis, the results are displayed in **Simulation Results** environment. Various tools in this environment are discussed next.

## Probing Results

The **Probe** tool is used to check analysis results at selected nodes. The procedure to use this tool is given next.

- Click on the **Probe** tool from the **Probe** group in the **Home** tab of **Ribbon**. The **Probe Table** dialog box will be displayed; refer to Figure-40.
- Select the **Node** radio button from the **Select** area of the dialog box to check the analysis results of selected node. Select the **Face** radio button from the area to check results at selected face. If you have selected **Face** radio button then all the nodes of selected face will get selected for checking results on clicking **Update** button; refer to Figure-41.
- Select the **Show displacement** check box to also check displacement results of nodes.
- Press **ESC** to exit the tool.

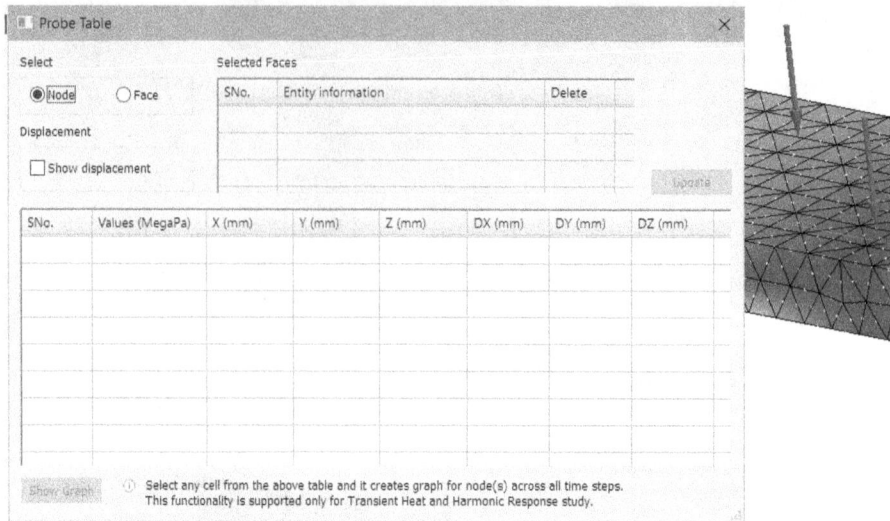

*Figure-40. Probe Table dialog box*

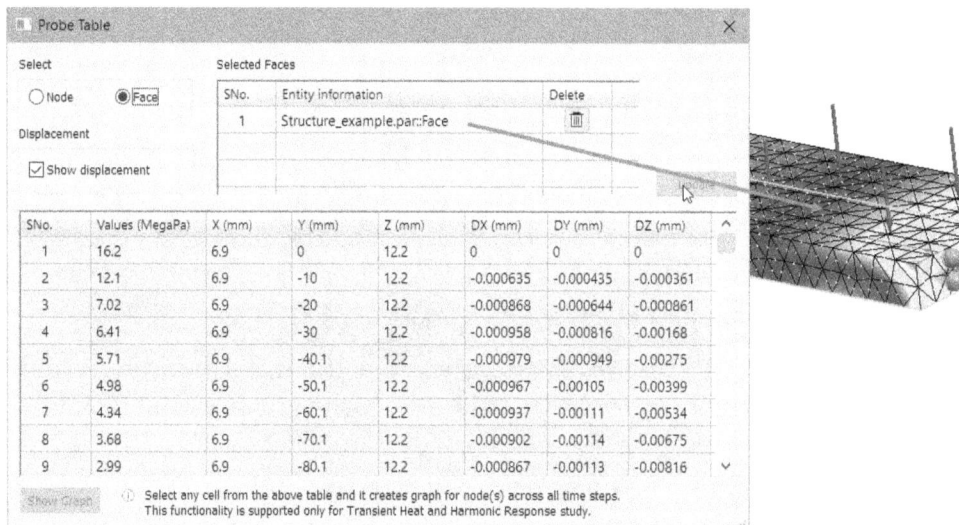

*Figure-41. Checking results on nodes*

## Display Options

The **Display Options** tool is used to show/hide various elements of result. On clicking this tool, the box to define which elements to be shown/hidden in analysis; refer to Figure-42. Select the desired check boxes from the box and click on the **Close** button.

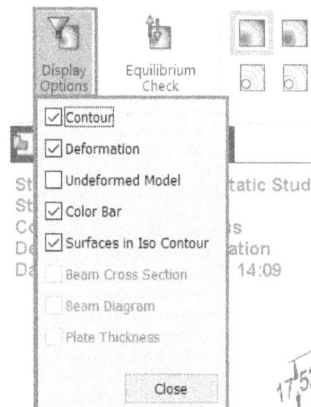

*Figure-42. Display Options*

## Equilibrium Check

The **Equilibrium Check** tool is used to check whether forces are balanced on all nodes in the model. On clicking this tool, the **Equilibrium Check** dialog box will be displayed; refer to Figure-43. Click on the **Export** button to export all the results of equilibrium check in the form of a csv file. After checking the results, click on the **OK** button.

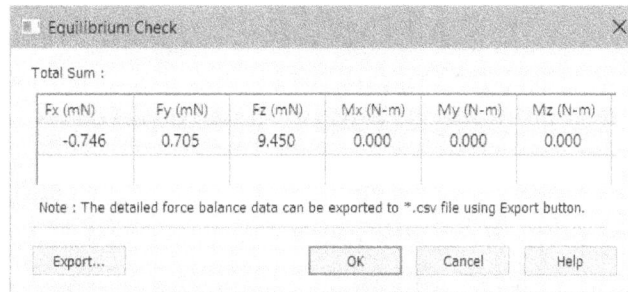

Figure-43. Equilibrium Check dialog box

## Contour Style

The buttons in the **Contour Style** group are used to display the results in contour styles like smooth contour, banded contour, element contour, and so on. Click on the **Dynamic Iso Contour** tool from the **Contour Style** group in the **Ribbon** to move iso contour lines in the result model.

## Deformation

The options in the **Deformation** group of **Home** tab in **Ribbon** are used to define how deformation results will be displayed in the drawing. Select the **Percentage** radio button from the **Deformation** group to display deformation in model as percentage of model size. Select the **Normalize** radio button from the **Deformation** group to set a maximum deformation value for result and adjust rest of the deformation in model accordingly. Select the **Actual** radio button from the **Deformation** group to display real deformation of model.

## Animating Results

The **Animate** tool is used to animation the deformation of model. Click on the **Animate** tool from the **Animate** group in the **Home** tab of **Ribbon**. The **Animate Command Bar** will be displayed; refer to Figure-44.

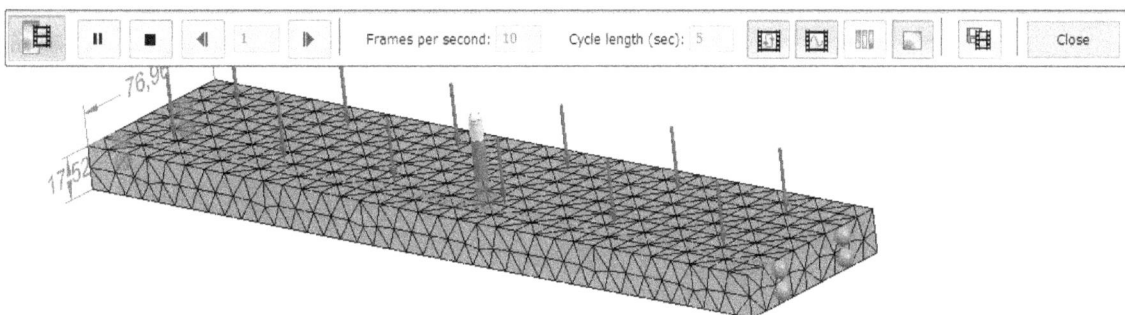

Figure-44. Animate Command Bar

You can use the **Play/Pause**, **Stop**, **Previous Frame**, **Next Frame**, and **Current Frame** options in the **Command Bar** to surf through the animation. Select the **Animate Full Cycle**, **Animate Sinusoidal Cycle**, **Animate Across Modes**, and **Animate Contours** buttons to set different parameters of animation. Click on the **Save as Movie** button from the **Command Bar** to save animation as video clip in AVI format. Click on the **Close** button to exit the **Command Bar**.

# CREATING REPORT

The **Create Report** tool from the **Output Results** group is used to generate report of analysis as html, doc, or docx file. The procedure to use this tool is given next.

- Click on the **Create Report** tool from the **Output Results** group in the **Home** tab of **Ribbon**. The **Create Report** dialog box will be displayed; refer to Figure-45.

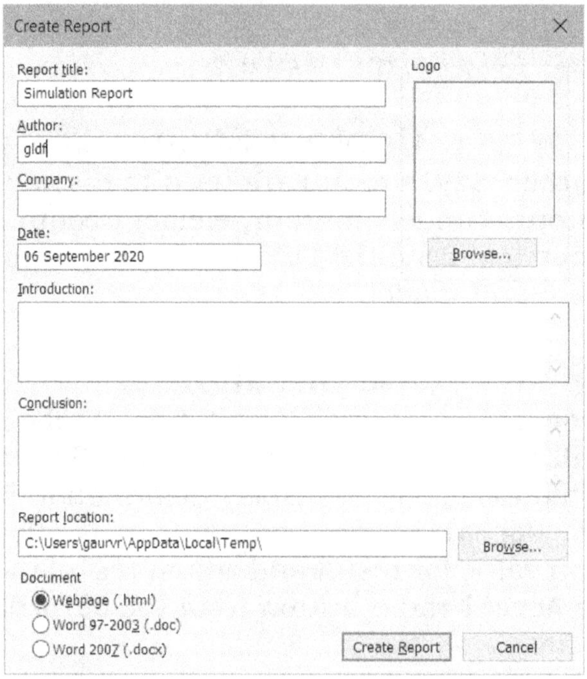

*Figure-45. Create Report dialog box*

- Specify desired parameters in various edit boxes of the dialog box.
- Select the desired radio button from the **Document** area of dialog box to define format of document.
- After setting all parameters, click on the **Create Report** button. The report will be generated; refer to Figure-46.

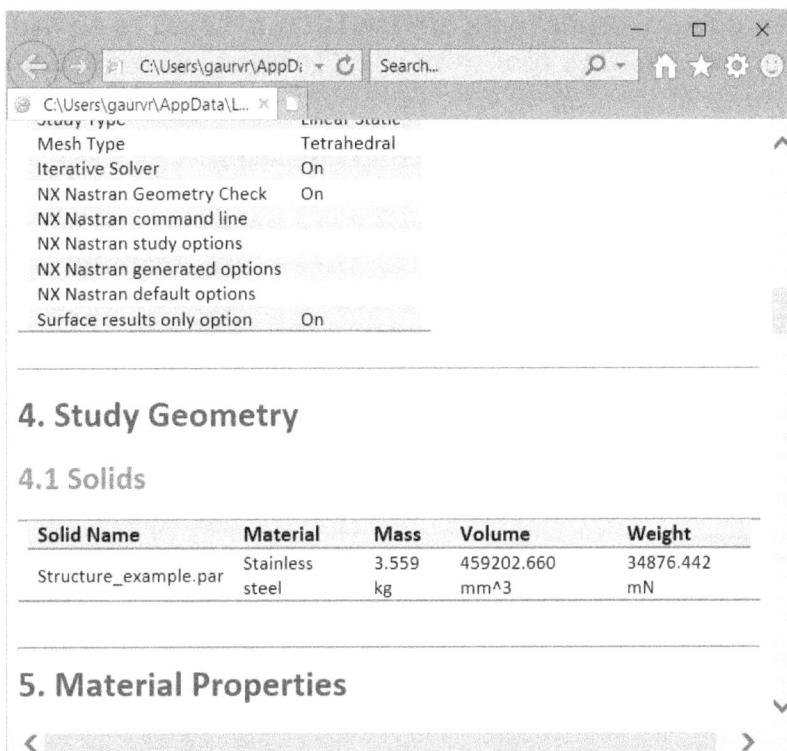

*Figure-46. Report created*

You can use the **Save As Movie** and **Save As Image** tools in the **Output Results** group of **Ribbon** as discussed earlier.

# DATA SELECTION FOR RESULTS

The options in the **Data Selection** group of **Ribbon** are used to set the parameters to be displayed in the results; refer to Figure-47.

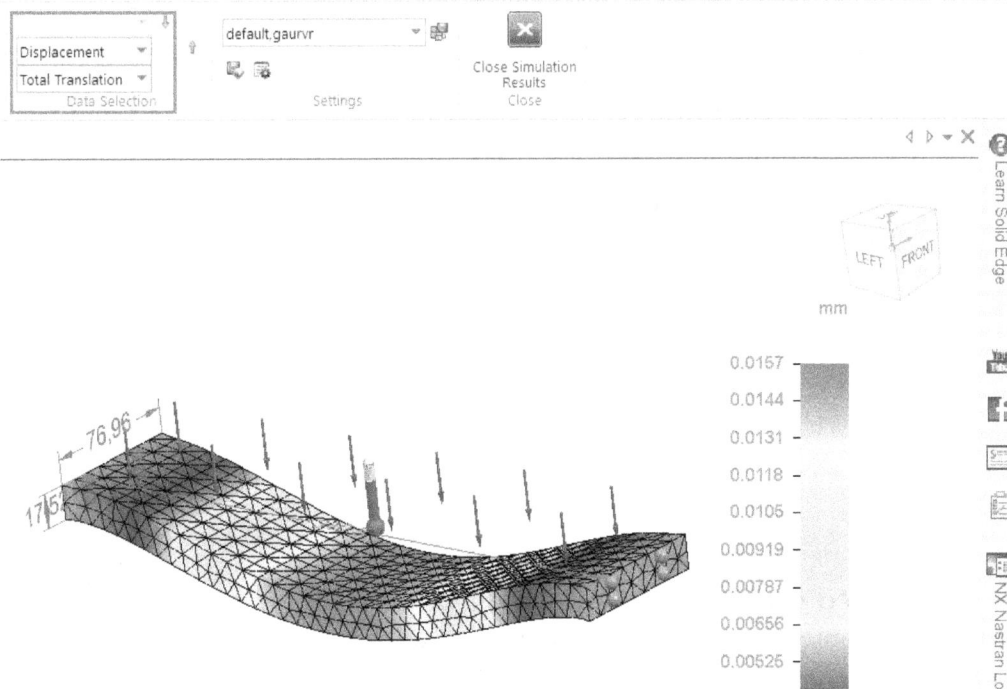

*Figure-47. Data selection for results*

Similarly, you can use the other tools of **Simulation Results** environment. Click on the **Close Simulation Results** tool from the **Close** group to exit the environment.

## Optimization Study

The Optimization study is used to change the geometry of model based on results of linear static analysis. Using this study, you can get the model with minimum mass which passes the linear static analysis. The procedure to create optimization study is given next.

- After running the linear static analysis on model, click on the **New Optimization** tool from the **Study** group in the **Simulation** tab of **Ribbon**. The **Optimization** dialog box will be displayed; refer to Figure-48.

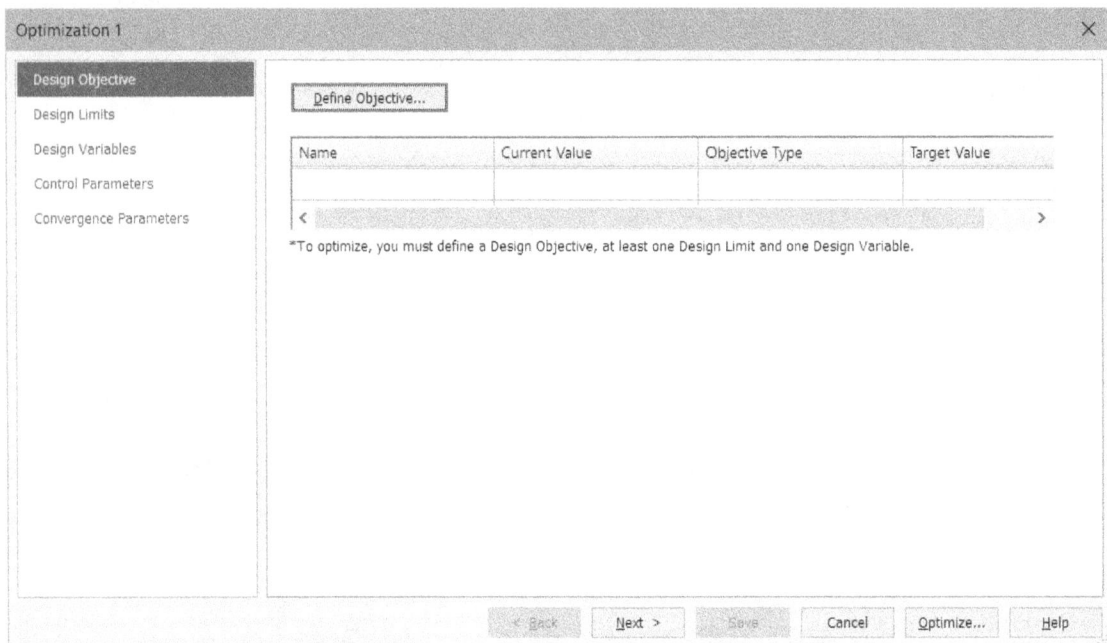

*Figure-48. Optimization dialog box*

- Click on the **Define Objective** button from the dialog box. The **Define Objective** dialog box will be displayed; refer to Figure-49.

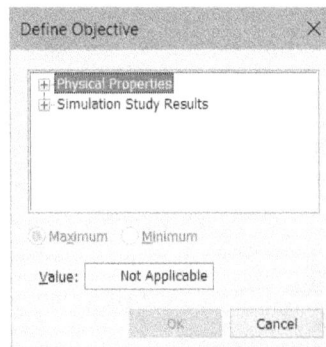

*Figure-49. Define Objective dialog box*

- Select the desired option from the dialog box to define which physical property or simulation result will be used for optimization. For example, select the **Mass** option from the dialog box to get an optimized model with minimum mass which passes the linear static analysis.
- After selecting the desired property, click on the **OK** button from the dialog box. The property will be added in the **Design Objective** list; refer to Figure-50.

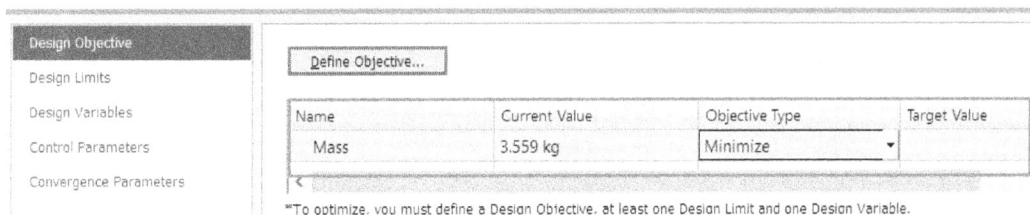

*Figure-50. Objective added in the list*

- Select the desired option from the **Objective Type** column to specify whether selected property will be minimized, maximized, or need to achieve specified target value.
- Click on the **Design Limits** option from the dialog box to define limit within which the selected property should be when optimizing the model based on analysis.
- Click on the **Design Variables** option from the dialog box to select the design variables to be used for optimizing the model. After selecting this option, click on the click on the **Add Model Variable** button. The **Variable Table** dialog box will be displayed; refer to Figure-51.

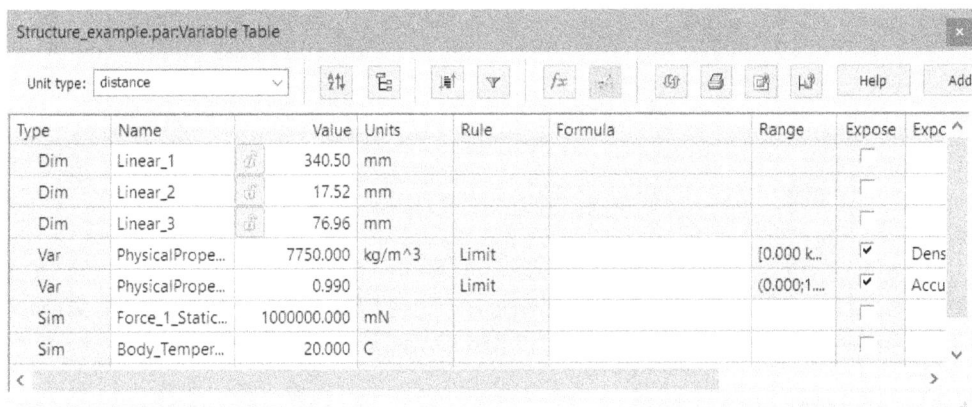

*Figure-51. Variable Table dialog box*

- Select the desired properties (while holding the **CTRL** key) you want to add in the list and click on the **Add** button from the dialog box. The list of variables will be added. Click on the button under **Rule** column for desired variable to define rules for properties during analysis. The **Design Variable Rule Editor** dialog box will be displayed; refer to Figure-52. Select the desired options from the **Minimum limit** and **Maximum limit** drop-downs, and specify the desired values for the parameters.

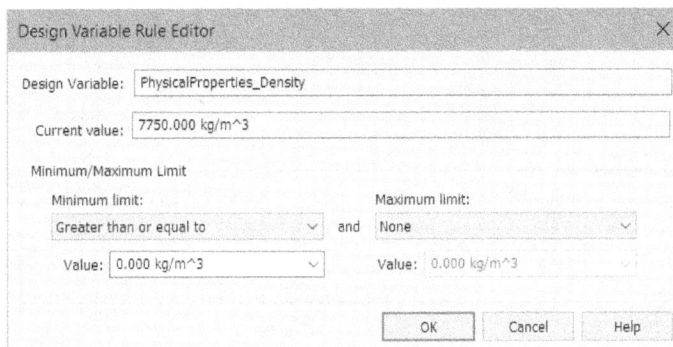

*Figure-52. Design Variable Rule Editor dialog box*

- Select the **Control Parameters** option from the dialog box to specify maximum number of iterations to be performed for optimization.
- Select the **Convergence Parameters** option from the dialog box to specify the convergence limit at which analysis will be complete; refer to Figure-53.

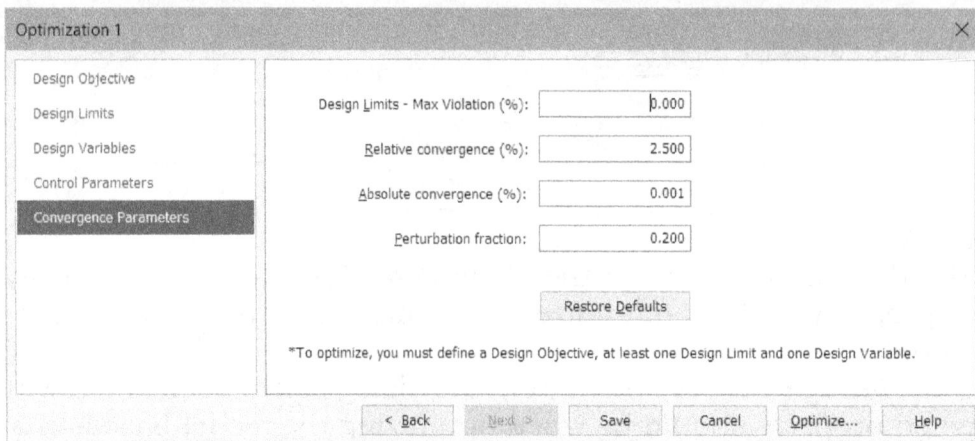

*Figure-53. Convergence Parameters options*

- After setting the parameters, click on the **Optimize** button to run optimization study. Once the study is complete, an **Optimization** dialog box will be displayed; refer to Figure-54.

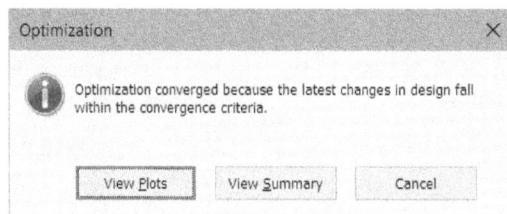

*Figure-54. Optimization dialog box*

- Click on the **View Plots** button to check plots. Click on the **View Summary** button to check analysis result summary; refer to Figure-55.

| | A | B | C | D | E | |
|---|---|---|---|---|---|---|
| 10 | | | | | | |
| 11 | Design Variables | | Type | Name | Value | Range |
| 12 | | mN | Sim | Force_1_Static_Study_1 | 1.000E+06 | [1,00,000.00 m |
| 13 | | | | | | |
| 14 | | | | | | |
| 15 | Optimization Results | Units | | | | |
| 16 | Iteration | | 1 | 2 | 3 | |
| 17 | Design Objective | | | | | |
| 18 | Mass | kg | 3.559 | 3.559 | 3.559 | |
| 19 | | | | | | |
| 20 | Design Variable | | | | | |
| 21 | Force_1_Static_Study_1 | mN | 1.000E+06 | 7.800E+05 | 7.816E+05 | |
| 22 | | | | | | |
| 23 | Design Limit | | | | | |
| 24 | Total Translation | mm | 0.016 | 0.012 | 0.012 | |
| 25 | | | | | | |
| 26 | | | | | | |
| 27 | Processed Results | Units | | | | |
| 28 | Iteration | | 1 | 2 | 3 | |
| 29 | Total Translation-Minimum | mm | 0 | 0 | 0 | |
| 30 | Total Translation-Maximum | mm | 0.016 | 0.012 | 0.012 | |

*Figure-55. Optimization summary*

Right-click on the iteration of optimization study whose geometry you want to be checked and select the **Show Geometry** option; refer to Figure-56.

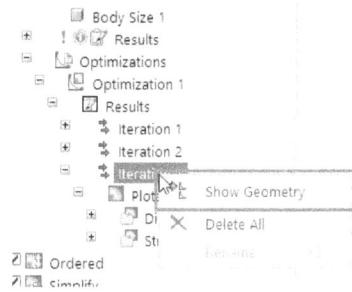

*Figure-56. Show Geometry option in Simulation Tree*

# PERFORMING GENERATIVE STUDY

The Generative Study is performed to generate the shape and size of model based on analysis results. The major goal of performing generative study is to reduce mass of model. The procedure to perform the study is given next.

* Click on the **Create Generative Study** tool from the **Generative Study** group in the **Generative Design** tab of **Ribbon**. The tools to perform generative study will be active in the **Ribbon**; refer to Figure-57.

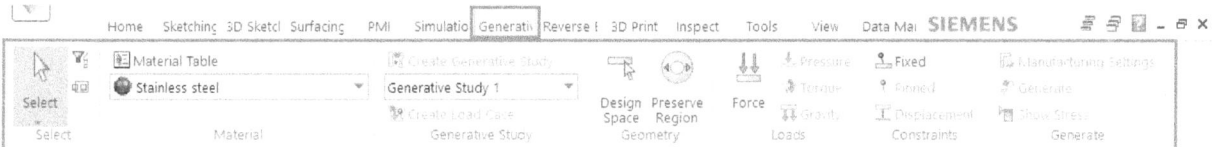

*Figure-57. Generative Design options*

* Click on the **Design Space** button from the **Geometry** group in the **Generative Design** tab of **Ribbon** to select the bodies to be used for generative study. The **Design Space Command Bar** will be displayed. Select the desired bodies and click on the **OK** button.
* Click on the **Preserve Region** button from the **Geometry** group in the **Generative Design** tab of **Ribbon**. The **Preserve Region Command Bar** will be displayed; refer to Figure-58. Select the faces of regions you want to preserve. You will be asked to specify offset value up to which the region will be preserved from selected faces; refer to Figure-59. Right-click in the drawing area to create preserve region.

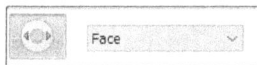

*Figure-58. Preserve Region Command Bar*

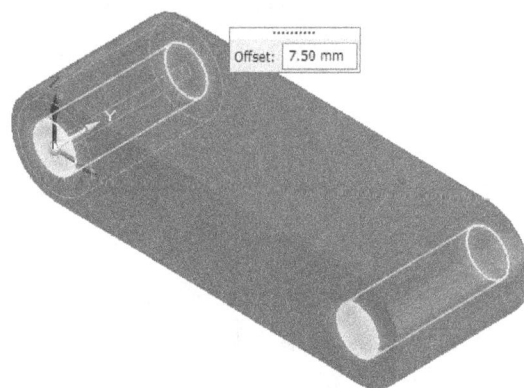

*Figure-59. Offset for preserve region*

* Apply the desired loads like force, pressure, torque, and gravity on the model as discussed earlier.

- Apply the desired constraints to the model like fixed, pinned, and displacement.
- After setting all the desired parameters, click on the **Generate** button from the **Generate** group in the **Generative Design** tab of the **Ribbon**. The **Generate Study** dialog box will be displayed; refer to Figure-60.

*Figure-60. Generate Study dialog box*

- Use the **Study Quality** slider to define the quality of analysis. The higher you set quality of analysis, the more time it takes to perform.
- Use the slider in **Mass Reduction** area to define maximum reduction in mass of model to be applied.
- Click in the **Factor of safety** edit box to specify factor of safety to be achieved by model generated.
- After setting the desired parameters, click on the **Generate** button. The generated model will be displayed; refer to Figure-61.

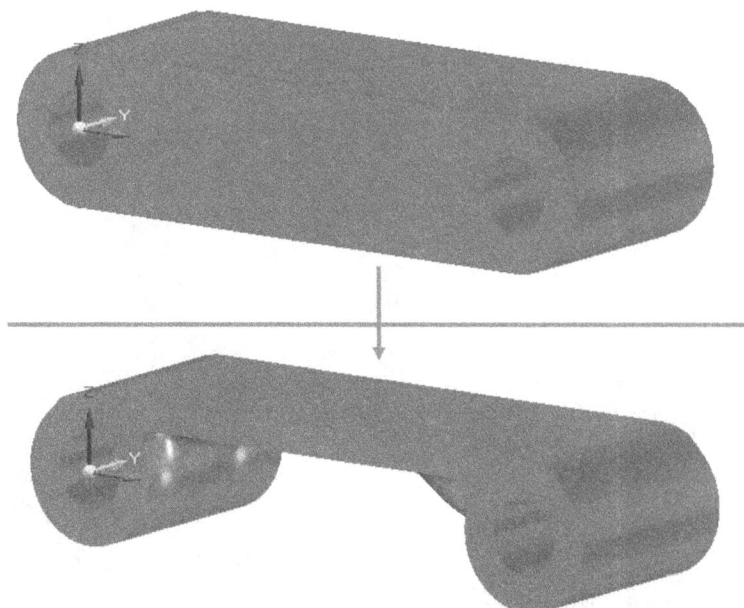

*Figure-61. Model generated by generative design study*

# FOR STUDENT NOTES

FOR STUDENT NOTES

# Index

**Symbols**

3D Builder button 1-23
3D Print cascading menu 1-21
3D Printing 5-4
3D Printing Processes 5-5
3D Print tab 5-11
3D Sketch 3-2

**A**

Add Body tool 4-21
Add drop-down 4-2
Add-Ins tool 1-36
Alignment Indicator tool 2-58
Along Curve Pattern 4-29
Along Curve tool 8-27
Alpha Tolerance option 7-7
Angle Between tool 2-44
Angled tool 3-16
Angle tool 8-8
Angular Coordinate Dimension tool 2-46
Animate tool 13-21
Animation Editor tool 9-23
Animation Properties button 9-24
Animation Properties dialog box 9-24
Arc by 3 Points tool 2-16
Arc by Center Point tool 2-16
Arc Segment tool 10-4
Arrange Dimensions tool 11-16
Assemble button 8-5
Assembly Relationship Assistant tool 8-14
Assembly Relationships Manager tool 8-13
Assembly Statistics tool 8-36
Auto-Dimension tool 2-48
Auto Explode tool 9-27
Auxiliary tool 11-8
Axial Align tool 8-6

**B**

Background tab 5-24
Back View button 5-17
Balloon tool 7-16
Base Reference Planes check box 2-2
Basic option 7-8
Bead tool 12-12

Bearing tool 13-11
Bend Angle tool 12-23
Bend Bulge Relief tool 12-21
Bend Radius tool 12-23
Bend Table tool 11-13
Bend tool 12-18
Bevel Gear Designer 9-21
Binder option 1-49
Bind tool 9-31
Blank option 7-9
Blank Surface tool 6-25
Blend Round 3-32
BlueDot tool 6-35
BlueSurf tool 6-13
Body Size tool 13-18
Body Temperature tool 13-11
Border tab 7-14
Bottom View button 5-18
Bounded tool 6-16
Break Corner tool 12-22
Broken-Out tool 11-11
Broken tool 11-9
Browse button 1-13

**C**

Cable tool 10-11
Callout tool 7-10
Cam Designer 9-14
Camera Path button 9-25
Cam tool 8-10
Capture Fit tool 8-16
Center check box 2-54
Center-Plane tool 8-13
Centrifugal tool 13-11
Chamfer Dimension tool 11-15
Chamfer tool 2-20, 3-33
Check Interference tool 8-35
Circle by 3 Points tool 2-13
Circle by Center Point tool 2-12
Circular Pattern 4-28
Class option 7-7
Close 2-Bend Corner tool 12-16
Close 3-Bend Corner tool 12-17
Close Curve button 2-17
Coincident Plane tool 3-15
Collapse tool 9-30
Collinear constraint 2-37
Color Manager tool 5-22
Colors option 1-29

Command Finder 1-8
Command Log tool 1-25
Compare Models option 1-42
Component Structure Editor tool 9-7
Component Tracker tool 9-4
Concentric constraint 2-36
Configuration Manager tool 9-34
Configuration Options tool 9-35
Configurations group 9-33
Connect tool 2-33, 8-7
Constant Radius Round 3-29
Constraints group 13-12
Construction Display button 5-15, 8-4
Construction tool 2-31
Contour Flange tool 12-5
Contour Style group 13-21
Contour tool 6-9
Convert to Curve tool 2-49
Coordinate Dimension tool 2-45
Coordinate System tool 3-18
Copy Attributes tool 11-16
Copy Current Display tool 9-36
Copy tool 6-23
Counterbore Holes 3-25
Create As Construction tool 2-32
Create Drawing dialog box 11-2
Create Generative Study tool 13-27
Create Geometry button 2-8
Create Report tool 13-22
Cross Brake tool 12-14
Cross tool 6-8
Curve by Table tool 6-3
Curve Control Vertex tool 2-56
Curve Options button 2-18
Curve Segment tool 10-5
Curve tool 2-17
Customize option 1-7
Customize tool 1-37
Cutting Plane tool 11-10
Cut tool 3-12
cycle 13-3
Cylindrical tool 13-14

**D**

Data Selection group 13-23
Datum Frame tool 7-22
Datum Target tool 7-24
Decrease PMI Font tool 2-64

Define tool 13-7
Deformation group 13-21
Delete Animation button 9-24
Delete Faces drop-down 4-37
Delete tools 4-37
Delete Voids tool 5-12
Derived tool 6-11
Design Space button 13-27
Design Variable Rule Editor dialog box 13-25
Detail tool 11-9
Dimension Axis tool 2-47
Dimension Prefix dialog box 7-6
Dimension Style page 1-33
Dimetric View button 5-18
Dimple tool 12-8
Disperse tool 8-23
Displacement tool 13-10
Display Configurations tool 9-33
Display Options tool 13-20
Distance Between tool 2-44
Draft tool 3-34
Drag Component tool 8-17
Draw button 9-32
Drawing of Active Model option 11-2
Drawing Standards page 1-36
Drawing View Layout button 11-5
Drawing View Wizard dialog box 11-4
Drawn Cutout tool 12-11
Draw tool 2-8
Drop tool 9-32
Duplicate Pattern 4-32
Dynamic Analysis 13-3

**E**

Edge Condition tool 7-19
Edge Size tool 13-16
Edit Links option 1-49
Electrical Routing tool 10-10
Ellipse by 3 Point tool 2-15
Ellipse by Center Point tool 2-14
Emboss tool 12-15
Enclosure tool 4-22
End Point check box 2-53
Engineering Reference drop-down 9-9
Equal constraint 2-35
Equilibrium Check tool 13-21
ERA tool 9-23
Error Assistant dialog box 9-6

Errors tool 9-6
Etch tool 12-14
Explode - Render - Animate environment 9-23
Explode tool 9-29
Export tool 5-14
Extend to Next tool 2-21
Extend tool 6-30
Extrude - Crown 3-8
Extruded tool 6-20
Extrude tool 3-3
Extrude - Treatment Step button 3-7

**F**

Feature Callout option 7-9
Feature Callout tab 7-14
Feature Control Frame tool 7-20, 7-23
File Locations option 1-31
File Properties tool 1-47
File Units 1-47
Fillet tool 2-19
Fillet Weld tool 4-13
Fill tool 2-50
Finite Extent button 3-5
Fixed tool 13-13
Flange tool 12-3
FlashFit tool 8-5
Floor Reflection tool 5-21
Floor Shadow tool 5-22
Flow Lines group 9-31
Force tool 13-8
Frame Design environment 10-12
Frame tool 10-12
FreeSketch tool 2-9
From/To Extent 3-6
Front View button 5-17

**G**

Gear tool 8-11
Geometry Inspector tool 8-36
Gravity tool 13-12
Grid Option tool 2-51
Groove Weld tool 4-14
Gusset tool 12-13

**H**

Helical Curve tool 6-5
Helix Cut tool 4-20
Helix tool 4-9
Helpers page 1-34

Hem tool 12-8
High-Quality Rendering button 5-22
Hole Table tool 11-12
Hole tool 3-22
Hooke's Law 13-2
Horizontal or Vertical check box 2-56

**I**

IGES Export Wizard 1-17
Increase PMI Font tool 2-63
Info cascading menu 1-45
Inquire Element tool 8-33
Insert Component tool 8-3
Insert tool 8-7
Inspection button 7-9
IntelliSketch 2-53
IntelliSketch Options tool 2-57
Inter-Part Manager tool 9-7
Inter-Part page 1-32
Intersection check box 2-55
Intersection Point tool 6-13
Intersection tool 6-4
Intersect tool 4-25, 6-26
Isocline tool 6-10
iso metric assembly.asm template 8-2
ISO View 5-18

**J**

Jog tool 12-20

**K**

Keyboard tab 1-37
Keypoint Curve Segment tool 10-6
Keypoint Curve tool 6-2
Keypoints button 7-9
KeyShot Animate button 9-27
KeyShot Render tool 9-27

**L**

Label Weld tool 4-17
Layout tab 1-41
Leader tool 7-25
Learn option 1-10
Left View button 5-17
Lights tab 5-24
Limit option 7-8
Linear Buckling Analysis 13-3
Linear Motor tool 8-24
Linear Static Analysis 13-2

Line Command Bar  2-7
Line drop-down  2-7
Line Segment tool  10-3
Line tool  2-7
Line Up Text tool  11-15
Lip tool  3-43
Local Edit button  2-18
Lock Dimension Plane button  7-9
Lock Plane tool  7-2
Lock tool  2-38
Loft Cutout tool  4-19
Lofted Flange tool  12-7
Loft tool  4-8
Look at Face tool  5-19
Louver tool  12-9

**M**

Mail Recipient tool  1-25
Maintain Relationships tool  2-38
Match Coordinate Systems tool  8-11
Material List drop-down  13-7
Material Table tool  1-46
Mate tool  8-5
Measure Angle tool  8-32
Measure Distance tool  8-31
Measure Minimum Distance tool  8-31
Measure Normal Distance tool  8-32
Measure tool  8-30
Mesh tool  13-15
Mesh type drop-down  13-5
Microsoft 3D Builder app  1-22
Midpoint check box  2-53
Mirror Copy Feature tool  4-33
Mirror Copy Part tool  4-34
Mirror tool  2-28, 8-28
Model Size PMI radio button  7-10
Modes group  9-36
Modify Surfaces group  6-26
Modify tool  9-33
Mounting Boss tool  3-46
Move Components Option button  8-19
Move Components tool  8-18
Move Faces tool  4-34
Move on Select tool  8-17
Move Segment tool  10-4
Move tool  2-26
Multi Section Sweep  4-4

**N**

New 3D Sketch tool  3-2
New Animation button  9-24
New dialog box  1-4
New Optimization tool  13-24
New option  1-3
New Study tool  13-4
Nominal option  7-7
Non-Symmetric Extent  3-7
Normal Cutout tool  4-20
Normal Modes Analysis  13-3
Normal tool  4-11
No Rotation tool  13-13

**O**

Occurrence Properties dialog box  8-3
Offset Edge tool  6-34
Offset Faces tool  4-36
Offset tool  2-24, 6-22
On Element check box  2-54
Optimization dialog box  13-26
Optimization study  13-24
Options dialog box  8-4
Options tool  1-25
Ordered environment  2-4
Overhang tool  5-14
Overlapping toggle button  5-4

**P**

Pack and Go tool  1-25
Panes drop-down  5-15
Paper Print cascading menu  1-19
Parallel constraint  2-35
Parallel tool  2-56, 3-15, 8-10
Parting Split tool  6-33
Parting Surface tool  6-34
Part Library box  8-2
Part Painter tool  5-25
Parts List tool  11-12
Path Finder  1-6
Path tool  8-9
PathXpres tool  10-2
Pattern by Table  4-30
Pattern tool  4-27, 8-26
Peer Variables tool  9-2
Perpendicular constraint  2-37
Perpendicular tool  2-57, 3-17
Perspective button  5-22
Physical Thread tool  5-11
Pinned tool  13-13

Piping Route tool 10-8
Pixel Size PMI radio button 7-10
Planar Align tool 8-6
PMI 7-2
Point tool 2-8
Polygon by Center tool 2-11
Preserve Region button 13-27
Pressure tool 13-9
Previous View tool 5-19
Principal tool 11-8
Print to file check box 1-20
Probe tool 13-19
Product Manufacturing Information 7-2
Project tool 4-10, 6-5
Project To Sketch tool 3-10
Prompt Bar 1-8
Properties Manager tool 8-34
Properties tool 8-33
Property Manager tool 1-48
Publish Terrain Models tool 9-9
Publish tool 10-14
Publish Virtual Components tool 9-8

**Q**

Quick Access tab 1-39
Quick Access toolbar 1-6
Quick View Cube 1-8
Quick View Cube Settings dialog box 1-8

**R**

Radial Menu 1-40
Rebend tool 12-20
Recognize Holes tool 3-28
Record 1-8
Record Video dialog box 1-9
Recover Bodies tool 13-8
Rectangle by 2 Points tool 2-10
Rectangle by 3 Points tool 2-10
Rectangle by Center tool 2-9
Rectangular Pattern 4-27
Redefine tool 6-18
Reference Text button 7-12
Reflection Box tab 5-25
Reflective Plane option 6-38
Relationship Assistant tool 2-39
Relationship Colors tool 2-58
Relationships Handles tool 2-38
Remove from Alignment Set tool 11-16
Remove tool 9-30

Rendering tab 5-24
Reorient tool 5-12
Replace Face tool 6-29
Replace Part drop-down 8-20
Replace Part tool 8-20
Replace Part with Copy tool 8-23
Replace Part with New Part tool 8-21
Replace Part with Standard Part tool 8-21
Reports tool 9-5
Reposition Origin tool 2-53
Reposition tool 9-29
Resize Holes tool 4-39
Resize Round tool 4-40
resonance 13-3
Retrieve Dimensions tool 11-15
Revolve Cut tool 3-21
Revolved tool 6-21
Revolve tool 3-19
Ribbon 1-6
Rib tool 3-39
Right View button 5-18
Rigid Set tool 2-38, 8-13
Rip Corner tool 12-17
Rotate Components Option button 8-19
Rotate Face tool 4-35
Rotate tool 2-27
Rotational Motor tool 8-23
Round tool 3-29
Route tool 10-6
Ruled tool 6-23
Run Macro tool 1-44

**S**

Save All tool 1-14
Save Animation button 9-24
Save As dialog box 1-13
Save As Image tool 1-15
Save as Movie button 9-24
Save As Translated tool 1-16
Save Current View option 5-18
Saved Views box 5-17
Save for Tablet tool 1-18
Save page 1-30
Save tool 1-13
Scale Body tool 4-26
Scale tool 2-29
Section by Plane tool 7-25
Section Curvature Settings tool 6-37
Section tool 7-26, 11-11

Select drop-down  5-2
Selection Filter tool  5-4
Set Axis tool  7-2
Set Planes tool  5-16
Settings option  1-25
Shaded tool  5-20
Shaded with Visible Edges tool  5-21
Shaft Designer tool  9-10
Shape Edit button  2-18
Share cascading menu  1-24
Sharpen button  5-22
Show Camera Path button  9-26
Show Geometry option  13-26
Show Grid tool  2-51
Show Non-Stitched Edges tool  6-32
Silhouette check box  2-55
Simple Hole  3-23
Simple Hole button  3-23
Simplify Curve button  2-18
Simulate Motor tool  8-25
Simulation  13-2
Single Edge Color button  5-22
Sketch tool  2-5, 2-49
Sliding Along Surface tool  13-13
Slot tool  3-27
Smart Depth tab  7-14
Smart Dimension tool  2-40, 7-3
Smart Measure tool  8-29
Snap to Grid tool  2-51
Solid Edge Options dialog box  1-25
Solid Edge Portal tool  1-25
Solid Sweep Cutout tool  4-18
Solid Sweep tool  4-7
Solve tool  13-18
Spin About tool  5-19
Split Path tool  10-5
Split tool  2-21, 4-25, 6-12, 6-31
Spur Gear Designer  9-16
Static Analysis  13-2
Stitched tool  6-31
Stitch Weld tool  4-15
Stretch tool  2-30
Study List drop-down  13-7
Styles tool  2-58, 7-10
Subtract tool  4-23
Surface Blend  3-33
Surface Size tool  13-17
Surface Texture Symbol tool  7-17
Surface Visualization tool  6-36

Surfacing tab  6-2
Sweep tool  4-2
Swept Cutout tool  4-17
Swept tool  6-19
Symmetric constraint  2-36
Symmetric Diameter tool  2-47, 11-14
Symmetric Extent  3-7
Symmetric Offset tool  2-24
Symmetry Axis button  2-38
Synchronous environment  2-4

**T**

Tab tool  12-3
Take Snapshot tool  9-36
Tangent Arc tool  2-15
Tangent check box  2-57
Tangent Circle tool  2-13
Tangent constraint  2-34
Tangent tool  8-9
Tear-Off tool  3-11
Text and Leader tab  7-13
Themes tool  1-41
Thicken tool  4-12
Thin Region tool  3-38
Thin Wall tool  3-36
Threaded Hole  3-24
Thread tool  3-25
Through All  3-5
Through Next  3-6
Tools tab  9-2
Top View button  5-18
Torque tool  13-10
Transfer tool  8-23
Transition to Ordered option  2-3
Trim Corner tool  2-22
Trimetric View button  5-18
Trim tool  2-22, 6-29
Tube tool  10-7
Twist Sweep  4-6

**U**

Unbend tool  12-19
Unbind tool  9-31
Unexplode tool  9-31
Union tool  4-23
Unite Bodies tool  13-8
Unit page  1-33
Unit Tolerance option  7-7
Unload Hidden Parts tool  9-36

Update Active Level tool 9-5
Update All Open Documents tool 9-5
Update Relationships tool 9-5
Update Retrieved Dimensions tool 11-16
Update Structure tool 9-9
Update Views tool 11-8
Upload to YouTube button 1-10
User-Defined tool 13-14
User Profile page 1-32

**V**

Variable Radius Round 3-31
Variable Table dialog box 13-25
Variable Table Motor tool 8-26
Vent tool 3-43
View Manager option 5-18
View option 1-28
View Orientation button 11-6
View Overrides tool 5-23
View Plots button 13-26
View tab 5-15
View tool 7-28
View Toolbar 2-5
View Wizard Command Bar 11-3
View Wizard tool 11-3
Visible and Hidden Edges tool 5-20
Visible Edges tool 5-20

**W**

Wall Thickness tool 5-13
Web Network tool 3-41
Weld Symbol tool 7-18
Where Used option 1-48
Wire Frame tool 5-19
Wire tool 10-10

**X**

Xpresroute tool 10-2

## Ethics of an Engineer

- Engineers shall hold paramount the safety, health, and welfare of the public and shall strive to comply with the principles of sustainable development in the performance of their professional duties.

- Engineers shall perform services only in areas of their competence.

- Engineers shall issue public statements only in an objective and truthful manner.

- Engineers shall act in professional manners for each employer or client as faithful agents or trustees, and shall avoid conflicts of interest.

- Engineers shall build their professional reputation on the merit of their services and shall not compete unfairly with others.

- Engineers shall act in such a manner as to uphold and enhance the honor, integrity, and dignity of the engineering profession and shall act with zero-tolerance for bribery, fraud, and corruption.

- Engineers shall continue their professional development throughout their careers, and shall provide opportunities for the professional development of those engineers under their supervision.

www.ingramcontent.com/pod-product-compliance
Lightning Source LLC
Chambersburg PA
CBHW081757200326
41597CB00023B/4060